AF782692

VOLUME FIVE HUNDRED AND EIGHTY ONE

METHODS IN ENZYMOLOGY

Single-Molecule Enzymology: Fluorescence-Based and High-Throughput Methods

METHODS IN ENZYMOLOGY

Editors-in-Chief

ANNA MARIE PYLE
Departments of Molecular, Cellular and Developmental Biology and Department of Chemistry
Investigator, Howard Hughes Medical Institute
Yale University

DAVID W. CHRISTIANSON
Roy and Diana Vagelos Laboratories
Department of Chemistry
University of Pennsylvania
Philadelphia, PA

Founding Editors

SIDNEY P. COLOWICK and NATHAN O. KAPLAN

VOLUME FIVE HUNDRED AND EIGHTY ONE

METHODS IN ENZYMOLOGY

Single-Molecule Enzymology: Fluorescence-Based and High-Throughput Methods

Edited by

MARIA SPIES

University of Iowa, Iowa City, IA, United States

YANN R. CHEMLA

University of Illinois at Urbana-Champaign, Urbana, IL, United States

ELSEVIER

AMSTERDAM • BOSTON • HEIDELBERG • LONDON
NEW YORK • OXFORD • PARIS • SAN DIEGO
SAN FRANCISCO • SINGAPORE • SYDNEY • TOKYO
Academic Press is an imprint of Elsevier

Academic Press is an imprint of Elsevier
50 Hampshire Street, 5th Floor, Cambridge, MA 02139, United States
525 B Street, Suite 1800, San Diego, CA 92101–4495, United States
The Boulevard, Langford Lane, Kidlington, Oxford OX5 1GB, United Kingdom
125 London Wall, London, EC2Y 5AS, United Kingdom

First edition 2016

Copyright © 2016 Elsevier Inc. All rights reserved.

No part of this publication may be reproduced or transmitted in any form or by any means, electronic or mechanical, including photocopying, recording, or any information storage and retrieval system, without permission in writing from the publisher. Details on how to seek permission, further information about the Publisher's permissions policies and our arrangements with organizations such as the Copyright Clearance Center and the Copyright Licensing Agency, can be found at our website: www.elsevier.com/permissions.

This book and the individual contributions contained in it are protected under copyright by the Publisher (other than as may be noted herein).

Notices
Knowledge and best practice in this field are constantly changing. As new research and experience broaden our understanding, changes in research methods, professional practices, or medical treatment may become necessary.

Practitioners and researchers must always rely on their own experience and knowledge in evaluating and using any information, methods, compounds, or experiments described herein. In using such information or methods they should be mindful of their own safety and the safety of others, including parties for whom they have a professional responsibility.

To the fullest extent of the law, neither the Publisher nor the authors, contributors, or editors, assume any liability for any injury and/or damage to persons or property as a matter of products liability, negligence or otherwise, or from any use or operation of any methods, products, instructions, or ideas contained in the material herein.

ISBN: 978-0-12-809267-5
ISSN: 0076-6879

For information on all Academic Press publications
visit our website at https://www.elsevier.com/

ELSEVIER Book Aid International Working together to grow libraries in developing countries

www.elsevier.com • www.bookaid.org

Publisher: Zoe Kruze
Acquisition Editor: Zoe Kruze
Editorial Project Manager: Helene Kabes
Production Project Manager: Magesh Kumar Mahalingam
Cover Designer: Mark Rogers

Typeset by SPi Global, India

CONTENTS

CONTRIBUTORS

J. Andrecka
Physical and Theoretical Chemistry Laboratory, University of Oxford, Oxford, United Kingdom

N.A. Bailey
Columbia University, New York, NY, United States

M.R. Blanco
Single Molecule Analysis Group and Center for RNA Biomedicine, University of Michigan, Ann Arbor, MI, United States

E.M. Boehm
University of Iowa Carver College of Medicine, Iowa City, IA, United States

M. Brehove
The Ohio State University, Columbus, OH, United States

L.A. Campos
Centro Nacional de Biotecnología, Consejo Superior de Investigaciones Científicas, Madrid, Spain

B.C. Case
Wesleyan University, Middletown, CT, United States

R. Chadda
The University of Iowa, Iowa City, IA, United States

D.A. Erie
University of North Carolina at Chapel Hill, Chapel Hill, NC, United States

J.C. Exell
University of California, Davis, CA, United States

H. Flyvbjerg
Department of Micro- and Nanotechnology, Technical University of Denmark, Kongens Lyngby, Denmark

J.W. Gauer
University of North Carolina at Chapel Hill, Chapel Hill, NC, United States

M. Ghoneim
Faculty of Science, Cairo University, Giza, Egypt

M.D. Gibson
The Ohio State University, Columbus, OH, United States

Y.E. Goldman
Pennsylvania Muscle Institute, Perelman School of Medicine, University of Pennsylvania, Philadelphia, PA, United States

R.L. Gonzalez Jr.
Columbia University, New York, NY, United States

M. Götz
Institute of Physical Chemistry, University of Freiburg, Freiburg, Germany

W.O. Hancock
Pennsylvania State University; Intercollege Graduate Degree Program in Bioengineering, Pennsylvania State University, University Park, PA, United States

S.R. Hansen
University of Wisconsin–Madison, Madison, WI, United States

P. Hao
North Carolina State University, Raleigh, NC, United States

S. Hartmann
Institute for Physical Chemistry, University of Muenster, Muenster, Germany

N.S. Hatzakis
Department of Chemistry & Nanoscience Center, University of Copenhagen, Copenhagen, Denmark

P. Hedegård
Niels Bohr Institute & Nanoscience Center, University of Copenhagen, Copenhagen, Denmark

M.M. Hingorani
Wesleyan University, Middletown, CT, United States

J. Hohlbein
Laboratory of Biophysics; Microspectroscopy Centre, Wageningen University and Research, Wageningen, The Netherlands

S. Hohng
National Center for Creative Research Initiatives; Institute of Applied Physics; Seoul National University, Seoul, Republic of Korea

A.A. Hoskins
University of Wisconsin–Madison, Madison, WI, United States

T. Hugel
Institute of Physical Chemistry, University of Freiburg, Freiburg, Germany

Y. Jang
Advanced Photonics Research Institute, Gwangju Institute of Science and Technology, Gwangju, Republic of Korea

A.N. Kapanidis
Clarendon Laboratory, University of Oxford, Oxford, United Kingdom

C.D. Kinz-Thompson
Columbia University, New York, NY, United States

D. Klostermeier
Institute for Physical Chemistry, University of Muenster, Muenster, Germany

S.C. Kowalczykowski
University of California, Davis, CA, United States

P. Kukura
Physical and Theoretical Chemistry Laboratory, University of Oxford, Oxford, United Kingdom

S. LeBlanc
University of North Carolina at Chapel Hill, Chapel Hill, NC, United States

S.-J. Lee
National Center for Creative Research Initiatives, Seoul National University, Seoul, Republic of Korea

S. Lee
Advanced Photonics Research Institute, Gwangju Institute of Science and Technology, Gwangju, Republic of Korea

L.G. Lippert
Pennsylvania Muscle Institute, Perelman School of Medicine, University of Pennsylvania, Philadelphia, PA, United States

Y. Luo
The Ohio State University, Columbus, OH, United States

K.J. Mickolajczyk
Pennsylvania State University; Intercollege Graduate Degree Program in Bioengineering, Pennsylvania State University, University Park, PA, United States

P. Mogalisetti
Tufts University, Medford, MA, United States

K.I. Mortensen
Department of Micro- and Nanotechnology, Technical University of Denmark, Kongens Lyngby, Denmark

M.E. Moses
Department of Chemistry & Nanoscience Center, University of Copenhagen, Copenhagen, Denmark

V. Muñoz
Bioengineering Program, School of Engineering, University of California at Merced, Merced, CA, United States; Centro Nacional de Biotecnología, Consejo Superior de Investigaciones Científicas, Madrid, Spain

S. Myong
Johns Hopkins University, Baltimore, MD, United States

J. North
The Ohio State University, Columbus, OH, United States

T.L. Pavankumar
University of California, Davis, CA, United States

M.G. Poirier
The Ohio State University, Columbus, OH, United States

R. Qiu
North Carolina State University, Raleigh, NC, United States

Y. Qiu
Johns Hopkins University, Baltimore, MD, United States

J.L. Robertson
The University of Iowa, Iowa City, IA, United States

M.L. Rodgers
University of Wisconsin–Madison, Madison, WI, United States

C.E. Rohlman
Albion College, Albion, MI, United States

M. Sakato
Wesleyan University, Middletown, CT, United States

S. Schmid
Institute of Physical Chemistry, University of Freiburg, Freiburg, Germany

J.R. Sellers
Laboratory of Molecular Physiology, National Heart, Lung and Blood Institute, National Institutes of Health, Bethesda, MD, United States

M. Spies
University of Iowa Carver College of Medicine, Iowa City, IA, United States

J.A. Spudich
Department of Biochemistry, Stanford University School of Medicine, Stanford, CA, United States

S. Subramanyam
University of Iowa Carver College of Medicine, Iowa City, IA, United States

J. Sung
Department of Cellular and Molecular Pharmacology, The Howard Hughes Medical Institute, University of California, San Francisco, CA, United States

Y. Takagi
Laboratory of Molecular Physiology, National Heart, Lung and Blood Institute, National Institutes of Health, Bethesda, MD, United States

M. Todd Washington
University of Iowa Carver College of Medicine, Iowa City, IA, United States

D.R. Walt
Tufts University, Medford, MA, United States

N.G. Walter
Single Molecule Analysis Group and Center for RNA Biomedicine, University of Michigan, Ann Arbor, MI, United States

Z. Wang
Bioengineering Program, School of Engineering, University of California at Merced, Merced, CA, United States

D. Weidlich
Institute for Physical Chemistry, University of Muenster, Muenster, Germany

K.R. Weninger
North Carolina State University, Raleigh, NC, United States

P. Wortmann
Institute of Physical Chemistry, University of Freiburg, Freiburg, Germany

PREFACE

Single-molecule biophysics is a burgeoning field that encompasses studies of individual molecules and macromolecular machines in isolation, within the cells, and even inside living organisms. The last two decades witnessed an explosion of single-molecule research. The technological advances made to visualize and investigate biologically important molecules individually, in real time, and under physiological conditions have fundamentally changed our understanding of how cells work and communicate, how the molecular machines of the cell assemble and function, and, in general, how physiological functions emerge from the chaos of stochastic molecular events and interactions. New insights have been gained into the mechanisms of molecular motors, cell division, DNA repair and replication, RNA transcription, translation, protein folding, enzymatic catalysis, and assembly and function of membrane proteins.

The time when the novelty of observing a single molecule was a cause for celebration has passed. Single-molecule techniques have grown more sensitive and sophisticated, capable of multidimensional or high-throughput measurements of biomolecular dynamics. Once arcane, single-molecule methodologies are also becoming more mainstream. With the technologies becoming increasingly available to a broader community of biophysicists, biochemists, and molecular biologists, what lags behind is a comprehensive understanding of the power of single-molecule techniques when combined with rigorous data analysis. Thus, we believe the time is right for a comprehensive text describing the most cutting-edge single-molecule methods. The two volumes in this "Single-Molecule Enzymology" set are not intended to replace the previous set (volumes 472 and 475 on "Single-Molecule Tools")—all the methods and analyses described in it are still highly relevant. Instead, our aim is to expand the previous volumes by providing a comprehensive toolkit of up-to-date methods and expert advice on their utilization and data analysis.

The first volume in the set (volume 581, "Single-Molecule Enzymology, Part A") focuses on advances in the fluorescence-based techniques including single-molecule fluorescence in single or multiple colors, FRET, multidimensional FRET, interferometry, and massively parallel single-molecule enzymatic analyses in nanoliter volumes. The second volume in the series (volume 582, "Single-Molecule Enzymology, Part B") focuses on the force-based single-molecule techniques and hybrid approaches that combine manipulation of single molecules by force with fluorescence-based observation.

Our philosophy when we selected the topics and contributors was to offer an authoritative practical guide on the most popular and newly emerging aspects of single-molecule enzymology. Our hope is that the two volumes in this series will provide valuable information on everything that goes into a successful single-molecule experiment from sample preparation, to measurements, data acquisition, and rigorous data analysis routines, and will be well used by everyone in the lab, from undergraduates and rotation students to postdoctoral fellows and faculty who wish to expand their research programs.

Maria Spies
University of Iowa
Yann R. Chemla
University of Illinois, Urbana-Champaign

CHAPTER ONE

Direct Fluorescent Imaging of Translocation and Unwinding by Individual DNA Helicases

T.L. Pavankumar[1], J.C. Exell[1], S.C. Kowalczykowski[2]
University of California, Davis, CA, United States
[2]Corresponding author: e-mail address: sckowalczykowski@ucdavis.edu

Contents

[1] Both authors contributed equally to this work.

Methods in Enzymology, Volume 581
ISSN 0076-6879
http://dx.doi.org/10.1016/bs.mie.2016.09.010
© 2016 Elsevier Inc.
All rights reserved.

Abstract

The unique translocation and DNA unwinding properties of DNA helicases can be concealed by the stochastic behavior of enzyme molecules within the necessarily large populations used in ensemble experiments. With recent technological advances, the direct visualization of helicases acting on individual DNA molecules has contributed significantly to the current understanding of their mechanisms of action and biological functions. The combination of single-molecule techniques that enable both manipulation of individual protein or DNA molecules and visualization of their actions has made it possible to literally see novel and unique biochemical characteristics that were previously masked. Here, we describe the execution and use of single-molecule fluorescence imaging techniques, focusing on methods that include optical trapping in conjunction with epifluorescent imaging, and also surface immobilization in conjunction with total internal reflection fluorescence visualization. Combined with microchannel flow cells and microfluidic control, these methods allow individual fluorescently labeled protein and DNA molecules to be imaged and tracked, affording measurement of DNA unwinding and translocation at single-molecule resolution.

1. INTRODUCTION

In this chapter, we describe methods to visualize DNA and proteins at the single-molecule level. Ensemble biochemical analyses have contributed considerably to an understanding of DNA unwinding and translocation behavior of helicases and motor proteins (Byrd & Raney, 2012; Gilhooly, Gwynn, & Dillingham, 2013; Lohman, Tomko, & Wu, 2008; Pyle, 2008; Singleton, Dillingham, & Wigley, 2007; Spies, Dillingham, & Kowalczykowski, 2005). Though important, these methods measure the average behavior of an ensemble of molecules. A complete understanding requires direct measurement of individual events that make up the distribution of behaviors that comprise the ensemble. In recent decades, we have used single-molecule imaging to visualize the movement and actions of DNA motor proteins (Amitani, Baskin, & Kowalczykowski, 2006; Bianco et al., 2001; Handa, Bianco, Baskin, & Kowalczykowski, 2005; Liu, Baskin, & Kowalczykowski, 2013; Nimonkar, Amitani,

Baskin, & Kowalczykowski, 2007; Rad, Forget, Baskin, & Kowalczykowski, 2015; Spies, Amitani, Baskin, & Kowalczykowski, 2007; Spies et al., 2003; Amitani, Liu, Dombrowski, Baskin, & Kowalczykowski, 2010; Forget, Dombrowski, Amitani, & Kowalczykowski, 2013; Hilario & Kowalczykowski, 2010; Sun & Wang, 2016). Here, we limit ourselves to a description of some of those methods. However, the study of protein–DNA interactions at the single-molecule level is supported by a wide range of complementary techniques, including, but not limited to, atomic force microscopy, electron microscopy, force spectroscopy, fluorescence correlation spectroscopy, Förster resonance energy transfer, magnetic tweezers, optical traps combined with fluorescence microscopy, and total internal reflection fluorescence (TIRF) microscopy (Bustamante, Bryant, & Smith, 2003; Greenleaf, Woodside, & Block, 2007; Ha, Kozlov, & Lohman, 2012; Kapanidis & Strick, 2009; Lionnet et al., 2012; Moffitt, Chemla, Smith, & Bustamante, 2008; Neuman & Nagy, 2008; Qi & Greene, 2016; Roy, Hohng, & Ha, 2008; Spies, 2013; Yodh, Schlierf, & Ha, 2010). Mechanical manipulation of individual DNA molecules by optical trapping of attached microspheres or by attachment to a surface has been a useful platform for studying the mechanisms of DNA helicases and translocases (Amitani et al., 2006, 2010; Bianco et al., 2001; Forget et al., 2013; Hilario & Kowalczykowski, 2010; Liu et al., 2013; Rad et al., 2015; Sun & Wang, 2016).

Imaging that employs optical trapping typically uses a laser trap to physically restrain a polystyrene bead that is attached to one end of a DNA molecule. The normally coiled DNA is extended either by force generated from solution flow (Perkins, Quake, Smith, & Chu, 1994; Perkins, Smith, & Chu, 1994) or by attachment of the free end to a second bead, which is captured by a second distal optical trap (Forget et al., 2013; Forget & Kowalczykowski, 2012). The now extended DNA is visualized by fluorescent dyes whose quantum yield is enhanced by binding to double-stranded DNA (dsDNA) (Bianco et al., 2001; Perkins, Quake, et al., 1994; Perkins, Smith, & Chu, 1994).

In TIRF microscopy, DNA–protein interactions are studied by attachment of either the DNA or the protein to a functionalized surface. The use of TIRF allows selective illumination of a region extending ~100–200 nm from the glass surface, enabling excitation of molecules only within this region (Axelrod, 1989). This selectivity greatly improves visualization of fluorescently labeled DNA or protein molecules attached, or near, to the surface, minimizing background fluorescence from the bulk solution, thereby increasing the signal over background.

In this review, we describe the instrumentation and experimental techniques required to image DNA helicases and translocases at the single-molecule level. In particular, we describe methods that allow visualization of the movement of these motor proteins, or their unwinding activity, using direct or indirect fluorescent labeling procedures. The usefulness of the aforementioned techniques was demonstrated in studies of *Escherichia coli* RecBCD and RecQ helicases. These two enzymes will serve as the prototypes for elaborating the procedures required for fluorescent visualization using optical trapping and TIRF, respectively. Although described specifically for RecBCD and RecQ, these approaches can be, and have been, successfully adapted for use with other DNA helicases and translocases (Amitani et al., 2006; Nimonkar et al., 2007; Reuter, Parry, Dryden, & Blakely, 2010).

1.1 RecBCD Function and Properties

RecBCD enzyme is a 330 kDa heterotrimeric protein complex involved in homologous recombination (Dillingham & Kowalczykowski, 2008). It is a combined helicase/nuclease that catalyzes DNA end resection by producing single-stranded DNA (ssDNA) during recombinational repair of double-strand DNA breaks. It is a bipolar helicase containing two DNA motor subunits (RecB and RecD) with opposite translocation polarities (Dillingham, Spies, & Kowalczykowski, 2003; Taylor & Smith, 2003). RecBCD is an extremely fast (1–2 kb/s) and highly processive enzyme (~30 kb per binding event) (Roman, Eggleston, & Kowalczykowski, 1992). During DNA end resection, RecBCD recognizes an octameric DNA sequence, 5′-GCTGGTGG-3′, called Chi (crossover hot spot instigator). Recognition of Chi alters the enzyme's biochemical properties, attenuating its nuclease activity (Dillingham & Kowalczykowski, 2008; Dixon & Kowalczykowski, 1993).

Single-molecule studies using an optical tweezer combined with fluorescence microscopy have revealed interesting and unprecedented biochemical properties of RecBCD that could not be determined using traditional ensemble biochemical methods. In brief, an optical trap was used to capture a single-DNA molecule to which RecBCD was bound (Bianco et al., 2001; Handa et al., 2005; Spies et al., 2003, 2007), and DNA unwinding was imaged by using fluorescent reporter molecules that either were bound to duplex DNA and were displaced upon unwinding (Bianco et al., 2001) (Fig. 1 and Movie 1 (http://dx.doi.org/10.1016/bs.mie.2016.09.010)), or were attached to the enzyme itself as it moved along the DNA

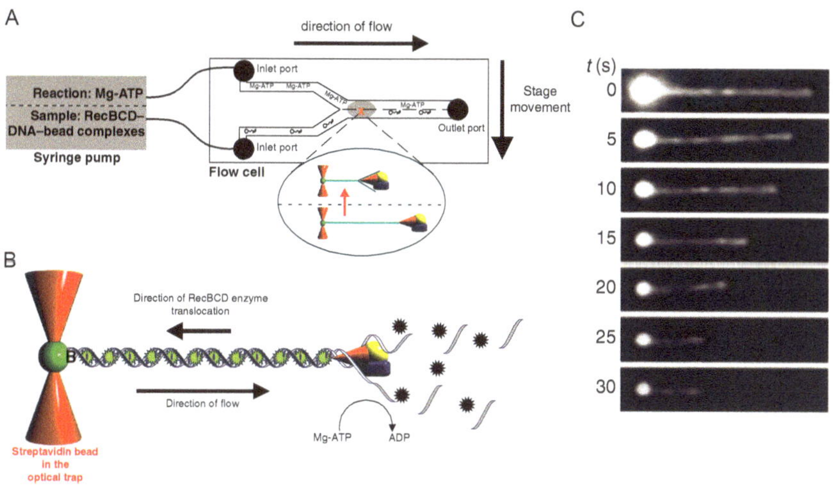

Fig. 1 Schematic representation of a multichannel flow cell used for visualization of DNA unwinding by an individual RecBCD enzyme. (A) Two-channel flow cell connected to a syringe pump. First channel contains RecBCD–DNA–bead complex and the second channel, the "reaction channel" contains ATP. "X" indicates the position of the optical trap, and this region is magnified to show movement of the enzyme-DNA–bead complex to the reaction channel. The RecBCD–DNA–bead complex is moved into the reaction channel by moving the stage by 400 μm. (B) The optically trapped fluorescently labeled dsDNA is extended by flow and its observed length decreases as RecBCD unwinds and degrades it. (C) Sequential images of DNA shortening as a function of time (*t*, in seconds) by RecBCD. *Panels (A) and (B): From Bianco, P. R., Brewer, L. R., Corzett, M., Balhorn, R., Yeh, Y., Kowalczykowski, S. C., & Baskin, R. J. (2001). Processive translocation and DNA unwinding by individual RecBCD enzyme molecules.* Nature, *409(6818), 374–378. doi:10.1038/35053131. Panel (C): From Liu, B., Baskin, R. J., & Kowalczykowski, S. C. (2013). DNA unwinding heterogeneity by RecBCD results from static molecules able to equilibrate.* Nature, *500(7463), 482–485. http://dx.doi.org/10.1038/nature12333.*

(Handa et al., 2005) (Fig. 2 and Movie 2 (http://dx.doi.org/10.1016/bs.mie.2016.09.010)). Using phage λ DNA containing a Chi sequence as a DNA substrate, RecBCD enzyme was seen to pause at Chi (for several seconds) but then resumed translocation afterward at a reduced rate, approximately one-half the pre-Chi rate (Spies et al., 2003). Furthermore, single-molecule analysis of RecBCD enzymes mutated in individual motor subunits revealed that the change in translocation rate was indeed due to switching of lead motor utilization, from RecD to RecB, upon Chi recognition (Spies et al., 2007). The heterogeneity in translocation velocities manifested by the RecBCD helicase is attributed to ergodic behavior by the enzyme (Liu et al., 2013).

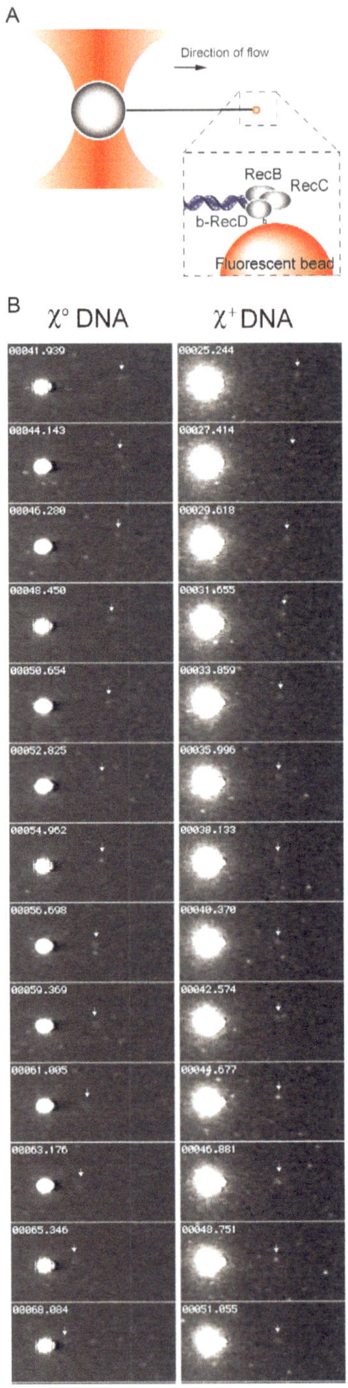

Fig. 2 See legend on opposite page.

1.2 RecQ Function and Properties

A different approach was used to monitor DNA unwinding by *E. coli* RecQ helicase, which is the founding member of the ubiquitous class of RecQ-like helicases (Chu & Hickson, 2009). The unwinding behavior of RecQ is dynamic (Harmon & Kowalczykowski, 2001; Rad & Kowalczykowski, 2012a, 2012b), requiring high concentrations of RecQ and ssDNA-binding (SSB) proteins. Consequently, surface immobilized DNA and visualization by TIRF were used to permit rapid and continuous exchange of DNA-bound proteins with those in the bulk solution. DNA unwinding by RecQ was visualized using a fluorescently modified SSB that served as a biosensor to directly detect ssDNA formation (Fig. 3). TIRF imaging was used to see the ssDNA unwinding tracts and the progression of unwinding forks in real time (Rad et al., 2015) (Fig. 4).

RecQ is a 3′–5′ helicase that participates in DNA-break repair by the RecF pathway of homologous recombination (Umezu, Nakayama, & Nakayama, 1990). In vivo, RecQ and RecJ, a 5′–3′ exonuclease, predominantly enlarge ssDNA gaps (Courcelle & Hanawalt, 1999), but can also resect dsDNA breaks into ssDNA needed for subsequent steps of homologous recombination (Handa, Morimatsu, Lovett, & Kowalczykowski, 2009; Morimatsu & Kowalczykowski, 2014). In vitro, RecQ can unwind a multitude of DNA substrates, requiring neither an ssDNA tail nor a dsDNA end from which to initiate unwinding (Harmon & Kowalczykowski, 1998, 2000, 2001; Umezu et al., 1990). Translocation occurs along ssDNA and this translocation is tightly coupled to DNA unwinding (Manosas, Xi, Bensimon, & Croquette, 2010; Rad & Kowalczykowski, 2012a; Sarlos, Gyimesi, & Kovacs, 2012). Single-molecule analysis of RecQ helicase

Fig. 2 Direct visualization of translocation by fluorescently labeled RecBCD enzyme. (A) Illustration of an optically trapped DNA molecule with fluorescently labeled RecBCD bound to the free end. The optically trapped dsDNA is extended by flow. In presence of ATP, the RecBCD enzyme (with fluorescent nanoparticle attached) translocates along the DNA from right to left, opposite to the direction of flow. (B) Sequential images of the RecBCD–nanoparticle complex translocating on two different DNA molecules, one devoid of Chi (*left*) and the other containing Chi (*right*). The intense bright spot on the left of each frame is the streptavidin-coated polystyrene bead, which is fluorescent due to nonspecific binding of the nanoparticles. The position of the RecBCD nanoparticle (*faint spot* to the *right*) is shown by an *arrow* in each frame. The numbers in image are an arbitrary time in seconds. *Panels (A) and (B): From Handa, N., Bianco, P. R., Baskin, R. J., & Kowalczykowski, S. C. (2005). Direct visualization of RecBCD movement reveals cotranslocation of the RecD motor after χ recognition.* Molecular Cell, 17*(5), 745–750. http://dx.doi.org/10.1016/j.molcel.2005.02.011.*

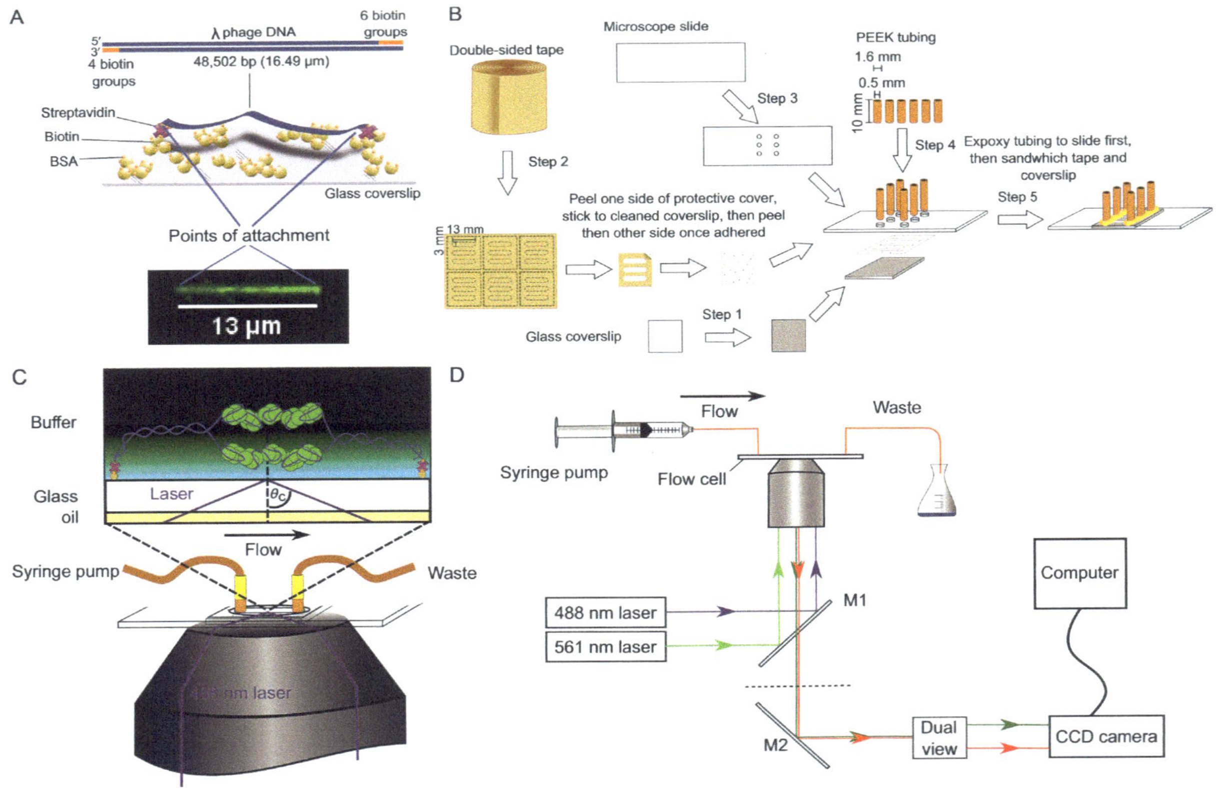

Fig. 3 See legend on opposite page.

activity showed that unwinding initiated with a dimer of RecQ locally melting dsDNA, and that processive unwinding of several kilobase pairs at maximal rates of ~40–60 bp/s involved multiple monomers at the DNA unwinding fork (Rad et al., 2015) (Movie 3 (http://dx.doi.org/10.1016/bs.mie.2016.09.010)). A microfluidic injection system, in combination with single-molecule TIRF, permitted rapid exchange of reaction components. As such, the unwinding rate could be measured in the absence of free RecQ (Movie 4 (http://dx.doi.org/10.1016/bs.mie.2016.09.010)). In this case, tracks of fluorescent SSB up to 2 kb were visualized from a single initiation event. Furthermore, it was demonstrated that unwinding could be reinitiated at a stalled unwinding fork. Such data, combined with the sigmoidal dependence of initiation events on protein concentration, suggested that multiple cooperating monomers of RecQ are required to efficiently unwind DNA (Byrd & Raney, 2015).

Using fluorescently modified SSB as a natural reporter of DNA unwinding activity has several advantages over the use of fluorescent dsDNA-binding dyes. Often, the fluorophore reduces the DNA helicase activity.

Fig. 3 Visualizing unwinding of individual DNA molecules using fluorescent SSB protein and TIRF microscopy. (A) *Top*: Diagram of phage λ DNA molecule with the indicated number of biotin groups incorporated in each 12-nt *cos* overhang. *Middle*: Illustration of biotinylated λ DNA attached at both ends via biotin–streptavidin linkage. *Bottom*: Image of an actual λ DNA molecule attached to the glass surface, stained with YO-PRO-1 (100 n*M*), and illuminated with a 488 nm laser. The image is false colored in *green* and the attachment points to the glass surface are indicated. (B) The process required to construct a flow cell containing three separate single-channels. The steps highlighted are equivalent to those described in Section 4.1. (C) The flow cell from (B) mounted onto the objective; biotinylated lambda DNA was injected under buffer flow, permitting attachment of both ends to the surface. Unwinding tracks are visualized by binding of AF488-SSB^{G26C} (*green*) to ssDNA regions. (D) Schematic representation of a TIRF microscope capable of visualizing DNA unwinding by RecQ by monitoring signal from both DNA and fluorescent SSB simultaneously. As shown in (C) the flow cell is mounted onto a 100× oil-immersion objective. The fluorescent SSB and DNA are excited by two lasers; 488 and 561 nm, respectively, and emission measured, via dichroic mirrors (M1 and M2). The deconvoluted emission is then directed onto different areas of a CCD camera generating a signal corresponding to either SSB or DNA. The lasers are operated using a custom LABview VI program to coordinate excitation with image acquisition so that the sample is illuminated only during the exposure times. *Panels (A) and (D): From Rad, B., Forget, A. L., Baskin, R. J., & Kowalczykowski, S. C. (2015). Single-molecule visualization of RecQ helicase reveals DNA melting, nucleation, and assembly are required for processive DNA unwinding.* Proceedings of the National Academy of Sciences of the United States of America, 112*(50), E6852–E6861. http://dx.doi.org/10.1073/pnas.1518028112.*

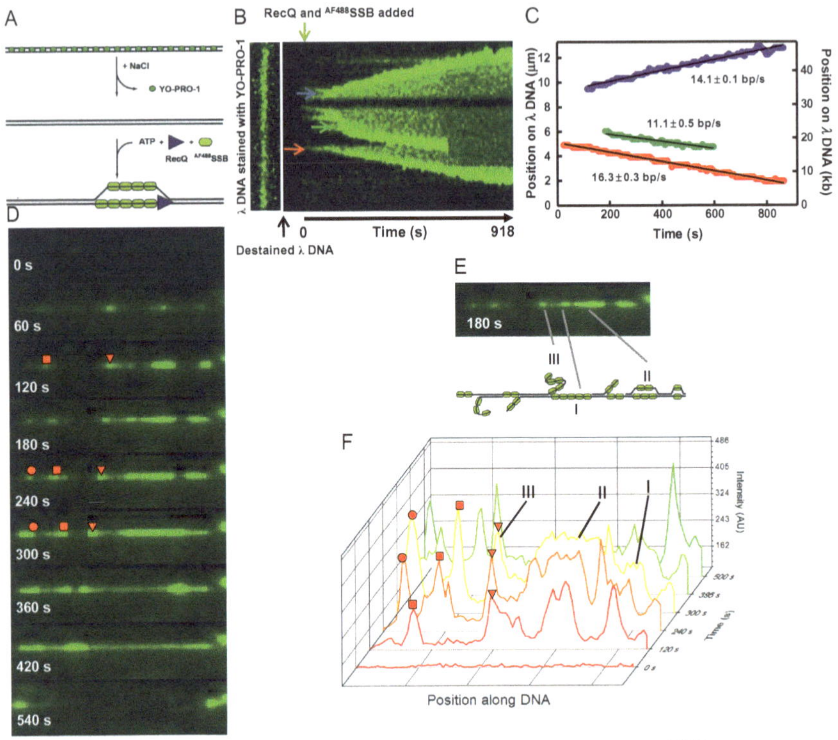

Fig. 4 Monitoring DNA unwinding by RecQ helicase using AF488-SSB^{G26C} as a sensor of ssDNA production. (A) Schematic of experiment employed to observe DNA unwinding by RecQ using signal from AF488-SSB^{G26C} bound to ssDNA. (B) Attachment to the surface at both ends of the λ DNA molecules is initially confirmed by staining dsDNA with YO-PRO-1. The stain is removed by washing with SMB2 containing 200 m*M* NaCl. *Left*, A doubly tethered λ DNA molecule stained with YO-PRO-1 (13.3 μm). *Right*: After destaining, RecQ (100 n*M*), ATP (1 m*M*), $Mg(OAc)_2$ (1 m*M*), and AF488-SSB^{G26C} (60 n*M*) in SMB2 are introduced into the flow cell; kymograph depicts three unwinding events on the same DNA molecule, each identified by AF488-SSB^{G26C} fluorescence. (C) The kymograph is analyzed using steps outlined in Section 7.2 to produce a graph showing position of unwinding forks in *B* vs time; *black line* is a fit to a linear region to obtain the unwinding rate. Error bars represent standard deviation. (D) Time-lapse images showing individual unwinding events on a single λ DNA molecule marked by bound AF488-SSB^{G26C}, in the presence of RecQ (80 n*M*). *Red circle*, *square*, and *triangle* highlight three unwinding forks tracked in F. (E) *Top*: The intensity per pixel due to AF488-SSB^{G26C} bound to unwound DNA can be used to determine different types of ssDNA products. Image shown is molecule in D after 180 s. *Bottom*: Interpretative diagram showing how different classes of unwound regions bound by AF488-SSB^{G26C} result in differences in their fluorescent intensity: class I, one strand of ssDNA; class II, unwinding bubbles with two strands of ssDNA; and class III, nicked ssDNA condensed by AF488-SSB^{G26C}. (F) The three unwinding forks highlighted in (D), *red circle*, *square*, and *triangle* are plotted as a

The level of inhibition is dependent on the dye and helicase, although judicious choice can yield unwinding rates that are essentially unaffected (Eggleston, Rahim, & Kowalczykowski, 1996; George, Ghate, Matson, & Besterman, 1992). The most significant drawback of fluorescent dyes is that they catalyze photocleavage of DNA (Akerman & Tuite, 1996). Although the newer generation of dyes (SYTOX or YO-PRO-1) has a significantly lower propensity to cause photocleavage compared to, for example, YOYO-1, they induce cleavage nonetheless.

1.3 SSB Function and Properties

In all cells, ssDNA is protected by the binding of a ssDNA-binding protein (Richard, Bolderson, & Khanna, 2009; Shereda, Kozlov, Lohman, Cox, & Keck, 2008). SSB proteins not only bind ssDNA but also interact with many different proteins to orchestrate DNA replication and repair processes. One group of these interacting partners is the DNA helicases. SSB proteins can stimulate the unwinding activity of helicases by direct interaction, by preventing DNA reannealing, or by sequestering ssDNA to which helicases can bind, preventing product inhibition (Cejka & Kowalczykowski, 2010; Harmon & Kowalczykowski, 2001; Umezu & Nakayama, 1993).

SSB from *E. coli* is a homotetramer. The DNA can bind and wrap around each monomer, and the extent of wrapping is dependent on the salt concentration, temperature, pH, and the ratio of SSB to ssDNA. As a result, the stoichiometry of ssDNA binding by SSB, referred to as "binding modes," is highly dependent on these variables (Lohman & Ferrari, 1994; Raghunathan, Kozlov, Lohman, & Waksman, 2000; Shereda et al., 2008). Two binding modes predominate. If conditions favor a binding mode where

function of pixel intensities along the contour of the unwound molecule. These traces are shown for six of the frames along the time-course in (D). The intensities of the three different classes of ssDNA bound by AF488-SSB^{G26C} are highlighted by *Roman numerals* and correspond to the structures shown in (E). Other details from RecQ unwinding can also been seen from these plots, the unwinding forks corresponding to the *red, triangle*, and *square* converge, as seen at 420 s in (D). In addition, the DNA molecule breaks internally at 500 s (*blue* trace). *Panels (A–F): From Rad, B., Forget, A. L., Baskin, R. J., & Kowalczykowski, S. C. (2015). Single-molecule visualization of RecQ helicase reveals DNA melting, nucleation, and assembly are required for processive DNA unwinding.* Proceedings of the National Academy of Sciences of the United States of America, 112*(50), E6852–E6861. http://dx.doi.org/10.1073/pnas.1518028112.*

DNA is bound to all four SSB subunits, then the tetramer occupies ~65 nucleotides (Lohman & Ferrari, 1994). Alternatively, if only two of the four monomers are bound to the DNA, then the tetramer binds ~35 nucleotides. Regardless of the binding mode, association of SSB and ssDNA is rapid and diffusion limited (Lohman & Ferrari, 1994). This is of paramount importance for their use in measuring DNA unwinding rates.

The first example of using SSB as a biosensor to follow DNA unwinding by a helicase took advantage of the decreased intrinsic tryptophan fluorescence of SSB upon binding to ssDNA (Roman & Kowalczykowski, 1989). The concept was then adapted to incorporate an extrinsic fluorophore into phage T4 ssDNA-binding protein, gp32 (Liu, Qian, & Morrical, 2006). This innovation improved the signal change upon binding and shifted the spectrum into the visible region. SSB was modified to directly measure unwinding by placing fluorescent dyes on residues near the ssDNA-binding domain (Dillingham et al., 2008). Each monomer contains oligosaccharide/oligonucleotide binding folds (OB-folds) that form groves along the protein's surface allowing ssDNA to bind. Taking advantage of the lack of natural cysteine residues in wild-type SSB, amino acid residues along these grooves were mutated individually to cysteine to permit coupling to fluorescent dyes. Placement of certain fluorophores at these sites produced a relatively large enhancement in the fluorescent signal upon binding to ssDNA. The initial study described a mutant protein, SSB^{G26C}, which had been reacted with a coumarin fluorophore, IDCC (*N*-[2-(iodoacetamido)ethyl]-7-diethylaminocoumarin-3-carboxamide) (Dillingham et al., 2008). The fluorescently labeled protein, DCC-SSB, was used to accurately determine the DNA unwinding rates of AddAB, RecBCD, and PcrA helicases. In a subsequent study, another variant, SSB^{W88C}, was coupled to Cy3b (Fili et al., 2010). This particular dye has significantly improved quantum yield and photostability, compared to previously used dyes, permitting the use of the fluorescent SSB in single-molecule experiments. Again, the rates of AddAB, RecBCD, and PcrA were measured, and it was demonstrated that Cy3b-SSB^{W88C} association with product ssDNA accurately reported on their rates of unwinding.

Here, this review concentrates on the use of fluorescently labeled SSB to image DNA unwinding by RecQ (Rad et al., 2015). The reagent is SSB^{G26C} conjugated with either Alexa Fluor 488 (AF488) or Alexa Fluor 546 (AF546) (Section 3.2). DNA unwinding is visualized by SSB binding and the concomitant appearance of fluorescence. By immobilizing λ dsDNA that is elongated on a surface using microfluidics, tracks due to SSB binding can be seen initiating and physically growing as a product of RecQ activity. The formation and progression of individual unwinding forks is readily visualized. The rate, direction, and processivity of unwinding at individual forks can be determined using this method. The steps required to observe and quantify unwinding by RecQ using fluorescent SSB are outlined below.

2. PREPARATION OF DNA SUBSTRATES

In practical terms, fluorescence imaging techniques, such as the two described herein (epifluorescence and TIRF), require immobilization of either protein or DNA substrate to visualize their behaviors by optical microscopy. Typically, immobilization is achieved by modification of the DNA to permit its attachment to a bead or a surface. Bacteriophage λ DNA (48,502 bp in length) is a widely used dsDNA substrate to study the DNA unwinding and translocating properties of proteins in single-molecule experiments. The DNA ends are modified to allow binding of the DNA to a functionalized surface of choice. The modifications can be introduced at one or both ends of the λ DNA, which conveniently has unique 3′ recessed ends that are 12 nucleotides in length (the *cos* site). These *cos* sites permit annealing and subsequent ligation of modified oligonucleotides (Sections 2.1 and 2.2) or incorporation of modified nucleotides by a DNA polymerase (Section 2.3).

2.1 Preparation of λ DNA With a Biotinylated End by Ligation

Bacteriophage λ DNA (New England Biolabs (NEB), Ipswich, MA) has complementary 12 nucleotide ssDNA cohesive sites at each terminus (*cos* ends). Depending on which end needs to be biotinylated, a 3′-biotinylated 12-mer oligonucleotide (5′-GGGCGGCGACCT-3′ or 5′-AGGTCGCCGCCC-3′; Integrated DNA Technology, Coralville, IA) is ligated to an end.

1. Phosphorylate oligonucleotide by incubating 5 μ*M* 3′-biotinylated 12-mer oligonucleotides with 1 m*M* ATP and 0.2 U/μL polynucleotide kinase (PNK) in 50 μL PNK buffer (70 m*M* Tris–HCl (pH 7.6), 10 m*M* $MgCl_2$, and 5 m*M* DTT) at 37°C for 1 h.
2. Heat-inactivate the enzyme by incubating at 75°C for 10 min.
3. Add 15 n*M* (molecules) phosphorylated oligonucleotides to 25 ng/μL (0.75 n*M* molecules) of λ DNA (20:1 ratio) in a 90-μL reaction volume containing 100 m*M* NaCl.

Note: Care must be taken while handling phage λ DNA. Due to its long length, λ DNA is highly susceptible to shearing. Hence, the use of cut-off pipette tips (wide-mouth tips), gentle mixing (no vortexing), and low-speed centrifugation are recommended.

4. Carry out the annealing reaction with an initial incubation at 75°C for 15 min, followed by a stepwise decrease in temperature at 1°C/min using a PCR machine. Alternatively, annealing can be carried out by starting at 75°C for 15 min in a heating block. Later, by simply turning off the heated block, allow it to cool down slowly to room temperature (over at least 1 h).
5. Ligate the annealed oligo-λ DNA complex by adding 10 μL of 10× T4 DNA ligase buffer (500 m*M* Tris–HCl (pH 7.5), 100 m*M* $MgCl_2$, 10 m*M* ATP, and 100 m*M* DTT), and 1 μL of T4 DNA ligase (400 units) to the annealed (90 μL) reaction.
6. Incubate the reaction at 16°C overnight, or room temperature for 1 h.
7. Stop reaction by adding ethylenediaminetetraacetic acid (EDTA) to a final concentration of 10 m*M* and heat-inactivating the T4 DNA ligase for 10 min at 65°C.
8. Remove excess oligonucleotides and other components by gel filtration using a prepacked MicroSpin column (e.g., Illustra MicroSpin S-400 HR Columns, GE Healthcare, Piscataway, NJ) as per manufacturer's protocol.

2.2 Preparation of λ DNA With Biotinylated and Digoxiginated (DIG) Ends by Ligation

The two cohesive ends (cos sites) can be treated differently, as needed. In this protocol, the attachment of complementary biotinylated oligonucleotide to the *cosB* (*cosL*) end and complementary digoxiginated oligonucleotide to the *cosA* (*cosR*) end of the DNA is explained.

1. Incubate 25 ng/μL of λ DNA (~750 p*M*, DNA molecules) with 15 n*M* of phosphorylated (see Section 2.1, steps 1 and 2) and digoxiginated oligonucleotide, 5′-GGGCGGCGACCT-3′ (*cosA*) (Operon Technologies, Huntsville, AL) in a 90-μL reaction mixture containing 100 m*M* NaCl at 75°C for 15 min in a heat block.
2. Allow the heat block slowly to cool to room temperature (over at least 1 h).
3. Add 10 μL of 10× T4 DNA ligase and T4 DNA ligase to a final concentration 4 U/μL and incubate at room temperature for 1 h.
4. Heat-inactivate T4 DNA ligase for 10 min at 65°C.

5. Remove free oligonucleotides using MicroSpin S-400 spin column as per manufacturer's protocol.
6. Add 37.5 n*M* (50-fold molar excess over the λ DNA end and more than 2-fold molar excess over the *cosA* oligonucleotide) biotinylated oligonucleotide, 5′-AGGTCGCCGCCC-3′ (*cosB*), and 4 U/μL T4 DNA ligase; incubate for 1 h at room temperature.
7. Stop the reaction by adding EDTA to a final concentration of 10 m*M* and heat-inactivating T4 DNA ligase for 10 min at 65°C.
8. Remove excess of digoxiginated oligonucleotides using MicroSpin S-400 column.
9. Modified oligonucleotides ligated to λ DNA can be stored at 4°C for several months (do not freeze).

   Note: The same protocol can be used to either biotinylate or digoxiginate both DNA ends, using appropriate modified oligonucleotides.

2.3 Preparation of λ DNA With Biotinylated Ends by DNA Synthesis

Both *cos* ends of λ DNA can be modified with multiple biotin moieties at each DNA end. This protocol describes biotinylation by synthetic fill-in of the 3′ recessed ends using exo$^-$ T4 DNA polymerase (NEB, Ipswich, MA).

1. In 30 μL of 1× NEB buffer-2.1, set up a reaction mixture containing 5 μg of λ DNA and 100 μ*M* each of dATP, dTTP, dCTP, and biotin-11-dGTP; add three units of T4 DNA polymerase.
2. Incubate the reaction for 15 min at 12°C.
3. Terminate the reaction by adding EDTA to a final concentration of 10 m*M* and heat-inactivating the enzyme for 20 min at 75°C.
4. Dilute the reaction mixture to a final volume of 100 μL with water and remove free dNTPs using MicroSpin S-400 HR column.
5. Store biotinylated λ DNA at 4°C.

2.4 Construction of λ DNA Containing a Unique Sequence

In this protocol, we briefly explain the steps involved in construction of a λ DNA substrate containing a unique sequence, using the Chi sequence as an example. The phage λgt10 vector is used. Lambda gt10 DNA is 43.34 kb in length with a unique *Eco*RI restriction site within the λ repressor (*cI*) gene. It can accept DNA restriction fragments up to 7.5 kb in length with *Eco*RI ends.

1. Prepare 30 μL of reaction mixture containing 2 μg of λgt10 DNA, 40% polyethylene glycol (PEG) 8000, 1× T4 DNA ligase buffer, and 4 U/μ L T4 DNA ligase. Incubate reaction mixture overnight at 16°C. This step helps to prevent spurious ligation in later steps of the protocol.
2. Precipitate ligated product by ethanol precipitation, dry, and resuspend in 20 μL of water.
3. Add ligated λ gt10 DNA to a reaction mixture containing 1× CutSmart buffer (20 m*M* Tris–acetate (pH 7.9), 50 m*M* KOAc, 10 m*M* magnesium acetate, and 100 μg/mL bovine serum albumin (BSA)) and 0.5 U/μL *Eco*RI-HF restriction enzyme (NEB). Incubate for 2 h at 37°C.
4. Precipitate the DNA with ethanol, air dry, and suspend in 10 μL of 10 m*M* Tris–HCl (pH 8.0).
5. Prepare dsDNA with the desired sequence in any *Eco*RI-digested DNA fragment (e.g., for a Chi sequence, use *Eco*RI-digested pSNH33 (Spies et al., 2003)).

   Note: dsDNA containing any sequence of interest, up to 7.5 kb can be inserted.
6. Perform a ligation reaction, as described in Section 2.3, step 1, using a ratio of vector to insert of 1:10.
7. The ligated recombinant λ DNA product is packaged into a λ capsid in vitro using the Packagene® Lambda DNA Packaging System (Promega, Madison, WI).
8. Screen for the desired λ DNA recombinants using *E. coli* C600 *hflA* or any suitable host strain.

   Note: An *E. coli* strain inactivated for *hflA* is critical for maintaining the lysogenic mode of phage λ and is important for the recombinant screening process.
9. Make recombinant λ DNA stock using standard λ DNA isolation protocol (e.g., using the QIAGEN Lambda Purification Kit per manufacturer's instructions (Spies et al., 2003)).
10. Store the purified recombinant λ DNA at 4°C.

3. FLUORESCENT LABELING OF PROTEINS

Single-molecule imaging of individual proteins requires fluorescent modification to permit direct visualization. The ability to label proteins with a fluorophore without altering function is paramount for single-molecule imaging. Various techniques and methods are available to covalently link fluorophores to the protein of interest—too many to be considered here (see, e.g., Chakraborty, Wang, Ebright, & Ebright, 2010; Hein et al., 2010; Kapanidis, Ebright, & Ebright, 2001; Shaner, Steinbach, & Tsien, 2005; Witte et al., 2013; Wu, Piatkevich, Lionnet, Singer, & Verkhusha, 2011). Here, we provide procedures for attachment of nanoparticles via biotinylated protein and for chemical labeling directly via amino or sulfhydryl groups.

3.1 Labeling of RecBCD Enzyme With a Fluorescent Nanoparticle

Here, we describe the technique used to directly visualizing individual RecBCD molecules translocating on single molecules of DNA. Real-time visualization of translocation was achieved by incorporating a biotin tag into a genetically modified RecBCD enzyme in vivo, and, then binding a streptavidin-coated fluorescent nanoparticle to the purified protein (Handa et al., 2005). As an alternative to nanoparticles, quantum dots can be substituted (Finkelstein, Visnapuu, & Greene, 2010).

1. Mix 4.8 μL of 1.22 μ*M* biotinylated RecBCD enzyme (stored in 20 m*M* Tris–HCl (pH 7.5), 0.1 m*M* EDTA, 0.1 m*M* DTT, 100 m*M* NaCl, and 50%, v/v glycerol) with 3 μL of streptavidin-coated florescent nanoparticle (~18 n*M*) (40 nm TransFluoSpheres; excitation 488 nm; emission 645 nm; Molecular Probes, Carlsbad, CA) in 50 m*M* sodium phosphate (pH 7.5), 50 m*M* NaCl, and 0.02% (v/v) Tween 20. Expression and purification of biotinylated RecBCD is elaborated elsewhere (Handa et al., 2005).

   Note: An excess of RecBCD over the bead concentration is necessary due to nonspecific loss of protein in the procedure.
2. Incubate for 10 min at 37°C and RecBCD–nanoparticles are ready to use with DNA–bead complex (proceed to Section 5.2, step 4).

3.2 Fluorescent Labeling of SSB, Used to Detect ssDNA

Proteins can be labeled relatively specifically at their N-terminus in a pH-controlled reaction that occurs between primary amines and a fluorophore that has been functionalized with a succinimidyl ester. This "N-terminal" labeling procedure has proven successful for covalently attachment of a fluorophore to a variety of proteins, RecA (Bell, Plank, Dombrowski, & Kowalczykowski, 2012; Galletto, Amitani, Baskin, Kowalczykowski, 2006), RAD51 (Hilario, Amitani, Baskin, & Kowalczykowski, 2009), and RPA2 (*Ferroplasma acidarmanus* SSB) (Honda, Park, Pugh, Ha, & Spies, 2009). However, the same method could not be used with *E. coli* SSB because the resultant protein had attenuated DNA-binding properties. Therefore, an alternative strategy was adopted whereby residues at various positions were mutated to cysteine. The sulfhydryl group of cysteine was then coupled to iodoacetamide or maleimide functionalized dyes, resulting in formation of a thioether linkage. Here, we describe fluorescent labeling of SSB^{G26C}, with either AF 488 or AF 546 using maleimide chemistry.

1. Prepare 1 L of buffer containing 20 m*M* Tris–acetate (pH 7.5), 500 m*M* NaCl, 1 m*M* EDTA, and 20% (v/v) glycerol. Before use, filter the buffer through 0.22 μm filter membrane.
2. Take the desired amount of SSB^{G26C} (1 mL of 1.7 mg/mL, ~90 μ*M*) and dialyze against a minimum of 100-fold volume of buffer, twice, each dialysis for at least 4 h. The DTT or 2-mercaptoethanol (2-ME) from the storage buffer will compete for reaction with the dye; hence the dialysis (or spin column) step is essential to remove the DTT or 2-ME (or any reactive thiols) from the storage buffer.
3. Measure the concentration of SSB^{G26C} after dialysis. Add a 10-fold molar excess of Alexa Fluor 488-C_5 maleimide or Alexa Fluor 546-C_5 maleimide (Molecular Probes, Eugene, OR) to the dialyzed SSB^{G26C}. Adjust the volume to keep the final volume of dimethyl sulfoxide below 5% (v/v) in the reaction.
4. Perform the labeling reaction at 4°C for 4 h in the dark. Stop the reaction by adding DTT to a final concentration of 1 m*M*.
5. Remove unreacted free dye from the reaction using a 20 mL P-10 gel filtration column (Bio-Rad, Hercules, CA), preequilibrated with 25 m*M* TrisOAc (pH 8.3), 0.5 *M* NaCl, 1 m*M* EDTA, 1 m*M* DTT and 20% (v/v) glycerol. The size of filtration column will depend on the scale of labeling reaction.
6. Collect the labeled SSB fractions, which elute first, and dialyze against buffer containing 25 m*M* TrisOAc (pH 8.3), 0.5 *M* NaCl, 1 m*M* EDTA, 1 m*M* DTT, and 50% (v/v) glycerol with at least two changes of at least a 10× volume each of buffer for at least 4 h each, or one change against at least a 100× volume.
7. Determine the concentration of labeled SSB by measuring absorbance (A) at 280 nm. Correct the measured absorbance for the contribution of absorbance from the dye at 280 nm by using the formula: $A_{\text{protein}} = A_{\text{measured}} - A_{\text{cf}} \times A_{\text{dye@max}}$. Use a molar extinction coefficient of 27,880 M^{-1} cm^{-1} at 280 nm for SSB protein. Similarly, use the molar extinction coefficients 72,000 M^{-1} cm^{-1} (at 493 nm) and 112,000 $M^{-1}cm^{-1}$ (at 556 nm) for AF488 and AF546 dyes, respectively, provided by the manufacturer. Apply a correction factor (A_{cf}) of 0.11

and 0.12 for AF 488 and AF 546, respectively, for dye absorption at 280 nm.

8. Calculate degree of labeling from the moles of dye incorporated per mole of protein and store protein at −80°C. Usually, the calculated degree of labeling for SSB ranges from ~0.8 to 1.0 dye/monomer. A comparison of unlabeled and labeled enzyme by SDS-PAGE is recommended to ensure that the concentration of protein is accurate, as the presence of free dye will interfere with any spectroscopic quantification.

4. INSTRUMENT

The flow cell design, flow cell fabrication, laser trapping, microfluidic system, and temperature control system are essential and unique components of every single-molecule instrument. A detailed description of these factors is beyond the scope of this chapter. However, this information is covered in detail elsewhere (Amitani et al., 2010; Forget et al., 2013). A brief explanation of the flow cell needed, and instrumentation used, for the TIRF imaging method described herein is provided.

4.1 TIRF Flow Cell Construction

One of the most important considerations when imaging fluorescent proteins by TIRF is to prevent nonspecific binding of the proteins to the surface to which the DNA is attached. Unfortunately, the extent of nonspecific binding by fluorescent proteins can only be empirically determined. Nonetheless, common to every procedure is the scrupulous cleaning of the surface that will be imaged by TIRF. There are several published methods for cleaning, functionalizing, and passivating cover glass that can be used to immobilize DNA and minimize protein adsorption to the surface (Bell, Liu, & Kowalczykowski, 2015; Bell et al., 2012; Chandradoss et al., 2014; Forget & Kowalczykowski, 2012; Gibb, Silverstein, Finkelstein, & Greene, 2012; Jain, Liu, Xiang, & Ha, 2012; Rad et al., 2015; Rasnik, McKinney, & Ha, 2005; Zhuang et al., 2000). These methods differ in their labor intensity and longevity (Rasnik et al., 2005). Alternative surface treatments include biotinylated BSA (Bell et al., 2012, 2015; Forget & Kowalczykowski, 2012; Rad et al., 2015), PEG (Chandradoss et al., 2014; Jain et al., 2012; Zhuang et al., 2000), and lipid bilayers (Gibb et al., 2012).

There are various methods to produce TIRF flowcells compatible with an inverted microscope (Amitani et al., 2010; Forget et al., 2013). The following procedure describes how to construct TIRF flowcells by sandwiching double-sided tape between a precleaned slide and a coverslip allowing objective-type TIRF to be conducted. Here, biotinylated BSA is used to coat the cover glass, providing stable anchor points for the biotinylated λ DNA when linked via streptavidin (Fig. 3A). The entire surface modification procedure is performed at room temperature (~22°C) (Fig. 3B).

1. Clean glass coverslips, thickness No. 1 (Fisher 22 × 22 × (0.13–0.17) mm), using a chemical etching process by sonicating them in a solution of 1 *M* KOH in methanol, for 1 h (Branson 1510 Ultrasonic Cleaner, 40 kHz), followed by sequential rinsing with reagent grade water (Barnstead Nanopure or Millipore Milli-Q) and methanol.
2. Cut channels into double-sided tape (3M™ Adhesive Transfer Tape 9437) using a RoboCutter (Craft ROBO, Graphtec). The number of channels will depend on the channel dimensions required. For this particular design, each flow channel measured 3 × 13 mm in width and length, respectively, and 58 μm (2.3 mils) deep (tape thickness).
3. Create holes for inlets and outlets in the glass slides (Fisher Scientific 25 × 75 × 1 mm) using a laser engraver (30-Watt mini, Epilog Laser). The slides were subsequently submerged in 2% (v/v) Hellmanex II (Hellma) overnight. Rigorous cleaning of the slide is unnecessary as it is relatively far from the imaging plane.
4. Attach PEEK tubing (0.5-mm inner diameter, Upchurch Scientific) to each of the etched holes using a 5-min epoxy (Devcon) to create ports.
5. Align the channels on the tape with the inlets and outlets on the slide and then place the coverslip over the area to create channels.
6. Functionalize the flow cell by sequential exposure to several solutions, using either a pipette or a syringe.
7. Immediately prior to use, using a syringe pump (KD Scientific, Holliston, MA), wash each channel with 1 *M* NaOH (degassed) for 30 min and rinse with 1 mL reagent grade water (Barnstead Nanopure or Millipore Milli-Q) (degassed), before equilibration with the appropriate buffer (e.g., 25 m*M* TrisOAc (pH 7.5), and 50 m*M* NaCl—also degassed).
8. Subsequently, inject 100 μL of 1 mg/mL biotinylated BSA (Thermo Scientific), in the same buffer (25 m*M* TrisOAc (pH 7.5), and 50 m*M* NaCl), to coat the flow cell. At this relatively high concentration of BSA, it is particularly important to be careful to avoid bubbles. Wash with 1 mL of buffer.

9. Inject 100 μL of 0.1 mg/mL streptavidin (Promega, stored at −20°C as 5 mg/mL aliquots dissolved in 25 m*M* TrisOAc (pH 7.5)) diluted in the experimentation buffer. Incubate for 10 min, and wash out free streptavidin with desired reaction buffer (1 mL).
10. Finally, the flow channel can be further passivated using Roche Blocking Reagent at 1.5 mg/mL (100 μL), dissolved in the same buffer as step 9, for 5 min.
11. Multiple channels can be prepared for use on the same day, but ensure that the channels remain solvated. This can be achieved by placing a pipette tip into the inlet and outlet and filling with buffer to act as reservoir. Proceed to Section 6.2.

4.2 TIRF Instrument

Visualization of DNA unwinding by RecQ was performed on an Eclipse TE2000U inverted microscope with a TIRF attachment (Nikon). Excitation is by a 488-nm (Picarro, 25 mW) or 561-nm (Cobolt, 25 mW) lasers, and the power measured at the objective lens is ~2–3 and ~5.5 mW, respectively. Fluorescence emission was detected and separated into 515 nm (30-nm bandpass) and 600 nm (40-nm bandpass) components (Optical Insights). Images were captured on a DU-897E iXon CCD camera (iXon+, Andor) and processed using Andor iQ software v. 2.4 (Amitani et al., 2010) (Fig. 3D).

5. IMAGING DNA UNWINDING BY AN INDIVIDUAL RecBCD ENZYME ON A SINGLE MOLECULE OF DNA

Briefly, the steps involved in visualizing DNA helicases or translocases on single-DNA molecules in experiments using an optical trap are (1) preparing DNA–bead complexes typically with bound protein; (2) delivery of appropriate buffers into the individual channels of a multichannel flowcell; (3) capturing a DNA–bead complex within the optical trap; and (4) moving DNA–bead complex to a channel with ATP to start the reaction and to capture images in real time.

5.1 DNA Unwinding Measured by Displacement of a Fluorescent Dye From DNA

To visualize DNA unwinding by an individual enzyme (Bianco et al., 2001; Liu et al., 2013; Spies et al., 2003), a YOYO-1-stained DNA–bead complex, to which RecBCD enzyme was bound, is optically trapped in the absence of ATP (Fig. 1B). To initiate DNA unwinding, the complex is then moved to the reaction channel that contains ATP (Fig. 1A). DNA unwinding and degradation, and concomitant RecBCD translocation, are seen as a shortening of DNA length with time (Fig. 1C).

1. Prepare single-molecule buffer containing 50 m*M* $NaHCO_3$ (pH 8.3), 15% sucrose (w/v), and 50 m*M* DTT (SMB1); degas at least for 1 h.

   Note: We routinely perform overnight degassing of buffer containing no DTT and degas for 1 h after addition of DTT. Removal of oxygen by degassing is to reduce oxygen-mediated photobleaching and cleavage of DNA.
2. Prepare DNA–bead complexes by mixing 1 μL (~35 p*M*) of streptavidin-coated polystyrene beads (1.0 μm, Bangs Laboratories, Fishers, IN), 1 μL of 100 m*M* $NaHCO_3$ (pH 8.3), and 2 μL of 100 p*M* biotinylated λ DNA for 30 min at 37°C. The ratio between the bead and DNA may need to be varied to obtain ~1 DNA molecule to each bead.
3. Stain DNA with fluorescent dye by adding 500 μL of SMB1 containing 20 n*M* YOYO-1 (Invitrogen, Carlsbad, CA) to the bead–DNA complexes. The dye to DNA (in base pairs) ratio can be varied from 1:1 to 1:5 (Eggleston et al., 1996).
4. Flush the syringes, tubing and, flow cell with water, followed by SMB1 containing 1 mg/mL BSA (typically, at a flow rate of 800 μL/h). BSA is used as a blocking agent to prevent nonspecific binding of reaction components to the syringe, tubing, and flowcell.
5. Add 2 m*M* $Mg(OAc)_2$ and 60 n*M* of RecBCD to the YOYO-1-stained DNA–bead complexes; immediately load this solution into the syringe for the first channel.
6. Prepare 500 μL SMB1 containing 2 m*M* $Mg(OAc)_2$, and ATP at desired concentration (typically, 1 m*M*); load this solution into the syringe for the second channel.
7. Start the syringe pump to inject solution into the flow cell at a flow rate of 800 μL/h; over the course of 5–10 min, decrease the flow rate stepwise to 50 μL/h.
8. Trap a DNA–bead–RecBCD complexes in first channel and move immediately into the ATP-containing reaction channel.
9. The unwinding of dsDNA is manifested by the shortening of YOYO-1 bound DNA (Movie 1 (http://dx.doi.org/10.1016/bs.mie.2016.09.010)).

5.2 Direct Observation of Translocation on DNA by an Individual RecBCD Enzyme Conjugated to a Nanoparticle

The translocation behavior of individual RecBCD molecules can also be monitored by direct visualization of fluorescently labeled RecBCD

(Handa et al., 2005). In this method, the optically trapped DNA–bead complex containing fluorescently labeled RecBCD enzyme at the free end is moved into the ATP-containing channel and the movement of fluorescently labeled RecBCD enzyme is tracked.

1. Prepare single-molecule buffer and wash all the channels as described in Section 5.1, step 4.
2. Prepare DNA–bead complexes as described in Section 5.1, step 2.
3. Label biotinylated RecBCD using a fluorescent nanoparticle as described in Section 3.1.
4. Mix 3 μL DNA–bead and 7.8 μL of RecBCD–nanoparticle complexes together, add 2 m*M* $Mg(OAc)_2$, and incubate for 2 min.
5. Add the nanoparticle–RecBCD–DNA–bead complex into 500 μL of degassed sample buffer containing 45 m*M* $NaHCO_3$ (pH 8.2), 20% (w/v) sucrose, 50 m*M* dithiothreitol, 2 m*M* magnesium acetate, and 0.5% (v/v) blocking solution (B-10710, Molecular Probes), and load into the first channel.
6. Load single-molecule buffer containing 1 m*M* ATP, 2 m*M* $Mg(OAc)_2$, and 0.5% (v/v) blocking solution (B-10710, Molecular Probes) into the second channel.
7. Trap a nanoparticle–RecBCD–DNA–bead complex and move into second channel. The fluorescent particle is seen to move toward the trapped bead as RecBCD enzyme translocates on the DNA (Fig. 2 and Movie 2 (http://dx.doi.org/10.1016/bs.mie.2016.09.010)).

6. MEASURING DNA UNWINDING BY IMAGING FORMATION OF FLUORESCENT SSB–ssDNA COMPLEXES

The use of AF488-SSB^{G26C} as an effective biosensor for tracking helicase activity was demonstrated recently for RecQ (Rad et al., 2015); alternatively, AF546-SSB^{G26C} can also be used. The following procedure describes the method used to measure DNA unwinding of doubly tethered λ DNA on a biotin-BSA surface using a fluorescent SSB to directly detect ssDNA formation (Fig. 4C).

6.1 Immobilization and Visualization of DNA

The concentration of DNA and time required to immobilize the desired number of DNA molecules in a single field of view (80 μm × 80 μm) will depend on the surface preparation method. The three methods mentioned, biotin-BSA PEGylation, and lipid bilayers (Sections 1.3 and 4.1) vary in their biotinylation surface density. Reactions are performed in single-molecule buffer 2 (SMB2) (25 m*M* TrisOAc (pH 7.5), 20% (w/v) sucrose, 50 m*M* DTT, and 0.1 mg/mL BSA), which had been degassed overnight in the absence of DTT (Section 5.1, step 1), unless stated otherwise.

As a guideline, for a flow channel prepared using biotin-BSA, 10–50 p*M* DNA is sufficient. The DNA (prepared as in Section 2.3) is diluted into SMB2 buffer, containing 100 n*M* YO-PRO-1 (or dye of choice, e.g., 75 n*M* Sytox Orange (Rad et al., 2015)).

1. Connect the inlet of a flow cell channel to a syringe pump (KD Scientific, Holliston, MA) and wash with SMB2, allowing the BSA to passivate the inner walls and syringe.
2. Inject the DNA solution in stages that allows association between the biotin and streptavidin but prevents both ends of the DNA being attached in close proximity. Initially, inject the DNA at a volumetric flow rate of 8000 μL/h (linear flow rate = 1.7 cm/s at center of channel, specific for channel with dimensions as described in Section 4.1). The flow is temporarily stopped for 2 min, and then the DNA is restretched by resuming flow (8000 μL/h) permitting attachment of the other end. This process is cycled until the desired number of DNA molecules has been reached.
3. Remove free DNA and stain the bound DNA by flowing in 100 vols (~200 μL) of 100 n*M* YO-PRO-1 in SMB2.
4. Acquire an image of each individual doubly tethered DNA molecule for subsequent measurement and normalization of DNA unwinding measurements.
5. The DNA can be destained if it interferes with helicase function. YO-PRO-1 is removed by washing with 200 m*M* NaCl in SMB2.
6. For a protocol specific to RecQ, the syringe is loaded with 50 μL of reaction mixture, 40–150 n*M* RecQ, 60 n*M* (monomer) AF488-SSB^{G26C}, 1 m*M* ATP, and 1 m*M* $Mg(OAc)_2$ in SMB2 to initiate the reaction. The flow required to fully stretch the DNA attached to the surface amounts to only ~0.1–0.5 pN (Bell et al., 2015).
7. For doubly tethered DNA, stop the flow.
8. Capture images every 2 s (or as appropriated for the rate of unwinding) at reduced laser power (e.g., 10–20% for our instrument; Section 4.2) to minimize photobleaching, using exposure of only ~200 ms.
9. Analyze data as described in Section 7.2.

10. Data using this method shows that RecQ can bind and start unwinding randomly on the tethered λ DNA (Movie 3 (http://dx.doi.org/10.1016/bs.mie.2016.09.010)). The individual events were each highlighted by tracks of fluorescently labeled SSB (Fig. 4A). Visualization shows different ssDNA intermediates, depending on whether RecQ had started unwinding from an end, a nick or melting of an intact duplex region (Fig. 4B). Using SYTOX Orange, instead of YO-PRO-1, and AF488-SSB^{G26C} enables both ssDNA and dsDNA to be visualized simultaneously, provided that the microscope is equipped with a Dual-View accessory (Optical Insights, LLC) (Movie 4 (http://dx.doi.org/10.1016/bs.mie.2016.09.010)).

6.2 Experimental Considerations Specific to Imaging With SSB Protein

Ordinarily, at the concentrations of SSB used, the rate of SSB association will be significantly faster than any helicase unwinding rate. Therefore, it is prudent to determine the effect of SSB concentration on the DNA unwinding rates, at each solution condition examined.

In general, the potential melting of duplex DNA by the specific SSB protein should always be considered when measuring initiation events in the presence of SSB. The eukaryotic ssDNA-bind protein, RPA (replication protein-A), induces DNA unwinding under low ionic strength conditions, especially at high concentrations of protein (Georgaki, Strack, Podust, & Hubscher, 1992; Treuner, Ramsperger, & Knippers, 1996); therefore, care must be taken to attribute initiation events to the helicase and not to the SSB. This can be easily tested with the fluorescent SSB by omitting the helicase.

It is also important to be aware that SSB-coated ssDNA is condensed compared to the contour length of naked ssDNA (Bell et al., 2015). However, by using double-tethers, thereby constraining the DNA, if the starting apparent length of the dsDNA is known, then the length of ssDNA produced can be calculated, to a good approximation, by assuming the lengths of dsDNA and SSB-coated ssDNA are equivalent (Section 7.2). A more accurate method is to calibrate the length of SSB–ssDNA complexes independently, either by fluorescent or by force measurements (Bell et al., 2015). Such steps become relevant when performing unwinding experiments in various solution conditions, as dsDNA length and the extent to which SSB-coated ssDNA condenses are dependent on the mono- and divalent-salt concentration (Bell et al., 2015).

One disadvantage of this assay, however, is that some helicases require a free ssDNA tail as a loading site to initiate unwinding. Clearly, if the SSB blocks this critical binding, then this approach for visualizing DNA unwinding cannot be used.

7. DATA ANALYSIS

7.1 Automatic DNA Length Measurement

To measure the length of fluorescent, dye-bound DNA or DNA bound with a fluorescent protein, we developed several plug-ins for automated analysis using Image J software. These were described previously (Amitani et al., 2010) and are available upon request.

7.2 Unwinding Analysis for the Fluorescent SSB-Binding Method, Using TIRF Microscopy

Time-lapse images collected during each experiment are first compiled into a stack. Then using Andor iQ software v. 2.4., to improve signal to noise, the stack is divided into groups of four frames, and each group is averaged using the built-in "lumped average" algorithm. The resultant stack contains images corresponding to 8 s time intervals, sufficient to define individual unwinding events for RecQ. For each molecule of interest, a kymograph is created in ImageJ using a line width of 1 pixel; an example is shown in Fig. 4B. Further analysis on the kymograph is conducted in ImageJ v. 1.44c (Schneider, Rasband, & Eliceiri, 2012), using the Point Picker plug-in to mark the position of the individual unwinding forks. The progression of multiple unwinding forks on the same DNA molecule can be monitored using this method. The distance from the point of initiation to the furthest point of continual fluorescent SSB signal defines unwinding track length in pixels. To determine the rate of unwinding in base pairs, a two-stage normalization procedure is required. First, for the particular camera used to acquire the images, measurements are converted into distance (in nanometers) using the calibrated camera resolution, in this instance 163.9 nm/pixel. Second, to convert nanometers into base pairs, the distance is normalized using the initial and expected lengths of the DNA molecule, outlined earlier; this can depend on the particular buffer condition. The measured length of the unwinding track is divided by the end-to-end distance of the doubly attached dsDNA molecule prior to starting the reaction. This serves as an internal normalization factor for each molecule. The length of the λ DNA (in nanometers) at the start of the reaction is equivalent to 48,502 bp. Therefore, the ratio of fluorescent SSB track and full-length λ DNA yields the unwound length in base pairs. Finally, the rate of unwinding is obtained from fitting to the linear region of a distance vs time trace.

ACKNOWLEDGMENTS

We particularly thank Ichiro Amitani, Nev Gilhooly, Anna Kalashnikova, Katsumi Morimatsu, and Artem Lada for their comments on this manuscript, and all of the lab members, including alumni of the Kowalczykowski lab, for their contributions to this research. Funding on this research work was supported by the grants from National Institutes of Health, GM-62653 and GM-64745, to S.C.K.

REFERENCES

Akerman, B., & Tuite, E. (1996). Single- and double-strand photocleavage of DNA by YO, YOYO and TOTO. *Nucleic Acids Research*, *24*(6), 1080–1090.

Amitani, I., Baskin, R. J., & Kowalczykowski, S. C. (2006). Visualization of Rad54, a chromatin remodeling protein, translocating on single DNA molecules. *Molecular Cell*, *23*(1), 143–148. http://dx.doi.org/10.1016/j.molcel.2006.05.009.

Amitani, I., Liu, B., Dombrowski, C. C., Baskin, R. J., & Kowalczykowski, S. C. (2010). Watching individual proteins acting on single molecules of DNA. *Methods in Enzymology*, *472*, 261–291. http://dx.doi.org/10.1016/S0076-6879(10)72007-3.

Axelrod, D. (1989). Total internal reflection fluorescence microscopy. *Methods in Cell Biology*, *30*, 245–270.

Bell, J. C., Liu, B., & Kowalczykowski, S. C. (2015). Imaging and energetics of single SSB-ssDNA molecules reveal intramolecular condensation and insight into RecOR function. *Elife*, *4*, e08646. http://dx.doi.org/10.7554/eLife.08646.

Bell, J. C., Plank, J. L., Dombrowski, C. C., & Kowalczykowski, S. C. (2012). Direct imaging of RecA nucleation and growth on single molecules of SSB-coated ssDNA. *Nature*, *491*(7423), 274–278. http://dx.doi.org/10.1038/nature11598.

Bianco, P. R., Brewer, L. R., Corzett, M., Balhorn, R., Yeh, Y., Kowalczykowski, S. C., & Baskin, R. J. (2001). Processive translocation and DNA unwinding by individual RecBCD enzyme molecules. *Nature*, *409*(6818), 374–378. http://dx.doi.org/10.1038/35053131.

Bustamante, C., Bryant, Z., & Smith, S. B. (2003). Ten years of tension: Single-molecule DNA mechanics. *Nature*, *421*(6921), 423–427. http://dx.doi.org/10.1038/nature01405.

Byrd, A. K., & Raney, K. D. (2012). Superfamily 2 helicases. *Frontiers in Bioscience (Landmark edition)*, *17*, 2070–2088.

Byrd, A. K., & Raney, K. D. (2015). Fine tuning of a DNA fork by the RecQ helicase. *Proceedings of the National Academy of Sciences of the United States of America*, *112*(50), 15263–15264. http://dx.doi.org/10.1073/pnas.1520119112.

Cejka, P., & Kowalczykowski, S. C. (2010). The full-length *Saccharomyces cerevisiae* Sgs1 protein is a vigorous DNA helicase that preferentially unwinds Holliday junctions. *Journal of Biological Chemistry*, *285*(11), 8290–8301. http://dx.doi.org/10.1074/jbc.M109.083196.

Chakraborty, A., Wang, D., Ebright, Y. W., & Ebright, R. H. (2010). Azide-specific labeling of biomolecules by Staudinger-Bertozzi ligation phosphine derivatives of fluorescent probes suitable for single-molecule fluorescence spectroscopy. *Methods in Enzymology*, *472*, 19–30. http://dx.doi.org/10.1016/S0076-6879(10)72018-8.

Chandradoss, S. D., Haagsma, A. C., Lee, Y. K., Hwang, J. H., Nam, J. M., & Joo, C. (2014). Surface passivation for single-molecule protein studies. *Journal of Visualized Experiments. 86.* , http://dx.doi.org/10.3791/50549.

Chu, W. K., & Hickson, I. D. (2009). RecQ helicases: Multifunctional genome caretakers. *Nature Reviews Cancer, 9*(9), 644–654. http://dx.doi.org/10.1038/nrc2682.

Courcelle, J., & Hanawalt, P. C. (1999). RecQ and RecJ process blocked replication forks prior to the resumption of replication in UV-irradiated Escherichia coli. *Molecular and General Genetics, 262*(3), 543–551.

Dillingham, M. S., & Kowalczykowski, S. C. (2008). RecBCD enzyme and the repair of double-stranded DNA breaks. *Microbiology and Molecular Biology Reviews, 72*(4), 642–671. http://dx.doi.org/10.1128/MMBR.00020-08.

Dillingham, M. S., Spies, M., & Kowalczykowski, S. C. (2003). RecBCD enzyme is a bipolar DNA helicase. *Nature, 423*(6942), 893–897. http://dx.doi.org/10.1038/nature01673.

Dillingham, M. S., Tibbles, K. L., Hunter, J. L., Bell, J. C., Kowalczykowski, S. C., & Webb, M. R. (2008). Fluorescent single-stranded DNA binding protein as a probe for sensitive, real-time assays of helicase activity. *Biophysical Journal, 95*(7), 3330–3339. http://dx.doi.org/10.1529/biophysj.108.133512.

Dixon, D. A., & Kowalczykowski, S. C. (1993). The recombination hotspot χ is a regulatory sequence that acts by attenuating the nuclease activity of the E. coli RecBCD enzyme. *Cell, 73*(1), 87–96.

Eggleston, A. K., Rahim, N. A., & Kowalczykowski, S. C. (1996). A helicase assay based on the displacement of fluorescent, nucleic acid-binding ligands. *Nucleic Acids Research, 24*(7), 1179–1186.

Fili, N., Mashanov, G. I., Toseland, C. P., Batters, C., Wallace, M. I., Yeeles, J. T., … Molloy, J. E. (2010). Visualizing helicases unwinding DNA at the single molecule level. *Nucleic Acids Research, 38*(13), 4448–4457. http://dx.doi.org/10.1093/nar/gkq173.

Finkelstein, I. J., Visnapuu, M. L., & Greene, E. C. (2010). Single-molecule imaging reveals mechanisms of protein disruption by a DNA translocase. *Nature, 468*(7326), 983–987. http://dx.doi.org/10.1038/nature09561.

Forget, A. L., Dombrowski, C. C., Amitani, I., & Kowalczykowski, S. C. (2013). Exploring protein-DNA interactions in 3D using in situ construction, manipulation and visualization of individual DNA dumbbells with optical traps, microfluidics and fluorescence microscopy. *Nature Protocols, 8*(3), 525–538. http://dx.doi.org/10.1038/nprot.2013.016.

Forget, A. L., & Kowalczykowski, S. C. (2012). Single-molecule imaging of DNA pairing by RecA reveals a three-dimensional homology search. *Nature, 482*(7385), 423–427. http://dx.doi.org/10.1038/nature10782.

Galletto, R., Amitani, I., Baskin, R. J., & Kowalczykowski, S. C. (2006). Direct observation of individual RecA filaments assembling on single DNA molecules. *Nature, 443*(7113), 875–878. http://dx.doi.org/10.1038/nature05197.

Georgaki, A., Strack, B., Podust, V., & Hubscher, U. (1992). DNA unwinding activity of replication protein A. *FEBS Letters, 308*(3), 240–244.

George, J. W., Ghate, S., Matson, S. W., & Besterman, J. M. (1992). Inhibition of DNA helicase II unwinding and ATPase activities by DNA-interacting ligands. Kinetics and specificity. *Journal of Biological Chemistry, 267*(15), 10683–10689.

Gibb, B., Silverstein, T. D., Finkelstein, I. J., & Greene, E. C. (2012). Single-stranded DNA curtains for real-time single-molecule visualization of protein-nucleic acid interactions. *Analytical Chemistry, 84*(18), 7607–7612. http://dx.doi.org/10.1021/ac302117z.

Gilhooly, N. S., Gwynn, E. J., & Dillingham, M. S. (2013). Superfamily 1 helicases. *Frontiers in Bioscience (Scholar Edition), 5*, 206–216.

Greenleaf, W. J., Woodside, M. T., & Block, S. M. (2007). High-resolution, single-molecule measurements of biomolecular motion. *Annual Review of Biophysics and Biomolecular Structure, 36*, 171–190. http://dx.doi.org/10.1146/annurev.biophys.36.101106.101451.

Ha, T., Kozlov, A. G., & Lohman, T. M. (2012). Single-molecule views of protein movement on single-stranded DNA. *Annual Review of Biophysics*, *41*, 295–319. http://dx.doi.org/10.1146/annurev-biophys-042910-155351.

Handa, N., Bianco, P. R., Baskin, R. J., & Kowalczykowski, S. C. (2005). Direct visualization of RecBCD movement reveals cotranslocation of the RecD motor after χ recognition. *Molecular Cell*, *17*(5), 745–750. http://dx.doi.org/10.1016/j.molcel. 2005.02.011.

Handa, N., Morimatsu, K., Lovett, S. T., & Kowalczykowski, S. C. (2009). Reconstitution of initial steps of dsDNA break repair by the RecF pathway of *E. coli*. *Genes and Development*, *23*(10), 1234–1245. http://dx.doi.org/10.1101/gad.1780709.

Harmon, F. G., & Kowalczykowski, S. C. (1998). RecQ helicase, in concert with RecA and SSB proteins, initiates and disrupts DNA recombination. *Genes and Development*, *12*(8), 1134–1144.

Harmon, F. G., & Kowalczykowski, S. C. (2000). Coupling of DNA helicase function to DNA strand exchange activity. *Methods in Molecular Biology*, *152*, 75–89. http://dx.doi.org/10.1385/1-59259-068-3:75.

Harmon, F. G., & Kowalczykowski, S. C. (2001). Biochemical characterization of the DNA helicase activity of the *Escherichia coli* RecQ helicase. *Journal of Biological Chemistry*, *276*(1), 232–243. http://dx.doi.org/10.1074/jbc.M006555200.

Hein, B., Willig, K. I., Wurm, C. A., Westphal, V., Jakobs, S., & Hell, S. W. (2010). Stimulated emission depletion nanoscopy of living cells using SNAP-tag fusion proteins. *Biophysical Journal*, *98*(1), 158–163. http://dx.doi.org/10.1016/j.bpj.2009.09.053.

Hilario, J., Amitani, I., Baskin, R. J., & Kowalczykowski, S. C. (2009). Direct imaging of human Rad51 nucleoprotein dynamics on individual DNA molecules. *Proceedings of the National Academy of Sciences of the United States of America*, *106*(2), 361–368. http://dx.doi.org/10.1073/pnas.0811965106.

Hilario, J., & Kowalczykowski, S. C. (2010). Visualizing protein-DNA interactions at the single-molecule level. *Current Opinion in Chemical Biology*, *14*(1), 15–22. http://dx.doi.org/10.1016/j.cbpa.2009.10.035.

Honda, M., Park, J., Pugh, R. A., Ha, T., & Spies, M. (2009). Single-molecule analysis reveals differential effect of ssDNA-binding proteins on DNA translocation by XPD helicase. *Molecular Cell*, *35*(5), 694–703. http://dx.doi.org/10.1016/j.molcel.2009.07.003.

Jain, A., Liu, R., Xiang, Y. K., & Ha, T. (2012). Single-molecule pull-down for studying protein interactions. *Nature Protocols*, 7(3), 445–452. http://dx.doi.org/10.1038/nprot.2011.452.

Kapanidis, A. N., Ebright, Y. W., & Ebright, R. H. (2001). Site-specific incorporation of fluorescent probes into protein: Hexahistidine-tag-mediated fluorescent labeling with Ni(2+):nitrilotriacetic acid (n)-fluorochrome conjugates. *Journal of the American Chemical Society*, *123*(48), 12123–12125.

Kapanidis, A. N., & Strick, T. (2009). Biology, one molecule at a time. *Trends in Biochemical Sciences*, *34*(5), 234–243. http://dx.doi.org/10.1016/j.tibs.2009.01.008.

Lionnet, T., Allemand, J. F., Revyakin, A., Strick, T. R., Saleh, O. A., Bensimon, D., & Croquette, V. (2012). Single-molecule studies using magnetic traps. *Cold Spring Harbor Protocols*, *2012*(1), 34–49. http://dx.doi.org/10.1101/pdb.top067488.

Liu, B., Baskin, R. J., & Kowalczykowski, S. C. (2013). DNA unwinding heterogeneity by RecBCD results from static molecules able to equilibrate. *Nature*, *500*(7463), 482–485. http://dx.doi.org/10.1038/nature12333.

Liu, J., Qian, N., & Morrical, S. W. (2006). Dynamics of bacteriophage T4 presynaptic filament assembly from extrinsic fluorescence measurements of Gp32-single-stranded DNA interactions. *Journal of Biological Chemistry*, *281*(36), 26308–26319. http://dx.doi.org/10.1074/jbc.M604349200.

Lohman, T. M., & Ferrari, M. E. (1994). Escherichia coli single-stranded DNA-binding protein: Multiple DNA-binding modes and cooperativities. *Annual Review of Biochemistry*, *63*, 527–570. http://dx.doi.org/10.1146/annurev.bi.63.070194.002523.

Lohman, T. M., Tomko, E. J., & Wu, C. G. (2008). Non-hexameric DNA helicases and translocases: Mechanisms and regulation. *Nature Reviews. Molecular Cell Biology*, *9*(5), 391–401. http://dx.doi.org/10.1038/nrm2394.

Manosas, M., Xi, X. G., Bensimon, D., & Croquette, V. (2010). Active and passive mechanisms of helicases. *Nucleic Acids Research*, *38*(16), 5518–5526. http://dx.doi.org/10.1093/nar/gkq273.

Moffitt, J. R., Chemla, Y. R., Smith, S. B., & Bustamante, C. (2008). Recent advances in optical tweezers. *Annual Review of Biochemistry*, 77, 205–228. http://dx.doi.org/10.1146/annurev.biochem.77.043007.090225.

Morimatsu, K., & Kowalczykowski, S. C. (2014). RecQ helicase and RecJ nuclease provide complementary functions to resect DNA for homologous recombination. *Proceedings of the National Academy of Sciences of the United States of America*, *111*(48), E5133–E5142. http://dx.doi.org/10.1073/pnas.1420009111.

Neuman, K. C., & Nagy, A. (2008). Single-molecule force spectroscopy: Optical tweezers, magnetic tweezers and atomic force microscopy. *Nature Methods*, *5*(6), 491–505. http://dx.doi.org/10.1038/nmeth.1218.

Nimonkar, A. V., Amitani, I., Baskin, R. J., & Kowalczykowski, S. C. (2007). Single molecule imaging of Tid1/Rdh54, a Rad54 homolog that translocates on duplex DNA and can disrupt joint molecules. *Journal of Biological Chemistry*, *282*(42), 30776–30784. http://dx.doi.org/10.1074/jbc.M704767200.

Perkins, T. T., Quake, S. R., Smith, D. E., & Chu, S. (1994). Relaxation of a single DNA molecule observed by optical microscopy. *Science*, *264*(5160), 822–826.

Perkins, T. T., Smith, D. E., & Chu, S. (1994). Direct observation of tube-like motion of a single polymer chain. *Science*, *264*(5160), 819–822.

Pyle, A. M. (2008). Translocation and unwinding mechanisms of RNA and DNA helicases. *Annual Review of Biophysics*, *37*, 317–336. http://dx.doi.org/10.1146/annurev.biophys.37.032807.125908.

Qi, Z., & Greene, E. C. (2016). Visualizing recombination intermediates with single-stranded DNA curtains. *Methods*, *105*, 62–74. http://dx.doi.org/10.1016/j.ymeth.2016.03.027.

Rad, B., Forget, A. L., Baskin, R. J., & Kowalczykowski, S. C. (2015). Single-molecule visualization of RecQ helicase reveals DNA melting, nucleation, and assembly are required for processive DNA unwinding. *Proceedings of the National Academy of Sciences of the United States of America*, *112*(50), E6852–E6861. http://dx.doi.org/10.1073/pnas.1518028112.

Rad, B., & Kowalczykowski, S. C. (2012a). Efficient coupling of ATP hydrolysis to translocation by RecQ helicase. *Proceedings of the National Academy of Sciences of the United States of America*, *109*(5), 1443–1448. http://dx.doi.org/10.1073/pnas.1119952109.

Rad, B., & Kowalczykowski, S. C. (2012b). Translocation of E. coli RecQ helicase on single-stranded DNA. *Biochemistry*, *51*(13), 2921–2929. http://dx.doi.org/10.1021/bi3000676.

Raghunathan, S., Kozlov, A. G., Lohman, T. M., & Waksman, G. (2000). Structure of the DNA binding domain of E. coli SSB bound to ssDNA. *Nature Structural Biology*, 7(8), 648–652. http://dx.doi.org/10.1038/77943.

Rasnik, I., McKinney, S. A., & Ha, T. (2005). Surfaces and orientations: Much to FRET about? *Accounts of Chemical Research*, *38*(7), 542–548. http://dx.doi.org/10.1021/ar040138c.

Reuter, M., Parry, F., Dryden, D. T., & Blakely, G. W. (2010). Single-molecule imaging of Bacteroides fragilis AddAB reveals the highly processive translocation of a single motor helicase. *Nucleic Acids Research*, *38*(11), 3721–3731. http://dx.doi.org/10.1093/nar/gkq100.

Richard, D. J., Bolderson, E., & Khanna, K. K. (2009). Multiple human single-stranded DNA binding proteins function in genome maintenance: Structural, biochemical and functional analysis. *Critical Reviews in Biochemistry and Molecular Biology, 44*(2–3), 98–116. http://dx.doi.org/10.1080/10409230902849180.

Roman, L. J., Eggleston, A. K., & Kowalczykowski, S. C. (1992). Processivity of the DNA helicase activity of Escherichia coli recBCD enzyme. *Journal of Biological Chemistry, 267*(6), 4207–4214.

Roman, L. J., & Kowalczykowski, S. C. (1989). Characterization of the helicase activity of the *Escherichia coli* RecBCD enzyme using a novel helicase assay. *Biochemistry, 28*(7), 2863–2873.

Roy, R., Hohng, S., & Ha, T. (2008). A practical guide to single-molecule FRET. *Nature Methods, 5*(6), 507–516. http://dx.doi.org/10.1038/nmeth.1208.

Sarlos, K., Gyimesi, M., & Kovacs, M. (2012). RecQ helicase translocates along single-stranded DNA with a moderate processivity and tight mechanochemical coupling. *Proceedings of the National Academy of Sciences of the United States of America, 109*(25), 9804–9809. http://dx.doi.org/10.1073/pnas.1114468109.

Schneider, C. A., Rasband, W. S., & Eliceiri, K. W. (2012). NIH Image to ImageJ: 25 years of image analysis. *Nature Methods, 9*(7), 671–675.

Shaner, N. C., Steinbach, P. A., & Tsien, R. Y. (2005). A guide to choosing fluorescent proteins. *Nature Methods, 2*(12), 905–909. http://dx.doi.org/10.1038/nmeth819.

Shereda, R. D., Kozlov, A. G., Lohman, T. M., Cox, M. M., & Keck, J. L. (2008). SSB as an organizer/mobilizer of genome maintenance complexes. *Critical Reviews in Biochemistry and Molecular Biology, 43*(5), 289–318. http://dx.doi.org/10.1080/10409230802341296.

Singleton, M. R., Dillingham, M. S., & Wigley, D. B. (2007). Structure and mechanism of helicases and nucleic acid translocases. *Annual Review of Biochemistry, 76*, 23–50. http://dx.doi.org/10.1146/annurev.biochem.76.052305.115300.

Spies, M. (2013). There and back again: New single-molecule insights in the motion of DNA repair proteins. *Current Opinion in Structural Biology, 23*(1), 154–160. http://dx.doi.org/10.1016/j.sbi.2012.11.008.

Spies, M., Amitani, I., Baskin, R. J., & Kowalczykowski, S. C. (2007). RecBCD enzyme switches lead motor subunits in response to c recognition. *Cell, 131*(4), 694–705. http://dx.doi.org/10.1016/j.cell.2007.09.023.

Spies, M., Bianco, P. R., Dillingham, M. S., Handa, N., Baskin, R. J., & Kowalczykowski, S. C. (2003). A molecular throttle: The recombination hotspot c controls DNA translocation by the RecBCD helicase. *Cell, 114*(5), 647–654.

Spies, M., Dillingham, M. S., & Kowalczykowski, S. C. (2005). DNA helicases. In *McGraw-Hill yearbook of science & technology* (pp. 95–98). New York, NY: McGraw-Hill.

Sun, B., & Wang, M. D. (2016). Single-molecule perspectives on helicase mechanisms and functions. *Critical Reviews in Biochemistry and Molecular Biology, 51*(1), 15–25. http://dx.doi.org/10.3109/10409238.2015.1102195.

Taylor, A. F., & Smith, G. R. (2003). RecBCD enzyme is a DNA helicase with fast and slow motors of opposite polarity. *Nature, 423*(6942), 889–893. http://dx.doi.org/10.1038/nature01674.

Treuner, K., Ramsperger, U., & Knippers, R. (1996). Replication protein A induces the unwinding of long double-stranded DNA regions. *Journal of Molecular Biology, 259*(1), 104–112. http://dx.doi.org/10.1006/jmbi.1996.0305.

Umezu, K., & Nakayama, H. (1993). RecQ DNA helicase of *Escherichia coli*. Characterization of the helix-unwinding activity with emphasis on the effect of single-stranded DNA-binding protein. *Journal of Molecular Biology, 230*(4), 1145–1150. http://dx.doi.org/10.1006/jmbi.1993.1231.

Umezu, K., Nakayama, K., & Nakayama, H. (1990). Escherichia coli RecQ protein is a DNA helicase. *Proceedings of the National Academy of Sciences of the United States of America*, *87*(14), 5363–5367. http://dx.doi.org/10.1073/pnas.87.14.5363.

Witte, M. D., Theile, C. S., Wu, T., Guimaraes, C. P., Blom, A. E., & Ploegh, H. L. (2013). Production of unnaturally linked chimeric proteins using a combination of sortase-catalyzed transpeptidation and click chemistry. *Nature Protocols*, *8*(9), 1808–1819. http://dx.doi.org/10.1038/nprot.2013.103.

Wu, B., Piatkevich, K. D., Lionnet, T., Singer, R. H., & Verkhusha, V. V. (2011). Modern fluorescent proteins and imaging technologies to study gene expression, nuclear localization, and dynamics. *Current Opinion in Cell Biology*, *23*(3), 310–317. http://dx.doi.org/10.1016/j.ceb.2010.12.004.

Yodh, J. G., Schlierf, M., & Ha, T. (2010). Insight into helicase mechanism and function revealed through single-molecule approaches. *Quarterly Reviews of Biophysics*, *43*(2), 185–217. http://dx.doi.org/10.1017/S0033583510000107.

Zhuang, X., Bartley, L. E., Babcock, H. P., Russell, R., Ha, T., Herschlag, D., & Chu, S. (2000). A single-molecule study of RNA catalysis and folding. *Science*, *288*(5473), 2048–2051.

CHAPTER TWO

Single-Molecule Imaging With One Color Fluorescence

Y. Qiu, S. Myong[1]
Johns Hopkins University, Baltimore, MD, United States
[1]Corresponding author: e-mail address: smyong@jhu.edu

Contents

Abstract

Single-molecule fluorescence imaging is a powerful tool that enables real-time observation of DNA–protein or RNA–protein interactions with a nanometer precision. Here, we provide a detailed procedure for a previously developed single-molecule fluorescence method, termed "single-molecule protein-induced fluorescence enhancement" (smPIFE). While smFRET (Förster resonance energy transfer) requires both donor and acceptor, protein-induced fluorescence enhancement (PIFE) employs a single dye and measures the increase in fluorescence intensity induced by protein binding near the dye. PIFE displays distance sensitivity within 0–4 nm, making it a powerful complementary or alternative tool to FRET method. In this chapter, we will discuss the various ways that PIFE has been utilized to study protein–nucleic acid interactions.

Methods in Enzymology, Volume 581
ISSN 0076-6879
http://dx.doi.org/10.1016/bs.mie.2016.08.011
© 2016 Elsevier Inc.
All rights reserved.

1. PROTEIN-INDUCED FLUORESCENCE ENHANCEMENT

Single-molecule Förster resonance energy transfer (FRET) assay is widely used to probe nucleic acid–protein interactions that cannot be resolved at an ensemble level (Abbondanzieri et al., 2008; Joo et al., 2006; Myong, Rasnik, Joo, Lohman, & Ha, 2005; Woodside et al., 2006). FRET occurs when an excitation of a donor fluorophore leads to a transfer of its energy to a nearby acceptor, leading to an anticorrelated change between the donor and acceptor intensity. FRET efficiency is calculated as the ratio of intensity emitted by the acceptor over the sum of acceptor and donor, hence can be used as a measure of distance between the two dyes. The two most common ways FRET is used to monitor protein–nucleic acid interactions include (i) an indirect measurement by placing two (FRET pair) dyes directly on DNA/RNA substrate at either end of single-strand DNA, for example and (ii) a direct measurement by applying singly labeling protein (donor, for example) to singly labeled DNA/RNA (acceptor, for example). While FRET is an excellent tool for determining relative distance change between the two dye-labeled molecules or molecular positions, it can be disadvantageous because fluorescent labeling of proteins is often inefficient, time consuming, and sometimes perturbs protein function. In addition, if the equilibrium binding constant of the protein is high, the high concentration of fluorescently labeled molecules may hinder single-molecule detection.

In single-molecule protein-induced fluorescence enhancement (smPIFE) assay, unlabeled protein is applied to fluorescently labeled DNA or RNA substrate. In this regard, the smPIFE presents a clear advantage over FRET because the protein-labeling process can be bypassed and the protein binding kinetic and movement are reported directly through the intensity of fluorescent dye attached to nucleic acid substrate (Hwang, Kim, & Myong, 2011; Myong et al., 2009). The intensity of a fluorophore is enhanced upon binding of a protein in its vicinity and the enhancement scales with the distance between the dye and protein. Due to the distance sensitivity of protein-induced fluorescence enhancement (PIFE) (0–4 nm) which complements that of FRET (3–8 nm) combined with the ease of use, PIFE is a powerful alternative single-molecule detection approach for protein–nucleic acid interactions (Hwang et al., 2011; Hwang & Myong, 2014).

1.1 How Does PIFE Work?

The maximum intensity emitted by a fluorophore is determined by the quantum yield of a particular dye. In 1984, Aramendia et al. observed that the viscosity of the local environment has a direct effect on the quantum yield of cyanine dyes (Aramendia, Negri, & San Roman, 1994). The change in the fluorescence quantum yield was attributed to the *cis–trans* isomerization of the cyanine dye from excited (fluorescent, photoactive) to dark (nonfluorescent) state. The isomerization entails a rotation of one carbon ring with respect to the other carbon ring which is connected through a carbon–carbon double bond (Fig. 1A). Protein binding near the fluorophore results in stabilizing the photoactive *trans* state, thus enhancing the fluorescence intensity, quantum yield, and lifetime (Aramendia et al., 1994; Levitus & Ranjit, 2011; Sanborn, Connolly, Gurunathan, & Levitus, 2007; Stennett, Ciuba, & Levitus, 2014).

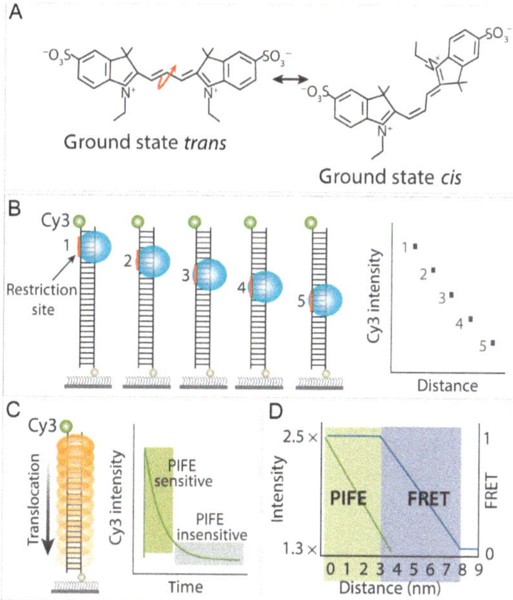

Fig. 1 (A) *cis–trans* Isomerization of a Cy3 dye. The *red arrow* indicates the rotation with respect to the carbon–carbon double bonds. (B) Schematic of the *Bam*HI protein binding to sequence-specific restriction sites positioned at five different base pair distances away from the Cy3 fluorophore. The closer the binding site, the brighter the Cy3 intensity. (C) Schematic of RIG-I protein translocates along the dsRNA axis. As RIG-I moves away from the Cy3 dye, the PIFE sensitivity decreases and movement is no longer detected beyond 4 nm. (D) Distance sensitivity of PIFE and FRET.

In addition, the location of the dye can also change the rigidity of *cis–trans* isomerization (Sanborn et al., 2007).

Cy3 is a popular choice of an organic fluorescent dye for PIFE and FRET experiments due to its high absorption coefficient, photostability, and modest quantum yield (Ha, 2001). The PIFE effect has also been observed with several other dyes including DY547, Cy5, and Alexa dyes (Hwang et al., 2011; Myong et al., 2009). In contrast, Cy3B which cannot undergo *cis–trans* isomerization exhibits no increase in quantum yield or fluorescent lifetime (Hwang et al., 2011). If a dye has the same type of chemical structure as the cyanine dyes, where two rings are interconnected by carbon–carbon double bonds that can undergo *cis–trans* isomerization, it is expected to exhibit the PIFE effect.

The fluorescence lifetime of the chosen dye can significantly influence the PIFE effect it produces. Sorokina et al. employed time-correlated single photon counting technique to demonstrate a stepwise increase in both the fluorescence intensity and lifetime when T7 RNA polymerase binds to a fluorescently labeled DNA (Sorokina, Koh, Patel, & Ha, 2009). A similar experiment (Fig. 1B) done using a restriction enzyme, *Bam*HI, binding 1–10 bp away from the Cy3 dye also reported corresponding intensity and lifetime changes (Hwang et al., 2011).

1.2 Early Uses of PIFE

Before PIFE was characterized in single-molecule experiments, the same photophysical effect had been utilized in stopped-flow ensemble measurements to investigate kinetics of DNA motor proteins (Fischer & Lohman, 2004; Fischer, Tomko, Wu, & Lohman, 2012; Tomko, Fischer, & Lohman, 2010). Additionally, a similar method was used to probe the binding of T7 Polymerase to DNA by tracking the intensity change (Luo, Wang, Konigsberg, & Xie, 2007). In 2009, we reported translocation activity of a human antiviral protein, RIG-I on double-stranded (ds) RNA, where the movement of RIG-I along dsRNA axis was visualized as gradual intensity increase and decrease in an ATP-dependent manner. Such movement was greatly accelerated on the viral mimic RNA, suggesting a viral recognition mechanism of RIG-I (Myong et al., 2009).

1.3 Distance Sensitivity of PIFE at Single-Molecule Resolution

The distance sensitivity of PIFE was determined by using a restriction enzyme, *Bam*HI, and the RIG-I translocation data (Hwang et al., 2011;

Myong et al., 2009). The binding site for *Bam*HI was specifically engineered to be 1–10 base pairs (bp) away from the Cy3 label positioned on one end of the double-stranded DNA (dsDNA), and the increase in Cy3 intensity was recorded to map out the distance sensitivity of PIFE (Fig. 1B). The PIFE effect was found to be the greatest (2.6 × enhancement) at 1 bp and the lowest (1.3 × enhancement) at 10 bp, with measurements in between showing an apparent linear relationship between the intensity and distance (Hwang & Myong, 2014). It is interesting to note that the lowest intensity is slightly higher than 1 ×. The 1.3 × enhancement is even observed when *Bam*HI binds 12 and 15 bp away from the fluorophore. We confirmed that this is not due to buffer but due to the protein binding; hence, it raises a possibility about an "action at a distance." This needs to be followed up by future studies.

Single-molecule traces obtained from RIG-I translocation showed repetitive translocation events that can be separated into two phases, the PIFE-sensitive and -insensitive region observed as a linear decline and a plateau in fluorescence intensity, respectively. The linear decrease corresponds to the PIFE-sensitive distance range when RIG-I translocates away from the fluorophore, and the plateaued region is interpreted as the PIFE-insensitive distance range where the protein movement is not detectable by PIFE. Assuming the rate of translocation occurs at a constant velocity, the RIG-I data indicated that the first 10–12 bp of movement was detectable by PIFE, which translates to 0–4 nm distance range (Fig. 1C). In light of the FRET-sensitive distance of 3–8 nm, PIFE fills in the 0–3 nm gap, thereby making it a powerful complementary tool to FRET (Fig. 1D).

2. EXPERIMENTAL PREPARATION FOR PIFE

As with single-molecule FRET measurements, PIFE can be conducted with either the prism- and objective-type total internal reflection fluorescence (TIRF) microscope. A detailed guide on TIRF microscope building has been previously published (Roy, Hohng, & Ha, 2008). Similar to FRET, the laser-excited fluorescence emissions from the DNA molecules on imaging surface are transmitted through the objective lens and captured by an electron multiplying charge-coupled device (EMCCD) camera. One major difference is that a dichroic beam splitter is unnecessary since the only one signal needs to be detected unless two-color PIFE is performed. Two- or multicolor PIFE is feasible if the dyes do not crosstalk or are positioned at a distance from one another.

Cy3 (donor) fluorophore on DNA is excited using an Nd:YAG laser (532 n*M*) via total internal reflection. The fluorescence signals from Cy3 are collected through an objective (Olympus Uplan S-Apo, 100×; 1.4 numerical aperture; oil immersion) and detected at 100 ms time resolution using an EMCCD camera (Andor iXon DU-897ECS0-#BV). The EMCCD camera is controlled using custom-made C++ program. Single-molecule traces can be extracted from the recorded video file using IDL software and analyzed with custom-made Matlab codes available for free on the CPLC software website (https://cplc.illinois.edu/software/).

2.1 Materials for Imaging

T50 buffer (a general wash buffer, also used for DNA or RNA immobilization)

10 m*M* Tris–HCl (pH 8.0) and 50 m*M* NaCl

NeutrAvidin

10 mg of NeutrAvidin powder is dissolved in 2 mL of T50 Buffer. The stock solution is diluted by adding 20–980 μL of T50 buffer. This can be store at 4°C for 2–3 months

Glucose oxidase–catalase

1 mg/mL glucose oxidase and 0.03 mg/mL catalase are dissolved in T50 buffer and stored at 4°C

Trolox

3–6 m*M* Trolox (6-hydroxy-2,5,7,8-tetramethylchromane-2-carboxylic) can be used instead of β-mercaptoethanol as shown in Section 2.4. Although the effect of trolox on smPIFE has not been studied, trolox does not seem to interfere with smPIFE imaging

2.2 Imaging Surface Preparation

We follow the same surface passivation protocol used for smFRET (Roy et al., 2008). Quartz slides and coverslips are cleansed by washing steps involving soap, water (Milli-Q), methanol, acetone, KOH, and burning. Then the slides are coated with a biotinylated PEG (1%) and mPEG (methoxypolyethylene glycol) (99%) mixture to form a passivation layer that prevents nonspecific binding of molecules to surface. The slides are then vacuum-packed (a pair of one cover slip and one quartz slide in one 50-mL Corning tube, stored in a plastic sandwich bag vacuum sealed) and stored at −80°C for future use.

On the day of experiment, a pair of cover slip and quartz slide are thawed at room temperature for 15 min, and assembled to create reaction chambers (3–4 chambers with inlet and outlet pores for each chamber) to which all reaction mixtures will be applied (Hwang & Myong, 2014). NeutrAvidin is added to the chamber in preparation for tethering biotinylated and fluorescently labeled nucleic acid substrate.

2.3 Oligo Substrate Preparation

DNA or RNA oligonucleotide labeled with fluorescent dye can be purchased from commercial vendors such as IDT DNA. For single-strand substrate, one end needs to be conjugated with fluorescent dye and the other end with biotin. For an internal labeling, oligos containing internal amino modifier deoxy-thymine (IDT DNA) can be labeled using Cy3 mono-functional NHS esters (Thermo Fisher). Briefly, 10 nmol of amino-modified oligonucleotides in 20 μL of 50 m*M* sodium tetraborate buffer (pH 8.5) and 100 nmol of Cy3 NHS ester dissolved in dimethylsulfoxide are mixed and incubated on a rotator overnight at room temperature. Since the dye is light sensitive, the labeling procedure should be performed in a dark room and/or the reaction tubes should be completely covered with aluminum foil. The labeled oligonucleotides are purified by ethanol precipitation repeated at least two times (Joo & Ha, 2012).

In case of preparing double-stranded or partial duplex DNA, the two complementary strands can be annealed by mixing the biotinylated and non-biotinylated oligonucleotides in a 1:2 molar ratio at 10 μ*M* in T50 buffer. Oligonucleotide mixtures are incubated at 95°C for 2 min followed by slow cooling to room temperature to complete the annealing reaction.

2.4 Imaging/Reaction Buffer Preparation

The imaging solution consists of a salt of choice (commonly used is NaCl or KCl) and a buffering reagent (e.g., TRIS, PBS, MOPS, HEPES). To reduce photobleaching of fluorescent dyes, the imaging buffer is supplemented with the oxygen scavenging system composed of 0.8% (v/v) dextrose, glucose oxidase, catalase, and 1% (v/v) β-mercaptoethanol (Roy et al., 2008).

2.5 Single-Molecule Fluorescence Assay

Annealed DNA or RNA substrates are diluted to 50–100 p*M* concentration and are applied to the NeutrAvidin coated slide surface. In this concentration range, we observe ~300–500 fluorescent spots in one field of view of

25 × 75 μm area (Joo, Balci, Ishitsuka, Buranachai, & Ha, 2008). If the number of spots is too low (below 100) or too high (above 600), PEG-biotin concentration can be modulated accordingly. It is important to have a clear separation between diffraction limited fluorescent spots for single-molecule detection in general. This is even more critical for smPIFE measurement because the PIFE signal is solely dependent on the fluorescence intensity of single molecules, i.e., even slightly overlapping signal between neighboring molecules may obscure the observable PIFE effect.

FRET value is determined by obtaining a ratio between donor and acceptor signals; hence, the noise in fluorescence signal due to thermal fluctuation or vibration is efficiently canceled out. In contrast, the readout from PIFE is inherently noisier because it is dependent on signals from single fluorophores. Therefore, it is crucial to establish a uniform intensity level before applying proteins for PIFE measurement. This can be best achieved by having an evenly illuminated field of view so that the resulting intensity histogram from all molecules yields one sharp peak. This will allow protein-induced intensity change to be clearly distinguished from DNA or RNA-only signals. We note that depending on the substrate design, the fluorescence signal may be noisier. For example, single-stranded DNA/RNA or structured substrate such as G-quadruplex (GQ) exhibits signal fluctuation due to the conformational flexibility within the substrate structure. In comparison, a fluorophore located at the end of duplexed substrate is less prone to noise due to the structural rigidity provided by the duplex.

2.6 Substrate Design

As with single-molecule FRET studies, there is a concern that the presence of fluorescent dye on substrate may perturb protein binding and kinetics. Therefore, it is critical to perform an orthogonal functional test to check for the potential dye effect. In the *Bam*HI study used to characterize PIFE distance sensitivity (Hwang et al., 2011), the *Bam*HI bound and cleaved the dsDNA regardless of the dye position. The DNA cleavage, detected by the disappearance of Cy3 dye, occurred immediately following the addition of magnesium. The similar rates of digestion observed in different constructs suggested that the fluorescent dye did not interfere with the function of the protein. Many previous single-molecule FRET studies also showed that the protein's biochemical activity (i.e., translocation, unwinding) is preserved when the DNA/RNA substrates are labeled with dyes Joo et al., 2006; Myong, Bruno, Pyle, & Ha, 2007; Myong et al., 2005;

Qiu et al., 2013; Tang, Roy, Bandwar, Ha, & Patel, 2009). However, it should be noted that the method of dye attachment can be important for the activity of the protein. For example, it is undesirable to attach a dye to the backbone of the DNA/RNA for helicases because many helicases use the phosphate backbone as a walking track (Myong et al., 2007, 2005; Myong & Ha, 2010; Park et al., 2010). In addition, dye can sometimes affect the folding of DNA secondary structures needed for the study. Commonly studied secondary structures such as hairpin junctions and GQ may not fold correctly if the fluorescent dye is located at important positions. This can be checked by using single-molecule FRET, as a control experiment, to see if the desirable FRET level is present. The GQ DNA structure can also be tested by using circular dichroism.

3. STATIC AND TRANSIENT PROTEIN BINDING

The most obvious advantage of PIFE measurement lies in its ability to directly visualize binding of protein to a fluorescently labeled substrate without the need to label or modify the protein in anyway. Any purified protein can be tested simply by applying to the fluorescently labeled substrate immobilized to surface. Once binding is confirmed, protein concentration can be titrated to obtain the protein's binding affinity. Fluorescence intensity histogram generated from collecting thousands of single-molecule traces can be quantified and compared before and after the addition of proteins. If the illumination field is uneven, it is still possible to construct and compare intensity histograms by using the relative value instead of the absolute intensity value. This is done by normalizing the individual protein-bound intensity by their respective protein-unbound intensity found within individual single-molecule traces. Once constructed, the area under the histogram curves can be quantified to estimate the amount bound and unbounded molecules, which can then be used to calculate K_d, the protein dissociation constant (Hwang & Myong, 2014).

Recent study employed smPIFE method to study DNA mismatch recognition/repair process (Jeon et al., 2016). Duplexed DNA with a single mismatch was prepared with Cy3, biotin, and digoxigenin (dig) attached as shown (Fig. 2A). In the presence on ATP, the mismatch recognition protein MSH2–MSH6 heterodimer forms a sliding clamp that undergoes diffusion of the dsDNA. The binding of MSH2–MSH6 to the mismatched DNA was visualized by PIFE signal (Fig. 2B) and the lifetime, T_{on}, was determined by fitting the individual dwell times (t_{on}) to exponential decay function

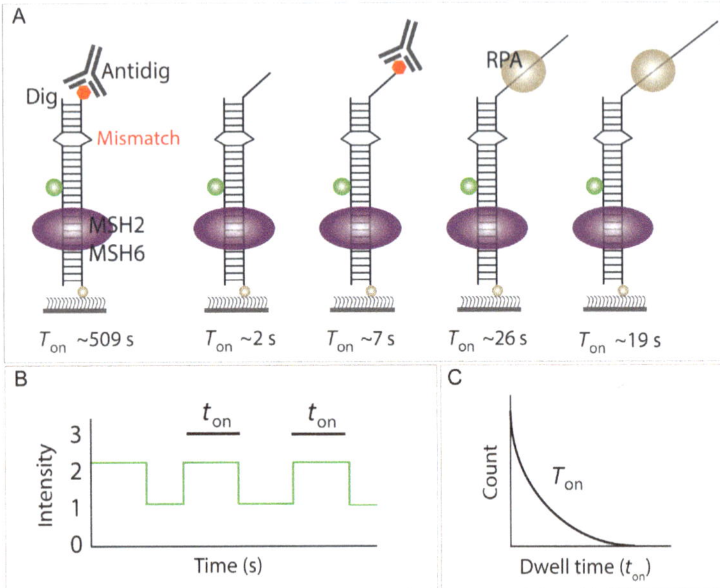

Fig. 2 (A) Schematic of MSH2–MSH6 heterodimer sliding on mismatched DNA. The substrate is labeled with a Cy3 dye near the mismatch as well as a digoxigenin at the one end and biotin on the other end to allow the immobilization of the DNA. (B) The binding of the heterodimer is detected via PIFE. Cy3 intensity increases when protein binds, and the duration of this increase is measured as t_{on}. (C) The overall lifetime T_{on} is determined by fitting all the t_{on} values collected to an exponential decay function.

(Fig. 2C). When the end was blocked with dig–antidig to trap freely diffusing sliding clamps, the T_{on} was extremely long (~509 s). In contrast, sliding clamp rapidly dissociated from DNA substrate containing a single-strand (ss) DNA tail, deduced from the short lifetime (Fig. 2A). The dissociation kinetic was similar regardless of the dig–antidig or RPA blocking, suggesting that the MSH2–MSH6 dissociation is stimulated upon encountering ssDNA–dsDNA junction.

While there are other conventional assays that could also detect the binding of protein to nucleic acid, it is impossible for ensemble technique to detect transient binding of the proteins to their substrate. Since its characterization in 2011, single-molecule PIFE assay has been used as the primary source of measurement for comparing the binding constant of a variety of proteins. Some examples include RIG-I and its truncation mutant (CARDless RIG-I) (Myong et al., 2009), Pol1A and its concentration-dependent binding constant (Markiewicz, Vrtis, Rueda, & Romano, 2012), and the heterodimeric and homodimeric forms of HIV-1 reverse transcriptase protein to DNA–DNA substrates (Marko et al., 2013). It is also often used as a complementary tool to single-molecule FRET, either to confirm the binding of a protein or complex structure to other substrates (Craggs, Hutton, Brenlla, White, & Penedo, 2014), or to estimate base pair distance based on the intensity increase.

4. FILAMENT FORMATION AND PROTEIN BINDING TO LONG DNA STRANDS

We have previously reported that the intensity of the Cy3 dye can be enhanced up to 2.6 × upon the binding of a single protein. This is the most striking advantage of PIFE over FRET, allowing the observation of protein binding kinetics without the need to label the protein. When multiple proteins bind as a group, however, the PIFE effect can reach as high as 3–4 times fluorescence enhancement. This effect was demonstrated when proteins such as RecA and Rad51 assemble into filaments around the DNA substrate (Hwang et al., 2011; Qiu et al., 2013). In our previous study, we used smFRET to determine the step size of Rad51 monomer (3 nt) that leads to filament formation. While stepwise FRET decrease reported on monomer Rad51 binding, it did not inform us about the directionality of filament formation. This was complemented by smPIFE which defined 5′ to 3′ filament growth direction (Hwang & Myong, 2014; Qiu et al., 2013). A partial duplex DNA with 18 bp of dsDNA with a 3′ ssDNA tail composed of polydeoxythymine was labeled with the Cy3 dye at either end of the ssDNA tail. After immobilizing the substrate DNA on single-molecule surface, 1 μ*M* of Rad51 and 1 m*M* ATP were added and the rate of PIFE increase was measured. The rate of intensity increase differed between the two dye positions since the Rad51 filament will grow toward the dye in one case and away from the dye in the other. The rate difference can only be measured by PIFE, and allows us to determine the direction of filament growth.

In a recent study by Song et al., smPIFE was utilized to monitor protein binding and DNA remodeling process (Song, Graham, & Loparo, 2016). A 20 kilobase (kb) of dsDNA was sparsely labeled with Cy3 fluorophores and immobilized to single-molecule surface by biotin–streptavidin (Fig. 3A, step 1). Flow was applied to add the DNA binding proteins, which resulted in PIFE effect (Fig. 3A and B, step 2). Subsequently, DNA compaction/shortening was monitored (Fig. 3A and B, step 3) as reduction in DNA length. In contrast, a mutant protein defective in DNA remodeling activity

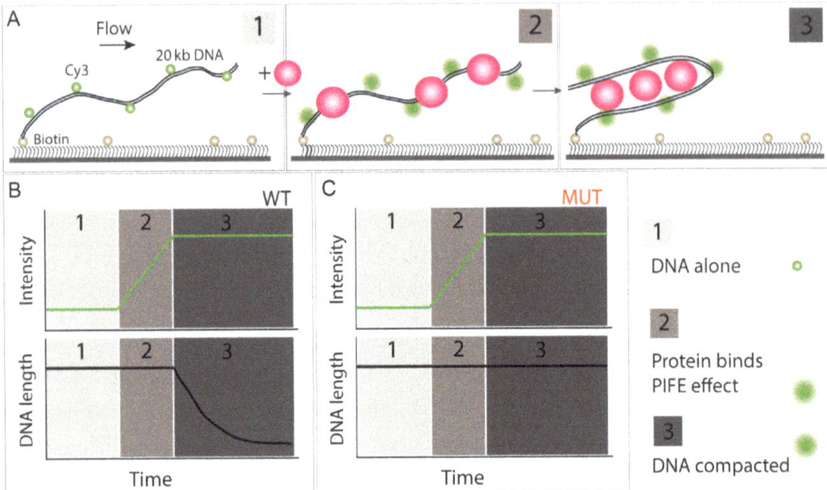

Fig. 3 (A) Schematic of multiple proteins binding on a long dsDNA sparsely labeled with Cy3 fluorophores. (B) As each protein binds, the Cy3 intensity increases until (C) DNA compaction is finished and maximum PIFE intensity has increased.

resulted in a similar PIFE effect followed by no change in DNA length, in agreement with the known defect in DNA compaction activity (Fig. 3C). This unique imaging capability allowed the authors to determine the protein binding kinetic and the rate of DNA compaction based on the dwell times collected from single-molecule traces.

5. PROTEIN BINDING AND MOTILITY KINETICS

While PIFE is extremely useful as a direct indicator of single or even multiple protein binding, it can also be an efficient tool to measure movement of motor proteins or intrinsically dynamic proteins that slide on DNA/RNA substrates. Many proteins exhibit sliding activity on DNA/RNA after binding. PIFE signal increases as protein translocates toward the fluorophore. In cases where protein traverses through DNA substrates, different DNA lengths can be tested. If the rate of PIFE fluctuation is correlated with the length of DNA/RNA substrate, it indicates that the protein traverses through the entire length of the substrate rather than a defined length. This rate of translocation or sliding can then be estimated from the speed of fluorescence increase based on the PIFE distance sensitivity. The rate is determined assuming a linear intensity–distance dependence based on our previous calibration, i.e., we assume the maximum and minimum intensity to correspond to 0 and approximately 4 nm from the dye (Hwang et al., 2011; Hwang & Myong, 2014). In cases where the protein activity is stimulated by ATP, the same kinetic experiment can be performed with varying ATP concentrations, which yields the rate of protein translocation at different ATP concentrations. The rate can be fitted to the Michaelis–Menten equation for obtaining K_m and V_{max} for the given protein (Hwang & Myong, 2014).

Recent study by Zhou et al. implemented smPIFE in conjunction with smFRET to determine the mechanism of ssDNA translocase, Pif1 (Zhou, Zhang, Bochman, Zakian, & Ha, 2014). The repetitive translocation activity was detected by smFRET which displayed periodic FRET increase and decrease, consistent with repetitive "reeling-in" activity of the protein. By performing smPIFE measurement on the same DNA with Cy3 situated in different locations, i.e., either tip of ssDNA, Pif1 produced a periodic PIFE signal fluctuation on one substrate (Fig. 4A) but not on the alternate substrate (Fig. 4B). Based on this result, the authors confirmed that Pif1 remains at ssDNA/dsDNA junction while periodically reeling in ssDNA powered by ATP hydrolysis.

Our recent study extended the usage of PIFE technique by testing proteins extracted from mammalian cell lysate (Wang, Vukovic, Koh, Schulten, & Myong, 2015). The antibody-based single-molecule pull-down technique was developed by Jain et al. (2011). Previously, we reported that a double-strand RNA binding protein (dsRBP), called Tar RNA binding protein (TRBP), exhibits ATP-independent sliding activity on dsRNA (Koh, Kidwell, Ragunathan, Doudna, & Myong, 2013). To examine if other dsRBPs also display such motility, we conducted a comprehensive study by expanding the array of dsRBPs including ADAD2, Staufen1, and ADAR1 in addition to TRBP. The four dsRBPs were chosen based on the presence of highly conserved dsRBD-I (type I dsRBD) domains even though they differed in total length and overall domain composition. The dsRBP protein was applied to single-molecule surface by adding cell lysate containing overexpressed dsRBP (Fig. 4C). Each protein was tagged with yellow fluorescence protein (YFP), which enabled visualizing the protein expression in cells and pulling down of dsRBP to single-molecule surface coated with anti-YFP antibody (Fig. 4C). Due to an extremely fast photobleaching (<3 s), the YFP does not interfere with the PIFE measurement. On this platform, nonbiotinylated, Cy3-labeled dsRNA (>25 bp) was added to check for the diffusion activity of the four dsRBPs. We performed such experiment on six structural variants of RNA substrates. The PIFE

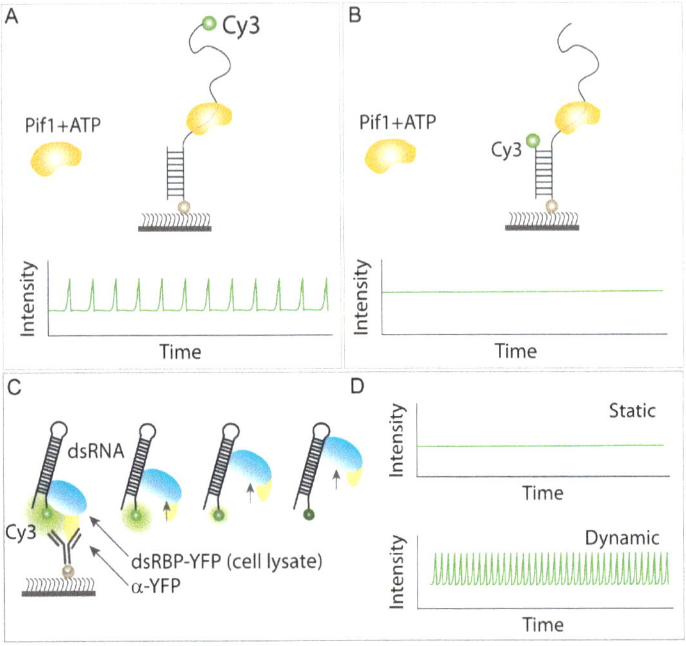

Fig. 4 (A) Schematic of Pif1 translocating on a DNA substrate that has a Cy3 dye at the free end of the ssDNA. For example, single-molecule trace below shows periodic intensity fluctuations corresponding to Pif1 coming near the dye. (B) Schematic of Pif1 translocating on a DNA substrate that has a Cy3 dye at the junction of the duplex DNA. For example, single-molecule trace below shows no increase in PIFE intensity, suggesting that Pif1 movement is not within the PIFE sensitivity range of the Cy3 dye. (C) Schematic of dsRBP binding experiments using mammalian cell lysate, dsRNA, and detection by PIFE. Two types of single-molecule traces are commonly seen in (D). A static intensity trace (*upper*) indicates protein binding only, and a dynamic intensity trace (*lower*) featuring periodic fluctuations indicates protein diffusion along the dsRNA.

signal fluctuation indicated sliding activity of proteins near the dye. This is a reverse configuration as the protein, instead of the dsRNA, was immobilized to the surface and movement of the RNA along dsRBP was monitored (Fig. 4D). The fraction of sliding molecules was quantified and TRBP showed the highest sliding activity (70%) out of the four, followed by Staufen1 (20–30%). Dwell time analysis for RNA length variants displayed a clear length dependence, indicating that the dsRBP likely slides on the entire length of dsRNA (Wang et al., 2015). Briefly, for each RNA tested, the rate of sliding, or dwell time, can be calculated from the peak-to-peak values of the PIFE fluctuations. The peak-to-peak dwell time represents the time in which the protein undergoes sliding away from the dye. Therefore, the dwell time indicates how long it takes for the protein to complete one cycle (round trip) of sliding. The dye is positioned at one end of the RNA substrates to minimize interfering with protein binding or sliding and also for comparing dwell times in different RNA lengths.

6. PIFE IN THE PRESENCE OF FRET

If the experiment is designed for FRET measurement, it is the most direct to interpret the result by FRET alone rather than decoupling PIFE and FRET effect. Nevertheless, there are some cases in which PIFE effect can be observed or deduced in FRET experiment. First, the PIFE effect can be seen after the photobleaching of an acceptor dye. This occurs frequently because Cy5 (acceptor) typically undergoes photobleaching more rapidly than Cy3 (donor). If the repetitive translocation activity of a motor protein such as PcrA (Park et al., 2010) and Pif1 (Zhou et al., 2014) was monitored on FRET–DNA construct with Cy3 and Cy5 attached at either end of ssDNA, the reeling-in of ssDNA can be visualized by the periodic FRET fluctuation. After the photobleaching of Cy5 dye at the ssDNA/dsDNA junction, the same activity can be displayed by PIFE signal change (Fig. 5A). In such cases, the confirmatory information obtained from PIFE experiment is that the protein is mediating the reeling-in motion by physically bringing in the end of ssDNA. Second, the PIFE effect can be observed when the protein binding occurs near both donor and acceptor dyes. We took advantage of this effect in our previous study in which RNA helicase A (RHA) protein binding to partially duplexed RNA substrate was visualized as a simultaneous increase in both Cy3 and Cy5 intensity (Fig. 5B). Since both dyes exhibit the *cis–trans* isomerization, it is reasonable to expect PIFE effect on Cy3 and Cy5. We note that we are not comparing PIFE effect between the two dyes quantitatively. We are interpreting the presence of PIFE effect on both dyes as a signature of protein binding near both dyes. This is particularly useful because while the protein binding does not induce FRET change, leaving the PIFE signal as the only means to observe the protein binding (Fig. 5C) (Koh, Xing, Kleiman, & Myong, 2014). The PIFE effect is more pronounced in Cy3 than in Cy5 because RHA binding occurs at the ssRNA/dsRNA junction more proximal to Cy3. Such effect can also be seen in the total intensity (sum of Cy3 and Cy5) trajectory in which the protein binding induces a clear stepwise increase in the total intensity. The time interval between the addition of protein (Fig. 5C, black arrow) and binding of protein (Fig. 5C, red arrow) was collected from many molecules to obtain the

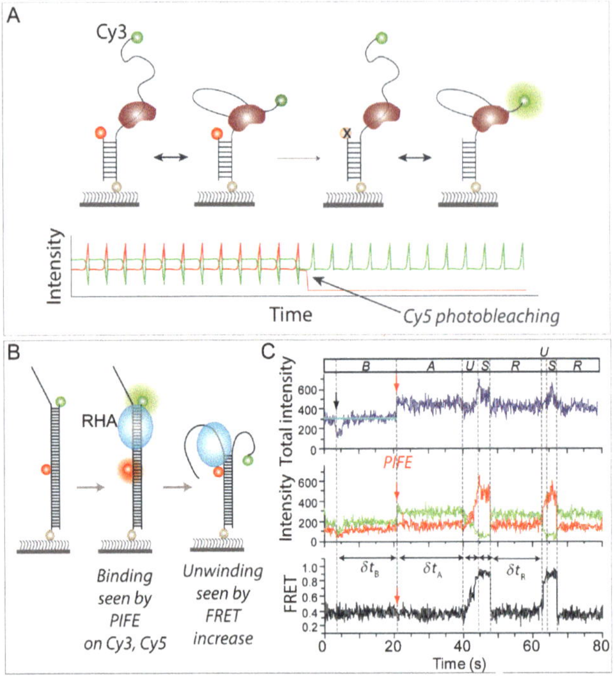

Fig. 5 (A) Schematic of protein interaction with a FRET DNA structure that has the Cy5 dye photobleaches after some time. Corresponding intensity traces show anticorrelated Cy3–Cy5 intensity traces that changes to Cy3 intensity only as Cy5 photobleaches. Note that the periodic intensity fluctuations remain even after Cy5 photobleaches. (B, C) Schematic and single-molecule trace of RHA protein binding to partially duplex RNA substrate labeled with FRET and its eventual unwinding. Binding of protein does not induce change in FRET level but can be seen in the PIFE increase.

binding kinetic. This enabled us to distinguish the binding rate (*B*) and the off rate (*A*), which cannot be discerned based on FRET alone (Fig. 5C). PIFE and FRET can be combined as long as the PIFE effect occurs without affecting FRET. In addition to the example given above, one can design a substrate in which two dyes are outside of FRET detection range. PIFE effect on one side can be distinguished from FRET change since only one intensity will increase without the anticorrelated decrease in the other dye.

7. CONCLUSIONS

Since its initial characterization in 2011, the smPIFE has become an increasingly useful tool for researchers to conduct single-molecule protein binding study without protein modification. The multitude of proteins for which PIFE technique has been applied strongly reflects the universality of PIFE method. Exceptions can be for proteins that possess nickel or manganese metal binding pocket and FeS-containing proteins (Spies, 2014) which may quench the fluorescence when it comes in contact with the fluorophore, although we have never encountered such a case. The high sensitivity of PIFE enables detection of very transient interactions that may not be resolved by biochemical means such as EMSA. In addition, as demonstrated by our previous study on Rad51, the directionally biased intensity increase (always toward dye) makes PIFE an ideal tool for determining the direction of filament growth. Most of the proteins studied so far are involved in the DNA and RNA processing pathways; therefore, it will be exciting to foray into proteins that modify other proteins such as kinases, phosphorylases, and methyl transferases in the near future.

REFERENCES

Abbondanzieri, E. A., Bokinsky, G., Rausch, J. W., Zhang, J. X., Le Grice, S. F., & Zhuang, X. (2008). Dynamic binding orientations direct activity of HIV reverse transcriptase. *Nature*, *453*(7192), 184–189. http://dx.doi.org/10.1038/nature06941.

Aramendia, P. F., Negri, R. M., & San Roman, E. (1994). Temperature dependence of fluorescence and photoisomerization in symmetric carbocyanines. Influence of medium viscosity and molecular structure. *The Journal of Physical Chemistry*, *98*, 3165–3173.

Craggs, T. D., Hutton, R. D., Brenlla, A., White, M. F., & Penedo, J. C. (2014). Single-molecule characterization of Fen1 and Fen1/PCNA complexes acting on flap substrates. *Nucleic Acids Research*, *42*(3), 1857–1872. http://dx.doi.org/10.1093/nar/gkt1116.

Fischer, C. J., & Lohman, T. M. (2004). ATP-dependent translocation of proteins along single-stranded DNA: Models and methods of analysis of pre-steady state kinetics. *Journal of Molecular Biology*, *344*(5), 1265–1286. http://dx.doi.org/10.1016/j.jmb.2004.10.004.

Fischer, C. J., Tomko, E. J., Wu, C. G., & Lohman, T. M. (2012). Fluorescence methods to study DNA translocation and unwinding kinetics by nucleic acid motors. *Methods in Molecular Biology*, *875*, 85–104. http://dx.doi.org/10.1007/978-1-61779-806-1_5.

Ha, T. (2001). Single-molecule fluorescence methods for the study of nucleic acids. *Current Opinion in Structural Biology*, *11*(3), 287–292. S0959-440X(00)00204-9 [pii].

Hwang, H., Kim, H., & Myong, S. (2011). Protein induced fluorescence enhancement as a single molecule assay with short distance sensitivity. *Proceedings of the National Academy of Sciences of the United States of America*, *108*(18), 7414–7418. http://dx.doi.org/10.1073/pnas.1017672108.

Hwang, H., & Myong, S. (2014). Protein induced fluorescence enhancement (PIFE) for probing protein-nucleic acid interactions. *Chemical Society Reviews*, *43*(4), 1221–1229. http://dx.doi.org/10.1039/c3cs60201j.

Jain, A., Liu, R., Ramani, B., Arauz, E., Ishitsuka, Y., Ragunathan, K., … Ha, T. (2011). Probing cellular protein complexes using single-molecule pull-down. *Nature*, *473*(7348), 484–488. http://dx.doi.org/10.1038/nature10016.

Jeon, Y., Kim, D., Martin-Lopez, J. V., Lee, R., Oh, J., Hanne, J., … Lee, J. B. (2016). Dynamic control of strand excision during human DNA mismatch repair. *Proceedings of the National Academy of Sciences of the United States of America*, *113*(12), 3281–3286. http://dx.doi.org/10.1073/pnas.1523748113.

Joo, C., Balci, H., Ishitsuka, Y., Buranachai, C., & Ha, T. (2008). Advances in single-molecule fluorescence methods for molecular biology. *Annual Review of Biochemistry*, 77, 51–76. http://dx.doi.org/10.1146/annurev.biochem.77.070606.101543.

Joo, C., & Ha, T. (2012). Labeling DNA (or RNA) for single-molecule FRET. *Cold Spring Harbor Protocols*, *2012*(9), 1005–1008. http://dx.doi.org/10.1101/pdb.prot071027.

Joo, C., McKinney, S. A., Nakamura, M., Rasnik, I., Myong, S., & Ha, T. (2006). Real-time observation of RecA filament dynamics with single monomer resolution. *Cell*, *126*(3), 515–527. http://dx.doi.org/10.1016/j.cell.2006.06.042.

Koh, H. R., Kidwell, M. A., Ragunathan, K., Doudna, J. A., & Myong, S. (2013). ATP-independent diffusion of double-stranded RNA binding proteins. *Proceedings of the National Academy of Sciences of the United States of America*, *110*(1), 151–156. http://dx.doi.org/10.1073/pnas.1212917110.

Koh, H. R., Xing, L., Kleiman, L., & Myong, S. (2014). Repetitive RNA unwinding by RNA helicase A facilitates RNA annealing. *Nucleic Acids Research*, *42*(13), 8556–8564. http://dx.doi.org/10.1093/nar/gku523.

Levitus, M., & Ranjit, S. (2011). Cyanine dyes in biophysical research: The photophysics of polymethine fluorescent dyes in biomolecular environments. *Quarterly Reviews of Biophysics*, *44*(1), 123–151. http://dx.doi.org/10.1017/S0033583510000247.

Luo, G., Wang, M., Konigsberg, W. H., & Xie, X. S. (2007). Single-molecule and ensemble fluorescence assays for a functionally important conformational change in T7 DNA polymerase. *Proceedings of the National Academy of Sciences of the United States of America*, *104*(31), 12610–12615. http://dx.doi.org/10.1073/pnas.0700920104.

Markiewicz, R. P., Vrtis, K. B., Rueda, D., & Romano, L. J. (2012). Single-molecule microscopy reveals new insights into nucleotide selection by DNA polymerase I. *Nucleic Acids Research*, *40*(16), 7975–7984. http://dx.doi.org/10.1093/nar/gks523.

Marko, R. A., Liu, H. W., Ablenas, C. J., Ehteshami, M., Gotte, M., & Cosa, G. (2013). Binding kinetics and affinities of heterodimeric versus homodimeric HIV-1 reverse transcriptase on DNA-DNA substrates at the single-molecule level. *The Journal of Physical Chemistry B*, *117*(16), 4560–4567. http://dx.doi.org/10.1021/jp308674g.

Myong, S., Bruno, M. M., Pyle, A. M., & Ha, T. (2007). Spring-loaded mechanism of DNA unwinding by hepatitis C virus NS3 helicase. *Science*, *317*(5837), 513–516. http://dx.doi.org/10.1126/science.1144130. 317/5837/513 [pii].

Myong, S., Cui, S., Cornish, P. V., Kirchhofer, A., Gack, M. U., Jung, J. U., … Ha, T. (2009). Cytosolic viral sensor RIG-I is a 5'-triphosphate-dependent translocase on double-stranded RNA. *Science*, *323*(5917), 1070–1074. http://dx.doi.org/10.1126/science.1168352.

Myong, S., & Ha, T. (2010). Stepwise translocation of nucleic acid motors. *Current Opinion in Structural Biology*, *20*(1), 121–127. http://dx.doi.org/10.1016/j.sbi.2009.12.008. S0959-440X(09)00196-1 [pii].

Myong, S., Rasnik, I., Joo, C., Lohman, T. M., & Ha, T. (2005). Repetitive shuttling of a motor protein on DNA. *Nature*, *437*(7063), 1321–1325. http://dx.doi.org/10.1038/nature04049. nature04049 [pii].

Park, J., Myong, S., Niedziela-Majka, A., Lee, K. S., Yu, J., Lohman, T. M., & Ha, T. (2010). PcrA helicase dismantles RecA filaments by reeling in DNA in uniform steps. *Cell*, *142*(4), 544–555. http://dx.doi.org/10.1016/j.cell.2010.07.016.

Qiu, Y., Antony, E., Doganay, S., Koh, H. R., Lohman, T. M., & Myong, S. (2013). Srs2 prevents Rad51 filament formation by repetitive motion on DNA. *Nature Communications*, *4*, 2281. http://dx.doi.org/10.1038/ncomms3281.

Roy, R., Hohng, S., & Ha, T. (2008). A practical guide to single-molecule FRET. *Nature Methods*, *5*(6), 507–516. http://dx.doi.org/10.1038/nmeth.1208.

Sanborn, M. E., Connolly, B. K., Gurunathan, K., & Levitus, M. (2007). Fluorescence properties and photophysics of the sulfoindocyanine Cy3 linked covalently to DNA. *The Journal of Physical Chemistry B*, *111*(37), 11064–11074. http://dx.doi.org/10.1021/jp072912u.

Song, D., Graham, T. G., & Loparo, J. J. (2016). A general approach to visualize protein binding and DNA conformation without protein labelling. *Nature Communications*. 7, 10976. http://dx.doi.org/10.1038/ncomms10976.

Sorokina, M., Koh, H. R., Patel, S. S., & Ha, T. (2009). Fluorescent lifetime trajectories of a single fluorophore reveal reaction intermediates during transcription initiation. *Journal of the American Chemical Society*, *131*(28), 9630–9631. http://dx.doi.org/10.1021/ja902861f.

Spies, M. (2014). Two steps forward, one step back: Determining XPD helicase mechanism by single-molecule fluorescence and high-resolution optical tweezers. *DNA Repair*, *20*, 58–70. http://dx.doi.org/10.1016/j.dnarep.2014.01.013.

Stennett, E. M., Ciuba, M. A., & Levitus, M. (2014). Photophysical processes in single molecule organic fluorescent probes. *Chemical Society Reviews*, *43*(4), 1057–1075. http://dx.doi.org/10.1039/c3cs60211g.

Tang, G. Q., Roy, R., Bandwar, R. P., Ha, T., & Patel, S. S. (2009). Real-time observation of the transition from transcription initiation to elongation of the RNA polymerase. *Proceedings of the National Academy of Sciences of the United States of America*, *106*(52), 22175–22180. http://dx.doi.org/10.1073/pnas.0906979106.

Tomko, E. J., Fischer, C. J., & Lohman, T. M. (2010). Ensemble methods for monitoring enzyme translocation along single stranded nucleic acids. *Methods*, *51*(3), 269–276. http://dx.doi.org/10.1016/j.ymeth.2010.03.010.

Wang, X., Vukovic, L., Koh, H. R., Schulten, K., & Myong, S. (2015). Dynamic profiling of double-stranded RNA binding proteins. *Nucleic Acids Research*, *43*(15), 7566–7576. http://dx.doi.org/10.1093/nar/gkv726.

Woodside, M. T., Anthony, P. C., Behnke-Parks, W. M., Larizadeh, K., Herschlag, D., & Block, S. M. (2006). Direct measurement of the full, sequence-dependent folding landscape of a nucleic acid. *Science*, *314*(5801), 1001–1004. http://dx.doi.org/10.1126/science.1133601.

Zhou, R., Zhang, J., Bochman, M. L., Zakian, V. A., & Ha, T. (2014). Periodic DNA patrolling underlies diverse functions of Pif1 on R-loops and G-rich DNA. *eLife*, *3*, e02190. http://dx.doi.org/10.7554/eLife.02190.

CHAPTER THREE

Measuring Membrane Protein Dimerization Equilibrium in Lipid Bilayers by Single-Molecule Fluorescence Microscopy

R. Chadda, J.L. Robertson[1]
The University of Iowa, Iowa City, IA, United States
[1]Corresponding author: e-mail address: janice-robertson@uiowa.edu

Contents

Abstract

Dimerization of membrane protein interfaces occurs during membrane protein folding and cell receptor signaling. Here, we summarize a method that allows for measurement of equilibrium dimerization reactions of membrane proteins in lipid bilayers, by measuring the Poisson distribution of subunit capture into liposomes by single-molecule

Methods in Enzymology, Volume 581
ISSN 0076-6879
http://dx.doi.org/10.1016/bs.mie.2016.08.025
© 2016 Elsevier Inc.
All rights reserved.

photobleaching analysis. This strategy is grounded in the fact that given a comparable labeling efficiency, monomeric or dimeric forms of a membrane protein will give rise to distinctly different photobleaching probability distributions. These methods have been used to verify the dimer stoichiometry of the Fluc F^- ion channel and the dimerization equilibrium constant of the ClC-ec1 Cl^-/H^+ antiporter in lipid bilayers. This approach can be applied to any membrane protein system provided it can be purified, fluorescently labeled in a quantitative manner, and verified to be correctly folded by functional assays, even if the structure is not yet known.

1. INTRODUCTION

Studies of membrane protein dimerization have been of keen interest in biological research for two reasons. First, there are situations where protein dimerization is physiologically significant, for instance, in the case of receptor activation (Ferré et al., 2014; Kovacs, Zorn, Huang, Barros, & Kuriyan, 2015). Second, dimerization serves as an equilibrium model for studying the physical forces underlying protein folding. In fact, for multihelix membrane proteins, where reversible assembly is practically intractable, dimerization offers the simplest strategy to obtain a thermodynamic understanding of how protein interactions are stabilized in the nonpolar solvent environment of the lipid bilayer. While equilibrium dimerization reactions have been measured in detergent micelles (Fleming, Ackerman, & Engelman, 1997; MacKenzie & Fleming, 2008), numerous technical challenges have limited our ability to study these reactions in lipid bilayers.

Dimerization involves the self-association of monomers (M) to form assembled dimers (D):

$$M + M \rightleftharpoons D \tag{1}$$

Since membrane proteins are solvated in a two-dimensional lipid bilayer, the reaction is typically measured on a subunit/lipid mole fraction scale, χ, with the following equilibrium constant, K_{eq}:

$$K_{eq} = \frac{\chi_D}{\chi_M^2} \tag{2}$$

where χ_M and χ_D are the monomer and dimer mole fraction densities, respectively (Fleming, 2002; White & Wimley, 1999). For soluble proteins in aqueous solution, the equilibrium constant can be measured by simply diluting the protein with buffer and measuring the shift in the fraction of dimer (F_{Dimer}) as a function of the total protein concentration. However, dilution is not an easy task for membrane proteins, as the solvent forms a lipid bilayer structure that does not readily exchange with other membranes. Another technical challenge arises from the fact that we work with relatively low concentrations of lipids ~35 m*M* during reconstitution, a ~2000-fold reduction in the solvent phase compared to 55.5 *M* water, resulting in a proportional decrease in protein signal. For example, to measure Glycophorin A dimerization by traditional dilution (K_{eq} ~10^9 POPC lipid/subunit (Hong, Blois, Cao, & Bowie, 2010)) one would have to detect the protein at picomolar concentrations. Other methods have been developed to extract the equilibrium constant while examining the protein at high densities in liposomes (Hong et al., 2010; Mathiasen et al., 2014); however, this may introduce other technical challenges, as many membrane proteins undergo aggregation at high densities. Therefore, macroscopic approaches for measuring membrane protein dimerization are inherently challenged as high affinity dimers must be diluted to ridiculously low concentrations in order to observe the dissociated state.

Single-molecule fluorescence microscopy offers a solution to this issue as it can detect single molecules within the membrane of proteoliposome. Furthermore, photobleaching analysis allows for rigorous determination of protein stoichiometry on the single-particle level (Stockbridge, Robertson, Kolmakova-Partensky, & Miller, 2013). To solve the practical problem of diluting the protein in the membrane, we fuse lipid bilayers through multiple freeze/thaw cycles creating large ~10 μm diameter multilamellar vesicles (MLVs) (Pozo Navas et al., 2005), our model of an infinite membrane whereupon the equilibrium reaction occurs. To prepare this system for single-molecule total internal reflection fluorescence (TIRF) microscopy, the MLVs are fractioned into liposomes, each carrying a small area of the membrane along with the protein subunits captured in that patch. Membrane protein capture into liposomes follows a Poisson distribution (Maduke, Pheasant, & Miller, 1999; Walden et al., 2007), wherein the number of protein particles per liposome depends on the protein–protein proximity in the large membrane state. Therefore, measurement of liposome occupancy by single-molecule photobleaching analysis allows us to follow the changing monomer–dimer population in the large membrane and extract the equilibrium constant that pertains to the statistical mechanical free energy of dimerization. This method (Fig. 1 and Table 1) requires knowledge of the extruded liposome size distribution and experimental determination of subunit fluorescent labeling yields. Protein/liposome colocalization microscopy allows for quality checks on

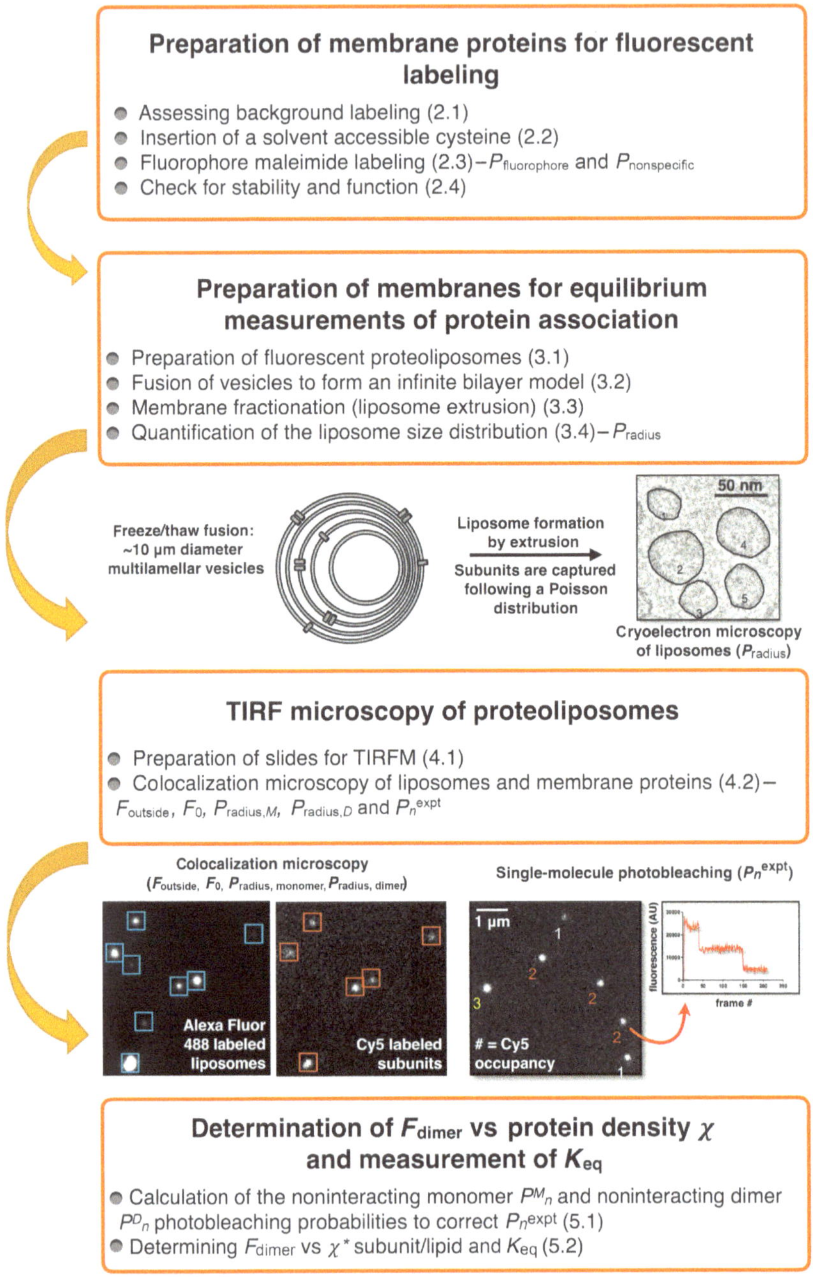

Fig. 1 Flowchart outlining the steps involved in measuring equilibrium membrane protein dimerization in lipid bilayers by subunit capture followed by single-molecule photobleaching analysis. See Table 1 for definition of experimental parameters.

Table 1 List of Essential Experimental Parameters for Determining the Fraction of Dimerization of a Protein in the Lipid Bilayer as a Function of the Subunit/Lipid Mole Fraction χ

Method/ Parameter	Description	Section
UV–visible spectrophotometry		
$P_{\text{fluorophore}}$	Overall labeling yield (fluorophore/subunits)	2.3
$P_{\text{nonspecific}}$	Nonspecific labeling yield (fluorophore/subunits), also referred to as P_{ns}	2.3
P_{cysteine}	Labeling yield at specific cysteine, $P_{\text{cysteine}} = P_{\text{fluorophore}} - P_{\text{nonspecific}}$, also referred to as P_{c}	2.3
Cryo-EM imaging		
P_{radius}	The liposome radius distribution from cryoelectron imaging of extruded liposomes	3.4
Colocalization microscopy		
F_{outside}	The fraction of protein spots that do not colocalize with liposomes. Measures fidelity of reconstitution	4.2.1
F_0	Fraction of unoccupied vesicles, i.e., liposomes that do not colocalize with protein	4.2.2
$P_{\text{radius},M}$	Renormalized liposome size distribution for monomer accessible liposomes	4.2.2
$P_{\text{radius},D}$	Renormalized liposome size distribution for dimer accessible liposomes	4.2.2
P_n^{expt}	The experimental photobleaching probabilities	4.2.3
Simulation		
P_n^M	The calculated photobleaching probabilities for noninteracting monomers. Requires experimental determination of P_{radius}, $P_{\text{radius},M}$, $P_{\text{fluorophore}}$, and $P_{\text{nonspecific}}$	5.1.1
P_n^D	The calculated photobleaching probabilities for noninteracting dimers. Requires experimental determination of P_{radius}, $P_{\text{radius},D}$, $P_{\text{fluorophore}}$, and $P_{\text{nonspecific}}$	5.1.2
F_{Dimer}	The fraction of dimer determined by least-squares fitting of the P_n^{expt} to $(1 - F_{\text{Dimer}})P_n^M + F_{\text{Dimer}} {*} P_n^D$	5.2

reconstitution, inspection of protein aggregation or nonspecific reactions, determination of differences in liposome accessibility, and, finally, the measurement of the photobleaching probability distribution from which the monomer–dimer equilibrium constant can be obtained. This method has recently been used to measure K_{eq} of the ClC-ec1 Cl^-/H^+ antiporter in 2:1 POPE/POPG lipid bilayers (Chadda et al., 2016) and can be adapted to study new types of membrane proteins systems using a similar approach.

2. PREPARATION OF MEMBRANE PROTEINS FOR FLUORESCENT STUDIES

In this section, we will discuss the strategy for subunit specific labeling of a membrane protein for single-molecule fluorescence microscopy experiments in lipid bilayers. We will assume that purification of stable histidine-tagged protein in detergent micelles has already been established, along with reconstitution methods into liposomes. We recommend that there be a quantitative method for assaying functional activity (Stockbridge & Tsai, 2015) as this is the most rigorous measure of whether the protein is properly folded.

In order to measure a dimerization reaction in the membrane using single-molecule photobleaching, each subunit should be labeled at a unique position with a single fluorophore. Hundred percent labeling is desirable for photobleaching experiments but it is unrealistic in a laboratory setting. In addition, fluorophores can irreversibly bleach or switch to a dark state (Ha & Tinnefeld, 2012) lowering the effective labeling yield. Increasing the reaction time or rate may improve the yield, but it also increases non-specific labeling. In reality, both of these situations can be tolerated within reason and corrected for if the overall labeling yield, $P_{\text{fluorophore}}$, and the nonspecific labeling yield, $P_{\text{nonspecific}}$, can be experimentally determined. Using the following procedure, we obtained labeling yields for ClC-ec1 $P_{\text{Cy5}}=0.72$ and $P_{\text{nonspecific}}=0.14$, which allowed for adequate discrimination between the monomer and dimer photobleaching probabilities.

2.1 Assessing the Protein for Background Labeling

We use the standard method of cysteine–maleimide conjugation because it is rapid and fairly specific, but there are other means to attach fluorophores to proteins (Hermanson, 2013). It is important to have only one fluorophore per subunit, so native cysteines should be removed and replaced with a single-reactive cysteine. If removal of all cysteines affects thé stability of the protein, then it is possible to screen and only remove the ones that are reactive when the protein is in a folded state. For example, ClC-ec1 has three endogenous cysteines, but only one is reactive in the folded structure (Chadda et al., 2016). The cysteine accessibility Ellman's reagent assay (Riddles, Blakeley, & Zerner, 1983) measures the amount of reactive thiol groups present, and then step-by-step mutagenesis will identify the cysteines that must be removed. Note that the sulfhydryl groups can be oxidized in vivo during the expression of the protein, so TCEP–HCl (tris(2-carboxyethyl)phosphine) is added to maintain the reduced state of the thiol (5 m*M* in the initial cell suspension, followed by 1 m*M* in metal affinity or ion affinity buffers). The pH of the buffers must be adjusted after addition of TCEP–HCl. The final purification step (size exclusion chromatography) does not contain TCEP as it can interfere with the maleimide reaction. We find that the cysteines on the purified protein remain reactive for weeks; however a slow loss of reactivity is observed, possibly due to irreversible oxidation (Reddie & Carroll, 2008).

2.1.1 Measurement of Cysteine Accessibility by Ellman's Reagent

In this reaction, the thiolate form of the cysteine sulfhydryl group reacts with Ellman's reagent (DNTB, 5,5′-dithiobis(2-nitrobenzoic acid)) to form TNB^{2-}, which absorbs light at 412 nm, A_{412}. Following the absorbance in a UV–visible light spectrophotometer (Nanodrop 2000c, ThermoFisher Scientific, Waltham, MA, USA) allows for direct monitoring of the reactivity of the cysteines. Baseline absorbance at 750 nm should also be measured to correct for drift during the acquisition.

1. Prepare a fresh stock of Ellman's reagent in reaction buffer (0.1 *M* sodium phosphate, 1 m*M* EDTA, pH 8.0) at 10 m*M* concentration. This master stock can be stored in the dark at room temperature (RT) for a day. Dilute the master stock to a 5 m*M* working solution with purification buffer (e.g., 150 m*M* NaCl, 20 m*M* MOPS, 5 m*M* *n*-decyl β-D-maltopyranoside (DM), pH 7.0 for ClC-ec1).

   Note. The purification buffer can be changed to suit the protein of interest, but it should not contain KCl if SDS is to be added at a later stage.
2. Take 300 μL of 10 μ*M* protein in a clean quartz cuvette and monitor A_{412} for a period of 5 min to establish the baseline.
3. Add 20 μL of Ellman's reagent working solution to the cuvette, mix well by pipetting up and down, and monitor A_{412} for 20 min. Surface

exposed cysteines reacts with DTNB in a rapid fashion (i.e., an immediate step up in A_{412}), whereas partially accessible cysteines show a slow creep in the absorbance over time.

4. Add 40 μL of denaturing buffer (e.g., 20% SDS stock prepared in purification buffer for ClC-ec1) to unfold the protein and expose internal cysteines to react with DTNB. Allow the reaction to proceed for 45 min, or until a steady absorbance is reached. For ClC-ec1, we observe this as a gradual exponential increase in A_{412}, presumably due to the slow kinetics of the reaction since unfolding is expected to be fast.
5. Include a negative control of purification buffer alone to allow measurement of $(A_{412} - A_{750})_{\text{control}}$ in the presence of Ellman's reagent and with the addition of denaturing buffer. This is necessary as the spectrum of DNTB bleeds into 412 nm.
6. The TNB^{2-}-specific absorbance is calculated using $A_{412,\text{TNB}^{2-}} = (A_{412} - A_{750})_{\text{protein}} - (A_{412} - A_{750})_{\text{control}}$, where the control represents either the purification buffer + Ellman's reagent or the addition of denaturing buffer. The concentration of $[TNB^{2-}]$ is calculated using the Beer–Lambert law:

$$[\text{TNB}^{2-}] = \frac{A_{412,\text{TNB}^{2-}}}{\varepsilon \cdot \ell} \tag{3}$$

where ℓ is the path length in cm, and ε is the molar extinction coefficient. Note that this value has been reported in the literature from 11,400 to 14,140 M^{-1} cm^{-1} (Riddles et al., 1983). It is recommended that a standard curve be generated with $A_{412} - A_{750}$ as a function of cysteine concentration, freshly prepared in the desired buffer condition. The number of reactive cysteines per subunit is calculated as $[TNB^{2-}]/([\text{protein}]*d)$, where d is the dilution factor.

2.2 Insertion of a Solvent Accessible Cysteine for Subunit-Specific Labeling

Once the protein has been wiped clean of all the reactive endogenous cysteines, then a new cysteine residue can be inserted that allows for rapid conjugation with maleimides. The position of the new cysteine should be optimized to obtain high-quality photobleaching data. Since photobleaching relies on observing step-wise changes in fluorescent intensity, any fluctuations can make it difficult to interpret the traces. Fluorescent intensity depends on the solvent environment around a fluorophore (Lakowicz, 2006), so it is advisable to place the cysteine in a structured part of the protein far away from the protein–membrane interface. Inserting the cysteine at the end of a rigid helix, in the center of a multihelix protein may yield better quality data than placing the cysteine in a flexible loop or near one of the termini. It is important that the site be exposed to the aqueous solution so that reaction time can be minimized to prevent nonspecific labeling at lysines or the N-terminus (Brewer & Riehm, 1967). Since fluorophores partition into membranes (Hughes, Rawle, & Boxer, 2014), it is recommended that the cysteine be placed away from the protein–lipid interface and also away from the dimerization interface, though this may not be possible for smaller proteins. Note that other factors outside of aqueous exposure, such as steric constraints or the surrounding electrostatic environment, can also play a role in the reactivity of the cysteine with maleimides. While knowledge of the structure is useful in this situation, it is not necessary as information from hydropathy analysis can serve as a guide. We recommend preparing five cysteine mutants to start, comparing the reactivity of each using Ellman's reagent and choosing the site that yields an instantaneous reaction of a single cysteine.

2.3 Fluorophore-Maleimide Labeling of Membrane Protein Subunits

The following procedure has been optimized for labeling of ClC-ec1 with Cy5-maleimide, maximizing the overall labeling yield ($P_{\text{fluorophore}}$), while minimizing nonspecific labeling ($P_{\text{nonspecific}}$). The procedure here should suffice as a starting point for any new protein system, but optimization is recommended.

1. Dissolve lyophilized Cy5-maleimide in anhydrous DMSO (Molecular Probes, ThermoFisher Scientific, Waltham, MA, USA) to make 10 m*M* master stocks. This stock should be divided into smaller, workable aliquots to prevent exposure to light and stored at −80°C in a box with anhydrous $CaSO_4$ crystals (Drierite, Xenia, OH, USA) to prevent hydrolysis of the maleimide.
2. Dilute protein with purification buffer to a final concentration of 10 μ*M*. In dark, thaw a vial of frozen 10 m*M* Cy5-maleimide working stock and add the appropriate volume for a final concentration of 50 μ*M*. Mix well by pipetting up and down and then incubate for 5 min at RT in the dark. Stop the reaction with 1 m*M* cysteine from a 100 m*M* stock prepared freshly in purification buffer, and pH adjusted to ~7.0.
3. For a 1 mL reaction (~500 μg ClC-ec1), we make a 250 μL cobalt resin column (Talon, Clontech, Takara, Mountain View, CA, USA) in a

micro-bio spin column (Bio-Rad, Hercules, CA, USA) and equilibrate the resin with 15 column volumes (CVs) of cobalt wash buffer (100 m*M* NaCl, 20 m*M* Tris, 5 m*M* DM, pH 7.5). Bind the protein to the column and then wash extensively with 30 CVs of purification buffer, making sure that all inner surfaces of the column are washed to remove any lingering free dye. Since the protein is fluorescently labeled, it is visible as a colored band at the top of the column. The protein is then eluted with 4 CVs of 400 m*M* imidazole in purification buffer (Fig. 2A).

Note. Imidazole absorbs light at 280 nm, and it is difficult to control for this as the amount of absorbance depends on the preparation, and the final concentration in the elution fraction, which varies. Therefore, all imidazole should be removed from the elution fraction to enable proper quantification of the protein labeling yield.

4. Swell Sephadex G50 resin in ddH_2O. This can be carried out in advance and the resin stored at 4°C; however, it must be warmed to RT before pouring to prevent de-gassing inside the column. Set up a 5 mL disposable column (Pierce, ThermoFisher Scientific, Waltham, MA, USA) and pour the preswelled G50 resin to the top, letting it drip dry, for a final CV ~2.5 mL. Equilibrate the resin with 15 CVs of cobalt wash buffer and let the column drip dry. Add the protein elution fraction from the previous step and then follow with cobalt wash buffer. The blue Cy5-labeled protein can be visualized as it moves through the column, and collected as soon as it reaches the end of the column (Fig. 2B). The smaller imidazole will elute at a later volume and thus can be separated from the protein.
5. Place a sample of the eluted protein in a quartz cuvette and measure the UV–visible light absorbance spectrum from 260 to 750 nm. A peak will be observed at 280 nm, representing the protein, and 655 nm for Cy5. The concentration of protein, fluorophore, and the fluorophore/protein labeling yield $P_{\text{fluorophore}}$ is determined as follows:

$$[\text{subunit}] = \frac{A_{280} - \left(A_{\text{fluorophore}} {}^{*}\text{CF}_{\text{fluorophore}}\right)}{\varepsilon_{\text{subunit}} \cdot \ell} \qquad (4)$$

$$\frac{P_{\text{fluorophore}} = {}^{A_{\text{fluorophore}}} \big/ \left(\varepsilon_{\text{fluorophore}} \cdot \ell\right)}{[\text{subunit}]} \qquad (5)$$

where ℓ is the path length in cm, $A_{\text{fluorophore}}$ is the peak absorbance of the fluorophore, $\text{CF}_{\text{fluorophore}}$ is the correction factor for the absorbance of the fluorophore at 280 nm, $\varepsilon_{\text{protein}}$ is the molar extinction coefficient of the individual subunit of the protein, and $\varepsilon_{\text{fluorophore}}$ is the molar

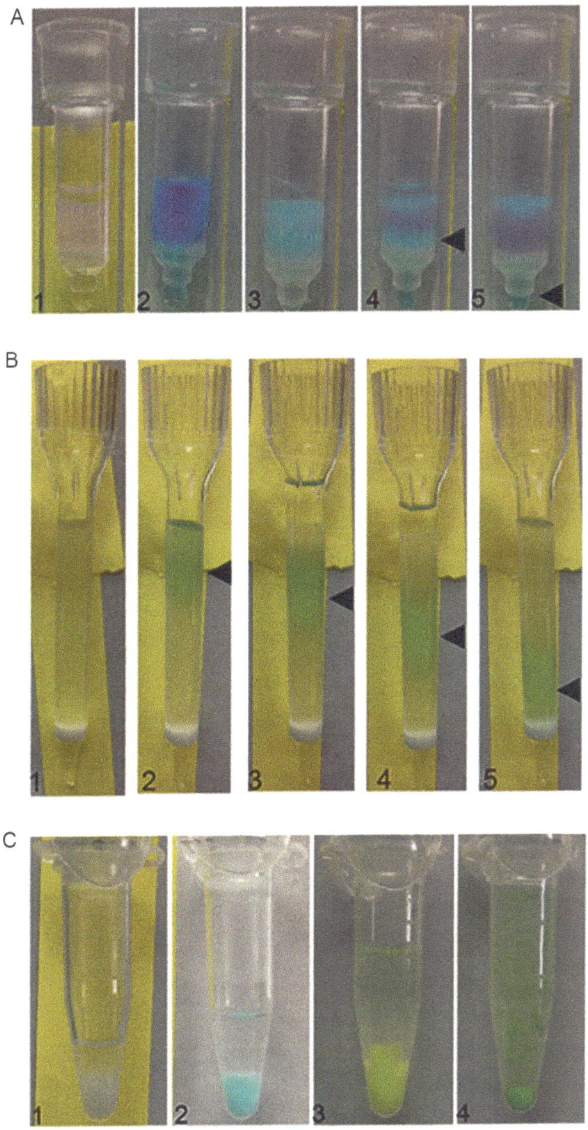

Fig. 2 Stages in preparation of fluorescently labeled protein and liposomes. (A.1) Equilibrated Bio-Rad mini-column with Talon His-tag purification resin. (A.2) Reaction containing 10 μ*M* ClC-ec1 + 50 μ*M* Cy5-maleimide from loaded onto mini-column. (A.3) ClC-ec1-Cy5 bound to the cobalt resin after free dye is washed off. (A.4) Appearance of column during elution of ClC-ec1-Cy5 with 400 m*M* imidazole. The eluted protein emerges at the bottom of mini-column for collection in an eppendorf, indicated by the *black triangle*. (A.5) Appearance of column near the end of elution. Note that some Cy5-labeled protein remains at the top of the column, indicating aggregation of protein

(*Continued*)

extinction coefficient of the fluorophore at the peak absorbance. For Cy5, A_{Cy5} is typically measured at 655 nm, with $\varepsilon_{Cy5} = 2.5 \times 10^5$ at 655 nm and $CF_{Cy5} = 0.02$.

Note. Maleimides will react with primary amines on lysines or the N-terminus, albeit at a much slower rate. Therefore, it is expected that there is a small population of subunits that will have more than one label, and thus artificially appear dimeric. The minimization of the reaction time to 5 min and the reduction of the reaction pH to 7.0 reduces nonspecific labeling. Still, the nonspecific labeling yield, $P_{nonspecific}$, must be measured experimentally to properly correct the photobleaching data. This can be done by carrying out the same reaction with a construct that lacks the reactive cysteine residue.

Note. We chose Cy5 because it is a far-red fluorophore that provides excellent signal to noise for photobleaching studies. It is also sufficiently far from Alexa Fluor 488 fluorescence that we use for lipid labeling. Other fluorophores have their advantages, for instance, Cy3 and Cy3B are far more stable allowing for long-lived photobleaching traces; however, liposome-only controls must be imaged to ensure that the signal from large liposomes do not bleed into this channel.

2.4 Check for Stability and Folding of the Fluorescently Labeled Protein

Adding a fluorophore to a membrane protein can alter its properties, as fluorophores are hydrophobic, and membrane proteins are situated in a hydrophobic solvent environment. Therefore, it is possible that the fluorophore will preferentially partition into the membrane and alter the structure and stability of the protein. It is imperative to confirm that fluorescent labeling does not affect the folding of the protein in both detergent micelles

Fig. 2—Cont'd during binding. (B.1) 5 mL disposable column filled with 2.5 mL G50 Sephadex resin equilibrated in buffer containing 5 m*M* DM for separation of imidazole from the eluted protein. Note that the tip of the column is cut at an angle using razor blade to increase the flow rate. (B.2) ClC-ec1-Cy5 eluted in (A.5) is applied to the column. (B.3–B.5) The movement of ClC-ec1-Cy5 in the column can be monitored visually (indicated by the *black triangle*). (C.1) Appearance of 7× freeze/thawed 1 μg/mg ClC-ec1-Cy5 liposomes after storage at room temperature. (C.2) Appearance of 7× freeze/thawed 10 μg/mg ClC-ec1-Cy5 liposomes after storage at room temperature. (C.3) Appearance of 7× freeze/thawed 1 μg/mg ClC-ec1-Cy5 in AF488-labeled liposomes after storage at room temperature. (C.4) Appearance of 7× freeze/thawed 10 μg/mg ClC-ec1-Cy5 AF488-labeled liposomes after storage at room temperature.

and liposomes, preferably by some type of functional assay (Stockbridge & Tsai, 2015).

3. PREPARATION OF MEMBRANES FOR EQUILIBRIUM MEASUREMENTS OF DIMERIZATION

Once the protein is fluorescently labeled, and verified to be functionally folded, the protein is reconstituted into lipid bilayers by standard dialysis methods (Stockbridge & Tsai, 2015). Reconstitution into membranes containing fluorescently labeled lipids allows for checks on the fidelity of reconstitution, and measurement of the liposome accessible population for both the monomer and dimer states. In the following steps, we describe how to prepare fluorescent liposomes, how to create the infinite model of the bilayer where equilibrium exchange can take place, and how to fractionate the membrane to quantify the monomer and dimer population by colocalization TIRF microscopy.

3.1 Preparation of Fluorescent Liposomes

It is prudent to check the fidelity of membrane protein reconstitution. Colocalization microscopy of Alexa Fluor 488 (AF488)-labeled liposomes and Cy5-labeled protein can readily provide this information. POPE lipids contain a primary amine on its head-group, which allows for covalent conjugation with NHS ester-linked fluorophores. We choose AF488 as it has been shown to exhibit significantly less nonspecific membrane incorporation (Hughes et al., 2014). Since the NHS esters are rapidly hydrolyzed in water, we use the more stable SDP ester form that is also available. The following protocol is for 1 mL of lipids at 20 mg/mL but the volumes can be scaled up as needed:

1. Under a fume hood, aliquot 18 mg POPE and 9 mg POPG in chloroform into a glass vial. Note that chloroform has a high vapor pressure and will drip from the pipette, leading to inaccuracies in lipid amounts. Make sure to gently pipette the solution up and down five times for accurate pipetting. Evaporate the chloroform from the 2:1 POPE/POPG lipids under N_2, to form a dried film at the bottom of the vial. To remove trace amounts of chloroform, resuspend the lipids in 500 μL of pentane and then dry the lipids again to obtain a thin film.
2. Immediately add 1.5 mL of lipid-labeling buffer: 300 m*M* KCl, 100 m*M* $NaHCO_3$ pH 8.3, to the lipid film in the glass vial. Sonicate briefly for 10 min to obtain a cloudy solution of lipid vesicles.

3. Add 33 mg of sol-grade CHAPS and continue sonication until the solution becomes translucent. Adjust the volume to 2 mL with labeling buffer, for a final concentration of 20 mg/mL 2:1 POPE/POPG and 35 m*M* CHAPS.

   Note. CHAPS detergent was chosen as it has a high critical micelle concentration and is thus readily removed during dialysis (Stockbridge & Tsai, 2015).

   Note. Different sonication methods have been used for solubilization of lipids, including vortexing, probe sonication, and cylindrical bath sonication. Recently, we have turned to sonication using a cup horn sonicator (Qsonica, Newtown, CT, USA) that allows for rapid solubilization of the lipids producing a transparent solution in 15–25 min.
4. Prepare 10 m*M* AF488 5-SDP ester stocks in anhydrous DMSO as described in Section 2.3. Aliquot the stock into single-use 16 μL volumes and store at −80°C to prevent exposure to light and hydrolysis of the reactive group during freeze/thaw.
5. When labeling, thaw an aliquot of 10 m*M* AF488 and add 15 μL to the solubilized lipids. The final mole ratio of AF488 to total lipids is 0.003.

   Note. A liposome with radius 25 nm will contain ~10^4 lipids, assuming a surface area per lipid of 0.6 nm^2 (Murzyn, Róg, & Pasenkiewicz-Gierula, 2005). Even if the reaction is only 50% successful, then this would predict ~15 fluorophores per liposome of that size. We confirmed that this saturated the liposomes by colocalization microscopy, examining the reconstitution as a function of the AF488/lipid mole fraction. If some liposomes were not labeled, then we would expect to see protein spots that do not colocalize with liposomes concomitant with a decrease in the fraction of unoccupied liposomes, F_0. In fact, we do not observe protein outside of the liposomes and there is no change in F_0 as the amount of AF488 is increased, indicating that there are few "dark" liposomes.
6. Incubate the reaction at RT on a cell mixer for 15 min in the dark. The reaction is stopped by addition of 85 μL 1.5 *M* of Tris–Cl, pH 8.3, which quenches the free, reactive fluorophore with its primary amine.
7. At this point, the fluorescently labeled protein can be combined with the lipid/fluorophore mixture for reconstitution of the protein. For ClC-ec1-Cy5, we reconstituted the protein at 1.5×10^{-9} to 7.5×10^{-4} subunit/lipid corresponding to observed mole fraction densities of $\chi=7.5\times10^{-10}$ to 3.8×10^{-4} subunit/lipid. Typically, Cy5-labeled protein is obtained at ~5 μ*M* during the purification, so the protein must be

serially diluted in 5 m*M* DM purification buffer before combining with the lipid mixture, for the low-density reconstitutions.

8. Liposomes are formed by dialysis using 10 kDa MWCO dialysis cassettes. This step does two things, it reconstitutes the protein in the membrane by dialyzing the CHAPS and DM detergent, and it also removes the free AF488 dye from the lipid labeling. For a 1 mL sample, we dialyze against 1 L of imaging buffer (300 m*M* KCl, 20 m*M* citrate pH 4.5) at 4°C, with buffer changes every 8–12 h for 48–60 h. When scaling up the sample volumes, we use a ratio of at least 1 sample volume to 500 volumes of dialysis buffer. The dialysis should be performed in the dark.

   Note. The imaging buffer was selected to match the functional buffer for ClC-ec1 chloride transport studies (Walden et al., 2007). For the protein of interest, a buffer should be selected to match conditions where the function can be measured.

   Note. The progress of the removal of free dye during dialysis can be followed by taking an aliquot of the spent dialysis buffer and measuring AF488 fluorescence in a fluorometer. We observe that using a 10 kDa MWCO dialysis membrane, all free dye is removed after four buffer changes, i.e., 48 h.

   Note. The MWCO of the dialysis cassette should be selected with the protein of interest in mind. For smaller molecular weight species, a 3 kDa MWCO cassette can be used, but the dialysis should be extended as it will take a longer amount of time and cycles to remove the detergent and free dye.

9. After dialysis, the liposomes should appear slightly cloudy in the 2:1 POPE/POPG lipid composition Fig. 2C, but this may also depend on the protein being reconstituted. The samples should be homogenized in the dialysis cassette by light vortexing, transferred to an eppendorf tube, and then supplemented with 0.02% NaN_3 to inhibit bacterial activity. The samples can be stored at −80°C until further use. Care must be taken to keep the samples away from light at all times.

   Note. To check that the detergent has been fully removed, the samples should be tested by freeze/thaw after removal from the dialysis cassette. They should fuse and become cloudy, indicating the formation of large MLVs. If the samples remain clear after several freeze/thaw cycles, then dialysis should be repeated to remove the remaining detergent.

   Note. We recommend carrying out the lipid labeling at the beginning of working with a new protein to check the fidelity of reconstitution and the liposome accessibility. However, once these parameters have been measured, the remaining photobleaching experiments can be carried out in unlabeled lipids. For ClC-ec1-Cy5, we do not observe any difference in the photobleaching probabilities from labeled or unlabeled lipid membranes.

3.2 Fusion of Vesicles to Form an Infinite Bilayer Model

During reconstitution, subunits can become trapped in liposomes with little possibility of exchange between vesicles. Since we are interested in measuring the equilibrium reaction of dimerization in the lipid bilayer, we must first create a large bilayer state where multiple copies of subunits exist and can exchange with one another, even at low densities. 2:1 POPE/POPG liposomes undergo fusion by freeze/thaw, forming large MLVs that are 10 μm in diameter, i.e., $\sim 5 \times 10^8$ lipids per bilayer (Pozo Navas et al., 2005). Our lowest observed mole fraction density of $\chi = 7.5 \times 10^{-10}$ subunits/lipid is expected to have 1 subunit per bilayer, just below the limit where equilibrium behavior can theoretically be observed. This large membrane state therefore creates a model of an infinite bilayer that allows for equilibrium behavior to be observed and measurement of the free energy of dimerization. The freeze/thaw procedure is carried out in the dark as follows:

1. Transfer dialyzed proteoliposomes to glass test tubes and place them in a floating tube rack.
2. Prepare a dry ice–100% ethanol bath in a sturdy ice bucket. Allow 30 min for the temperature to equilibrate. Set up a separate styrofoam bucket with 25–30°C water for thawing.
3. Immerse the tubes containing the proteoliposomes in the dry ice–ethanol bath for 5 min. At the end of incubation transfer tubes to the water bath for thawing for 15 min. Repeat this cycle seven times. The samples should appear opaque, and after sitting for some time, the MLVs will settle to the bottom of the tube. If the sample remains clear after freeze/thaw, then it is possible that there is still detergent in the solution preventing the formation of bilayers. If this occurs, then the dialysis should be repeated or preferably new samples prepared.

   Note. The seven times freeze/thaw procedure was developed by monitoring the Förster resonance energy transfer (FRET) signal after freeze/thaw of 2:1 POPE/POPG/PE-RhB and 2:1 POPE/POPG/PE-NBD. The FRET signal saturates after 4–7 cycles.

Note. Care must be taken to ensure that ethanol, or water, does not enter the tubes. Balance the rack well to prevent tipping over in water or ethanol bath.

4. After freeze–thaw, the samples should be stored at RT and away from ambient light.

The protein is now in the infinite membrane reaction state where subunits can exchange with one another. For ClC-ec1-Cy5, we find that the protein is stable, at least up to 27 days at RT. With this, we can examine the photobleaching probabilities as a function of incubation time after fusion of membranes and find that the monomer/dimer population comes to equilibrium during the freeze/thaw process, and that this is stable over the course of a month. Of course, this could be different for other protein systems so we advise investigating this experimentally.

3.3 Membrane Fractionation by Extrusion

To measure the population of monomers and dimers in the large membrane state by TIRF microscopy, we fractionate the MLVs by extrusion. This forms a population of smaller liposomes with a defined size distribution, and the encapsulation of protein into these liposomes follows a Poisson distribution (Maduke et al., 1999; Walden et al., 2007). The liposome extrusion is carried out as follows:

1. Assemble the extruder following manufacturer's guidance. We recommend using the Avestin's LiposoFast-Basic with a stabilizer support.
2. Using 0.5 mL gas-tight glass syringe pass ddH_2O, back and forth, through the extruder 4–5 times. Discard.
3. Now, using the same glass syringe, pass filtered imaging buffer through the extruder 4–5 times. Discard.

   Note. Imaging buffer can be prepared in advance and frozen in 10 mL aliquots and stored at −20°C. The imaging buffer is filtered by passing three times through same 0.22-μm syringe filter. The filtered buffer should be checked on the microscope for background fluorescence.
4. Wet a 0.4-μm Nucleopore membrane filter (Sigma-Aldrich, St. Louis, MO, USA) in ddH_2O and gently place the white membrane between the two O-rings on the Teflon cylinders. Be careful not to touch the membrane directly. Hold it at the edges with forceps or with clean gloves.
5. To equilibrate the membrane filter pass 0.4 mL imaging buffer through extruder 4–5 times. Discard.

6. Finally fill the glass syringe with 0.4 mL of the freeze/thawed proteoliposome sample and pass 21 times through the membrane filter. Apply steady but gentle pressure while moving back and forth between two syringes.

   Note. It is imperative that the extruder be properly cleaned in between the samples. We found that the most efficient way of cleaning the extruder, was to dismantle it completely (including O-rings) and soak in hot water/dilute detergent (0.1%, v/v Dial dish soap) followed by sonication in ddH_2O. "Extruding" buffer after this cleaning and checking on the microscope shows that there is no cross-contamination following this method.

3.4 Quantification of the Liposome Size Distribution

The size distribution of the extruded liposome population must be known in order to correct the photobleaching probability data and extract the monomer–dimer reaction. Extrusion yields reproducible liposome populations, but the radii are far from uniform (Walden et al., 2007). As such, liposome sizes must be measured experimentally, and cryoelectron microscopy (cryo-EM) provides a direct approach. The methods are well established (Almgren, Edwards, & Karlsson, 2000; Frederik & Hubert, 2005) and any cryo-EM imaging facility should be able to carry out these experiments, but the liposomes should be extruded immediately before freezing to prevent changes in the sample over time. The images can be analyzed in Fiji (Schindelin et al., 2012) using the polygon outline tool to obtain the radius of each liposome, and then a probability distribution, P_{radius}, can be determined from the normalized histogram of radii. If an internal control is desired for pixel-to-nm conversion, 0.1 mg/mL of purified tobacco mosaic virus (width = 1.8 nm; Namba & Stubbs, 1986) can be added to the liposome sample prior to freezing. We find that experimental distributions reported in the literature are quite reproducible even with small changes in composition, e.g., 0.4 μm extrusion of *Escherichia coli* polar lipids or 2:1 POPE/POPG, both in 300 m*M* KCl, 25 m*M* citrate pH 4.5, yield similar P_{radius} distributions (Walden et al., 2007), even though cardiolipin is lacking in the latter condition.

3.5 Quantification of the Protein/Lipid Mole Fraction χ

During dialysis, freeze/thaw, and extrusion, there can be loss of protein or lipids in different amounts leading to a deviation in the experimental mole fraction value compared to that set during reconstitution. To measure the actual mole fraction, we suggest quantifying the protein and lipid amounts in each sample as follows:

1. 10 μL of the vortexed proteoliposome sample is solubilized in 190 μL of 5 m*M* DM purification buffer supplemented with 35 m*M* CHAPS and 10% CTAB (Hexadecyltrimethylammonium bromide). CTAB is an ionic detergent used to denature the protein in potassium containing buffers, as opposed to SDS which will precipitate upon forming the potassium salt.
2. Prepare a standard curve using Cy5-labeled protein (with similar labeling yield) in the same buffer conditions used for the proteoliposome samples. Measure the fluorescence of the proteoliposome sample along side the standard curve to determine the moles of protein in the sample. We carry out our measurements in a 96-well plate on a Typhoon FLA 9500 gel imager (GE Healthcare Life Sciences, Little Chalfont, Buckinghamshire, UK), but a fluorometer or another high-sensitivity fluorescent plate reader can be used.
3. The exact same samples that were prepared for protein quantification can be used for lipid quantification by colorimetric analysis of phosphate using well-established procedures (Fiske & Subbarow, 1925). The samples are compared to a standard curve prepared with sodium phosphate dibasic for determination of the moles of lipid in each sample.
4. The ratio of the moles of protein to lipids gives the actual experimental mole fraction after freeze/thaw or extrusion. For ClC-ec1-Cy5, we find that the mole fraction is 50% of the reconstituted value and use this corrected mole fraction in all further analysis. Note that this may be different for different proteins and lipids and should be tested for each sample.

4. TIRF MICROSCOPY OF PROTEOLIPOSOMES

We use passive adsorption (Johnson, Ha, Chu, & Boxer, 2002) to attach 2:1 POPE/POPG liposomes to the glass slide. This is an extremely effective method that requires little slide preparation, outside of a basic cleaning procedure. While binding to the slide is expected to deform the liposome, we rarely observe rupture or fusion of the membrane in the 2:1 POPE/POPG condition (Nollert, Kiefer, & Jähnig, 1995). This is not expected for all lipid conditions, as we observe POPC liposomes forming blurry spots, indicating rupture and formation of membrane patches on the slide. At high densities of POPC liposomes, we observe the formation of a complete bilayer on the glass where single-molecule diffusion of lipids can be observed. Therefore, a liposome tethering method may be used, e.g., via PE-biotin (Diao et al., 2012) or lipid-DNA tethering (M. Chung, Lowe, Chan, Ganesan, & Boxer, 2009) for other types of lipid compositions. Our choice of 2:1 POPE/POPG to study ClC-ec1 was based on the fact that this protein is native to *E. coli*, and is functional in this membrane composition. When working with a new protein system, the lipid composition should be selected to match the condition where the functional activity can be measured. However, for preliminary studies, 2:1 POPE/POPG offers a quick and simple way of carrying out single-vesicle microscopy studies.

4.1 Preparation of Slides for TIRFM

The cleaning of glass slides and coverslips serves two purposes. First, it leads to a uniformly low background, and second, KOH treatment renders the surface hydrophilic enabling passive adhesion of liposomes to the surface (Chandradoss et al., 2014).

1. Place glass slides and coverslips (Gold-Seal slides were 24 × 60 mm no. 1.5 thickness and coverslips were 25 × 25 mm no. 1.0 thickness) in slide mailers 5 at a time. Fill up the container with 0.1% Micro-90 detergent (Cole-Parmer, Vernon Hills, IL, USA) and sonicate in bath sonicator for 30 min. At the end of 30 min, empty out detergent and rinse 10 times with deionized, distilled grade water (ddH_2O).
2. Next, fill slide mailers with 100% ethanol and sonicate for 30 min. At the end of 30 min, empty out ethanol and rinse 10 times with ddH_2O.
3. Add freshly prepared 0.2 *M* KOH and sonicate for 5 min followed by rinsing 10 times with ddH_2O.

   Note. Clean coverslips and slides can be stored in ddH_2O in a laminar flow bench for up to a week.

4.2 Colocalization Microscopy of Liposomes and Membrane Proteins

Our experiments were carried out on a multiwavelength single-molecule TIRF microscope that we built in our lab following the CoSMoS design (Friedman, Chung, & Gelles, 2006; Larson et al., 2014). Imaging is carried out in two channels, using 488 and 637 nm excitation lasers, while emitted fluorescence is collected and filtered using ET525/50m or ET650LP, respectively, and focused onto the CCD chip of an EM/CCD camera. Since liposomes adsorb onto the glass slide, and free liposomes can be washed away, epi-fluorescence or confocal microscopy (Mathiasen et al., 2014) could be used as long as the signal is sufficient to observe unambiguous photobleaching steps.

1. Apply extruded proteoliposomes to the flow cell for passive binding to the glass coverslip. For high protein densities, the samples must be serially diluted with filtered imaging buffer in low-adhesion tubes. The binding occurs quickly and excess liposomes can be washed off within 1–2 min of application. Low protein density liposomes can be extruded and applied directly to the chamber without further dilution, allowing 20–30 min incubation time to maximize liposome binding. The density of fluorescent spots on the glass slide was maintained at 0.02–0.09 spots/μm^2 to minimize overloading.

   Note. The binding of liposomes to the glass is practically irreversible (Johnson et al., 2002). We confirm this by imaging the same field of liposomes over time, demonstrating little change in the number of spots.

   Note. Cy5 triplet-state blinking (Ha & Tinnefeld, 2012) is avoided by using imaging buffers that do not contain oxygen scavengers.

2. Set EM gain to 300 and acquisition rate to 1 frame per second. The laser intensity used in these experiments depends on the sample, but we typically set the 488 nm laser to 15 μW to image the liposomes and the 637 nm laser to 240 μW for the Cy5-labeled subunits. The latter allows for long photobleaching traces while maintaining good signal to noise, but this must be optimized for different samples.

   Note. Our power meter is positioned between the broadband mirror that combines the different excitation paths and the focusing lens that focuses the laser onto the sample (Larson et al., 2014). Therefore, the actual laser power at the sample is expected to be less than the values we report.

3. On the same day of imaging, collect a mapping dataset to allow for pixel registration between 488 nm and 637 nm imaging channels, especially necessary for determining colocalization at low densities. This can be done independently with a separate high protein density sample of sonicated liposomes, or multicolor beads (Friedman & Gelles, 2015).

   Note. When carrying out AF488/Cy5 colocalization imaging, image the Cy5 channel first, followed by AF488. We observe Cy5 going to a dark state upon exposure to the 488 nm laser.

4.2.1 Measurement of $F_{outside}$, the Fraction of Protein That Does not Colocalize With Liposomes

F_{outside} is the fraction of Cy5 protein spots that do not colocalize with AF488 liposomes. This is a measurement of the fidelity of reconstitution, as appearance of protein outside of liposomes indicates aggregation during reconstitution (Mathiasen et al., 2014). If this value is more than 1–2%, then the protein does not properly reconstitute into the membrane or the lipid-labeling yield is too low. If increasing the amount of lipid labeling does not reduce $F_{outside}$, then it can be concluded that there is significant protein loss during the reconstitution preventing reliable measurement of the dimerization reaction.

4.2.2 Measurement of F_0, the Fraction of Unoccupied Liposomes

As reconstitution follows a Poisson distribution, the fraction of unoccupied vesicles, F_0, should decrease as the protein density increases. However, it has been shown that incorporation of protein in liposomes can depend on both size and curvature (Mathiasen et al., 2014), and so it is possible that there is subpopulation of liposomes that are not accessible to protein, either in the monomeric or in the dimeric state. Measuring F_0 as a function of protein density allows for direct determination of the accessible liposome population. For this, we recommend measuring the protein at high densities, $\chi=3.8\times10^{-5}$ to 3.8×10^{-4} subunit/lipid as this will reveal the plateau of the curve. For example, the dimeric wild-type ClC-ec1 exhibits a plateau at $F_0=0.4$, whereas the monomeric I201W/I422W plateau is ~0.1. This indicates that even at the highest protein density, dimers cannot access 40% of the liposomes, whereas the monomer occupies almost all of them. We assume that this is related to a conflict in protein vs liposome size, indicating that liposomes with radius <25 nm are excluded from the dimer population. With this, $P_{radius,D}$ can be calculated as the renormalized accessible liposome probability distribution, excluding the smaller radii bins. ClC-ec1 is a large protein (10 nm end-to-end), and so this may not be needed for smaller protein systems. It is worth checking but it requires having a positive dimer control, either through mutagenesis or by reconstitution of a cross-linked form of the complex.

4.2.3 Measurement of the Photobleaching Probabilities, P_n

Here, we describe the manual counting of photobleaching steps. While there are numerous methods that use algorithms for automated counting (Hines, Bankston, & Aldrich, 2015; McGuire, Aurousseau, Bowie, & Blunck, 2012), we find that these results often have to be checked against manual counting. We have compared the subjectivity of counting between different individuals and find no significant difference as long as the quality of the images has been optimized to maximize signal to noise (by increasing laser power) while maximizing the length of each photobleaching state (by decreasing laser power). This will be different for each protein system due to changes in photostability of the fluorophore on the protein near the membrane. Note that we do not observe Cy5 blinking behavior, as we do not include an oxygen scavenging system in our buffers. Since our goal is to measure a reaction in the membrane, we avoid triplet-state quenchers such as Trolox, which partitions into membranes and may affect protein properties (Alejo, Blanchard, & Andersen, 2013).

Image files can be stored as stacked tiff files (or some other binary image format) and analyzed in any image analysis program. We use a Matlab-based CoSMoS image analysis program (Friedman & Gelles, 2015; https://github.com/gelles-brandeis/CoSMoS_Analysis) as it has been customized to serve many needs in colocalization imaging analysis such as mapping, drift correction, and autodetection of fluorescent spots. We used raw images directly for analysis, as conditions were optimized for maximum signal to noise and relatively long photobleaching traces. An area of interest (AOI) around each fluorescent spot can be detected manually or by an autodetection method. Pixel intensity within an AOI is then intergrated over time to obtain a photobleaching trajectory, where photodestruction steps can be clearly counted.

The photobleaching probability distribution, (P_n^{expt}), is calculated as the number of Cy5 occupied liposomes that bleach in n steps, over the total number of Cy5 occupied liposomes. On our microscope, we can count up to nine discrete photobleaching steps, while higher occupied liposomes exhibit a macroscopic decay in bulk fluorescence. In general, it is sufficient to calculate probabilities for $n=1$, 2, and >3 to follow the monomer–dimer reaction.

5. CALCULATION OF CORRECTION FACTORS FOR DETERMINATION OF F_{DIMER} VS χ

5.1 Correcting the Experimental Photobleaching Probabilities

The photobleaching probabilities describe the monomer–dimer equilibrium in the membrane, but to extract the fraction of dimer (F_{Dimer}) due to the equilibrium dimerization reaction, the data must be corrected to account for the random capture of nonreactive monomer or dimer species. In addition, fluorescence microscopy detects the fluorophores attached to subunits, and so the data must also be corrected for the fluorescent labeling yields, as different amounts of labeling will yield different results. Here, we describe the steps involved for simulating the photobleaching probabilities for non-interacting monomers and noninteracting dimers as a function of the protein density. The procedure is depicted in Fig. 3 and described in full detail in Chadda et al. (2016).

5.1.1 The Noninteracting Monomer Distribution

The noninteracting monomer distribution represents the probability of coencapsulation of membrane proteins that do not associate in the membrane. It has been referred to as 'artifactual togetherness' for association reactions in detergent micelles (Fleming et al., 1997; Tanford & Reynolds, 1976). This can be simulated using the experimental parameters obtained so far (Table 1) and the following steps:

1. Set $N_{\mathrm{monomers}} = N_{\mathrm{subunits}}$ and $N_{\mathrm{liposomes}}$. The experimental mole fraction, χ, is calculated as:

$$\chi = \frac{N_{\mathrm{subunits}} SA_{\mathrm{lipid}}}{N_{\mathrm{liposomes}} 8\pi \sum_r P_{\mathrm{radius}}(r) r^2} \tag{6}$$

   where SA_{lipid} is the surface area per lipid in a bilayer, and P_{radius} is the liposome size probability distribution measured by cryo-EM (Section 3.4).
2. Define a matrix for each radius bin in the P_{radius} distribution. The number of rows in the matrix is equal to $P_{\mathrm{radius}}(r) * N_{\mathrm{liposomes}}$, where each individual row represents a single liposome.
3. Calculate the number of each fluorescently labeled species. We assume that each subunit has two sites for fluorescent labeling: (1) the reactive cysteine and (2) a nonspecific site. The probability of labeling at the cysteine is $P_{\mathrm{cysteine}} = P_{\mathrm{fluorophore}} - P_{\mathrm{nonspecific}}$, where $P_{\mathrm{fluorophore}}$ and $P_{\mathrm{nonspecific}}$ are the labeling yields measured from the absorbance spectrum (Section 2.3). For monomers, this results in subunits labeled with 0, 1, and 2 fluorophores with associated probabilities listed in Fig. 3C.
4. The proportion of each monomer species to be inserted into the liposomes is calculated using the fractional surface area from the accessible liposome population, $P_{\mathrm{radius},M}(r)$ determined from the measurements of F_0 (Section 4.2.2).

$$N_{\mathrm{monomers}}(r) = \frac{r^2 P_{\mathrm{radius},M}(r)}{\sum_r r^2 P_{\mathrm{radius},M}(r)} \cdot N_{\mathrm{monomers}} \tag{7}$$

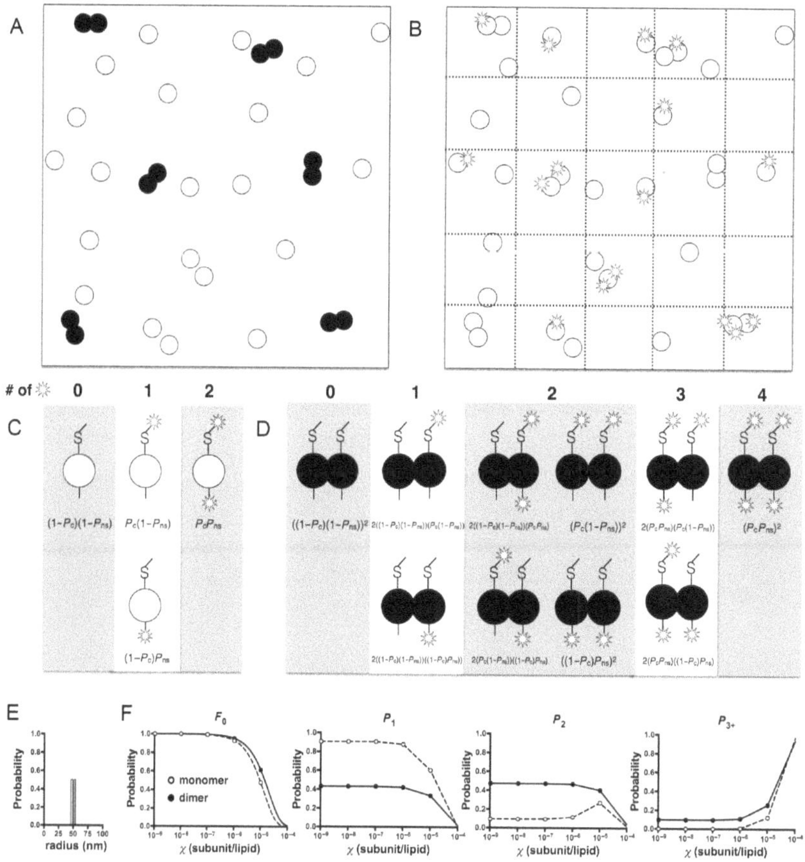

Fig. 3 Simulation of subunit capture into liposomes and photobleaching probabilities. (A) Large membrane (*square*) containing monomers (*white*) and dimers (*black*). (B) In the subunit capture method, the membrane is fractionated (*dotted lines*) by extrusion that forms liposomes, encapsulating monomers and/or dimers into each area of membrane. In the photobleaching analysis, only the fluorophores (*stars*) are detected, and the probabilities P_n^{expt} depend on protein density, χ subunit/lipid, fluorescent labeling yields $P_{fluorophore}$ and $P_{nonspecific}$, and monomer–dimer equilibrium constant K_{eq}. (C) Monomer fluorescent species and association probabilities. Each subunit has two sites for labeling, a high probability cysteine (-S site, $P_{cysteine}$ abbreviated to P_c) and a low probability nonspecific site (-site, $P_{nonspecific}$ abbreviated to P_{ns}). (D) Dimer fluorescent species with associated probabilities. Simulation of the photobleaching probabilities using a homogeneous liposome size distribution in (E) and labeling probabilities $P_{fluorophore}=0.75$ and $P_{nonspecific}=0.1$ (i.e., $P_{cysteine}=0.65$). (F) The fraction of unoccupied and invisible liposomes, F_0, and probability distributions of protein-occupied liposome that bleach in one (P_1), two (P_2), or more than three steps (P_{3+}) are shown for noninteracting monomer species (*white*) and noninteracting dimer species (*black*).

5. Define three columns in each matrix representing the occupancy of each of the different fluorescent species. Randomly insert the calculated number of the each monomer species into each subpopulation of liposomes.
6. At the end of the simulation, add up the number of fluorophores in each liposome row and generate a histogram that is normalized by $N_{\text{liposomes}}$. This gives the theoretical fluorophore occupancy probability distribution, $P^*(k)$, where $k=0$, 1, 2 and >3 represents the expected number of fluorophores per liposome, and $P^*(0)$ is equivalent to the experimentally measured F_0 (Section 4.2.2).
7. The photobleaching distribution is calculated from $P^*(k)$ by considering only the fluorophore occupied liposomes, i.e., $P^*(k)/(1-P^*(0))$ for $k=1$, 2, >3. This gives P_n, where $n=1$, 2, 3+ represents the number of photobleaching steps corresponding to the experimental photobleaching probabilities (Section 4.2.3).

   The simulation is repeated for different χ subunit/lipid densities by changing N_{subunits} and $N_{\text{liposomes}}$. Note that $N_{\text{subunits}} > 100$ and $N_{\text{liposomes}} > 1000$ should be used to avoid rounding errors.

5.1.2 The Nonnteracting Dimer Distribution

The noninteracting dimer distribution must also be calculated. For single dimer occupied liposomes with 100% labeling, only double steps would be observed. However, given typical labeling yields, i.e., ~75%, single and double steps are observed with equal probabilities. This is because there are now five different dimer fluorescent species with 0–4 fluorophores, where the 1 and 2 fluorophore labeling states are almost equal in probability (Fig. 3C). The rest of the calculation is almost identical to the monomer simulation described in Section 5.1.1, except now the user sets $N_{\text{dimer}} = N_{\text{subunit}}/2$, the dimer accessible liposome distribution $P_{\text{radius},D}(r)$ is used, and the simulation involves the five fluorescent dimer species.

5.2 Determination of F_{Dimer} vs χ^*

With the all-monomer and all-dimer probabilities calculated, the experimental photobleaching data (P_n^{expt}) can be fit to the noninteracting monomer (P_n^M) and noninteracting dimer (P_n^D) distributions by least-squares analysis of the sum of the squared residuals (R^2) to obtain F_{Dimer} as a function of χ:

$$R^2 = \sum_{n=1,2,3+} \left(P_n^{\text{expt}} - \left((1-F_{\text{Dimer}})\cdot P_n^M + F_{\text{Dimer}}\cdot P_n^D\right)\right)^2 \qquad (8)$$

The resulting data is fit to an equilibrium isotherm for extrapolation of the equilibrium constant, K_{eq}, of the reaction of the protein in the large membrane state

$$F_{\text{Dimer}} = \frac{1 + 4\chi^* K_{eq} - \sqrt{1 + 8\chi^* K_{eq}}}{4\chi^* K_{eq}} \tag{9}$$

Note that the reactive mole fraction density, $\chi^* = \chi/2$, must be used to account for random orientation of the protein in the membrane and that the reaction only occurs between proteins in the same direction. In order to establish that the reaction is at equilibrium, subunit exchange, reversibility and path independence must be experimentally tested.

5.3 Determining the Free Energy of Dimerization

With measurement of K_{eq}, the standard state dimerization free energy on the reactive mole fraction scale can be calculated using the following equation:

$$\Delta G^\circ = -RTln(\chi^\circ K_{eq}) \tag{10}$$

where χ° is the mole fraction standard state, set to an arbitrary reference of 1 subunit/lipid, R is the gas constant, and T is the temperature. For further discussion of the measurement of the free energy of dimerization of membrane proteins in lipids and detergent, we refer the readers to the following references: Chadda et al. (2016), Fleming (2002), Lemmon and Engelman (1994), and White and Wimley (1999).

6. SUMMARY

The methods in this chapter describe the general procedure for measuring subunit stoichiometry by subunit capture into liposomes followed by colocalization microscopy and photobleaching analysis. This approach can be used to determine protein stoichiometry in the membrane even if the structure is not yet known, as well as reactions of changing stoichiometry, such as equilibrium dimerization. Colocalization microscopy offers an opportunity to check the protein during the reconstitution process, to measure the success of protein incorporation, aggregation, and other higher-order oligomeric states. In principle, the same approach can be used to measure the equilibrium of higher-order oligomeric structures as well as heteromeric complexes. Finally, these methods map out the mole fraction range for single-molecule liposome experiments that can be applied to studies of conformational dynamics by FRET or functional experiments.

REFERENCES

Alejo, J. L., Blanchard, S. C., & Andersen, O. S. (2013). Small-molecule photostabilizing agents are modifiers of lipid bilayer properties. *Biophysical Journal*, *104*(11), 2410–2418.

Almgren, M., Edwards, K., & Karlsson, G. (2000). Cryo transmission electron microscopy of liposomes and related structures. *Colloids and Surfaces A: Physicochemical and Engineering Aspects*, *174*(1–2), 3–21.

Brewer, C. F., & Riehm, J. P. (1967). Evidence for possible nonspecific reactions between N-ethylmaleimide and proteins. *Analytical Biochemistry*, *18*(2), 248–255.

Chadda, R., Krishnamani, V., Mersch, K., Wong, J., Brimberry, M., Chadda, A., et al. (2016). The dimerization equilibrium of a ClC Cl^-/H^+ antiporter in lipid bilayers. *eLife*, *5*, e17438.

Chandradoss, S. D., Haagsma, A. C., Lee, Y. K., Hwang, J. H., Nam, J. M., & Joo, C. (2014). Surface passivation for single-molecule protein studies. *Journal of Visualized Experiments*, (86), e50549.

Chung, M., Lowe, R. D., Chan, Y.-H. M., Ganesan, P. V., & Boxer, S. G. (2009). DNA-tethered membranes formed by giant vesicle rupture. *Journal of Structural Biology*, *168*(1), 190–199.

Diao, J., Grob, P., Cipriano, D. J., Kyoung, M., Zhang, Y., Shah, S., et al. (2012). Synaptic proteins promote calcium-triggered fast transition from point contact to full fusion. *eLife*, *1*, e00109. http://dx.doi.org/10.7554/eLife.00109.

Ferré, S., Casadó, V., Devi, L. A., Filizola, M., Jockers, R., Lohse, M. J., et al. (2014). G protein–coupled receptor oligomerization revisited: Functional and pharmacological perspectives. *Pharmacological Reviews*, *66*(2), 413–434. http://dx.doi.org/10.1124/pr.113.008052.

Fiske, C. H., & Subbarow, Y. (1925). The colorimetric determination of phosphorous. *Journal of Biological Chemistry*, *66*, 374–389.

Fleming, K. G. (2002). Standardizing the free energy change of transmembrane helix–helix interactions. *Journal of Molecular Biology*, *323*(3), 563–571.

Fleming, K. G., Ackerman, A. L., & Engelman, D. M. (1997). The effect of point mutations on the free energy of transmembrane alpha-helix dimerization. *Journal of Molecular Biology*, *272*(2), 266–275.

Frederik, P. M., & Hubert, D. H. W. (2005). Cryoelectron microscopy of liposomes. In: *Methods in Enzymology: Vol. 391*, (pp. 431–448). Academic Press.

Friedman, L. J., Chung, J., & Gelles, J. (2006). Viewing dynamic assembly of molecular complexes by multi-wavelength single-molecule fluorescence. *Biophysical Journal*, *91*(3), 1023–1031.

Friedman, L. J., & Gelles, J. (2015). Multi-wavelength single-molecule fluorescence analysis of transcription mechanisms. *Methods*, *86*, 27–36.

Ha, T., & Tinnefeld, P. (2012). Photophysics of fluorescent probes for single-molecule biophysics and super-resolution imaging. *Annual Review of Physical Chemistry*, *63*(1), 595–617.

Hermanson, G. T. (2013). *Bioconjugate techniques*. Amsterdam: Academic Press, Elsevier Inc.

Hines, K. E., Bankston, J. R., & Aldrich, R. W. (2015). Analyzing single-molecule time series via nonparametric Bayesian inference. *Biophysical Journal*, *108*(3), 540–556.

Hong, H., Blois, T. M., Cao, Z., & Bowie, J. U. (2010). Method to measure strong protein–protein interactions in lipid bilayers using a steric trap. *Proceedings of the National Academy of Sciences of the United States of America*, *107*(46), 19802–19807.

Hughes, L. D., Rawle, R. J., & Boxer, S. G. (2014). Choose your label wisely: Water-soluble fluorophores often interact with lipid bilayers. *PLoS One*, *9*(2), e87649. Public Library of Science.

Johnson, J. M., Ha, T., Chu, S., & Boxer, S. G. (2002). Early steps of supported bilayer formation probed by single vesicle fluorescence assays. *Biophysical Journal*, *83*(6), 3371–3379.

Kovacs, E., Zorn, J. A., Huang, Y., Barros, T., & Kuriyan, J. (2015). A structural perspective on the regulation of the epidermal growth factor receptor. *Annual Review of Biochemistry*, *84*, 739–764.

Lakowicz, J. R. (2006). *Principle of fluorescence spectroscopy* (3rd ed.). Springer.

Larson, J., Kirk, M., Drier, E. A., O'Brien, W., MacKay, J. F., Friedman, L. J., et al. (2014). Design and construction of a multiwavelength, micromirror total internal reflectance fluorescence microscope. *Nature Protocols*, *9*(10), 2317–2328.

Lemmon, M. A., & Engelman, D. M. (1994). Specificity and promiscuity in membrane helix interactions. *Quarterly Review of Biophysics*, *27*(2), 157–218.

MacKenzie, K., & Fleming, K. G. (2008). Association energetics of membrane spanning α-helices. *Current Opinion in Structural Biology*, *18*(4), 412–419.

Maduke, M., Pheasant, D. J., & Miller, C. (1999). High-level expression, functional reconstitution, and quaternary structure of a prokaryotic ClC-type chloride channel. *The Journal of General Physiology*, *114*(5), 713–722.

Mathiasen, S., Christensen, S. M., Fung, J. J., Rasmussen, S. G. F., Fay, J. F., Jorgensen, S. K., et al. (2014). Nanoscale high-content analysis using compositional heterogeneities of single proteoliposomes. *Nature Methods*, *11*(9), 931–934.

McGuire, H., Aurousseau, M. R. P., Bowie, D., & Blunck, R. (2012). Automating single subunit counting of membrane proteins in mammalian cells. *The Journal of Biological Chemistry*, *287*(43), 35912–35921.

Murzyn, K., Róg, T., & Pasenkiewicz-Gierula, M. (2005). Phosphatidylethanolamine-phosphatidylglycerol bilayer as a model of the inner bacterial membrane. *Biophysical Journal*, *88*(2), 1091–1103.

Namba, K., & Stubbs, G. (1986). Structure of tobacco mosaic virus at 3.6 A resolution: Implications for assembly. *Science*, *231*(4744), 1401–1406.

Nollert, P., Kiefer, H., & Jähnig, F. (1995). Lipid vesicle adsorption versus formation of planar bilayers on solid surfaces. *Biophysical Journal*, *69*(4), 1447–1455.

Pozo Navas, B., Lohner, K., Deutsch, G., Sevcsik, E., Riske, K. A., Dimova, R., et al. (2005). Composition dependence of vesicle morphology and mixing properties in a bacterial model membrane system. *Biochimica et Biophysica Acta*, *1716*(1), 40–48.

Reddie, K. G., & Carroll, K. S. (2008). Expanding the functional diversity of proteins through cysteine oxidation. *Current Opinion in Chemical Biology*, *12*(6), 746–754.

Riddles, P. W., Blakeley, R. L., & Zerner, B. (1983). Reassessment of Ellman's reagent. In *Enzyme structure part I, Methods in enzymology: Vol. 91* (pp. 49–60). Elsevier, Amsterdam.

Schindelin, J., Arganda-Carreras, I., Frise, E., Kaynig, V., Longair, M., Pietzsch, T., et al. (2012). Fiji: An open-source platform for biological-image analysis. *Nature Methods*, *9*(7), 676–682.

Stockbridge, R. B., Robertson, J. L., Kolmakova-Partensky, L., & Miller, C. (2013). A family of fluoride-specific ion channels with dual-topology architecture. *eLife*, *2*, e01084.

Stockbridge, R. B., & Tsai, M.-F. (2015). Lipid reconstitution and recording of recombinant ion channels. *Methods in Enzymology*, *556*, 385–404. Elsevier.

Tanford, C., & Reynolds, J. A. (1976). Characterization of membrane proteins in detergent solutions. *Biochimica et Biophysica Acta*, *457*(2), 133–170.

Walden, M., Accardi, A., Wu, F., Xu, C., Williams, C., & Miller, C. (2007). Uncoupling and turnover in a Cl^-/H^+ exchange transporter. *The Journal of General Physiology*, *129*(4), 317–329. Rockefeller University Press.

White, S. H., & Wimley, W. C. (1999). Membrane protein folding and stability: Physical principles. *Annual Review of Biophysics and Biomolecular Structure*, *28*(1), 319–365.

CHAPTER FOUR

Fluorescent Labeling of Proteins in Whole Cell Extracts for Single-Molecule Imaging

S.R. Hansen[1], M.L. Rodgers[1], A.A. Hoskins[2]
University of Wisconsin–Madison, Madison, WI, United States
[2]Corresponding author: e-mail address: ahoskins@wisc.edu

Contents

Abstract

Cellular machines such as the spliceosome and ribosome can be composed of dozens of individual proteins and nucleic acids. Given this complexity, it is not surprising that many cellular activities have not yet been biochemically reconstituted. Such processes are often studied in vitro in whole cell or fractionated lysates. This presents a challenge for obtaining detailed biochemical information when the components being investigated may be only a minor component of the extract and unrelated processes may

[1] These authors contributed equally.

Methods in Enzymology, Volume 581
ISSN 0076-6879
http://dx.doi.org/10.1016/bs.mie.2016.08.018
© 2016 Elsevier Inc.
All rights reserved.

interfere with the assay. Single-molecule fluorescence microscopy methods allow particular biomolecules to be analyzed even in the complex milieu of a cell extract. This is due to the use of bright fluorophores that emit light at wavelengths at which few cellular components fluoresce, and the development of chemical biology tools for attaching these fluorophores to specific cellular proteins. Here, we describe a protocol for fluorescent labeling of endogenous, SNAP-tagged yeast proteins in whole cell extract. This method allows biochemical reactions to be followed in cell lysates in real time using colocalization single-molecule fluorescence microscopy. Labeled complexes can also be isolated from extract and characterized by SNAP tag single-molecule pull-down (SNAP-SiMPull). These approaches have proven useful for studying complex biological machines such as the spliceosome that cannot yet be reconstituted from purified components.

1. INTRODUCTION

Single-molecule fluorescence experiments can be used to simultaneously monitor the behavior of hundreds of molecules carrying out unsynchronized biochemical reactions in real time (Walter, Huang, Manzo, & Sobhy, 2008). It has been instrumental in describing the kinetics of transient or unstable interactions, detecting heterogeneity across populations, and characterizing intermediate steps in kinetic mechanisms (Larson, Rodgers, & Hoskins, 2014). While single-molecule techniques have been used extensively to study purified proteins and nucleic acids, it is often advantageous to study some processes in cellular extracts when a reconstituted system is unavailable. Complex cellular machines such as those involved in transcription, splicing, and translation can include dozens of interacting components. Not all of these components can be isolated in an active form and thus are often studied in vivo or in cell extracts.

Single-molecule fluorescence experiments in cell extracts can be carried out by using chemical tools to fluorescently label endogenous proteins. Often these tools combine a genetically encoded protein tag that can be functionalized with a small-molecule organic fluorophore. While fluorescent proteins (FPs) have often been used for single-molecule imaging, organic fluorophores offer distinct advantages. Experiments with FPs can be complicated due to low quantum yield of the FP, incomplete fluorophore maturation, and short observation lifetime due to fast photobleaching (Dickson, Cubitt, Tsien, & Moerner, 1997; Hendrix, Flors, Dedecker, Hofkens, & Engelborghs, 2008; Kubitscheck, Kückmann, Kues, & Peters, 2000). Small-molecule organic fluorophores can ameliorate many of these shortcomings to provide relatively bright, stable fluorescence. Additionally, by using genetically encoded tags that are not intrinsically fluorescent, the small-molecule fluorophore can easily be customized to the requirements of each experiment without introduction of a different tag or FP. For example, switching between green laser- and red laser-excitable small-molecule fluorophores may not require any additional genetic manipulations.

In this chapter we discuss the general workflow of preparing *Saccharomyces cerevisiae* (yeast) whole cell extract (WCE) with proteins labeled with small-molecule fluorophores (Fig. 1). We focus on labeling methods,

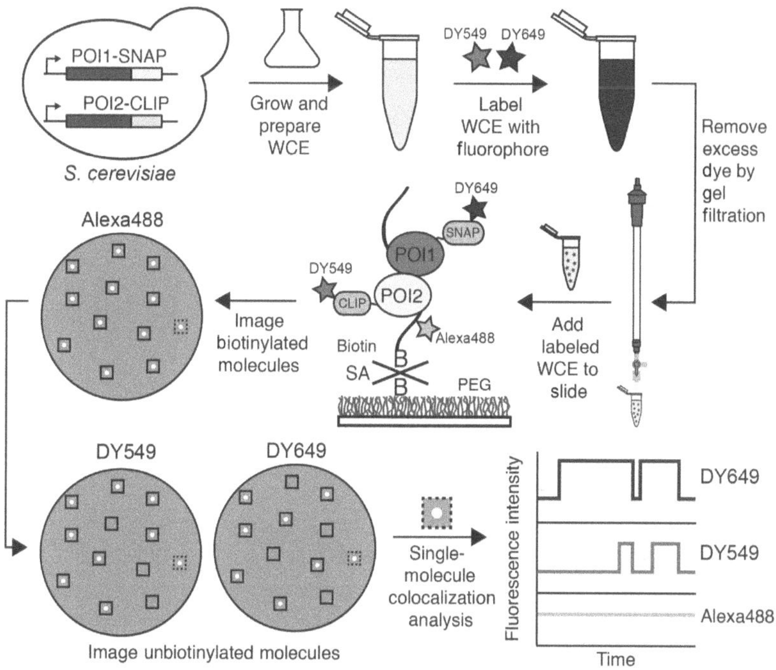

Fig. 1 The general workflow of preparing yeast WCE with fluorescently labeled proteins and conducting single-molecule colocalization experiments. The proteins of interest (POIs) are labeled using a genetically encoded tag (see Section 6.1). The WCE of this strain is labeled with a small-molecule organic fluorophore, and unreacted fluorophore is removed by gel filtration (as described in Section 4). Then the labeled WCE is applied to a PEG/PEG-biotin-coated slide on which a biotinylated species is immobilized via streptavidin. Two or more fluorophores (e.g., Alexa488, DY549, or DY649) can be imaged simultaneously by filtering the fluorescence emission by wavelength or imaged consecutively by using alternating laser excitation (ALEX). Single-molecule colocalization analysis of the fluorescence intensity from each fluorophore reveals the identity, stoichiometry, and lifetimes of complexes formed with the immobilized molecules.

removal of unreacted fluorophores, and preparation of the labeled extract for single-molecule fluorescence microscopy. As an example, we discuss the application of these techniques in the study of a multiprotein complex by SNAP tag single-molecule pull-down (SNAP-SiMPull).

2. GENETICALLY ENCODED TAGS FOR FLUORESCENT LABELING

Proteins in extract can be selectively labeled with a small-molecule organic fluorophore of choice by creating a genetically encoded fusion of that protein and a labeling tag such as one of those described in Table 1. Methods of genetic manipulation of yeast to create fusion genes are discussed in Section 6.1. The number and choice of tags is an important aspect of experimental design. The options described in Table 1 have been used by a number of laboratories and in general have high reactivity and specificity for their small-molecule fluorophore ligands.

The 20 kDa SNAP tag is an engineered version of the human DNA repair enzyme O^6-alkylguanine-DNA alkyltransferase (hAGT). The tag

Table 1 Common Protein Tags Used for Fluorophore Labeling

Tag	Approx. Size	Ligand	Reference	Notes
SNAP	20 kDa	O^6-Benzylguanine	Juillerat et al. (2003)	Modified human O^6-alkylguanine-DNA alkyltransferase (hAGT)
$SNAP_f$	20 kDa	O^6-Benzylguanine	Sun et al. (2011)	A variant of the SNAP tag with increased reactivity
CLIP	20 kDa	O^2-Benzylcytosine	Gautier et al. (2008)	Engineered from the SNAP tag to react with a different ligand
$CLIP_f$	20 kDa	O^2-Benzylcytosine	Sun et al. (2011)	A variant of the CLIP tag with increased reactivity
HaloTag	33 kDa	Chloroalkane ligands	Los et al. (2008)	Modified haloalkane dehalogenase
eDHFR	18 kDa	Trimethoprim (TMP)	Miller, Cai, Sheetz, and Cornish (2005)	*E. coli* dihydrofolate reductase; binds TMP ligands with K_D ~1 nM

quickly and efficiently forms a covalent bond between an active site cysteine and fluorescent derivatives of O^6-benzylguanine (Juillerat et al., 2003). The CLIP tag was generated by mutating the SNAP tag to react with derivatives of O^2-benzylcytosine instead of O^6-benzylguanine (Gautier et al., 2008). Fast SNAP and CLIP tags ($SNAP_f$ and $CLIP_f$) contain additional mutations that accelerate the rate of labeling compared with the original variants (Sun et al., 2011). The 33 kDa HaloTag is a modified haloalkane dehalogenase which has been engineered to form covalent adducts with fluorescent derivatives of chloroalkane ligands (Los et al., 2008). The 18 kDa *Escherichia coli* dihydrofolate reductase (eDHFR) tightly binds, but does not react with, fluorescent derivatives of trimethoprim (Miller et al., 2005).

Spectrally distinct fluorophores can be used simultaneously to differentiate between species with different protein tags in the same single-molecule experiment. Ideally, the excitation spectrum of each fluorophore should have a maximal peak near the wavelengths of lasers used by the microscope. Fluorophores can then be distinguished from one another by a combination of excitation with lasers of varying wavelength and passage of the emission light through appropriate filters. Many fluorescent ligands useful for these types of single-molecule experiments are commercially available or can be easily synthesized (Calloway et al., 2007; Corrêa, 2015; So, Yao, & Rao, 2008; Sun et al., 2011).

Since brightness and photobleaching behavior can vary dramatically between fluorophores, the final choice of which protein, tag, and fluorophore combination to use often depends on the goal of the experiment. When labeling a multiprotein complex, incorporating the same type of tag onto multiple protein components (e.g., simultaneously incorporating two or more SNAP tags into the same complex) is an effective way to overcome issues with incomplete labeling, dim fluorophores, and photobleaching during single-molecule experiments (Hoskins et al., 2011). In addition, it is often critical to ensure that incorporation of the protein tag or fluorescent label does not alter the function of the tagged protein. Bulk biochemical assays should be used to compare activities of tagged and untagged complexes. Finally, we note that many protein tags can also provide a secondary function in addition to fluorophore labeling. For example, the SNAP tag has been used to immobilize fluorescent complexes in single-molecule microscopy experiments via use of biotinylated ligands or by biotinylation of the fusion protein itself in vivo (Rodgers, Paulson, & Hoskins, 2015; Smith, Daugherty-Clarke, Goode, & Gelles, 2013).

3. COLOCALIZATION SINGLE-MOLECULE SPECTROSCOPY

Colocalization single-molecule spectroscopy (CoSMoS) is a technique for simultaneous monitoring of multiple species, labeled with spectrally distinct fluorophores, in real time by fluorescence microscopy (Larson, Rodgers, et al., 2014). Typically, CoSMoS experiments involve both immobilized and freely diffusing fluorescent biomolecules with binding interactions being monitored between the two. The immobilized species is usually tethered to a microscope slide surface with biotin/streptavidin or by using an antibody. The immobilized molecules appear as discrete spots on the slide surface, while the labeled species in solution diffuse rapidly in and out of an evanescent field established by total internal reflection (TIR) illumination and contribute only to background. If the free and immobilized molecules stably interact, then they both appear as discrete, colocalized spots of fluorescence. The dynamics of biomolecular association and dissociation can thus be monitored in real time to reveal the identity, stoichiometry, and lifetimes of complexes formed between the fluorescently labeled molecules (Friedman & Gelles, 2015). CoSMoS has been used extensively to study splicing, the process of intron removal from precursor mRNAs (pre-mRNAs) by the spliceosome. These colocalization measurements have proven useful for determining the order of spliceosome assembly, detecting spliceosome heterogeneity, and observing pre-mRNA conformational remodeling by the spliceosome (Crawford, Hoskins, Friedman, Gelles, & Moore, 2013; Hoskins et al., 2011; Hoskins, Rodgers, Friedman, Gelles, & Moore, 2016; Rodgers et al., 2015; Shcherbakova et al., 2013). Though not performed in cellular extract, CoSMoS has also been used to study the molecular features of bacterial transcription, branched actin network formation, and to examine processes involved in DNA repair (Boehm, Spies, & Washington, 2016; Friedman & Gelles, 2012; Ghoneim & Spies, 2014; Sanchez, Osborne, Friedman, Kondev, & Gelles, 2011; Smith et al., 2013). Thus, the CoSMoS approach is broadly applicable to a wide range of biological systems.

One practical consideration for any type of CoSMoS experiment is the concentration of the freely diffusing component. Even with TIR illumination, the concentration of freely diffusing fluorescent molecules must be at or below ~50 nM to maintain a high signal-to-noise ratio and distinguish colocalization from background fluorescence (Friedman, Chung, & Gelles, 2006). In cases where higher concentrations of fluorescent molecules are needed, the observation volume can be further reduced with the use of zero mode waveguides to confine illumination to a microenvironment of a few zeptoliters (10^{-21} L) (Levene et al., 2003). This reduces the background fluorescence even at the high concentrations of fluorophore-tagged components needed for studying systems such as translation (Tsai et al., 2012).

In the broadest sense, CoSMoS is a technique used to identify, count, and describe fluorescently labeled molecules. Since Förster resonance energy transfer (FRET) between fluorophores is not required for a CoSMoS-type experiment, use of genetically encoded tags fused to the N- or C-terminus of a protein is a pragmatic approach for labeling proteins with fluorophores in cell extracts. In CoSMoS, fluorophores attached to members of a biomolecular complex will be individually excited and monitored without requiring that the fluorophores be within a certain FRET proximity. This is useful for analysis of complexes whose structures are unknown and greatly facilitates fluorophore labeling of extract proteins.

4. LABELING SNAP$_f$-TAGGED PROTEINS IN CELL EXTRACT

4.1 Materials

- WCE containing SNAP$_f$-tagged proteins (1.2 mL for yeast WCE, extract preparation is described in Section 6.1)
- Labeling buffer (25 m*M* HEPES–KOH pH 7.9, 50 m*M* KCl, 1 m*M* DTT, 10% v/v glycerol)
- Peristaltic pump and tubing
- Syringe and Luer lock stopcocks
- Chromatography column (Kontes, 7×150 mm^2, 6 mL volume)
- G-25 Sephadex (DNA grade), equilibrated in water and resuspended in a 50:50 slurry
- SNAP tag fluorophore (e.g., SNAP-Surface 549®, New England Biolabs, 500 μ*M* in DMSO). For fluorophore considerations see Section 4.2.2 and Fig. 3
- Oxygen Scavenging System (OSS): final concentrations of 0.9 U/mL protocatechuate dioxygenase (PCD), 5 m*M* protocatechuic acid (PCA). See Section 6.4 for additional options for OSSs
- Triplet state quenchers (TSQs): Trolox, stock concentration of 100 m*M* in DMSO. See Section 6.4 for additional options for TSQs

4.2 Protocol

4.2.1 Assembling and Preparing the Gel Filtration Column

1. Set up the column apparatus with the Luer lock stopcock on the bottom and tubing connecting the top of the column to the pump (Fig. 2A). If gel filtration needs to be carried out at low temperatures, it is easiest to set up the column in a 4°C cold room.
2. The peristaltic pump should be calibrated for flow rates of ~0.38 and ~0.25 mL/min. Calibration can be done by adjusting the pump flow settings while measuring the amount of time it takes to flow a defined volume through the tubing.
3. Add 7–8 mL of G-25 Sephadex slurry to the capped column and allow the resin to settle. The settled bed volume should be at least 3.5–3.8 mL. It is often beneficial to add volume markings to the exterior of the column using known volumes of water prior to addition of resin.
4. Fill the rest of the column by carefully adding labeling buffer on top of the resin. Then, connect the column to the peristaltic pump (Fig. 2A).
5. Using a flow rate of ~0.38 mL/min, allow the resin to compact to a final bed volume between 3.5 and 3.8 mL. If too much resin is present, adjust the bed volume carefully using a needle and syringe to disrupt the surface of the resin and remove excess. Reconnect the pump and let the resin settle and compact under flow.

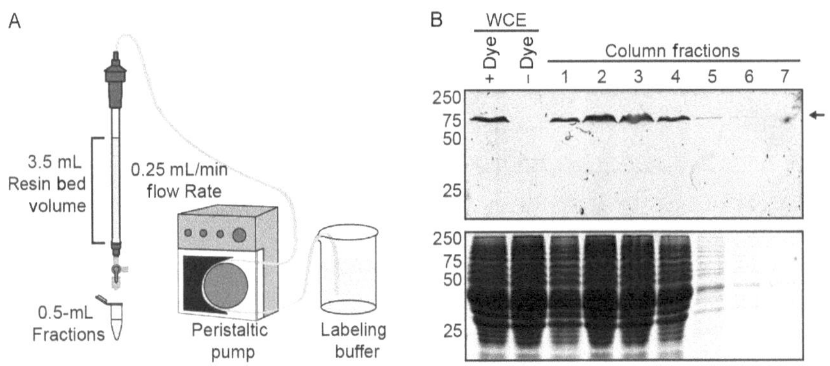

Fig. 2 (A) Diagram of the apparatus used to remove fluorophores from WCE by gel filtration described in Section 4.2.1. The labeled cell extract is added to the top of the equilibrated resin in the column and eluted at a flow rate of 0.25 mL/min while 0.5-mL fractions are collected. (B) An example denaturing SDS-PAGE, where the *S. cerevisiae* whole cell extract contains a 76 kDa SNAP tag fusion protein, imaged for SNAP-Surface 549® fluorescence (*top, arrow*) then stained with Coomassie dye (*bottom*). Extract containing the labeled protein should elute in the first few fractions.

6. Once the appropriate bed volume has been achieved, equilibrate the column in labeling buffer with a flow rate of ~0.38 mL/min for at least 1 h.

4.2.2 Labeling the SNAP$_f$ Tag in WCE With O^6-Benzylguanine Fluorophores

1. Care should be taken when choosing a fluorophore for single-molecule experiments. Some fluorophores or ligands may cross-label or tightly bind to extract proteins and increase background fluorescence (see Section 4.3 and Fig. 3).
2. For labeling SNAP$_f$ tag fusion proteins present at n*M* concentrations in extracts, add the desired O^6-benzylguanine fluorophore to a final concentration of 1 μ*M* to 1.2 mL of WCE. Other tags may require different dye concentrations for complete labeling. Orthogonal tags may be labeled simultaneously as discussed in Section 5.2.
3. Incubate the extract with dye for 30–45 min at room temperature while protected from light.

4.2.3 Excess Dye Removal by Gel Filtration Column Chromatography

1. Close the stopcock at the bottom of the prepared column and disconnect the column from the peristaltic pump. Remove excess buffer from the top of the column using a syringe while being careful to not disrupt the

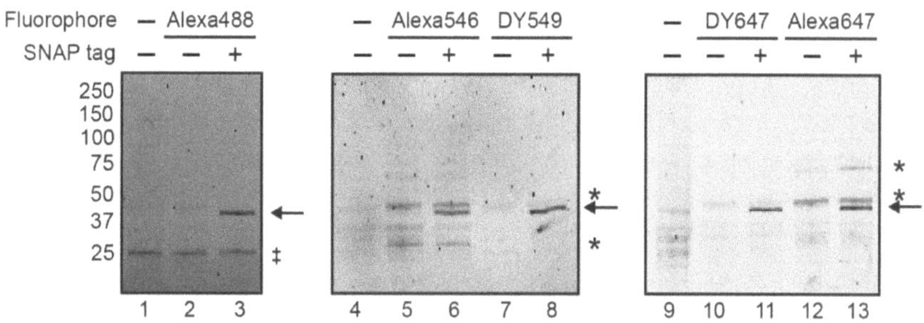

Fig. 3 SNAP tag ligands labeled with different fluorophores can show off-target labeling in whole cell extracts. Extracts were prepared with a SNAP-tagged protein or without to see if addition of ligand produced any off-target labeling. Extract samples were incubated with different color SNAP tag ligands (identified by the fluorophores noted in the figure) for 30 min at room temperature followed by 12% SDS-PAGE and imaging using a fluorescent gel imager. The *arrow* points to the correctly labeled SNAP protein. *Asterisks* (*) indicate off-target labeling of specific color SNAP tag ligands, and *double dagger mark* (‡, lanes 1–3) indicates an autofluorescent protein present in yeast WCE even in the absence of added fluorophore.

resin bed. Steps after this should be done quickly to avoid drying of the resin.

2. Apply the labeled WCE to the top of the resin bed slowly and carefully. The next steps should be done with reduced lighting to avoid photobleaching of the fluorophores.
3. Attach an air-filled 5- or 10-mL syringe to the top of the column with a stopcock, open the stopcock at the bottom of the column, and apply gentle pressure to push the labeled extract into the resin. The flow rate should be no more than 0.5 mL/min. When the WCE has completely entered the resin, release the pressure from the syringe (this is facilitated by using a three-way stopcock) and close the stopcock at the bottom of the column.
4. Carefully apply labeling buffer (~3 to 4 mL) to the top of the resin, filling the rest of the column, and reattach the column to the peristaltic pump.
5. Using a flow rate of ~0.25 mL/min, open the stopcock at the bottom of the column and begin collecting 0.5-mL fractions in 1.5-mL microfuge tubes. Usually 5–6 fractions are collected. The labeled extract should appear in fractions 2 and 3, while the bulk of the free fluorophore will begin eluting in fraction 6 or remain in the resin (Fig. 2B).
6. Fractions containing the WCE should appear clear with a yellow tint in comparison to earlier and later fractions.
7. Fractions can then be aliquoted, flash frozen in liquid N_2, and stored at −80°C.
8. To confirm that the $SNAP_f$ tag has been specifically labeled, denaturing polyacrylamide gel electrophoresis (SDS-PAGE) can be used. Fluorescently labeled proteins can be visualized using a fluorescence-based gel imager (e.g., a Typhoon imager), and total protein content can be visualized by Coomassie staining of the gel after fluorescence imaging (Fig. 2B).

4.2.4 Single-Molecule Experiments Using Labeled Cell Extracts

1. Since single-molecule experiments will vary greatly depending on the system being investigated, we provide here a broad outline of steps that will be common to many types of single-molecule experiments carried out using cell extracts.
2. The microscope setup for experiments in cell extracts should be chosen based on several factors such as background fluorescence, photon collection efficiency, and need for other components (e.g., syringe pumps or

stage/objective heating). See Section 6.2 for a discussion of microscopes that have been commonly used for experiments in cell extracts.

3. Slide preparation will also depend on particular microscope setups used. See Section 6.3 for protocols pertaining to slide preparation for experiments carried out using objective-based TIR microscopes.
4. Since many single-molecule experiments in cell extracts are designed to detect colocalization of an immobilized molecule with labeled molecules in solution, we focus here on that particular application. Other types of experiments (e.g., combined FRET and colocalization measurements) can follow a similar approach (Crawford et al., 2013).
5. We typically first incubate the prepared microscope slide with the molecule to be immobilized prior to addition of the labeled WCE. It is important that the immobilized molecules sparsely populate the surface so that single molecules can be observed. For attachment of biotinylated RNAs to a streptavidin-coated surface, we typically incubate the slide with the RNA at a low concentration (20–100 p*M*) and monitor immobilization in real time using the single-molecule microscope. Once the appropriate density has been obtained, the RNA solution is then washed off to prevent immobilization of additional molecules.
6. Before application of the labeled WCE to the slide, the WCE should be mixed with any buffer components and additional reagents. Single-molecule fluorescence in buffer and WCE benefits greatly from addition of an OSS and TSQs to extend fluorophore lifetimes. OSSs must be chosen wisely. For example, the commonly used system composed of glucose, glucose oxidase, and catalase cannot be used to study enzymes that require ATP in yeast cell extracts since endogenous hexokinase activity will rapidly consume the ATP to form glucose-6-phosphate from the added glucose. In this case, the protocatechuic acid/protocatechuate-3,4-dioxygenase system (PCA/PCD) can be used instead. Likewise, TSQs should be chosen with care and tested beforehand to determine if they alter the activity of the extract. See Section 6.4 for details on preparation of oxygen scavengers and triplet quenchers.
7. Cell extracts can then be applied to the slide and fluorescent molecules imaged. Laser power should be optimized such that cross-linking of fluorescent molecules to the slide surface is minimized. Typically, laser powers between 150 and 500 μW for 532- and 633-nm lasers can be used on a micromirror TIRF microscope. Appropriate laser powers should be determined empirically since they will vary setup to setup.

8. Imaging of two fluorophores simultaneously can be accomplished by spatially separating the fluorescence emission based on wavelength with a homebuilt or commercial "DualView" apparatus as described in Section 6.2. Additionally, multiple fluorophores can be imaged consecutively using alternating laser excitation (ALEX) (Kapanidis et al., 2005; Lee et al., 2007).
9. Single-molecule images obtained by the microscope are recorded using an electron multiplying charge-coupled device or complementary metal–oxide–semiconductor (CMOS) camera. Colocalization is often analyzed by correlating the locations of each fluorescent spot in different channels and integrating the fluorescent signal over time in each area of interest (AOI) to observe changes in fluorescence indicative of binding or dissociation events (see Section 6.5).

4.3 Troubleshooting

1. The $SNAP_f$ tag is not labeled or is not labeled efficiently in extracts. Labeling efficiency for the $SNAP_f$ tag can be near 90–100%, with higher labeling efficiencies greatly facilitating interpretation of single-molecule data. Efficiency can be determined using a time course assay relying on a nonfluorescent reagent that competes for labeling (SNAP-Cell Block®, NEB) (Rodgers et al., 2015). If labeling efficiency is low, the tag may be functionally impaired, and altering the tag orientation (moving to the N- or C-terminus) may help improve labeling.
2. Too much excess dye remains in the extract after gel filtration. In some cases, an additional dye removal step can be performed. Application of the labeled extract to a size exclusion spin column (Centri-Sep, Princeton Separations) after the gel filtration step has been shown to decrease the amount of excess dye an additional ~20% (Hoskins et al., 2011). Some fluorophores are also easier to remove from extract than others, and switching fluorophores can often remedy problems encountered during gel filtration (see later).
3. Off-target labeling is observed by SDS-PAGE. Fluorophores should be carefully chosen based on the experimental setup as well as potential for cross-labeling in cell extracts. For instance, in yeast WCE prepared for studies of pre-mRNA splicing, some fluorophores strongly associate with non-SNAP-tagged cellular proteins and cross-labeling can be observed by SDS-PAGE (Fig. 3). To avoid these issues, the dye with the least amount of cross-labeling should be chosen. We recommend

SNAP-Surface 549® (New England Biolabs, a benzylguanine derivative of the Dyomics fluorophore DY-549P1) as a good starting point since it routinely shows the least amount of cross-reactivity in yeast extracts.

5. SNAP-SiMPull OF THE YEAST U1 snRNP

5.1 $SNAP_{AP}$ Tag for Fluorescent Labeling and Biotin Immobilization

SNAP-SiMPull facilitates isolation of native complexes from yeast extracts and analysis by single-molecule fluorescence (Rodgers et al., 2015). Complexes are immobilized on a prepared microscope slide via a biotinylated and fluorescently labeled tag on a protein of interest ($SNAP_{AP}$ tag, Fig. 4A). The $SNAP_{AP}$ tag consists of the $SNAP_f$ protein fused to a short biotin acceptor peptide (AP), which can be site-specifically biotinylated by *E. coli* biotin ligase (BirA) in vivo or in vitro (Beckett, Kovaleva, & Schatz, 1999; Chen, Choi, & Ting, 2007). The $SNAP_{AP}$ tag can be incorporated onto endogenous proteins through the same genetic manipulations used for incorporation of other tags (for common procedures in yeast see Section 6.1).

The $SNAP_{AP}$ tag can be efficiently biotinylated in vivo by coexpression of BirA in yeast from a plasmid. In many cases, it is not necessary to include additional biotin in the yeast media since yeast provide sufficient biotin to carry out the biotinylation reaction to near completion. Biotinylation can be detected by incubating a labeled extract with streptavidin followed by SDS-PAGE (Fig. 4B). In our experience, this interaction is resistant to levels of SDS normally used for SDS-PAGE; however, it is important that the samples are not heated before loading the gel since this will denature the biotin/streptavidin interaction. A gel shift is observed for the biotinylated $SNAP_{AP}$ tag bound to streptavidin, and the biotinylation efficiency can be quantified by comparing the intensities of the shifted vs unshifted species.

5.2 Fluorescent Labeling of Orthogonal $SNAP_{AP}$ and $CLIP_f$ Tags in Yeast WCE

The SNAP and CLIP tags rely on different substrates for covalent labeling: benzylguanine and benzylcytosine, respectively (Sun et al., 2011). Cross-labeling is possible and must be optimized by adjusting fluorophore concentrations, incubation times, as well as testing sequential or simultaneous labeling strategies. For instance, fluorophore concentrations needed for efficient labeling can be lowered with the use of either $SNAP_f$ or $CLIP_f$

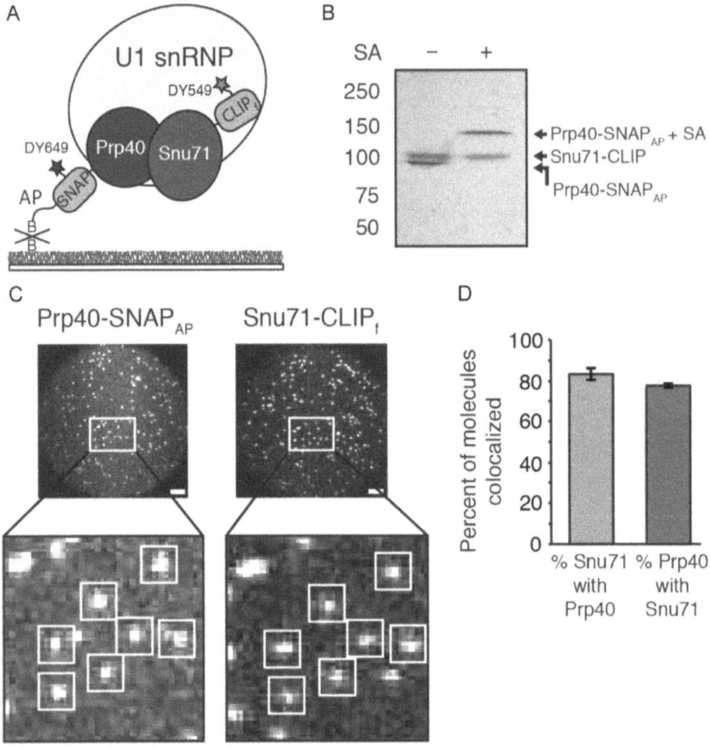

Fig. 4 SNAP-SiMPull of the U1 snRNP complex. (A) *Cartoon* depiction of pull-down and colocalization of the U1 snRNP proteins, Prp40-SNAP$_{AP}$ and Snu71-CLIP$_f$. Prp40-SNAP$_{AP}$ was labeled with SNAP-Surface 649® (New England Biolabs, labeled DY549), and Snu71-CLIP$_f$ was labeled with SNAP-Surface 549® (New England Biolabs, labeled DY649). (B) SDS-PAGE gel of the doubly labeled extracts incubated with and without streptavidin (SA). A gel shift can be observed specifically for Prp40-SNAP$_{AP}$ in the presence of SA, indicating biotinylation of the SNAP$_{AP}$ tag. (C) Representative microscope fields of view for Prp40-SNAP$_{AP}$ and Snu71-CLIP$_f$. *Zoomed-in regions* show colocalization of Prp40-SNAP$_{AP}$ and Snu71-CLIP$_f$ spots in each channel. (D) Quantification of colocalization between Prp40-SNAP$_{AP}$ and Snu71-CLIP$_f$ at the single-molecule level. Error bars represent the standard error from three replicates.

tags (Table 1). Often we have observed cross-reactivity of the SNAP$_f$ tag with benzylcytosine ligands (i.e., CLIP fluorophores). If this is observed, it is advantageous to adjust the order of labeling. For example, a SNAP$_f$ tag may be labeled first by adding the SNAP fluorophore to the extract, incubating for a period of time, and then adding the CLIP fluorophore to label the CLIP$_f$ tag. This allows for the faster labeling reaction (SNAP$_f$ tag with a SNAP fluorophore) to occur without the presence of a competing ligand (the CLIP fluorophore). While fluorophore concentrations necessary for labeling should be optimized for each system, we often find that 1 μ*M* SNAP ligand and 2 μ*M* CLIP ligand are sufficient to label $SNAP_f$- and $CLIP_f$-tagged proteins present at n*M* concentrations in yeast WCE in approximately 30 min at room temperature.

5.3 Pull-Down of U1 snRNP Complexes via SNAP-SiMPull

Extract containing the labeled $SNAP_{AP}$-tagged protein can be applied directly to a passivated, streptavidin-coated single-molecule slide for a single step pull-down. While yeast WCE contains a number of biotinylated proteins, we have found that these generally do not interfere with the ability to isolate $SNAP_{AP}$-tagged spliceosome components. Here, we will discuss an example of SNAP-SiMPull in which the U1 small nuclear ribonucleoprotein, an snRNP containing 17 proteins and an snRNA, was isolated. In order to assess colocalization within the U1 snRNP complex, we incorporated a $SNAP_{AP}$ and $CLIP_f$ tag on two U1 proteins, Prp40 and Snu71, respectively (Fig. 4A). Colocalization between these two proteins confirmed the presence of a complex, consistent with SNAP-SiMPull of the U1 snRNP. In these experiments, Prp40-$SNAP_{AP}$ was labeled with a red laser-excitable fluorophore (SNAP-Surface 649®, New England Biolabs) and Snu71-$CLIP_f$ was labeled with a green fluorophore (CLIP-Surface 549®, New England Biolabs) (Fig. 4).

Extract was first incubated on a prepared, streptavidin-coated microscope slide to pull-down the labeled snRNPs. The extract was then removed by repeated washing with buffer (25 m*M* HEPES–KOH pH 7.9, 50 m*M* KCl, 10% v/v glycerol, 1 m*M* DTT, 100 m*M* potassium phosphate, 3% w/v PEG-8000, 2.5 m*M* $MgCl_2$), and spots of fluorescence from both Prp40-$SNAP_{AP}$ and Snu71-$CLIP_f$ could be observed in images produced by a DualView apparatus in long- and short-wavelength channels (corresponding to Prp40-$SNAP_{AP}$ labeled with SNAP-Surface 649® and Snu71-$CLIP_f$ labeled with CLIP-Surface 549®, respectively). We determined the degree of colocalization by choosing spots in one channel, mapping those locations onto the other channel, and assessing overlap. Common procedures for mapping spot locations by creating a mathematical relationship between the two channels can be found in Section 6.5. In these experiments, two sets of colocalization analyses were performed in which the percent of Prp40-$SNAP_{AP}$ spots colocalized with Snu71-$CLIP_f$ was determined and vice versa. By carrying out both colocalization analyses particle heterogeneity could be measured: spots of fluorescence containing both Snu71-CLIP$_f$ and Prp40-SNAP$_{AP}$ could be identified as well as those containing one molecule but not the other.

Colocalization analyses revealed that $82 \pm 3\%$ Snu71-CLIP$_f$ molecules could be colocalized with Prp40-SNAP$_{AP}$ and similarly, $77 \pm 1\%$ of Prp40-SNAP$_{AP}$ molecules colocalized with Snu71-CLIP$_f$ (Fig. 4C and D). We determined that the maximum predicted colocalization was ~90% based on the observation that each tag was ~95% labeled. The discrepancy between the maximum predicted and observed colocalization between Prp40-SNAP$_{AP}$ and Snu71-CLIP$_f$ may be due to experimental factors, such as the presence of dark fluorophores, or it may have arisen from biologically relevant factors, such as incomplete or heterogeneous complexes. Overall, these data suggested that most of the isolated complexes contain both Prp40-SNAP$_{AP}$ and Snu71-CLIP$_f$ in agreement with other yeast U1 snRNP purification methods (Rigaut et al., 1999). Notably, these U1 snRNP particles were isolated by SNAP-SiMPull in only a few seconds and from <40 μL of yeast WCE. Bulk biochemical purification of U1 snRNP typically involves >40 mL of WCE prepared from 10 L of yeast culture and is carried out over several days (Rigaut et al., 1999). SNAP-SiMPull allows rapid capture and identification of complexes from yeast WCE and enables experiments in which other purification methods may become limiting.

6. COMMON PROTOCOLS

6.1 Genetic Manipulation in *S. cerevisiae* and Preparation of Yeast WCE

Endogenous yeast genes can be altered or fused to other genes by use of the native homologous recombination machinery. A linear PCR product containing the fusion gene (e.g., SNAP$_{AP}$ tag) flanked by 40–100 base pair homology regions to the gene of interest, and a genetic marker (e.g., *HIS3* or hygromycin resistance) can be introduced into yeast cells using lithium acetate transformation method (Gietz & Woods, 2002; Janke et al., 2004). Following transformation, yeast are grown on selective media, and transformants are confirmed by colony PCR and/or Southern blotting (Hoskins et al., 2011).

Yeast WCE can be prepared using the liquid nitrogen method (Stevens & Abelson, 2002). Briefly, 2 L of yeast culture can be grown to end log-phase, collected by centrifugation, and made into droplets using liquid nitrogen. A mortar and pestle or ball mill machine (Mayas, Maita, & Staley, 2006) can be used to grind the yeast droplets into a fine powder. Clarification of the extracts can then be performed by ultracentrifugation followed by dialysis (Ansari & Schwer, 1995) or gel filtration (Hoskins et al., 2011).

6.2 Construction of a Multiwavelength Micromirror TIR Microscope

TIR microscopy can be used to excite fluorophores nearest the slide surface, reducing background fluorescence from molecules in solution (Axelrod, 2008). TIR can be achieved through a prism-based or objective-based system both of which can be incorporated in a variety of microscope setups (Walter et al., 2008). Micromirror TIR fluorescence microscopes offer advantages such as increased photon collection efficiency for multiple wavelengths of light, making them well suited for multiwavelength colocalization analyses (Larson, Kirk, et al., 2014). If multiple fluorophores are to be viewed simultaneously, this is best achieved with a "DualView" apparatus that splits the fluorescence emission based on wavelength using a dichroic mirror (e.g., into short- and long-wavelength fluorescence emission images) and projects both images onto one or more cameras. DualView systems can be homebuilt or purchased commercially (Larson, Kirk, et al., 2014).

6.3 Preparation of Streptavidin-Coated Slides for Single-Molecule Studies

For single-molecule colocalization experiments using an objective-based TIR microscope, passivated glass slides are often used (Friedman et al., 2006). Detailed protocols for slide preparation have been described elsewhere (Anderson & Hoskins, 2014). There are several types of passivation methods that can be optimized for colocalization experiments carried out in WCE. To prevent nonspecific adsorbing of fluorescent molecules, glass slides may be derivatized with polyethyleneglycol (PEG) (Friedman et al., 2006; Rasnik, Myong, Cheng, Lohman, & Ha, 2004). In addition to PEG, the slide surface can be treated with polysiloxane and Tween20 to further reduce nonspecific binding (Revyakin et al., 2012). Other additives may also be included to help change the surface charge to reduce nonspecific interactions. These may include PEG conjugates (e.g., poly-L-lysine PEG) as well as coating the surface with cellular tRNA or heparin solutions (Salomon, Jolly, Moore, Zamore, & Serebrov, 2015). Care should be taken when choosing an appropriate surface coating, and there is not yet a one-size-fits-all solution. For example, heparin may reduce nonspecific binding in some cases while in others promote tight interactions with nucleic acid binding proteins.

6.4 Preparation of Oxygen Scavenging and Triplet Quenching Systems

Several OSS have been commonly used in single-molecule fluorescence experiments. Glucose oxidase/catalase is relatively inexpensive and has been applied in many different types of experiments (Yildiz et al., 2003). However, the glucose oxidase/catalase system is incompatible with yeast cell extract experiments involving ATP, since addition of glucose depletes the ATP from the extract. Alternative OSS such as galactose oxidase/catalase (Crawford, Hoskins, Friedman, Gelles, & Moore, 2008) or protocatechuate dioxygenase/protocatechuate acid (PCD/PCA) (Aitken, Marshall, & Puglisi, 2008; Crawford et al., 2008) can be used instead to prevent ATP depletion in yeast WCE. Protocols for their use have been described previously (Crawford et al., 2008).

There are several types of TSQs that can also be used in single-molecule fluorescence experiments. Reducing agents, such as DTT and β-mercaptoethanol, are often used in biochemical buffers and can also function as TSQs. Addition of Trolox has been shown to further prevent blinking of the fluorophores (Rasnik, McKinney, & Ha, 2006). Examples of other TSQs that have also been used in single-molecule experiments include cyclooctatetraene (COT), propyl gallate, and 4-nitrobenzyl alcohol (Dave, Terry, Munro, & Blanchard, 2009; Hoskins et al., 2011). We typically prepare TSQs together with DMSO in a 100–200 × stock and add them to WCE at a final concentration of 1 m*M* Trolox, 1 m*M* COT, 1 m*M* 4-nitrobenzyl alcohol, and 0.5 m*M* propyl gallate (Hoskins et al., 2011). COT produces very noxious fumes and stock solutions should be prepared in a fume hood.

6.5 Analysis of Colocalization Single-Molecule Data

Following data collection, we typically first identify well-separated spots of fluorescence representing the immobilized molecules. These are used to identify AOIs that are then mapped to other images, representing fluorescence collected from the freely diffusing fluorescent biomolecules. Mapping of AOI locations should be done with care, and functions for carrying this out have been described (Friedman & Gelles, 2015). A mapping function can be created using fluorescent beads that fluoresce at multiple wavelengths or by using a biotinylated DNA standard containing two or more distinct fluorophores. Mapping is not strictly linear since the optical properties of the microscope may lead to aberrations near the edges of the field of view.

Once AOIs in the images have been identified and mapped, the fluorescence intensity in that AOI can be integrated over the time course of the experiment. Colocalization can be determined by looking for peaks of fluorescence intensity appearing at AOIs corresponding to the same physical location on the microscope slide. Features of the biomolecular interaction such as lifetime kinetics, order of binding, and complex heterogeneity can then be determined by analyzing these single-molecule traces (Friedman & Gelles, 2015).

ACKNOWLEDGMENTS

This work was supported by the National Institutes of Health (R00 GM086471, R01 GM112735 to A.A.H., T32-GM08293 to M.L.R., T32-GM008505 to S.R.H.), Shaw Scientist and Beckman Young Investigator Awards, and startup funding from the University of Wisconsin–Madison, Wisconsin Alumni Research Foundation (WARF), and the Department of Biochemistry.

REFERENCES

Aitken, C. E., Marshall, R. A., & Puglisi, J. D. (2008). An oxygen scavenging system for improvement of dye stability in single-molecule fluorescence experiments. *Biophysical Journal*, *94*(5), 1826–1835.

Anderson, E. G., & Hoskins, A. A. (2014). Single molecule approaches for studying spliceosome assembly and catalysis. *Methods in Molecular Biology (Clifton, N.J.)*, *1126*, 217–241.

Ansari, A., & Schwer, B. (1995). SLU7 and a novel activity, SSF1, act during the PRP16-dependent step of yeast pre-mRNA splicing. *EMBO Journal*, *14*(16), 4001–4009.

Axelrod, D. (2008). Chapter 7: Total internal reflection fluorescence microscopy. *Methods in Cell Biology*, *89*, 169–221.

Beckett, D., Kovaleva, E., & Schatz, P. J. (1999). A minimal peptide substrate in biotin holoenzyme synthetase-catalyzed biotinylation. *Protein Science: A Publication of the Protein Society*, *8*(4), 921–929.

Boehm, E. M., Spies, M., & Washington, M. T. (2016). PCNA tool belts and polymerase bridges form during translesion synthesis. *Nucleic Acids Research*. pii: gkw563 [Epub ahead of print].

Calloway, N. T., Choob, M., Sanz, A., Sheetz, M. P., Miller, L. W., & Cornish, V. W. (2007). Optimized fluorescent trimethoprim derivatives for in vivo protein labeling. *ChemBioChem*, *8*(7), 767–774.

Chen, I., Choi, Y.-A., & Ting, A. Y. (2007). Phage display evolution of a peptide substrate for yeast biotin ligase and application to two-color quantum dot labeling of cell surface proteins. *Journal of the American Chemical Society*, *129*(20), 6619–6625.

Corrêa, I. R. (2015). Considerations and protocols for the synthesis of custom protein labeling probes. *Methods in Molecular Biology (Clifton, N.J.)*, *1266*, 55–79.

Crawford, D. J., Hoskins, A. A., Friedman, L. J., Gelles, J., & Moore, M. J. (2008). Visualizing the splicing of single pre-mRNA molecules in whole cell extract. *RNA*, *14*(1), 170–179.

Crawford, D. J., Hoskins, A. A., Friedman, L. J., Gelles, J., & Moore, M. J. (2013). Single-molecule colocalization FRET evidence that spliceosome activation precedes stable approach of 5′ splice site and branch site. *Proceedings of the National Academy of Sciences of the United States of America*, *110*(17), 6783–6788.

Dave, R., Terry, D. S., Munro, J. B., & Blanchard, S. C. (2009). Mitigating unwanted photophysical processes for improved single-molecule fluorescence imaging. *Biophysical Journal*, *96*(6), 2371–2381.

Dickson, R. M., Cubitt, A. B., Tsien, R. Y., & Moerner, W. E. (1997). On/off blinking and switching behaviour of single molecules of green fluorescent protein. *Nature*, *388*(6640), 355–358.

Friedman, L. J., Chung, J., & Gelles, J. (2006). Viewing dynamic assembly of molecular complexes by multi-wavelength single-molecule fluorescence. *Biophysical Journal*, *91*(3), 1023–1031.

Friedman, L. J., & Gelles, J. (2012). Mechanism of transcription initiation at an activator-dependent promoter defined by single-molecule observation. *Cell*, *148*(4), 679–689.

Friedman, L. J., & Gelles, J. (2015). Multi-wavelength single-molecule fluorescence analysis of transcription mechanisms. *Methods*, *86*(C), 27–36.

Gautier, A., Juillerat, A., Heinis, C., Corrêa, I. R., Jr., Kindermann, M., Beaufils, F., et al. (2008). An engineered protein tag for multiprotein labeling in living cells. *Chemistry & Biology*, *15*(2), 128–136.

Ghoneim, M., & Spies, M. (2014). Direct correlation of DNA binding and single protein domain motion via dual illumination fluorescence microscopy. *Nano Letters*, *14*(10), 5920–5931.

Gietz, R. D., & Woods, R. A. (2002). Transformation of yeast by lithium acetate/single-stranded carrier DNA/polyethylene glycol method. *Methods in Enzymology*, *350*, 87–96.

Hendrix, J., Flors, C., Dedecker, P., Hofkens, J., & Engelborghs, Y. (2008). Dark states in monomeric red fluorescent proteins studied by fluorescence correlation and single molecule spectroscopy. *Biophysical Journal*, *94*(10), 4103–4113.

Hoskins, A. A., Friedman, L. J., Gallagher, S. S., Crawford, D. J., Anderson, E. G., Wombacher, R., et al. (2011). Ordered and dynamic assembly of single spliceosomes. *Science*, *331*(6022), 1289–1295.

Hoskins, A. A., Rodgers, M. L., Friedman, L. J., Gelles, J., & Moore, M. J. (2016). Single molecule analysis reveals reversible and irreversible steps during spliceosome activation. *eLife*, *5*, e14166.

Janke, C., Magiera, M. M., Rathfelder, N., Taxis, C., Reber, S., Maekawa, H., et al. (2004). A versatile toolbox for PCR-based tagging of yeast genes: New fluorescent proteins, more markers and promoter substitution cassettes. *Yeast*, *21*(11), 947–962.

Juillerat, A., Gronemeyer, T., Keppler, A., Gendreizig, S., Pick, H., Vogel, H., et al. (2003). Directed evolution of O6-alkylguanine-DNA alkyltransferase for efficient labeling of fusion proteins with small molecules in vivo. *Chemistry & Biology*, *10*(4), 313–317.

Kapanidis, A. N., Laurence, T. A., Lee, N. K., Margeat, E., Kong, X., & Weiss, S. (2005). Alternating-laser excitation of single molecules. *Accounts of Chemical Research*, *38*(7), 523–533.

Kubitscheck, U., Kückmann, O., Kues, T., & Peters, R. (2000). Imaging and tracking of single GFP molecules in solution. *Biophysical Journal*, *78*(4), 2170–2179.

Larson, J., Kirk, M., Drier, E. A., O'Brien, W., MacKay, J. F., Friedman, L. J., et al. (2014). Design and construction of a multiwavelength, micromirror total internal reflectance fluorescence microscope. *Nature Protocols*, *9*(10), 2317–2328.

Larson, J. D., Rodgers, M. L., & Hoskins, A. A. (2014). Visualizing cellular machines with colocalization single molecule microscopy. *Chemical Society Reviews*, *43*(4), 1189–1200.

Lee, N. K., Kapanidis, A. N., Koh, H. R., Korlann, Y., Ho, S. O., Kim, Y., et al. (2007). Three-color alternating-laser excitation of single molecules: Monitoring multiple interactions and distances. *Biophysical Journal*, *92*(1), 303–312.

Levene, M. J., Korlach, J., Turner, S. W., Foquet, M., Craighead, H. G., & Webb, W. W. (2003). Zero-mode waveguides for single-molecule analysis at high concentrations. *Science*, *299*(5607), 682–686.

Los, G. V., Encell, L. P., McDougall, M. G., Hartzell, D. D., Karassina, N., Zimprich, C., et al. (2008). HaloTag: A novel protein labeling technology for cell imaging and protein analysis. *ACS Chemical Biology*, *3*(6), 373–382.

Mayas, R. M., Maita, H., & Staley, J. P. (2006). Exon ligation is proofread by the DExD/H-box ATPase Prp22p. *Nature Structural & Molecular Biology*, *13*(6), 482–490.

Miller, L. W., Cai, Y., Sheetz, M. P., & Cornish, V. W. (2005). In vivo protein labeling with trimethoprim conjugates: A flexible chemical tag. *Nature Methods*, *2*(4), 255–257.

Rasnik, I., McKinney, S. A., & Ha, T. (2006). Nonblinking and long-lasting single-molecule fluorescence imaging. *Nature Methods*, *3*(11), 891–893.

Rasnik, I., Myong, S., Cheng, W., Lohman, T. M., & Ha, T. (2004). DNA-binding orientation and domain conformation of the *E. coli* Rep helicase monomer bound to a partial duplex junction: Single-molecule studies of fluorescently labeled enzymes. *Journal of Molecular Biology*, *336*(2), 395–408.

Revyakin, A., Zhang, Z., Coleman, R. A., Li, Y., Inouye, C., Lucas, J. K., et al. (2012). Transcription initiation by human RNA polymerase II visualized at single-molecule resolution. *Genes & Development*, *26*(15), 1691–1702.

Rigaut, G., Shevchenko, A., Rutz, B., Wilm, M., Mann, M., & Séraphin, B. (1999). A generic protein purification method for protein complex characterization and proteome exploration. *Nature Biotechnology*, *17*(10), 1030–1032.

Rodgers, M. L., Paulson, J., & Hoskins, A. A. (2015). Rapid isolation and single-molecule analysis of ribonucleoproteins from cell lysate by SNAP-SiMPull. *RNA*, *21*(5), 1031–1041.

Salomon, W. E., Jolly, S. M., Moore, M. J., Zamore, P. D., & Serebrov, V. (2015). Single-molecule imaging reveals that argonaute reshapes the binding properties of its nucleic acid guides. *Cell*, *162*(1), 84–95.

Sanchez, A., Osborne, M. L., Friedman, L. J., Kondev, J., & Gelles, J. (2011). Mechanism of transcriptional repression at a bacterial promoter by analysis of single molecules. *The EMBO Journal*, *30*(19), 3940–3946.

Shcherbakova, I., Hoskins, A. A., Friedman, L. J., Serebrov, V., Corrêa, I. R., Xu, M.-Q., et al. (2013). Alternative spliceosome assembly pathways revealed by single-molecule fluorescence microscopy. *Cell Reports*, *5*(1), 151–165.

Smith, B. A., Daugherty-Clarke, K., Goode, B. L., & Gelles, J. (2013). Pathway of actin filament branch formation by Arp2/3 complex revealed by single-molecule imaging. *Proceedings of the National Academy of Sciences of the United States of America*, *110*(4), 1285–1290.

So, M.-K., Yao, H., & Rao, J. (2008). HaloTag protein-mediated specific labeling of living cells with quantum dots. *Biochemical and Biophysical Research Communications*, *374*(3), 419–423.

Stevens, S. W., & Abelson, J. (2002). Yeast pre-mRNA splicing: Methods, mechanisms, and machinery. *Methods in Enzymology*, *351*, 200–220.

Sun, X., Zhang, A., Baker, B., Sun, L., Howard, A., Buswell, J., et al. (2011). Development of SNAP-Tag fluorogenic probes for wash-free fluorescence imaging. *ChemBioChem*, *12*(14), 2217–2226.

Tsai, A., Petrov, A., Marshall, R. A., Korlach, J., Uemura, S., & Puglisi, J. D. (2012). Heterogeneous pathways and timing of factor departure during translation initiation. *Nature*, *487*(7407), 390–393.

Walter, N. G., Huang, C.-Y., Manzo, A. J., & Sobhy, M. A. (2008). Do-it-yourself guide: How to use the modern single-molecule toolkit. *Nature Methods*, *5*(6), 475–489.

Yildiz, A., Forkey, J. N., McKinney, S. A., Ha, T., Goldman, Y. E., & Selvin, P. R. (2003). Myosin V walks hand-over-hand: Single fluorophore imaging with 1.5-nm localization. *Science*, *300*(5628), 2061–2065.

CHAPTER FIVE

Quantifying the Assembly of Multicomponent Molecular Machines by Single-Molecule Total Internal Reflection Fluorescence Microscopy

E.M. Boehm*[,1], S. Subramanyam*, M. Ghoneim[†,2], M. Todd Washington*[,2], M. Spies*[,2]

*University of Iowa Carver College of Medicine, Iowa City, IA, United States
[†]Faculty of Science, Cairo University, Giza, Egypt
[2]Corresponding authors: e-mail addresses: moh.kar.ghoneim@cu.edu.eg; moh.kar.ghoneim@gmail.com; todd-washington@uiowa.edu; maria-spies@uiowa.edu

Contents

[1] Present address: Department of Biological Chemistry and Molecular Pharmacology, Harvard Medical School, Boston, MA, USA.

Methods in Enzymology, Volume 581
ISSN 0076-6879
http://dx.doi.org/10.1016/bs.mie.2016.08.019
© 2016 Elsevier Inc.
All rights reserved.

Abstract

Large, dynamic macromolecular complexes play essential roles in many cellular processes. Knowing how the components of these complexes associate with one another and undergo structural rearrangements is critical to understanding how they function. Single-molecule total internal reflection fluorescence (TIRF) microscopy is a powerful approach for addressing these fundamental issues. In this article, we first discuss single-molecule TIRF microscopes and strategies to immobilize and fluorescently label macromolecules. We then review the use of single-molecule TIRF microscopy to study the formation of binary macromolecular complexes using one-color imaging and inhibitors. We conclude with a discussion of the use of TIRF microscopy to examine the formation of higher-order (i.e., ternary) complexes using multicolor setups. The focus throughout this article is on experimental design, controls, data acquisition, and data analysis. We hope that single-molecule TIRF microscopy, which has largely been the province of specialists, will soon become as common in the tool box of biophysicists and biochemists as structural approaches have become today.

1. INTRODUCTION

Many essential cellular processes are carried out by heterogeneous macromolecular machines that dynamically assemble, disassemble, exchange components and undergo various rearrangements. On the most fundamental level, many questions regarding the activity and organization of these complex molecular machines may be reduced to questions of occupancy: whether and for how long do the molecules associate, what binary and higher-order (i.e., ternary, quaternary, quinary, etc.) complexes form, and whether the interactions between certain components are mutually exclusive. The ability to observe formation and decomposition of individual macromolecular complexes in real time and at physiologically relevant temperature permits direct investigation of the role of weak and transient interactions in the assembly and function of these complexes, as well as the heterogeneity in their composition over time and in the ensemble. These interactions are hard to investigate using bulk kinetics tools and almost always refractory to analysis by more traditional biochemical techniques such as electrophoretic gel mobility shift assays or pull-downs. Single-molecule analysis by total internal reflection fluorescence microscopy (TIRFM) has revolutionized our ability to monitor the assembly, function, and dynamics of these complexes in real time and under physiological conditions (Elenko, Szostak, & Van Oijen, 2009). The same molecule can be observed through multiple cycles of binding to and dissociation from its partner(s), a process visualized as an activity trajectory. Because of the stochastic nature of molecular interactions, every individual single-molecule binding/dissociation trajectory is unique. The overall statistical properties that are extracted from a set of trajectories, however, are reproducible and may offer unprecedented insight into the mechanism of the binding reaction and the stability of the complex.

Global analysis of these trajectories allows for accurate determination of the kinetic rate constants and interaction affinities (Chen et al., 2014; Elenko et al., 2009; Haghighat Jahromi, Honda, Zimmerman, & Spies, 2013; Huppa et al., 2010; Joshi et al., 2012; Wang, Guo, Golding, Cox, & Ong, 2009), helps to reveal populations of conformers with different binding or catalytic properties (Amrute-Nayak et al., 2014; Chen, Subramanyam, Elcock, Spies, & Wold, 2016; Honda et al., 2014), allows the investigator to distinguish different orders of complex assembly, and provides the tools to infer the mechanisms of competition or cooperation in the assembly of the higher-order complexes (Boehm, Spies, & Washington, 2016; Hoskins, Rodgers, Friedman, Gelles, & Moore, 2016). With the advent of commercially available TIRFM systems, the methodology is becoming less arcane and hopefully will become a routine technique in biophysics and biochemistry laboratories. Robust single-molecule data analysis protocols, as well as the methods for extracting meaningful mechanistic information from the collections of single-molecule dwell times, however, are needed for this methodology to become truly accessible. In this chapter, we will discuss TIRFM-enabled single-molecule colocalization experiments that offer a straightforward way to monitor and quantify the interactions between a surface-tethered macromolecule (protein or nucleic acid) and one or more fluorescently labeled freely diffusing interacting partner(s). We will focus on the details of the data acquisition and analysis for single-color and multicolor-binding experiments, constructing bound and unbound dwell-time histograms, and fitting these histograms to extract the mechanistic information about the macromolecular assembly and organization.

2. OBSERVING BINARY MACROMOLECULAR INTERACTIONS USING SINGLE-MOLECULE TIRFM

2.1 Total Internal Reflection Fluorescence Microscopes

Total internal reflection fluorescence (TIRF) microscopes provide one of the most convenient ways to visualize the molecular interactions between the surface-tethered species and freely diffusing molecules labeled with a fluorophore in real time and at the single-molecule level. The evanescent wave generated by total internal reflection of the illumination source selectively excites molecules very near to the surface of the slide ($\leq$100 nm), leaving the rest of the reaction solution dark (Axelrod, 2008). There exist numerous versions of home-built and commercially available configurations of TIRFM setups.

The data that we use in the examples later have been acquired on our prism-type TIRF microscope configured for the optimal detection of Cy3 and Cy5 dyes. The construction of the system is described in detail in Bain, Wu, and Spies (2016) and Ghoneim and Spies (2014). Briefly, this system was built on an Olympus IX-71 frame (Olympus America, Inc, IL, USA). The excitation sources for the Cy3 and Cy5 dyes in our system are the diode-pumped solid-state laser (532 nm; Coherent, Inc, CA, USA) and the diode laser (640 nm, Coherent, Inc, CA, USA), respectively. The most current equivalents of these lasers sold by Coherent are from the OBIS and CUBE series. A power output of at least 100 mW is desirable, though 50 mW can be sufficient for a majority of studies. In the two-color imaging scheme, the lasers are coaligned by a polarizing cube beam splitter (Melles Griot, NY, USA, Cat. No. PBSH-450-700-050), and then guided through Pellin–Broca prism (Eksma Optics, Vilnius, Lithuania, Cat. No. 325-1206) to generate two overlapping evanescent fields of illumination for the excitation of Cy3 and Cy5 fluorophores, respectively. Fluorescence signals of both fluorophores are collected by a water immersion 60 × objective (UPLANSAPO, numerical aperture 1.2, Olympus). Scattered excitation light is removed using Cy3/Cy5 dual band-pass filter (Semrock, NY, USA, FF01-577/690) in the emission optical pathway.

The same system configuration is used for both dual-illumination and single-color illumination imaging experiments to ensure accurate comparison of imaging conditions between the two types of experiments. Images originating from the two fluorophores are chromatically separated into Cy3 image and Cy5 image using 630 nm dichroic mirror inside the dual view system (DV2; Photometrics, AZ, USA) and then recorded using an EMCCD camera (Andor technology, Belfast, UK, DU-897-E-CSO-#BV) at a 100 or 30 ms per frame acquisition rate and amplification gain of 250 without binning. For dual-illumination experiments, the intensity of the two lasers is adjusted independently. First, we adjust the size and the focus of the green laser (532 nm Cy3 excitation), then select the laser power that provides suitable signal-to-noise ratio in the Cy3 channel. Second, we adjust the power of the red laser (640 nm Cy5 excitation) to achieve

Cy3 and Cy5 fluorescence signals of comparable levels. While exact values vary from one experiment to another, in most of our two-color excitation experiments, the intensities at the slide are set around 45 and 90 mW for the green and red laser, respectively, reflecting the difference in the quantum yield of the two dyes, differences in beam sizes at the slide–aqueous solution interface, penetration depth dependence on wavelength, and polarization of the two lasers.

2.2 Labeling and Surface Tethering of Macromolecules

To observe the binding reaction using a TIRFM-enabled single-molecule scheme (Fig. 1), one of the interacting partners needs to be tethered to the surface of the microscope imagine chamber, while the second is labeled with an appropriate fluorescent dye (we use Cy3 and Cy5 dyes in the experiments described herein).

2.2.1 Fluorescent Nucleic Acids

Oligonucleotides (DNA and RNA) site specifically labeled with a wide range of fluorophores compatible with most single-molecule TIRFM setups are readily available for purchase through commercial companies like IDT, MWG Operon, etc. These oligonucleotides are labeled during synthesis using a modified solid support (CPG) for 3′-modifications or a specialized phosphoramidite reagent for internal and 5′-modifications.

2.2.2 Fluorescent Labeling of Proteins

Labeling of proteins can be achieved through several methods depending on the nature of the protein as well as the requirement for site specificity. Direct chemical labeling of proteins involves the reaction of cysteine residues in the protein with maleimide–fluorophore conjugates (GE Healthcare Life Sciences, MA, USA). This reaction is carried out at moderate pH (optimally pH 7) and temperatures. Pretreatment of the protein with a reducing agent (DTT or TCEP) ensures that the cysteine residues are reduced prior to labeling. The presence of several reactive cysteines on the protein may lead to nonhomogenous labeling of the protein. One can replace the cysteines whose labeling is not desired with serine or threonine residues providing that this does not affect the structure, the stability, or the activities of the protein. N-terminal labeling of proteins can be carried out using amine reactive dyes such as succinimidyl esters (NHS) like CyDye™ (GE Healthcare Life Sciences, MA, USA) at a neutral pH. At high pH, this method can also be used to label selectively the ε-amino group on lysine residues; however, because

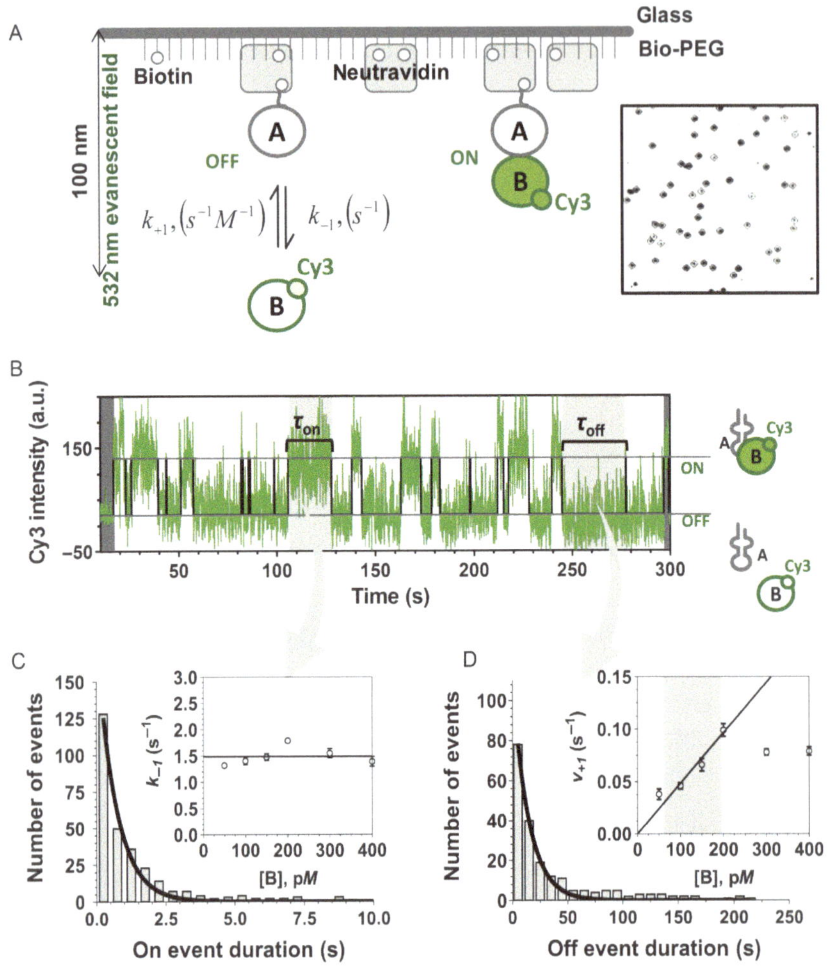

Fig. 1 Experimental scheme for the TIRFM-enabled single-molecule binding experiment. (A) The evanescent wave generated by TIR penetrates the reaction chamber to approximately 100 nm. The biotinylated macromolecules (A) are tethered to the surface of the passified, PEGylated microscope slide via interaction with the neutravidin, which bridges them with the sparsely biotinylated PEG. The Cy3-labeled molecule (B) is then flowed into the reaction chamber and the recording is initiated. Formation of the AB complex (governed by the association rate constant k_{+1} and the concentration of freely diffusing B) brings the Cy3-labeled B in the evanescent filed, which excites the dye. Decomposition of the AB complex (governed by the dissociation rate constant k_{-1}) leads to diffusion of B away from the evanescent field and in disappearance of the fluorescence signal. The *inset* shows a part of the camera field of view with dark spots corresponding to the points on the slide where the molecules are tethered and complexes are being formed during the experiment. The *diamonds* around the *dark spots* are the areas identified by the mapping software and selected as the origins of the signals.

lysine is a fairly common residue in proteins, this method may lead to heterogeneous labeling, which is undesirable.

A growing number of methods enable site-specific chemical labeling of proteins. The *Mycobacterium tuberculosis* formyl–glycine-generating enzyme (FGE) specifically catalyzes the cotranslational oxidation of the cysteine residue on the hexapeptide sequence LCTPSR to an aldehyde containing a Cα-formylglycine (fGly) residue, which can be then labeled in vitro (Appel & Bertozzi, 2015; Ghoneim & Spies, 2014; Liu et al., 2015; Rabuka, Rush, Dehart, Wu, & Bertozzi, 2012; Shi et al., 2012). The FGE recognition sequence can be genetically encoded into a protein, which is then coexpressed with FGE, both in eukaryotic as well as prokaryotic expression systems (Rabuka et al., 2012). Fluorescently labeled aminooxy- or hydrazide-functionalized molecules are readily available through GE healthcare (CyDye™ hydrazides) and other companies, which can be used directly in labeling reactions. Alternatively, a fluorescent dye can be irreversibly linked to the fGly via a Hydrazinyl-Iso-Pictet-Spengler (HIPS) ligation reaction (Liu et al., 2015). The FGE-mediated labeling is particularly useful when the location of the fluorophore on the protein is critical to observation or when the protein cysteines are important for the activity (Ghoneim & Spies, 2014). As with other labeling strategies, the appropriate controls should be employed to ensure that the addition of the hexapeptide FGE

(B) A representative fluorescence trajectory from the experiment where a surface-tethered RNA molecule (A) interacts with a fluorescently labeled protein (B), shows the evolution of the Cy3 signal in one of the spots on the slide. Raw data are shown in green overlaid with the idealized trajectory shown as a *black line.* High Cy3 signal corresponds to the AB complex (τ_{on} is the "ON" dwell time for the event), while the background fluorescence corresponds to the unbound A (and the corresponding "OFF" dwell time τ_{off}). The *dark gray rectangles* in the beginning and at the end of the trajectory indicate the events excluded from the analysis. (C) Fitting of the dwell-time distribution constructed from all "ON" dwell times with a single-exponential function provides the dissociation rate constant k_{-1}, which is independent of the concentration of B (*inset*). (D) Fitting of the dwell-time distribution constructed from all "OFF" dwell times with a single-exponential function provides the association rate v_{+1}, which increases with increasing concentration of B until reaching saturation (*inset*). At low concentrations of B, the slope $v_{+1}/[B]$ represents the association rate constant k_{+1}. The equilibrium dissociation constant can be then calculated as $K_D = k_{-1}/k_{+1}$. The range of the concentrations useful for calculating the association rate constant k_{+1} is marked by the *gray* rectangle in the *inset. The data in this figure are adapted with permission from Haghighat Jahromi, A., Honda, M., Zimmerman, S. C. & Spies, M. (2013). Single-molecule study of the CUG repeat-MBNL1 interaction and its inhibition by small molecules.* Nucleic Acids Research, 41, *6687–6697.*

recognition sequence and/or the labeling conditions do not alter the function or activity of the protein of interest.

The sortase system (Theile et al., 2013) is an alternative method that uses the *Staphylococcus aureus* sortase A enzyme, which recognizes an LPXTG motif in substrate proteins and cleaves the peptide bond between the threonine and glycine residues. A nucleophilic substitution by an α-amine of an incoming oligoglycine-based nucleophile conjugated to a fluorescent dye leads to a modified protein with the fluorophore at the N- or C-terminus.

2.2.3 Surface-Tethering of Macromolecules for TIRFM

Surface-tethering is commonly achieved by labeling the protein or nucleic acid with biotin and subsequent immobilization on a neutravidin-coated slide surface. Biotinylated oligonucleotides are readily available from many commercial sources, such as IDT. However, direct biotinylation may be impractical if a long RNA or DNA molecule is used in an experiment. To circumvent this, a biotinylated complementary DNA oligonucleotide can be annealed to a terminus of the RNA or DNA construct, or to an internal region that is not expected to affect the interaction being investigated.

For protein immobilization, a biotin acceptor peptide (BAP) sequence (GLNDIFEAQ<u>K</u>IEWHE or 5′-ggtcttaatgatatttttgaagctcag<u>**aag**</u>attgaatggcatgaa-3′) is engineered into the protein expression vector. The BAP is the most efficient recognition sequence for the *E. coli* BirA biotin ligase, which biotinylates the underlined lysine residue (Beckett, Kovaleva, & Schatz, 1999). It is typically incorporated at the N- or C-terminus of the protein or, if desired, in a solvent exposed loop. For bacterially expressed proteins, BAP and the presence of 100 μ*M* biotin in the growth medium is sufficient for the efficient biotinylation of the target proteins. For proteins expressed in human, baculovirus or yeast cells, the expression system must also contain a vector for the expression of BirA ligase (see Bain et al., 2016 for a protocol for the expression of biotinylated proteins in human cells; Honda et al., 2014 for the baculovirus expression, and Boehm, Powers, et al., 2016; Boehm, Spies, et al., 2016 for the yeast expression).

The biotinylated molecules are immobilized on the surface of the passivated imaging chamber coated with sparsely biotinylated (1:25) PEG polymer (MPEG-SVA-5000 and Biotin-PEG-SVA-5000, Laysan Bio), and 100 p*M* to 3 μ*M* neutravidin (Pierce). A detailed protocol describing the imaging chamber construction and preparation of the imaging buffer can be found in Bain et al. (2016). After a 3–5-min incubation, the excess untethered molecules are removed from the microscope imaging chamber by washing with the tethering buffer, which is then replaced with the imaging buffer supplemented with oxygen scavenging system (1 mg/mL glucose oxidase, 0.4% (w/v) D-glucose, 0.04 mg/mL catalase) and 12 mM Trolox (6-hydroxy-2,5,7,8-tetramethylchromane-2-carboxylic acid; note that Trolox concentration is based on the measurement before filtering the solution and can therefore be lower) and containing the fluorescently labeled molecules. The concentration of the biotinylated molecules during the tethering step ranges from 20 to 100 pM; the concentration of the fluorescently labeled molecules in the reaction is typically varied between 50 pM and 10 nM. Higher concentrations of the freely diffusing fluorescent molecules can be used, but will reduce the signal-to-noise ratio due to increase in the background fluorescence. The exact concentration of the fluorescently labeled molecule that gives signal-to-noise ratio of 1 depends on the optical alignment of the TIRFM system and the exact experimental design, and hence can vary from experiment to experiment. A more important reason to keep the concentration of the fluorescently labeled freely diffusing species low is that at high concentrations the concentration dependence of the association rate, v_{+1}, deviates from linear, which makes it difficult to calculate the association rate constant, k_{+1}, and consequently the equilibrium dissociation constant, K_D (see later).

Alternatively, surface-tethering of the proteins can also be achieved through interaction with surface-tethered biotinylated antibodies (e.g., biotin-conjugated anti-6xHis, anti-FLAG, or anti-HA antibodies), a technique known as "single-molecule pull down" (Aggarwal & Ha, 2014; Jain, Liu, Xiang, & Ha, 2012).

For all surface-tethering protocols it is essential to confirm that surface immobilization does not interfere with the biochemical properties of the protein. This can be achieved by carrying out the experiment in an alternative configuration where the surface-tethered and fluorescent molecules are switched.

2.3 Acquisition of Single-Molecule Binding Data

The binding experiment discussed here is designed so that the surface-tethered molecules are nonfluorescent and therefore invisible per se. Because of their capacity to bind the freely diffusing, fluorescently labeled partners, their presence on the surface of the TIRFM reaction chamber results in a sequence of high-intensity fluorescent events that appear, persist, and disappear in a single diffraction-limited spot. The time-based changes in the fluorescence in a designated location are referenced herein as a fluorescence trajectory (Fig. 1B). If surface-tethered molecules are separated further than the diffraction limit (ensured by the low ratio of biotinylated to nonbiotinylated PEG molecules and the concentrations of neutravidin and biotinylated molecules during surface immobilization reaction) and in the absence of a nonspecific binding, each fluorescence trajectory can be attributed to the binding activity of a single, surface-tethered molecule. Several hundreds of individual trajectories are extracted from each recorded video and visualized using the single-molecule FRET data acquisition and analysis suite https://cplc.illinois.edu/software (Joo & Ha, 2012) as described in detail in Bain et al. (2016). Similar to more established single-molecule FRET measurements, the raw data in single-molecule binding and colocalization studies are first processed using the peak-finder and signal integration algorithms implemented in IDL (Blanco & Walter, 2010; Roy, Hohng, & Ha, 2008). These algorithms allow one to select molecules from a wide field of view (FOV), to extract fluorescence trajectories, and, in the case of multicolor imaging, to match the peaks collected in different emission channels.

2.4 Selecting and Preparing the Trajectories for Dwell-Time Analysis

2.4.1 Selection Rules

The first step in the analysis of single-molecule trajectories is to establish an unbiased set of selection rules to separate the trajectories for analysis from all extracted trajectories. This allows for an unbiased analysis of the subset of the quality data, as the selection is typically done by eye and may therefore be subjectively biased. For simple one-color measurements of the association of fluorescently labeled macromolecule with the surface-tethered partner, the main selection criterion is the presence of at least two true binding events, i.e., the events whose dwell times are longer than two frames of the movie and the fluorescence increase is within a minimum predefined threshold of signal-to-noise ratio. These "ON" events define the boundaries of the "OFF" event in between, thus allowing to extract at least one "OFF" event from the trajectory. Moreover, to be recorded as a dwell time, the event should start after the beginning of the recording and end before the recording is terminated. For example, the first "OFF" and the last "ON" events in the trajectory shown in Fig. 1B (dark gray rectangles) do not adhere to this criterion and are removed from the analysis. If only one binding event is present in the trajectory, this event can still contribute to the analysis of the "ON" dwell times. Often, it is also advisable to exclude trajectories that show a monotonic decay of intensity of binding events across the total length of the recoded video, and to exclude trajectories that contain events with disproportionally large signal intensity compared to the dominant level of signal intensity in most of the trajectories.

2.4.2 Extracting the Dwell Times From Fluorescence Trajectories

Similar to ion channel open–close transition kinetics (Colquhoun & Sigworth, 1995) and times to completion of the catalytic cycle (Moffitt, Chemla, & Bustamante, 2010; Szoszkiewicz et al., 2008), the dwell times of the molecular association events are computed from the idealized trajectories. When the states are clearly separated and the transitions are obvious, a so-called "thresholding" method may be sufficient for extracting the dwell times that the trajectory spends at each state (Blanco & Walter, 2010; Sachs, Neil, & Barkakati, 1982; Zhou, Kunzelmann, Webb, & Ha, 2011). This description readily applies to the single-molecule binding trajectories where the labeled macromolecule is freely diffusing and is binding to a non-fluorescent surface-tethered partner. For such trajectories, the "bound" ("ON") state corresponds to a high fluorescence signal and the "free" ("OFF") state corresponds to background fluorescence level.

When the thresholding method is insufficient, for example, due to the presence of multiple, poorly separated states, or to the noise in the recording, other methods allow better quality idealization, and thereby a more accurate identification of the states and extraction of dwell times (Bronson, Fei, Hofman, Gonzalez, & Wiggins, 2009; Shuang et al., 2014; Watkins & Yang, 2005). A powerful set of methods for extracting kinetic information from noisy trajectories or in the presence of multiple ill-defined states is based on hidden Markov modeling (HMM) (Blanco & Walter, 2010; Chung, Moore, Xia, Premkumar, & Gage, 1990; Liu, Park, Dahmen, Chemla, & Ha, 2010; Talaga, 2007). HMM assumes that the process probed by the single-molecule analysis is a "Markovian," stochastic, memoryless process in which probability of the next event depends only on the current event and is independent from prior events (Talaga, 2007). Although this assumption is not always true in single-molecule data (Lu, Xun, & Xie, 1998), HMM still represents a practical method to identify the various signal states and to quantify the kinetics of transitions between these states (Blanco & Walter, 2010). A detailed and comprehensive description of the basic principles of HMM can be found in Rabiner (1989).

There are several HMM-based academic open access programs for idealization of the single-molecule trajectories (Bronson et al., 2009; Greenfeld, Pavlichin, Mabuchi, & Herschlag, 2012; Mckinney, Joo, & Ha, 2006;

Nicolai & Sachs, 2013; Qin & Li, 2004; Van De Meent, Bronson, Wiggins, & Gonzalez, 2014). Most of these programs are designed for analyzing single-molecule FRET trajectories, which are calculated from two sets of original trajectories for the FRET donor and the acceptor, respectively. To analyze the non-FRET single-molecule binding data derived from the trajectories exemplified in Fig. 1B, we prefer to use the program QuB ("Q" refers to a rate transition matrix, and "UB" refers to University of Buffalo) (Nicolai & Sachs, 2013; Qin & Li, 2004). This software package was originally developed for the analysis of single ion channel recordings. It can idealize multiple trajectories simultaneously and utilizes multistate models. It also can read data in many different types of formats. Using QuB, one can select a specific part of each trajectory and edit the data that is being analyzed (see Sections 2.4.3 and 2.5.5). It also can calculate rates and globally fit data obtained under different experimental conditions (e.g., ionic strength, concentrations of freely diffusing macromolecule, etc.). One of the advantages of the QuB package is that it can estimate the likelihood of different models (including all possible connection schemes between states).

2.4.3 Mapping the Positions of Surface-Tethered Molecules

When one-color single-molecule data are obtained using a TIRFM system capable of two-color imaging, Cy5-labeled streptavidin can be utilized to map the positions of biotinylated immobilized molecules on the slide surface.

- Replace neutravidin with the Cy5-labeled streptavidin (GE healthcare PA45001) or use a Cy5-labeled biotinylated target molecule. At the beginning of imaging, use 5 s excitation with the 640-nm laser, then switch the excitation source to the 532-nm laser for the rest of the recording. In this scheme, the MatLab program for extraction of the fluorescence trajectories only needs to consider the first 5 s of the recording to map the possible locations of surface-tethered molecules, aiding in the selection of trajectories (Boehm, Powers, et al., 2016; Boehm, Spies, et al., 2016; Haghighat Jahromi et al., 2013). When the unlabeled neutravidin is used for surface tethering, the MatLab code needs to be set to scan the entire movie in the Cy3 channel to identify the locations where the Cy3 signal associated with the binding of the freely diffusing Cy3-labeled ligand appears.
- The portion of the trajectory containing the Cy5 signal is trimmed prior to analysis. We have developed a software tool that allows for the

automated removal of this initial Cy5 excitation from the beginning of each trajectory in a given experiment folder (available upon request from Maria Spies or Todd Washington). The software will automatically truncate all files in a given folder, at a given row number and will save the updated file. Fig. 1B shows the trajectory after trimming to remove the Cy5 label.

- The trajectories for a given single-molecule experiment are then individually analyzed using QuB (Nicolai & Sachs, 2013). The one-color fluorescence trajectories are fit to a two-state model, wherein the "OFF" times are denoted as "0" and the "ON" times are denoted as "1". The "Amps" function in QuB is used to calculate the average values for both a low fluorescence population and a high fluorescence population for each individual trajectory. The "Idealize" command then identifies binding events utilizing these low and high fluorescence values, such that increases in fluorescence which are within one standard deviation of the average value for the high fluorescence population and persist for two or more consecutive camera frames will be identified as binding events. Increases in fluorescence signal which do not meet these requirements are considered noise and are ignored. An idealized data trajectory is then generated (Fig. 1B) along with the idealized data file containing durations of all "ON" and "OFF" events in a given trajectory in order of appearance.
- The QuB idealized data are then input into Microsoft Excel for further analysis (the automated software tool for the batch processing of QuB output files is available on request from Maria Spies or Todd Washington). This software opens each idealized QuB data file within a given folder, identifying them by the file extension .qwt and copies and pastes the information into the Excel document with a line between each idealized file to identify the trajectory number that it originated from. Binding events, whose beginning or whose end are truncated by the recording window of the movie (dark gray rectangles in the trajectory in Fig. 1B), are automatically omitted from analysis. Trajectories that have only one binding event are marked to be used only for consideration of the "ON" dwell time. The data are then sorted into "ON" times (given the designation of 1 by QuB) and "OFF" times (given the designation of 0 by QuB). The "ON" and "OFF" dwell-time data are then used in the histogram construction and kinetic analysis as described later in Section 2.5.

2.4.4 Errors Introduced Into Observed Binding Kinetics due to the Fluorophore Blinking and Photo-Bleaching

Reversible blinking (Rasnik, Mckinney, & Ha, 2006; Valeur, 2012) and irreversible photo-bleaching of fluorescent dyes (Diaspro, Usai, Ramoino, & Dobrucki, 2006; Lakowicz, 2006) are photophysical phenomena frequently observed in single-molecule fluorescence imaging experiments. In the case of blinking, the fluorescent dye visits a dark nonemissive state, stays in this state for some time (usually microseconds to milliseconds; Akyuz, Altman, Blanchard, & Boudker, 2013; Bartley, Zhuang, Das, Chu, & Herschlag, 2003), then returns to the electronic ground state and restarts the cycles of excitation and emissive deexcitation. Photo-bleaching is a process in which the dye encounters a photochemical reaction that leads to complete loss of its emissive property. Both phenomena are comprehensively discussed in a recent review (Ha & Tinnefeld, 2012). After a fluorescently labeled freely diffusing macromolecule binds to its surface-tethered partner ("ON"-state), its dissociation (indicated by the decrease in the fluorescence signal to the background level) can be confused with blinking or photo-bleaching (especially when dissociation is slow and the dwell times are long). Hence, blinking and photo-bleaching can introduce significant errors in the evaluation of rate of any process studied by single-molecule fluorescence imaging (Bartley et al., 2003; Rueda et al., 2004). To determine that the fluorophore blinking or photo-bleaching does not affect a single-molecule binding experiment, we recommend the following routine:

- Select the lowest possible laser excitation intensity that provides acceptable signal-to-noise ratio sufficient for extracting reliable kinetics (Haller, Rieder, Aigner, Blanchard, & Micura, 2011).
- Measure the photo-bleaching time (τ_{bleach}) of the fluorescent dye at the chosen excitation laser intensity using an immobilized form of the same fluorescently labeled macromolecule by plotting the distribution of lengths of the fluorescence trajectories and then fitting this distribution to an exponential function (Solomatin, Greenfeld, Chu, & Herschlag, 2010). Using the same fluorescently labeled species is important, because the dye photophysics are influenced by its chemical environment (Bartley et al., 2003).
- Carry out the binding experiment. Repeat the acquisition of the single-molecule binding data at least one more time, but using higher excitation intensity.

- Compare the binding kinetics (i.e., dwell-time distributions) evaluated from the data acquired using the two different excitation laser intensities.
- If no dependence on excitation intensity was observed, then this means that blinking and photo-bleaching did not induce significant errors in the observed binding kinetics. Using the current photostabilizing and oxygen scavenging strategies (Ha & Tinnefeld, 2012; Zheng et al., 2014), many recent single-molecule binding studies showed no significant errors induced in the observed binding kinetics due to blinking or photo-bleaching.
- In some cases, significant errors can still be introduced into the observed binding kinetics. This can be due to more than one reason; it can be because the dissociation rate is relatively slow which makes data acquisition limited by photo-bleaching (Honda et al., 2014), or because the fluorescence emission quantum yield of the dye is significantly reduced upon attachment to the target macromolecule sample, which consequently forces the experimenter to use very high excitation laser intensity in order to collect data with sufficient signal-to-noise ratio.
- If the observed kinetics show laser intensity dependence, it may be influenced by (i) both blinking and photo-bleaching, (ii) photo-bleaching only, or (iii) blinking only.
- In the first scenario, binding dwell-time distributions are expected to be of multiphasic nature; with one shorter phase whose fraction/amplitude and time constant depend on excitation intensity but does not depend on perturbations related to the biochemical reaction under study (e.g., ionic strength, etc.), and another longer phase whose time constant depends on both the biochemical reaction perturbations and excitation laser intensity. The former is usually due to blinking, while the latter originates from the binding/assembly process but has also a photo-bleaching associated error.
- In the second scenario, binding dwell-time distributions should be monophasic (i.e., can be fit to a single-exponential model).
- In the third scenario, binding dwell-time distributions should be again multiphasic, but in this case the time constant of the longer phase does not depend on excitation laser intensity, although it depends on the biochemical reaction perturbations.

The procedures for accounting for the errors induced due to these photophysical phenomena are described in detail in the next section.

2.5 Construction and Analysis of the Dwell-Time Distributions

2.5.1 Presenting the Dwell-Time Data as Ordinary Frequency and Cumulative Distributions

To make meaningful inferences into the mechanisms of biomolecular interactions, it is usually not enough merely to observe the binding events in the single-molecule trajectories. While every individual trajectory is unique, reflecting a stochastic nature of the macromolecular collisions and binding, the overall statistical properties extracted from a large collection of trajectories reflect on the macroscopic behavior of the process under investigation. The theory behind constructing, fitting, and statistical analysis of dwell-time distributions developed for single-channel recording data (Colquhoun & Sigworth, 1995) has been applied to analysis of many different single-molecule systems (Abelson et al., 2010; Munro, Altman, O'Connor, & Blanchard, 2007; Szoszkiewicz et al., 2008). Collectively, the single-molecule binding data may be presented as the histograms of the observed dwell times. Two types of distributions are most commonly used, ordinary or frequency distributions and cumulative distributions. Frequency distributions are constructed by plotting the number (or frequency) of the dwell times that fell in a certain time interval (bin) vs time (Figs. 1C and D, 2A and E). The resulting histogram can then be compared to a theoretical probability density function describing the process (e.g., exponential or multiple exponential). When normalized for the size of the data set, the height of each bin defines the probability that a random dwell-time falls within a particular interval. In cumulative distributions, the value on the y-axis represents the sum of all events that had a dwell time between zero and a corresponding time. When normalized, the cumulative distribution defines the probability that a random dwell-time falls below a given value. The number of points in the exact cumulative distribution is equal to the number of observed dwell times, which is especially useful for smaller data sets (see the discussion later).

2.5.2 Selecting a Bin Size of the Ordinary Frequency Histogram

Presenting the dwell-time data as frequency histograms requires large data sets (hundreds to thousands of data points) to yield accurate rate constants. Not surprisingly, the distributions built from a smaller number of dwell times show the greatest sensitivity to the effects of binning. The rules for the unbiased, objective choice of the optimal bin size have evolved from the historically earliest Sturges' rule (Sturges, 1926) to a number of more recent mathematical rules (Doane, 1976; Scott, 1979; Shimazaki & Shinomoto, 2007; Wand, 1997). However, the main problem with these rules of optimal

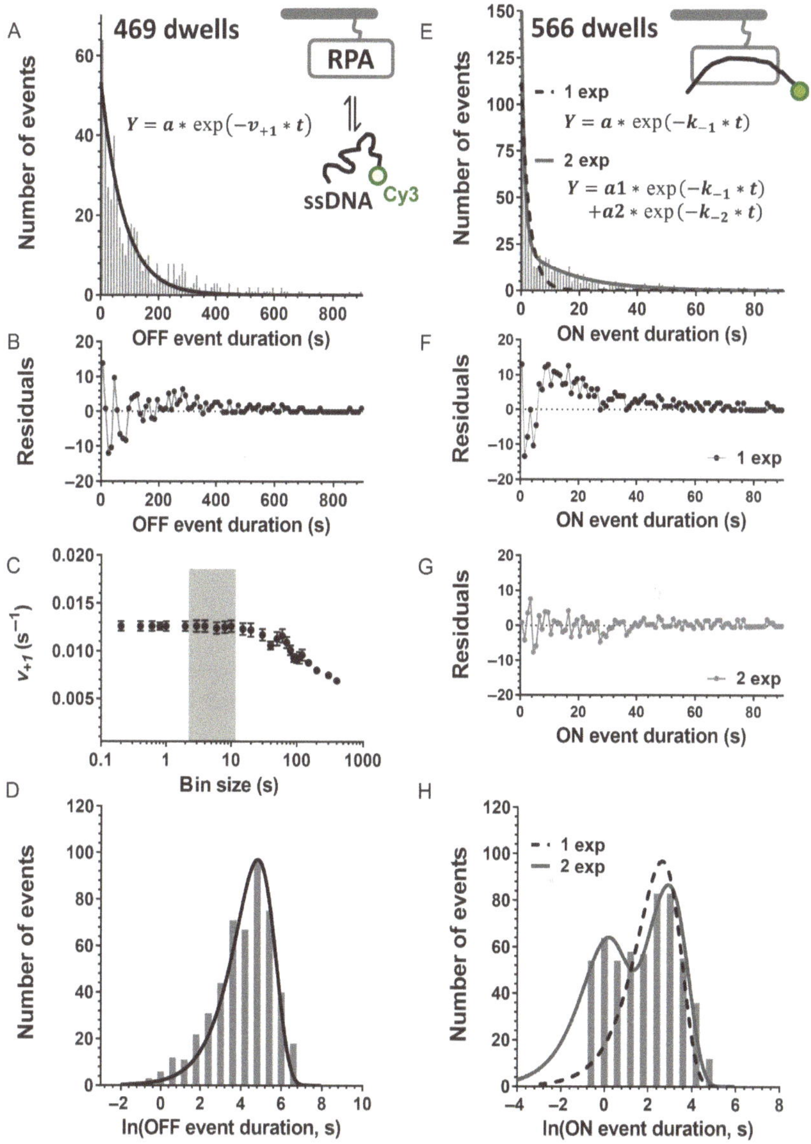

Fig. 2 Construction and analysis of the dwell-time distributions. (A) A histogram of the 469 individual "OFF" dwell times for the complex formation between the surface-tethered RPA protein and a Cy3-labeled 30-mer ssDNA constructed using 10 s bins (*gray bars*) is fitted with a single-exponential function (*black line*). (B) The residuals from the fit shown in (A) display a symmetric scatter around zero. (C) The same data set was binned with varying bin size and fitted with a single-exponential function. The resulting association rate is plotted as a function of bin size. It remains the same between the minimal reasonable bin size of 200 ms to about 20 s. Larger bin size combines most of the data in

(Continued)

bin size selection is that they are based on assumptions with regard to the nature of the distribution. Ideally, optimal bin size calculation should be model-independent, such as in the method developed by Shimazaki and Shinomoto (2007). In this method, the bin size associated with the minimum mean integrated squared error (MISE) is the optimal one. The MISE is calculated directly from the data, and the method is available as a web application at http://toyoizumilab.brain.riken.jp/hideaki/res/histogram.html. The only assumption in this method is that data are sampled independently of each other. Later this method was updated (Shimazaki & Shinomoto, 2010) to be able to find an optimal bin width in case of processes whose rates change over time (e.g., binding of two macromolecules, with one of them switching during the same trajectory between multiple conformers that have different association and dissociation rates). In our experience, this method requires a very large data set for best performance. Even with data set sizes in the range of a few hundred dwell times, the calculated optimal bin size did not permit good fitting of a multiphasic dwell-time distribution to multiexponential function because a significant amount of the data was spread over the longer side of the scale (unpublished results).

Practically, the optimum bin size depends on the number of recorded observations, the spread of the data, and the overall shape of the distribution. While a smaller bin size improves the accurate determination of the rate constants, the small number of events in narrow bins will result in a noisy signal that may be difficult to fit. Moreover, the 30 or 100 ms camera resolution used in our experiments provides the uncertainty in determining the beginning and the end of each event; i.e., if the camera resolution is set for 100 ms, and the threshold for recognizing the event in QuB is set as at least two consecutive frames, a bin size less than 200 ms is not meaningful. Intuitively, the

Fig. 2—Cont'd a very few bins (the half-life for this process is around 55 s) and consequently to an erroneous determination of the rate. The *gray rectangle* suggests the optimal bin size range. (D) The same data are plotted on a logarithmic scale. (E) A histogram of the 566 individual "ON" dwell times for the RPA-ssDNA constructed using 1 s bins (*gray bars*) is fitted with a single-exponential function (*black dashed line*) and double exponential (*gray continuous line*). (F) The residuals from the single-exponential fit shown in (D) display a systematic deviation from zero. (G) The residuals from the double-exponential fit display a symmetric scatter around zero and are much tighter. (H) The same data plotted on a logarithmic scale clearly reveal that the process is indeed a double exponential. *The data in this figure are adapted with permission from Chen, R., Subramanyam, S., Elcock, A. H., Spies, M. & Wold, M. S. (2016). Dynamic binding of replication protein A is required for DNA repair.* Nucleic Acids Research, 44, *5758–5772.*

bin size should be selected so that more than two bins are present before the half point of the fast phase of the exponential decay fitted to the distribution. This is because an ordinary frequency histogram is built of multiples of a sampling interval and the bin heights at the middle of each bin should approximate the probability density function for the distribution. This approximation becomes poor when the bin width approaches or exceeds the time constant of the probability density function. If the bin size is too large, it may skew the fitting, i.e., all meaningful data will be combined in a single bin, and the respective rate constants (the overall rate constant for a single-exponential distribution or the constant describing the rapid phase of the process) will be inaccurate. Fig. 2C provides an illustration how the association rate determined from the fitting of the "OFF" time distribution of the RPA–ssDNA complexation (Chen et al., 2016) is affected by the change in the bin size. In this particular case, the calculated rate remains the same until the bin size starts exceeding 20 s, which is about a half of a half-life of the distribution.

2.5.3 Logarithmic Histograms

One disadvantage of the ordinary histogram is that the bins containing the long dwell times have very few data points. The rare long events that may carry important biological consequences may therefore be overlooked during the fitting (Honda et al., 2014). It may also be difficult to determine unambiguously whether the distribution is monophasic (i.e., can be fit by a single-exponential function) or multiphasic. When distinguishing between single and multiple processes from the linear dwell-time distribution is problematic, the distribution can be replotted as a logarithmic histogram (Fig. 2D and H). Note that it is sometime advantageous to use a square root ordinate instead of the histogram with logarithmic x-axis and linear y-axis depicted in Fig. 2D and H. Numerous reports from the ion channel field (McManus, Blatz, & Magleby, 1987; Sigworth & Sine, 1987) and more recent single-molecule work (Szoszkiewicz et al., 2008) showed that logarithmic histograms can robustly discriminate between the processes characterized by a single rate vs multiple rates. Another advantage of log-dwell-time histogram is that it is more suitable in case of data with a range that covers multiples scales (e.g., from microseconds to 10s of seconds) (Colquhoun & Sigworth, 1995). Fig. 2D and H shows the data from Fig. 2A and E plotted on the logarithmic scale. Because the bins of such a histogram contain logarithmic values of the dwell-time durations, the bins for longer and shorter times are better balanced. Recently, Fernandez and colleagues suggested a method to achieve two targets at the same time: (i) evaluating optimal bin size and (ii) judging whether a process should be described with single or multiple rates (Szoszkiewicz et al., 2008). This method is based on combining a logarithmic time abscissa type of histogram with using reduced χ^2 as an objective bin size estimation parameter. An optimal bin size can be determined through an iterative process starting from the smallest reasonable bin size. For each iteration, the bin size is increased, the resulting distribution is fitted and the reduced χ^2 is calculated. The χ^2 value decreases with the increase in the bin size until it reaches a global minimum or levels off (Szoszkiewicz et al., 2008). It is worth mentioning that this method depends on the model (e.g., single-exponential or multiexponential function) that the experimenter uses to calculate the reduced χ^2 value for each bin size tested.

2.5.4 Cumulative Distributions

Both linear and logarithmic histograms require large data sets (i.e., several hundreds to several thousands of data points) to yield accurate rate constants. Not surprisingly, the distributions built from a smaller number of dwell times show the greatest sensitivity to the effects of binning. Cumulative representation of distributions helps to solve a great deal of this problem (Sung, Sivaramakrishnan, Dunn, & Spudich, 2010). In addition, it avoids the complexities associated with finding the optimal bin size, as it is independent of binning (Gebhardt, Clemen, Jaud, & Rief, 2006). One of the common forms of the cumulative distribution displays the fraction of dwell times that are longer than a given time "*t*" (Ghoneim & Spies, 2014; Strick, Croquette, & Bensimon, 2000; Sung et al., 2010). Each event in this type of representation is given an equal weight, which increases the ratio of data points to parameters allowing for more robust analysis (Sung et al., 2010) and increases the weight of the rare long events (Bartley et al., 2003). Cumulative representations of dwell-time distributions are especially useful for estimation of the number of rate constants. However, care must be taken when using this representation of distributions (Friedman & Gelles, 2015). Successive points in this type of data representation are correlated, which may bias curve fitting (Colquhoun & Sigworth, 1995). Least-squares fitting of cumulative distributions using commercial software may lead to an underestimation of fit uncertainty if the software did not consider covariance of fit parameters (Sung et al., 2010), and if an independent error in successive points on the graph is assumed (Friedman & Gelles, 2015). It is suggested that fitting cumulative distributions using maximum-likelihood methods combined with an estimation of fit uncertainty using a bootstrapping method can provide a more reliable strategy to evaluate rates of processes from single-molecule data. In the bootstrapping approach, a data set identical in size to the original data set is constructed by randomly selecting the dwell times with replacement, and the new dataset is fitted to obtain the rate constants. The process is repeated several hundred times. The average for the resulting set of values for each rate constant and the standard deviation are then used to assess the quality of the fit.

2.5.5 Correcting the Observed Rates for the Fluorophore Photo-Bleaching and Blinking

Several methods allow one to correct for photo-bleaching errors. In one popular method, the photo-bleaching rate (inverse of τ_{bleach}) is subtracted from the observed rate to calculate the actual corrected rate (Honda et al., 2014; Miyanaga & Ueda, 2011; Rueda et al., 2004; Watanabe & Mitchison, 2002). Another method uses a semiempirical approach where the relation between change in the observed dissociation rate is plotted and the change in laser exposure time, using the same excitation intensity and period of data recording (Bombardier et al., 2015; Friedman & Gelles, 2015; Huppa et al., 2010). Assuming a linear dependence of these two parameters, actual binding kinetics can be evaluated by extrapolating the curve to zero laser exposure.

Errors associated with blinking are more difficult to handle. Some studies ignore the fast, laser power dependent component of the dwell-time distributions (Akyuz et al., 2013). In these studies, the slower phase of the dwell-time distribution is used to measure the rate of the process after confirming that the time constant of this phase depends on biochemical perturbations (e.g., changing the ionic strength). A data-screening method was developed in other studies (Bartley et al., 2003; Rueda et al., 2004). Here, the fluorescent dye is immobilized, so that the sudden decrease in fluorescence signal to background level is due to blinking only, not because of dissociation from the surface of the microscope slide. The kinetics of blinking of the dye is then characterized. By combining this rough characterization of the blinking kinetics with a selection criteria related to emission intensity, a subpopulation of short dark intervals in the trajectories is ignored, then the analysis of dynamics of fluctuations of fluorescence trajectories is repeated and the dwell-time distributions are reconstructed and refitted. It is important to note that the design of the selection criteria mentioned earlier differs from one experimental system to another.

2.5.6 Mechanistic Interpretation of the Rates Obtained from the Dwell-Time Distributions

Finding the best-fitting probability distribution for the single-molecule data does not automatically generate a reasonable model of the actual process under investigation. For a sufficiently large data set, a mathematical shape of a distribution function can be identified by curve fitting followed up by a statistical analysis as exemplified by the two cases earlier. To understand the underlying molecular process, however, it is essential to explain the shape of a given distribution function by an underlying generative principle. Many macromolecular association processes discussed in this chapter are simple Poissonian processes with a single dominant energy barrier. As result, their respective probability density functions are single exponentials (see for example the dwell-time distributions for both "ON" and "OFF" dwell times in Fig. 1C and D and the "OFF" time distribution in Fig. 2A). Similar to the bulk associated kinetics described by the Debye-Smoluchowski equation, the probability of the surface-tethered molecule to form an observable complex or the frequency of appearance of the fluorescent signal in a particular location on the microscope slide (reflected in the "OFF" time distributions) is proportional to the frequency of the collisions between the molecule on the surface and the freely diffusing counterpart adjusted by a unitless factor that accounts for the fraction of productive collisions. If the "OFF" time distribution deviates from single exponential, this may signify the presence of distinct subpopulations of interconverting conformers or distinct subpopulations among the surface-tethered molecules with distinct binding properties. The "ON" time distribution reflects stability of the complex, which in the case of a homogenous system decomposes in a single step leading to a single-exponential distribution. Multiple exponentials are frequently observed and are likely to reflect the formation of distinct complexes with different stabilities (Amrute-Nayak et al., 2014; Chen et al., 2016) (Fig. 2E).

2.5.7 Interaction Stoichiometry

If the fluorescent labeling is homogenous (i.e., all freely diffusing molecules contain no more than a single dye) and only the 1:1 complexes are formed, each binding event generates a signal of similar intensity, and the process can be modeled as two-state, "bound" ("ON") and "free" ("OFF"). More complex binding patterns can be observed if the surface-tethered molecule contains multiple binding sites for the fluorescently labeled binding partner(s), which may bind independently or oligomerize upon binding to the surface-tethered substrate. An example of the interaction involving multiple independent binding sites can be found in a study examining the interaction of muscleblind-like 1 protein (MBNL1) with RNA molecules containing CUG repeats (Haghighat Jahromi et al., 2013). The protein bound the surface-tethered $(CUG)_4$ construct with 1:1 stoichiometry. In contrast, more complex trajectories were observed with the longer $(CUG)_{12}$ construct, and the intensity histograms suggested the presence of three independent binding sites for MBNL1 on this RNA, reflecting the expectation for the etiology of Myotonic dystrophy type 1 where MBNL1 is sequestered by the $(CUG)_n$ repeats transcribed from the expanded $(CAG)_n$ repetitive DNA. The presence of multiple binding sites on the surface-tethered macromolecule does not necessarily complicate the analysis of macromolecular complexation as it would in the analogous bulk experiment. This is because each binding event is observed independently at sufficiently low concentration of the fluorescently labeled freely diffusing molecules. Another elegant study by Pond, Ridgeway, Robertson, Wang, and Millar (2009) is an example of single-molecule enabled analysis that revealed oligomerization of the protein on the nucleic acid. This study probed the stoichiometry and the mechanism of assembly of the HIV-1 Rev protein on the Rev Response Element (RRE), a highly conserved region of the viral mRNA. The intensity histograms assembled from ~1000 individual 20-s trajectories and subjected to Hidden Markov analysis using the HAMMY program (Mckinney et al., 2006) showed the presence of four distinct bound states with 1–4 Rev proteins bound to the RRE RNA. The stepwise changes in the fluorescence intensity revealed the sequential addition and dissociation of the monomer from the RNA with higher-order complex formation on the RRE RNA driven primarily by protein–protein interaction.

3. BINARY INTERACTIONS IN THE PRESENCE OF INHIBITORS

3.1 Inhibitors of Macromolecular Interactions

Even in its simplest one-color version, TIRFM-based single-molecule localization provides a suite of powerful approaches to develop a comprehensive understanding of the inhibition mechanism and the kinetics of macromolecular interactions in the presence of an inhibitor (Haghighat Jahromi et al., 2013). In this section, we provide a guide on how to design and carry this type of analysis. We will use a study of the noncompetitive inhibition of the interaction between MBNL1 and the CUG repeat containing RNA

(Haghighat Jahromi et al., 2013), but the overall approach can be generally applied to competitive and noncompetitive inhibitors, as well as the allosteric effectors of protein–protein and protein–nucleic acid interactions. The major advantage of this approach is that it is model independent, i.e., we do not need to assume the mechanism a priori. The ability to detect and measure the individual binding events in real time and under equilibrium conditions at multiple inhibitor concentrations provides all the information one needs not only to access the inhibitor potency but also to infer its mechanism of action.

The overall experimental scheme starts from establishing and optimizing the conditions for the analysis of macromolecular interaction (as described in Section 2).

- One of the interacting molecules should be biotinylated for surface-tethering, and the other fluorescently labeled and freely diffusing. The inhibitor (effector) is unlabeled and should not have significant absorbance or fluorescence in the experimental excitation/emission windows. If the inhibitor produces a significant signal that interferes with the observation of the binding trajectories, we recommend labeling the freely diffusing macromolecule with a fluorophore whose spectral properties do not overlap with those of the inhibitor. This will require using a different laser for the excitation along with optimizing the excitation and emission filters.
- If feasible, the system should be selected or simplified to ensure a 1:1 binding stoichiometry. In the case of the MBNL1–RNA interaction, for example, we determined that a stem-loop RNA (5′-bio-GCUGCUGUGCGCUGCUG-3′), referred herein as $(CUG)_4$, provides a minimal substrate to which a single MBNL1 molecule binds with the affinity similar to that of the expanded CUG repeats. Moreover, this RNA molecule is expected to form a 1:1 complex with the bioactive dimeric inhibitor, which the study investigated (Haghighat Jahromi et al., 2013).
- The conditions are then established and optimized for collecting the fluorescence binding trajectories and for extracting the most reliable on and off rates, respective rate constants, and the equilibrium dissociation constant.
- The binding experiments are then repeated at several concentrations of the inhibitor. The range of the inhibitor concentrations should be greater than twice the expected K_i (equilibrium inhibition constant for the competitive inhibitor) or K_D (equilibrium dissociation constant for the inhibitor–target complex). For example, for the inhibitor which

binds $(CUG)_4$ with $K_D = 2.9 \pm 0.2 \mu M$, we carried out the single-molecule binding reactions in the presence of 0, 0.5. 1, 2, 4, and 6 μM inhibitor, while 0, 4, 20, 100, and 200 μM inhibitor was used in the case of another compound whose affinity for $(CUG)_4$ was $71 \pm 27 \mu M$.

- The "ON" and "OFF" dwell times are extracted from the fluorescence trajectories as described earlier, binned and plotted as ordinary frequency distributions, which are then globally fitted to determine the rate and equilibrium constants and the inhibition mechanism (see later).

3.2 Determining the Inhibition Mechanism: Competitive Inhibition

Competitive inhibitors (Fig. 3A) compete with the ligand (B) by occupying the same site on the target macromolecule (A). In the experimental

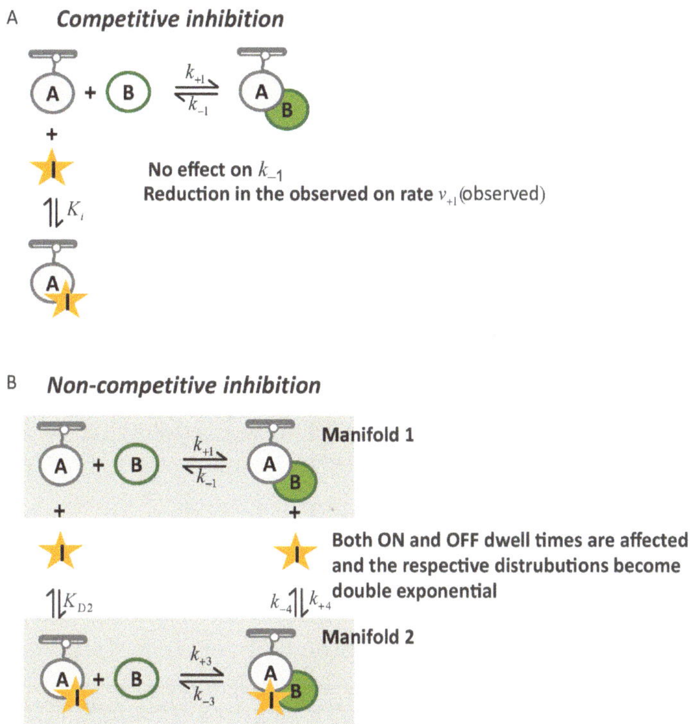

Fig. 3 Single-molecule analysis of the binary macromolecular interactions in the presence of unlabeled inhibitors. (A) Competitive inhibition. The experimental scheme is essentially the same as in the single-molecule analysis of binary interaction. The inhibitor that interacts with A is depicted as a *star*. K_i is the equilibrium inhibition constant. (B) Noncompetitive inhibition. Both A–B and A–B–I complexes are observed.

scheme outlined in Fig. 3A, macromolecule A is tethered to the surface, B is fluorescently labeled and is free in solution, and I is a competitive inhibitor that binds A. In the absence of the inhibitor, formation of the A–B complex is observed as the appearance of fluorescence signal. A competitive inhibitor should have no effect on the "ON" dwell-time distributions and therefore on the dissociation rate constant k_{-1}. In contrast, while the A–I complex is undetectable on its own, its formation precludes A–B complexation resulting in reduced frequency of binding events, and eventually no detectable binding at high inhibitor concentrations. When detectable, the distributions of the "OFF" dwell times will remain single exponential

$$Y = a * \exp\left(-\nu_{+1}^{\text{obs}} * t\right) \tag{1}$$

where ν_{+1}^{obs} is the observed association rate at the chosen inhibitor concentration [I].

$$\nu_{+1}^{\text{obs}} = \nu_{+1} \frac{[\text{B}]}{K_{\text{D}}\left(1 + \frac{[\text{I}]}{K_i}\right) + [\text{B}]}, \tag{2}$$

where K_i is the inhibition constant that reflects the affinity of the inhibitor for A, $K_{\text{D}} = \frac{k_{-1}}{k_{+1}}$ is the affinity of the ligand B for A, and [B] is the concentration of freely diffusing fluorescently labeled B in the reaction chamber. Constant a in Eq. (1) allows to fit the distribution without the need for normalization and accounts for all binding events, including those too short to be detected due to the temporal resolution of the measurement. If the inhibitor is determined to be competitive, there are two possible procedures to obtain K_i. One can choose to fit the "OFF" time distributions at each inhibitor concentration individually using Eq. (1), plot the observed association rate as a function of [I] and fit the resulting graph using Eq. (2). The accurate determination of K_i in this case requires at least 5 ν_{+1}^{obs} values in the broad range of the inhibitor concentrations. Alternatively, one can determine K_{D} from the binding experiments in the absence of the inhibitor, and then globally fit the "OFF" time distribution from several inhibitor concentrations to the combined equation:

$$Y = a*\exp\left(-\nu_{+1}\frac{[\mathrm{B}]}{K_{\mathrm{D}}\left(1+\frac{[\mathrm{I}]}{K_i}\right)+[\mathrm{B}]}*t\right), \qquad (3)$$

where K_i is a shared value for all data sets, and a is unique for each set allowing to fit the distributions globally without normalization. Note that there is no need to select the same bin size or the same size of the data set for each distribution.

3.3 Unlabeled Ligand as a Competitive Inhibitor

An unlabeled macromolecule as a competitive inhibitor may be used to determine if labeling affects its affinity for the surface-tethered partner. We recommend carrying out the binding reaction at several concentrations of the unlabeled molecule and analyzing them as described in Section 3.2. If labeling has no effect on the affinity, the calculated K_i will equal K_{D}; if $K_i > K_{\mathrm{D}}$, the labeling has a negative effect on the binding affinity.

Labeling of the macromolecules for single-molecule analysis is not always 100% efficient, and the freely diffusing species contain a mixture of labeled and unlabeled molecules of B. If this is the case, an accurate determination of the binding properties by single-molecule TIRFM is still possible. If both labeled and unlabeled species display the same binding properties, the actual association rate can be calculated from the observed rate as

$$\nu_{+1} = \frac{\nu_{+1}^{\mathrm{obs}} + \sqrt{\left(\nu_{+1}^{\mathrm{obs}}\right)^2 + 4\alpha\nu_{+1}^{\mathrm{obs}}k_{-1}}}{2\alpha[\mathrm{B}]}, \qquad (4)$$

where α is a fraction of labeled B. The binding analysis here should be carried out in the linear range of the association rate dependence of [B]. This equation can be derived from Eq. (2) by substituting [I] with $(1-\alpha)[\mathrm{B}]$, [B] with $\alpha[\mathrm{B}]$, K_i with $K_{\mathrm{D}} = k_{-1}/k_{+1}$, and then solving the resulting quadratic equation.

3.4 Determining the Inhibition Mechanism: Noncompetitive Inhibition

In a more complex noncompetitive inhibition scenario, a ternary complex A–B–I containing the inhibitor, ligand, and the target molecule may form (Fig. 3B). This complex is expected to have a different stability from the A–B complex and may be assembled by either binding of the ligand to the A–I complex, or by binding of the inhibitor to the A–B complex. Similarly, the A–B–I complex may decompose to either A–B or A–I complexes. The presence of the noncompetitive inhibitor does not completely abolish the A–B complex formation at any concentration, and both the "ON" and "OFF" dwell times are affected. For simplicity, we will consider a scenario where, in the absence of the inhibitor, both "ON" and "OFF" dwell-time distributions fit well to the single-exponential functions. The kinetic scheme for the reaction in the presence of noncompetitive inhibitor and all relevant rate constants are depicted in Fig. 3B. Appearance of the fluorescent signal in the single-molecule trajectories in this case may signify the formation of the A–B complex by binding of B to the surface-tethered A (k_{+1}), or by the formation of A–B–I complexes by binding of B to the A–I complex (k_{+3}). The interconversion between A–B and A–B–I complexes (k_{+4} and k_{-4}, respectively) is not detected, nor does binding of I to A (the affinity if I for A is defined by K_{D2}; respective rate constants cannot be determined from this analysis). Similarly, A–B–I may decompose to A–I with the rate constant k_{-3}, which is observed as a loss of the signal. A–B–I decomposition may also proceed through the A–B complex, which then dissociates with the rate constant k_{-1}.

- The binding experiments are carried out as described in Section 3.1 in the presence of multiple concentrations of the noncompetitive inhibitor.
- The "OFF" time data are then binned and plotted as ordinary frequency distributions, which will have a double-exponential shape.
- The "OFF" dwell-time distributions for all inhibitor concentrations are globally fit to the following equation:

$$Y = A[K_{D2}\exp(-k_{+1}[B]t) + [I]\exp(-k_{+3}[B]t)], \tag{5}$$

  where k_{+1} is obtained from the analysis of A–B complexation in the absence of the inhibitor, K_{D2} is a shared value for all data sets and the constant is allowed to vary to reflect different sizes of the datasets at different inhibitor concentrations. $A * K_{D2}$ is the weight of manifold (1) whereas $A * [I]$ is the weight of manifold (2) adjusted by the number of observed events. See Haghighat Jahromi et al. (2013) for the detailed explanations and the full list of the assumptions underlying this analysis.
- Similarly, the "ON" dwell-time distributions are globally fit to equation:

$$Y = A[K_{D2}\exp(-(k_{-1} - k_{+4}[I])t) + [I]\exp(-(k_{-3} - k_{-4})t)] \tag{6}$$

  yielding K_{D2}, k_{+4} and $(k_{-3} - k_{-4})$. Parameters k_{-3} and k_{-4} are linked as $(k_{-3} - k_{-4})$ and therefore cannot be determined individually by fitting but can be calculated from the linked equilibria as $k_{-3} = \frac{(k_{-3} - k_{-4})k_{+3}K_{D1}}{k_{+3}K_{D1} - k_{+4}K_{D2}}$

This analysis was initially developed to describe and compare the two inhibitors of the MBNL1 binding to the CUG repeat containing RNA (Haghighat Jahromi et al., 2013), but may be readily extended to the analysis of any noncompetitive inhibitors of the macromolecular assembly.

While powerful and straightforward in their implementation, the analysis detailed in this section fails to address the kinetics of the inhibitor–target interaction. We show earlier how to obtain the K_I and K_{D2}, but not individual k_{+2} and k_{-2}. To access these parameters directly, the smTIRFM study may be complemented by an SPR (surface plasmon resonance) experiment, which follows the kinetics of the inhibitor binding to the surface-tethered target molecule and detects the A–I complex formation in a label-independent manner. Availability of the biotinylated macromolecule developed for the smTIRFM studies greatly streamlines setting up the SPR experiment.

In some cases, the competitor is a macromolecule (nucleic acid or protein) that can be fluorescently labeled and binds to the target molecule with sufficiently high affinity to be compatible with the single-molecule TIRFM experiment. Such a competitor may be analyzed using two-color imaging scheme described in the next section.

4. ANALYSIS OF BINARY AND TERNARY COMPLEXES BY SINGLE-MOLECULE TIRFM

4.1 TIRFM-Enabled Colocalization Single-Molecule Imaging of Binding Processes

Single-molecule optical colocalization at a physiologically relevant temperature is a powerful method that can be dated to the mid-1990s (Funatsu, Harada, Tokunaga, Saito, & Yanagida, 1995). It can be achieved using many types of imaging modes or configurations (Churchman, Okten, Rock, Dawson, & Spudich, 2005; Enderle et al., 1997; Friedman & Gelles, 2012; Schutz, Trabesinger, & Schmidt, 1998; Selvin et al., 2008; Uemura et al., 2010). Conceptually, it also includes techniques like single-molecule FRET (Ha et al., 1996). Here we will specifically focus on single-molecule colocalization binding studies (with diffusion in 2D or/and 3D) using TIRFM, in the absence of FRET or any type of quenching (Boehm, Spies, et al., 2016; Friedman, Chung, & Gelles, 2006; Koyama-Honda et al., 2005). In a colocalization experiment, one has to obtain multiple spectrally distinct images of the same FOV, and, in the excitation pathway, the illumination fields of multiple wavelengths should overlap in space. Ideally, fluorophores should be selected so that they (i) have a minimal spectral overlap (to minimize signal leakage between the channels) and (ii) are of comparable photophysical properties, such as photo-bleaching rates, blinking dynamics and emission quantum yield. Colocalization can be achieved using multiple emission collection schemes, for example, (i) single detector (e.g., CCD camera) whose sensor area is spatially split into multiple sections (i.e., channels), recording signals from multiple dyes simultaneously; (ii) single detector alternating and quickly recording signals from multiple dyes and (iii) multiple detectors, each one recording signal from one dye separately (see Larson et al., 2014; Roy et al., 2008; Selvin et al., 2008, for comprehensive reviews). During analysis of videos, mathematical mapping is used to identify corresponding locations in all detection channels (Friedman & Gelles, 2015; Ghoneim & Spies, 2014; Roy et al., 2008). False positives (i.e., detection of colocalized spots when they are not actually present) are minimized using a combination of optimum mapping and spot detection algorithms. A practical tip to minimize false negatives is to use comparable excitation rates for all fluorophores so that emission rates are also comparable in addition to the selection of dyes with similar photophysical properties.

4.2 Detecting Binary and Ternary Complexes With Two-Color Imaging

This section will focus on two-color fluorescent imaging, which affords the direct visualization of two fluorescently labeled macromolecules binding to a nonfluorescent, immobilized macromolecule. These studies can be particularly informative when studying the architecture of the ternary complexes formed by these three macromolecules, as well as the kinetics of the assembly and disassembly of these complexes. This approach will be illustrated with a recent example in which the formation and dynamics of the ternary complexes that form among the translesion synthesis proteins DNA polymerase eta (pol η), Rev1, and proliferating cell nuclear antigen (PCNA) was studied (Boehm, Spies, et al., 2016). In these experiments, biotinylated pol η was immobilized on the surface of a quartz slide and the binding of Cy3-labeled PCNA and Cy5-labeled Rev1 to pol η was observed.

The binding events involving single immobilized pol η molecules are represented as fluorescence trajectories with time on the *x*-axis and fluorescence intensity on the *y*-axis (Fig. 4A). Cy3-labeled PCNA fluorescence is shown in green and Cy5-labeled Rev1 fluorescence is shown in red.

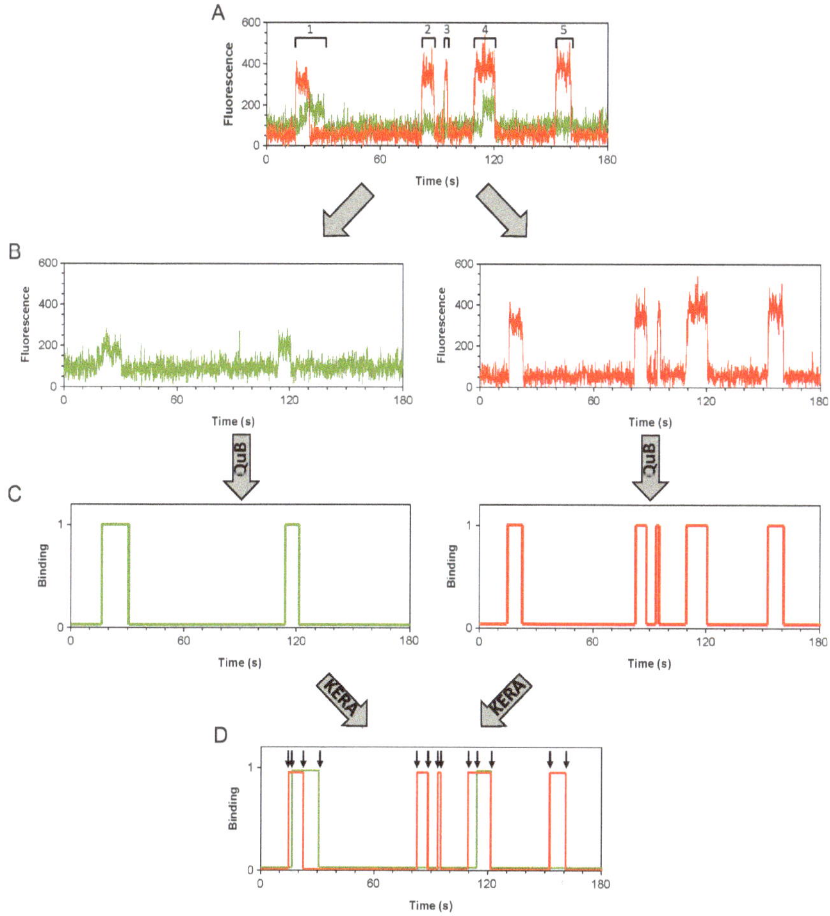

Fig. 4 Single-molecule analysis of ternary macromolecular complexes. (A) Sample fluorescence trajectory shows the evolution of the Cy3 (*green*) and Cy5 (*red*) signals in one of the spots on the slide. High Cy3 signal corresponds to the PCNA binding to the immobilized pol η molecule. High Cy5 signal represents Rev1 bound to the immobilized pol η. The five binding events are indicated. (B) The Cy3 and Cy5 signals are separated into two trajectories. (C) Idealized Cy3 and Cy5 trajectories are obtained from QuB. (D) The idealized Cy3 and Cy5 trajectories are recombined and analyzed by KERA. *The data in this figure are adapted with permission from Boehm, E.M., Spies, M., & Washington, M.T. (2016). PCNA tool belts and polymerase bridges form and interconvert in translesion synthesis complexes. Nucleic Acids Research, gkw563.*

Baseline fluorescence levels signify that no proteins are bound to the immobilized pol η molecule. An increase in Cy3 fluorescence above the baseline indicates the binding of PCNA to pol η, and an increase in Cy5 fluorescence indicates Rev1 binding. Likewise, a decrease in fluorescence back to the baseline indicates the dissociation of either PCNA or Rev1 from pol η. In this sample trajectory, there are five binding events: three events (2, 3, and 5) are binary complexes in which Cy5-labeled Rev1 binds to pol η and two events (1 and 4) contain ternary complexes in which Cy3-labeled PCNA and Cy5-labeled Rev1 simultaneously bind to pol η. Binding events truncated by the beginnings or the endings of the recordings are omitted from analysis because the exact timing of Cy3-PCNA or Cy5-Rev1 association or dissociation cannot be determined.

4.3 Kinetic Event Resolving Algorithm

Using two-color fluorescence imaging experiments, one can observe the formation of a variety of binary and ternary complexes. Thus, it is necessary to categorize the individual binding events. Because of the size of these single-molecule data sets, it is helpful to automate the categorization process in order to promote more rapid and efficient analysis. We have developed software in the form of a plug-in for Microsoft Excel for categorizing binding events from two-color fluorescence imaging experiments, which we have named kinetic event resolving algorithm (KERA).

The fluorescence trajectories from the two channels, Cy3 and Cy5, are analyzed separately by QuB to identify the Cy3 and Cy5 binding events (Fig. 4B and C). The two normalized, idealized trajectories corresponding to a given immobilized macromolecule are then recombined by KERA (Fig. 4D). The binding events for each channel are imported and enumerated in a table, which contains the beginning time, ending time, and duration of each Cy3 and Cy5 binding event. KERA then examines the binding events in the Cy3 channel and determines whether there are coincident binding events (i.e., occurring in the same time frame) in the Cy5 channel utilizing the Index function in Excel. If there is no coincident binding, then any events that occur are categorized as Cy3 binding only or Cy5 binding only binary events.

If there is coincident binding, KERA executes a series of logic statements to classify the binding event. We will use the first binding event in the idealized trajectory shown in Fig. 4D as an example of how KERA classifies binding events. In this example, we have four transitions (indicated by arrows). The first transition occurs at time $t = 15.3$ s and corresponds to the binding of a Cy5-labeled Rev1 to the immobilized pol η to form a pol η–Rev1 binary complex. The second transition occurs at time $t = 17.1$ s and corresponds to the binding of a Cy3-labeled PCNA to form a pol η-Rev1–PCNA ternary complex. The next transition occurs at time $t=22.7$ s and corresponds to the dissociation of the Cy5-labeled Rev1 leaving a pol η-PCNA binary complex. The final transition occurs at time $t=30.6$ s and corresponds to the dissociation of Cy3-labeled PCNA from the immobilized pol η. Table 1 shows the logic statements that KERA applies in order to categorize these binding events.

By applying such logic statements to all Cy3 and Cy5 binding data, KERA categorizes events into one of the following eleven categories (Fig. 5). Two of these are binary complexes:

(a) Cy3 binding only
(b) Cy5 binding only

Nine of these are ternary complexes:

(c) Cy3 binding first and Cy3 dissociating first
(d) Cy3 binding first and Cy5 dissociating first
(e) Cy3 binding first and Cy3 and Cy5 dissociating simultaneously
(f) Cy5 binding first and Cy3 dissociating first
(g) Cy5 binding first and Cy5 dissociating first
(h) Cy5 binding first and Cy3 and Cy5 dissociating simultaneously
(i) Cy3 and Cy5 binding simultaneously and Cy3 dissociating first
(j) Cy3 and Cy5 binding simultaneously and Cy5 dissociating first

Table 1 Categorization of Binding Events by KERA

Event	Logical Statement	Event Classification	Duration (s)
1	At $t=15.3$ (Cy3 = 0, Cy5 = 1) and at $t=17.1$ (Cy3 = 1, Cy5 = 1) and at $t=22.7$ (Cy3 = 1, Cy5 = 0) and at $t=30.6$ (Cy3 = 0, Cy5 = 0)	Cy5 binding first and Cy5 dissociating first	15.3
2	At $t=81.8$ (Cy3 = 0, Cy5 = 1) and at $t=88.5$ (Cy3 = 0, Cy5 = 0)	Cy5 binding only	6.7
3	At $t=93.8$ (Cy3 = 0, Cy5 = 1) and at $t=95.4$ (Cy3 = 0, Cy5 = 0)	Cy5 binding only	1.6
4	At $t=109.2$ (Cy3 = 0, Cy5 = 1) and at $t=114.4$ (Cy3 = 1, Cy5 = 1) AND at $t=120.6$ (Cy3 = 0, Cy5 = 0)	Cy5 binding first and Cy3 and Cy5 dissociating simultaneously	11.4
5	At $t=152.4$ (Cy3 = 0, Cy5 = 1) and at $t=160.3$ (Cy3 = 0, Cy5 = 0)	Cy5 binding only	7.9

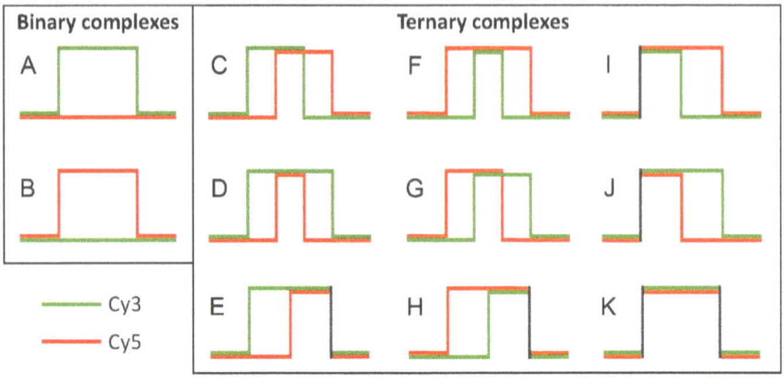

Fig. 5 Idealized trajectories of binary and ternary complexes. Idealized trajectories of (A) the binary complex between Cy3-labeled PCNA binding to pol η and (B) the binary complex between Cy5-labeled Rev1 binding to pol η. Idealized trajectories of (C) the ternary complex in which Cy3-labeled PCNA binds pol η first and the Cy3-labeled PCNA releases from pol η first, (D) the ternary complex in which Cy3-labeled PCNA binds pol η first and the Cy5-labeled Rev1 releases from pol η first (i.e., the hallmark of a PCNA tool belt), (E) the ternary complex in which Cy3-labeled PCNA binds pol η first and both proteins release from pol η simultaneously, (F) the ternary complex in which Cy5-labeled Rev1 binds pol η first and the Cy3-labeled PCNA releases from pol η first (i.e., the hallmark of a Rev1 bridge), (G) the ternary complex in which Cy5-labeled Rev1 binds pol η first and the Cy5-labeled Rev1 releases from pol η first, (H) the ternary complex in which Cy5-labeled Rev1 binds pol η first and both proteins release from pol η simultaneously, (I) the ternary complex in which both proteins bind pol η simultaneously and Cy3-labeled PCNA releases from η first, (J) the ternary complex in which both proteins bind pol η simultaneously and Cy5-labeled Rev1 releases from η first, and (K) the ternary complex in which both proteins bind pol η simultaneously and both proteins release from η simultaneously.

(k) Cy3 and Cy5 binding simultaneously and Cy3 and Cy5 dissociating simultaneously

KERA can be expanded to categorize and analyze data from more complicated experimental systems. Additional fluorescence channels can be added to analyze the formation of higher-order complexes (quaternary, quinary, etc.). Because KERA was built using VBA in Microsoft Excel, it is an approachable and intuitive software tool to both use and edit for a variety of applications. KERA is available upon request from Maria Spies and Todd Washington.

4.4 Determining the Order of Assembly and Architecture of Multicomponent Complexes

We used two-color fluorescence imaging experiments combined with an analysis facilitated by KERA to study the architecture and dynamics of the ternary complex that forms between the translesion synthesis proteins pol η, Rev1, and PCNA (Boehm, Spies, et al., 2016). The interaction between pol η and PCNA and the interaction between pol η and Rev1 are both mediated by the pol η (PCNA-interacting protein) PIP motif. Consequently, pol η cannot directly bind to both PCNA and Rev1 simultaneously. This implies that the architectures of the ternary complexes formed by these three proteins must be either "PCNA tool belts" or "Rev1 bridges" (Fig. 6). In a PCNA tool belt, pol η directly interacts with PCNA and PCNA directly interacts with Rev1, but pol η and Rev1 do not directly interact. In a Rev1 bridge, pol η directly interacts with Rev1 and Rev1 directly interacts with PCNA, but pol η and PCNA do not directly interact. The architecture of individual ternary complexes can be deduced in several ways.

One way to determine the architecture of each ternary complex is by examining their order of assembly and disassembly. For example, if the immobilized pol η first binds PCNA to form a PCNA–pol η binary complex and then this binary complex binds Rev1 to form a ternary complex, one can conclude that the ternary complex assembles as a PCNA tool belt. By contrast, if pol η first binds Rev1 and then binds PCNA, one can conclude that the resultant ternary complex assembles as a Rev1 bridge. Likewise, if Rev1 first dissociates from a ternary complex leaving behind a pol η–PCNA binary complex and then PCNA dissociates next, one can conclude that the ternary complex disassembled as a PCNA tool belt. Last, if PCNA dissociates first from a ternary complex and Rev1 dissociates next, one can conclude that the ternary complex dissembled as a Rev1 bridge.

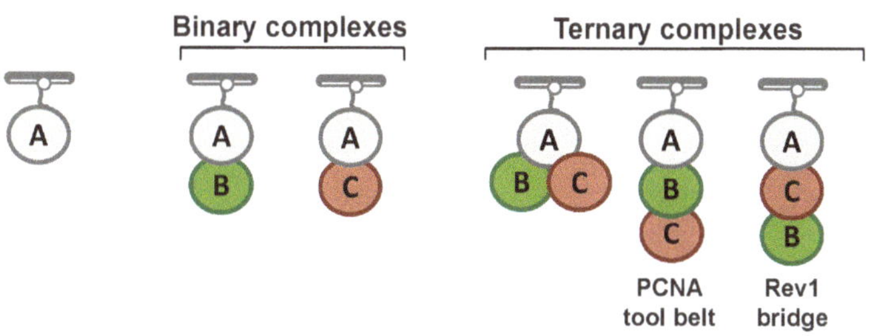

Fig. 6 Types of binary and ternary complexes. Binary complexes are shown between A and B and between A and C. Ternary complexes are shown among A, B, and C. In the first ternary complex (*left*), B and C both simultaneously bind A. In the second (*middle*), A directly binds B and B directly binds C, but A does not directly bind C. When A is pol η, B is PCNA, and C is Rev1, this arrangement is a PCNA tool belt. In the third (*right*) A directly binds C and C directly binds B, but A does not directly bind B. When A is pol η, B is PCNA, and C is Rev1, this arrangement is a Rev1 bridge.

We observed approximately 1000 ternary complexes formed by Cy3-labeled PCNA, Cy5-labeled Rev1, and immobilized, unlabeled pol η. Based on their orders of assembly and disassembly, we found that about 40% of the ternary complexes were PCNA tool belts and roughly 20% were Rev1 bridges. We were unable to unambiguously determine the architecture of about 20% of the ternary complexes. This was because PCNA and Rev1 bound pol η together as a preformed complex, obscuring their order of assembly or because both PCNA and Rev1 dissociated from pol η together, obscuring their order of disassembly. Surprisingly, we found that approximately 10% of the ternary complexes assembled as a PCNA tool belt but disassembled as a Rev1 bridge and that about 10% of the complexes assembled as a Rev1 bridge but disassembled as a PCNA tool belt. This suggests that these ternary complexes are dynamic and are capable of switching architectures (from a PCNA tool belt to a Rev1 bridge and vice versa) without any proteins dissociating from the complex.

Another way to determine the architecture of a ternary complex is to use mutant proteins that are defective in binding one of the other binding partners. For example, the interaction between pol η and Rev1 requires the C-terminal domain (CTD) of Rev1. We made a series of amino acid substitutions in the Rev1 CTD that interfere with the ability of Rev1 to bind pol η, but not the ability of Rev1 to bind PCNA. As a result, mutant Rev1 protein would be unable to engage in the formation of Rev1 bridges, but will not interfere with the formation of PCNA tool belts. When identical two-color fluorescence imaging experiments were carried out with this mutant Rev1 protein, we observed a greatly reduced number of Rev1 bridges, but not PCNA tool belts. Studies with mutant proteins such as these can help unambiguously define the architecture ternary complexes.

5. CONCLUSIONS AND FUTURE OUTLOOK

Single-molecule fluorescence microscopy is emerging as a powerful tool for the analysis of the dynamic assembly of various molecular machines. The smTIRFM-enabled workflow for the analysis presented in this chapter illustrates the power of this straightforward single-molecule assay to provide a comprehensive mechanistic description of macromolecular interactions ranging in scope from the binary interactions, to the formation and dynamics of ternary complexes, to the application of single-molecule fluorescence to study the inhibitors of protein–protein and protein–nucleic acid interactions. We discussed general considerations of the experimental design, controls, and data acquisition and analysis.

Analysis of the assembly and dynamics of larger multicomponent complexes, and the ability to combine multicolor imaging with studies of the inhibitors will require further development of more complicated data analysis routines as well as the experimental schemes that will allow one to evaluate the sources of heterogeneity. The basic routines, however, are in place and hopefully will become as common in the tool box of many biophysicists and biochemists as structural biology approaches and presteady state kinetics analysis are today.

ACKNOWLEDGMENTS

The work in Spies and Washington lab is supported by the National Institute of Health R01 GM108027, R01 GM081433 (M.T.W.), NIH R01 GM108617, R01 GM101167 (M.S.), and T32 GM067795 (E.M.B.).

REFERENCES

Abelson, J., Blanco, M., Ditzler, M. A., Fuller, F., Aravamudhan, P., Wood, M., et al. (2010). Conformational dynamics of single pre-mRNA molecules during in vitro splicing. *Nature Structural & Molecular Biology*, *17*, 504–512.

Aggarwal, V., & Ha, T. (2014). Single-molecule pull-down (SiMPull) for new-age biochemistry: Methodology and biochemical applications of single-molecule pull-down (SiMPull) for probing biomolecular interactions in crude cell extracts. *Bioessays*, *36*, 1109–1119.

Akyuz, N., Altman, R. B., Blanchard, S. C., & Boudker, O. (2013). Transport dynamics in a glutamate transporter homologue. *Nature*, *502*, 114–118.

Amrute-Nayak, M., Lambeck, K. A., Radocaj, A., Huhnt, H. E., Scholz, T., Hahn, N., et al. (2014). ATP turnover by individual myosin molecules hints at two conformers of the myosin active site. *Proceedings of the National Academy of Sciences of the United States of America*, *111*, 2536–2541.

Appel, M. J., & Bertozzi, C. R. (2015). Formylglycine, a post-translationally generated residue with unique catalytic capabilities and biotechnology applications. *ACS Chemical Biology*, *10*, 72–84.

Axelrod, D. (2008). Chapter 7: Total internal reflection fluorescence microscopy. *Methods in Cell Biology*, *89*, 169–221.

Bain, F. E., Wu, C. G., & Spies, M. (2016). Single-molecule sorting of DNA helicases. *Methods*, *S1046–2023*(16), 30132–30133.

Bartley, L. E., Zhuang, X., Das, R., Chu, S., & Herschlag, D. (2003). Exploration of the transition state for tertiary structure formation between an RNA helix and a large structured RNA. *Journal of Molecular Biology*, *328*, 1011–1026.

Beckett, D., Kovaleva, E., & Schatz, P. J. (1999). A minimal peptide substrate in biotin holoenzyme synthetase-catalyzed biotinylation. *Protein Science*, *8*, 921–929.

Blanco, M., & Walter, N. G. (2010). Analysis of complex single-molecule FRET time trajectories. *Methods in Enzymology*, *472*, 153–178.

Boehm, E. M., Powers, K. T., Kondratick, C. M., Spies, M., Houtman, J. C., & Washington, M. T. (2016). The proliferating cell nuclear antigen (PCNA)-interacting protein (PIP) motif of DNA polymerase eta mediates its interaction with the C-terminal domain of Rev1. *Journal of Biological Chemistry, 291*, 8735–8744.

Boehm, E. M., Spies, M., & Washington, M. T. (2016). PCNA tool belts and polymerase bridges form and interconvert in translesion synthesis complexes. *Nucleic Acids Research*. gkw563.

Bombardier, J. P., Eskin, J. A., Jaiswal, R., Correa, I. R., Jr., Xu, M. Q., Goode, B. L., et al. (2015). Single-molecule visualization of a formin-capping protein 'decision complex' at the actin filament barbed end. *Nature Communications, 6*, 8707.

Bronson, J. E., Fei, J., Hofman, J. M., Gonzalez, R. L., Jr., & Wiggins, C. H. (2009). Learning rates and states from biophysical time series: A Bayesian approach to model selection and single-molecule FRET data. *Biophysical Journal, 97*, 3196–3205.

Chen, R., Subramanyam, S., Elcock, A. H., Spies, M., & Wold, M. S. (2016). Dynamic binding of replication protein a is required for DNA repair. *Nucleic Acids Research, 44*, 5758–5772.

Chen, J., Zhang, Z., Li, L., Chen, B. C., Revyakin, A., Hajj, B., et al. (2014). Single-molecule dynamics of enhanceosome assembly in embryonic stem cells. *Cell, 156*, 1274–1285.

Chung, S. H., Moore, J. B., Xia, L. G., Premkumar, L. S., & Gage, P. W. (1990). Characterization of single channel currents using digital signal processing techniques based on Hidden Markov Models. *Philosophical Transactions of the Royal Society of London. Series B, Biological Sciences, 329*, 265–285.

Churchman, L. S., Okten, Z., Rock, R. S., Dawson, J. F., & Spudich, J. A. (2005). Single molecule high-resolution colocalization of Cy3 and Cy5 attached to macromolecules measures intramolecular distances through time. *Proceedings of the National Academy of Sciences of the United States of America, 102*, 1419–1423.

Colquhoun, D., & Sigworth, F. (1995). Fitting and statistical analysis of single-channel records. In E. Neher & B. Sakmann (Eds.), *Single-channel recording*. New York: Plenum Press.

Diaspro, A. C. G., Usai, C., Ramoino, P., & Dobrucki, J. (2006). Photobleaching. In J. Pawley (Ed.), *Handbook of biological confocal microscopy*. (3rd ed.). New York: Springer Science+Business Media, LLC.

Doane, D. P. (1976). Aesthetic frequency classification. *American Statistician, 30*, 181–183.

Elenko, M. P., Szostak, J. W., & Van Oijen, A. M. (2009). Single-molecule imaging of an in vitro-evolved RNA aptamer reveals homogeneous ligand binding kinetics. *Journal of the American Chemical Society, 131*, 9866–9867.

Enderle, T., Ha, T., Ogletree, D. F., Chemla, D. S., Magowan, C., & Weiss, S. (1997). Membrane specific mapping and colocalization of malarial and host skeletal proteins in the Plasmodium falciparum infected erythrocyte by dual-color near-field scanning optical microscopy. *Proceedings of the National Academy of Sciences of the United States of America, 94*, 520–525.

Friedman, L. J., Chung, J., & Gelles, J. (2006). Viewing dynamic assembly of molecular complexes by multi-wavelength single-molecule fluorescence. *Biophysical Journal, 91*, 1023–1031.

Friedman, L. J., & Gelles, J. (2012). Mechanism of transcription initiation at an activator-dependent promoter defined by single-molecule observation. *Cell, 148*, 679–689.

Friedman, L. J., & Gelles, J. (2015). Multi-wavelength single-molecule fluorescence analysis of transcription mechanisms. *Methods, 86*, 27–36.

Funatsu, T., Harada, Y., Tokunaga, M., Saito, K., & Yanagida, T. (1995). Imaging of single fluorescent molecules and individual ATP turnovers by single myosin molecules in aqueous solution. *Nature, 374*, 555–559.

Gebhardt, J. C., Clemen, A. E., Jaud, J., & Rief, M. (2006). Myosin-V is a mechanical ratchet. *Proceedings of the National Academy of Sciences of the United States of America, 103*, 8680–8685.

Ghoneim, M., & Spies, M. (2014). Direct correlation of DNA binding and single protein domain motion via dual illumination fluorescence microscopy. *Nano Letters*, *14*, 5920–5931.

Greenfeld, M., Pavlichin, D. S., Mabuchi, H., & Herschlag, D. (2012). Single molecule analysis research tool (SMART): An integrated approach for analyzing single molecule data. *PloS One*, 7, e30024.

Ha, T., Enderle, T., Ogletree, D. F., Chemla, D. S., Selvin, P. R., & Weiss, S. (1996). Probing the interaction between two single molecules: Fluorescence resonance energy transfer between a single donor and a single acceptor. *Proceedings of the National Academy of Sciences of the United States of America*, *93*, 6264–6268.

Ha, T., & Tinnefeld, P. (2012). Photophysics of fluorescent probes for single-molecule biophysics and super-resolution imaging. *Annual Review of Physical Chemistry*, *63*, 595–617.

Haghighat Jahromi, A., Honda, M., Zimmerman, S. C., & Spies, M. (2013). Single-molecule study of the CUG repeat-MBNL1 interaction and its inhibition by small molecules. *Nucleic Acids Research*, *41*, 6687–6697.

Haller, A., Rieder, U., Aigner, M., Blanchard, S. C., & Micura, R. (2011). Conformational capture of the SAM-II riboswitch. *Nature Chemical Biology*, 7, 393–400.

Honda, M., Okuno, Y., Hengel, S. R., Martin-Lopez, J. V., Cook, C. P., Amunugama, R., et al. (2014). Mismatch repair protein hMSH2-hMSH6 recognizes mismatches and forms sliding clamps within a D-loop recombination intermediate. *Proceedings of the National Academy of Sciences of the United States of America*, *111*, E316–E325.

Hoskins, A. A., Rodgers, M. L., Friedman, L. J., Gelles, J., & Moore, M. J. (2016). Single molecule analysis reveals reversible and irreversible steps during spliceosome activation. *eLife*, *5*, e14166.

Huppa, J. B., Axmann, M., Mortelmaier, M. A., Lillemeier, B. F., Newell, E. W., Brameshuber, M., et al. (2010). TCR-peptide-MHC interactions in situ show accelerated kinetics and increased affinity. *Nature*, *463*, 963–967.

Jain, A., Liu, R., Xiang, Y. K., & Ha, T. (2012). Single-molecule pull-down for studying protein interactions. *Nature Protocols*, 7, 445–452.

Joo, C., & Ha, T. (2012). Single-molecule FRET with total internal reflection microscopy. *Cold Spring Harbor Protocols*. *2012*, http://dx.doi.org/10.1101/pdb.top072058.

Joshi, C. P., Panda, D., Martell, D. J., Andoy, N. M., Chen, T. Y., Gaballa, A., et al. (2012). Direct substitution and assisted dissociation pathways for turning off transcription by a MerR-family metalloregulator. *Proceedings of the National Academy of Sciences of the United States of America*, *109*, 15121–15126.

Koyama-Honda, I., Ritchie, K., Fujiwara, T., Iino, R., Murakoshi, H., Kasai, R. S., et al. (2005). Fluorescence imaging for monitoring the colocalization of two single molecules in living cells. *Biophysical Journal*, *88*, 2126–2136.

Lakowicz, J. (2006). Single-molecule detection. In J. Lakowicz (Ed.), *Principles of fluorescence spectroscopy* (3rd ed.). New York: Springer Science+Business Media, LLC.

Larson, J., Kirk, M., Drier, E. A., O'Brien, W., MacKay, J. F., Friedman, L. J., et al. (2014). Design and construction of a multiwavelength, micromirror total internal reflectance fluorescence microscope. *Nature Protocols*, *9*, 2317–2328.

Liu, J., Hanne, J., Britton, B. M., Shoffner, M., Albers, A. E., Bennett, J., et al. (2015). An efficient site-specific method for irreversible covalent labeling of proteins with a fluorophore. *Scientific Reports*, *5*, 16883.

Liu, Y., Park, J., Dahmen, K. A., Chemla, Y. R., & Ha, T. (2010). A comparative study of multivariate and univariate hidden Markov modelings in time-binned single-molecule FRET data analysis. *Journal of Physical Chemistry B*, *114*, 5386–5403.

Lu, H. P., Xun, L., & Xie, X. S. (1998). Single-molecule enzymatic dynamics. *Science*, *282*, 1877–1882.

Mckinney, S. A., Joo, C., & Ha, T. (2006). Analysis of single-molecule FRET trajectories using hidden Markov modeling. *Biophysical Journal*, *91*, 1941–1951.
McManus, O. B., Blatz, A. L., & Magleby, K. L. (1987). Sampling, log binning, fitting, and plotting durations of open and shut intervals from single channels and the effects of noise. *Pflügers Archiv*, *410*, 530–553.
Miyanaga, Y., & Ueda, M. (2011). Single-molecule kinetic analysis of stochastic signal transduction mediated by G-protein coupled chemoattractant receptors. In Y. Sako & M. Ueda (Eds.), *Cell signaling reactions: Single-molecular kinetic analysis* (1st ed.). The Netherlands: Springer Science+Business Media B.V.
Moffitt, J. R., Chemla, Y. R., & Bustamante, C. (2010). Methods in statistical kinetics. *Methods in Enzymology*, *475*, 221–257.
Munro, J. B., Altman, R. B., O'Connor, N., & Blanchard, S. C. (2007). Identification of two distinct hybrid state intermediates on the ribosome. *Molecular Cell*, *25*, 505–517.
Nicolai, C., & Sachs, F. (2013). Solving Ion channel kinetics with the QuB software. *Biophysical Reviews and Letters*, *8*, 1–21.
Pond, S. J., Ridgeway, W. K., Robertson, R., Wang, J., & Millar, D. P. (2009). HIV-1 Rev protein assembles on viral RNA one molecule at a time. *Proceedings of the National Academy of Sciences of the United States of America*, *106*, 1404–1408.
Qin, F., & Li, L. (2004). Model-based fitting of single-channel dwell-time distributions. *Biophysical Journal*, *87*, 1657–1671.
Rabiner, L. (1989). A tutorial on hidden Markov models and selected applications in speech recognition. *Proceeding of the IEEE*, *77*, 257–286.
Rabuka, D., Rush, J. S., Dehart, G. W., Wu, P., & Bertozzi, C. R. (2012). Site-specific chemical protein conjugation using genetically encoded aldehyde tags. *Nature Protocols*, *7*, 1052–1067.
Rasnik, I., Mckinney, S. A., & Ha, T. (2006). Nonblinking and longlasting single-molecule fluorescence imaging. *Nature Methods*, *3*, 891–893.
Roy, R., Hohng, S., & Ha, T. (2008). A practical guide to single-molecule FRET. *Nature Methods*, *5*, 507–516.
Rueda, D., Bokinsky, G., Rhodes, M. M., Rust, M. J., Zhuang, X., & Walter, N. G. (2004). Single-molecule enzymology of RNA: Essential functional groups impact catalysis from a distance. *Proceedings of the National Academy of Sciences of the United States of America*, *101*, 10066–10071.
Sachs, F., Neil, J., & Barkakati, N. (1982). The automated analysis of data from single ionic channels. *Pflügers Archiv*, *395*, 331–340.
Schutz, G. J., Trabesinger, W., & Schmidt, T. (1998). Direct observation of ligand colocalization on individual receptor molecules. *Biophysical Journal*, *74*, 2223–2226.
Scott, D. W. (1979). On optimal and data-based histograms. *Biometrika*, *66*, 605–610.
Selvin, P., Lougheed, T., Hoffman, M., Park, H., Balci, H., Blehm, B., et al. (2008). In vitro and in vivo FIONA and other acronyms for watching molecular motors walk. In P. Selvin & T. Ha (Eds.), *Single-molecule techniques* (pp. 37–71). Cold Spring Harbor, NY: Cold Spring Harbor Laboratory Press.
Shi, X., Jung, Y., Lin, L. J., Liu, C., Wu, C., Cann, I. K., et al. (2012). Quantitative fluorescence labeling of aldehyde-tagged proteins for single-molecule imaging. *Nature Methods*, *9*, 499–503.
Shimazaki, H., & Shinomoto, S. (2007). A method for selecting the bin size of a time histogram. *Neural Computation*, *19*, 1503–1527.
Shimazaki, H., & Shinomoto, S. (2010). Kernel bandwidth optimization in spike rate estimation. *Journal of Computational Neuroscience*, *29*, 171–182.
Shuang, B., Cooper, D., Taylor, J. N., Kisley, L., Chen, J., Wang, W., et al. (2014). Fast step transition and state identification (STaSI) for discrete single-molecule data analysis. *Journal of Physical Chemistry Letters*, *5*, 3157–3161.

Sigworth, F. J., & Sine, S. M. (1987). Data transformations for improved display and fitting of single-channel dwell time histograms. *Biophysical Journal, 52*, 1047–1054.

Solomatin, S. V., Greenfeld, M., Chu, S., & Herschlag, D. (2010). Multiple native states reveal persistent ruggedness of an RNA folding landscape. *Nature, 463*, 681–684.

Strick, T. R., Croquette, V., & Bensimon, D. (2000). Single-molecule analysis of DNA uncoiling by a type II topoisomerase. *Nature, 404*, 901–904.

Sturges, H. A. (1926). The choice of a class interval. *Journal of the American Statistical Association, 21*, 65–66.

Sung, J., Sivaramakrishnan, S., Dunn, A. R., & Spudich, J. A. (2010). Single-molecule dual-beam optical trap analysis of protein structure and function. *Methods in Enzymology, 475*, 321–375.

Szoszkiewicz, R., Ainavarapu, S. R., Wiita, A. P., Perez-Jimenez, R., Sanchez-Ruiz, J. M., & Fernandez, J. M. (2008). Dwell time analysis of a single-molecule mechanochemical reaction. *Langmuir, 24*, 1356–1364.

Talaga, D. S. (2007). COCIS: Markov processes in single molecule fluorescence. *Current Opinion in Colloid & Interface Science, 12*, 285–296.

Theile, C. S., Witte, M. D., Blom, A. E., Kundrat, L., Ploegh, H. L., & Guimaraes, C. P. (2013). Site-specific N-terminal labeling of proteins using sortase-mediated reactions. *Nature Protocols, 8*, 1800–1807.

Uemura, S., Aitken, C. E., Korlach, J., Flusberg, B. A., Turner, S. W., & Puglisi, J. D. (2010). Real-time tRNA transit on single translating ribosomes at codon resolution. *Nature, 464*, 1012–1017.

Valeur, B. (2012). Fluorescence correlation spectroscopy and single-molecule fluorescence spectroscopy. In B. Valeur (Ed.), *Molecular fluorescence: Principles and applications* (2nd ed.). Germany: Wiley-VCH Verlag GmbH & Co KGaA.

Van De Meent, J. W., Bronson, J. E., Wiggins, C. H., & Gonzalez, R. L., Jr. (2014). Empirical Bayes methods enable advanced population-level analyses of single-molecule FRET experiments. *Biophysical Journal, 106*, 1327–1337.

Wand, M. (1997). Data-based choice of histogram bin width. *The American Statistician, 51*, 59–64.

Wang, Y., Guo, L., Golding, I., Cox, E. C., & Ong, N. P. (2009). Quantitative transcription factor binding kinetics at the single-molecule level. *Biophysical Journal, 96*, 609–620.

Watanabe, N., & Mitchison, T. J. (2002). Single-molecule speckle analysis of actin filament turnover in lamellipodia. *Science, 295*, 1083–1086.

Watkins, L. P., & Yang, H. (2005). Detection of intensity change points in time-resolved single-molecule measurements. *Journal of Physical Chemistry B, 109*, 617–628.

Zheng, Q. S., Juette, M. F., Jockusch, S., Wasserman, M. R., Zhou, Z., Altman, R. B., et al. (2014). Ultra-stable organic fluorophores for single-molecule research. *Chemical Society Reviews, 43*, 1044–1056.

Zhou, R., Kunzelmann, S., Webb, M. R., & Ha, T. (2011). Detecting intramolecular conformational dynamics of single molecules in short distance range with subnanometer sensitivity. *Nano Letters, 11*, 5482–5488.

CHAPTER SIX

How to Measure Separations and Angles Between Intramolecular Fluorescent Markers

K.I. Mortensen*, J. Sung†, J.A. Spudich‡, H. Flyvbjerg*,[1]
*Department of Micro- and Nanotechnology, Technical University of Denmark, Kongens Lyngby, Denmark
†Department of Cellular and Molecular Pharmacology, The Howard Hughes Medical Institute, University of California, San Francisco, CA, United States
‡Department of Biochemistry, Stanford University School of Medicine, Stanford, CA, United States
[1]Corresponding author: e-mail address: henrik.flyvbjerg@nanotech.dtu.dk

Contents

Abstract

Structure and function of an individual biomolecule can be explored with minimum two fluorescent markers of different colors. Since the light of such markers can be spectrally separated and imaged simultaneously, the markers can be colocalized. Here, we describe the method used for such two-color colocalization microscopy. Then we extend it to fluorescent markers with fixed orientations and in intramolecular proximity. Our benchmarking of this extension produced two extra results: (a) we established short

Methods in Enzymology, Volume 581
ISSN 0076-6879
http://dx.doi.org/10.1016/bs.mie.2016.08.020
© 2016 Elsevier Inc.
All rights reserved.

double-labeled DNA molecules as probes of 3D orientation of anything to which one can attach them firmly; (b) we established how to map with super-resolution between color-separated channels, which should be useful for all dual-color colocalization measurements with either fixed or freely rotating fluorescent molecules. Throughout, we use only simple means: from each color-separated microscope image in a time-lapse movie, we simultaneously determine both the relative (*x*,*y*)-separation of the fluorophores and their individual orientations in space, both with accuracy and precision. The relative positions and orientations of two domains of the same molecule are thus time-resolved. Using short double-stranded DNA (dsDNA) molecules internally labeled with two fixed fluorophores, we (i) demonstrate the accuracy and precision of our localization- and mapping-methods, using the known structure of dsDNA as benchmark; (ii) resolve 10 base pair differences in fluorophore separations; (iii) determine the unique 3D orientation of each DNA molecule.

1. INTRODUCTION

In the investigation of the performance of a scientific method, it is essential to consider both its precision and accuracy. In colloquial use, precision and accuracy are often used interchangeably, and in the scientific literature, they are seen used in disagreement with their standard definitions. In science, however, it is paramount to discriminate between these two notions, and in the method development in this chapter, we will use the terms consistently as defined in Fig. 1: accuracy quantifies systematic errors, i.e., agreement of a measurement with its true value in the absence of stochastic errors. Precision quantifies the reproducibility of repetitions of measurements subject to stochastic errors. Information theory furthermore allows one to establish the optimal precision obtainable in a given situation, the *information limit* (Rao, 1973). The precision of any method can therefore be quantified in terms of its ability to reach this limit (Fig. 1).

Later, we detail how to use a fluorescence-based tool that can resolve the relative positions and orientations of two domains on the same molecule with precision and accuracy, as we recently reported (Mortensen, Sung, Flyvbjerg, & Spudich, 2015).

To probe structure and function, fluorophores are imaged in a microscope in localization-based single-molecule studies and super-resolution microscopy. Fundamental diffraction effects cause individual fluorophores to image as diffraction-limited spots. In order to directly resolve two such spots in one image, they must be separated by at least their half-width, typically ~200 nm, given by the wavelength of the light that produced them (Abbe's law) (Born & Wolf, 1999). One may circumvent this limit by

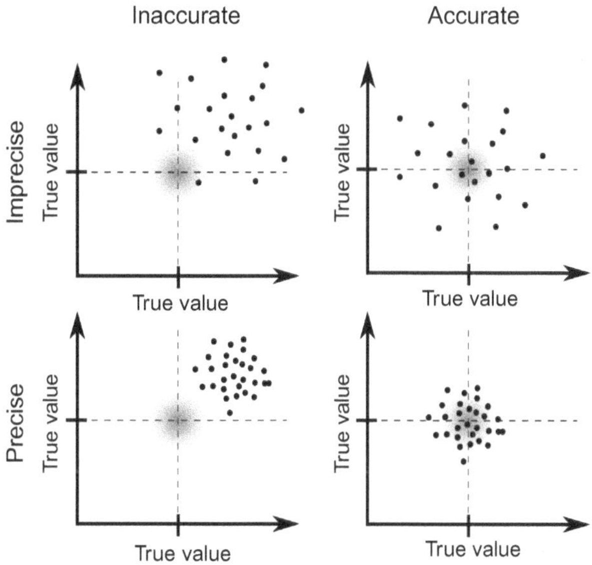

Fig. 1 Schematic illustration of accuracy and precision as they are used in science. Accuracy and precision are independent quality measures for measurable quantities. Accuracy quantifies the agreement of the measured mean value of a quantity with that quantity's true value in the absence of stochastic errors. Precision quantifies the reproducibility of repetitions of measurements subject to stochastic error. Furthermore, in the context of parametric statistics, a method is denoted precise if it achieves the maximum precision permitted by information theory (Mortensen, Churchman, Spudich, & Flyvbjerg, 2010; Rao, 1973). Here, the standard deviation corresponding to the maximum precision is indicated by the *shaded region*. The method in the *lower right panel* is accurate and precise, because its measurements scatter around the true values with an s.d. given by the information limit.

separating imaged spots either in time, by switching the fluorophores between active and inactive fluorescent states, as in the various super-resolution methods (STORM, PALM) (Betzig et al., 2006; Lippincott-Schwartz & Manley, 2009; Rust, Bates, & Zhuang, 2006), or in frequency space, by using fluorophores of different colors as in single-molecule high-resolution colocalization (Churchman, Okten, Rock, Dawson, & Spudich, 2005). Either way, the isolated spot has an intensity distribution that is a realization of the fluorophore's theoretical point spread function (PSF). Consequently, one may fit the latter to the measured intensity distribution and thus localize the fluorophore with a precision that increases with the number of photons in the imaged spot. With this method, fluorescent probes are routinely localized with nanometer *precision* (Betzig et al., 2006; Bobroff, 1986;

Churchman et al., 2005; Deschout et al., 2014; Lippincott-Schwartz & Manley, 2009; Mortensen et al., 2010; Ober, Ram, & Ward, 2004; Pertsinidis, Zhang, & Chu, 2010; Rust et al., 2006; Thompson, Larson, & Webb, 2002; Yildiz et al., 2003), i.e., nanometer reproducibility (Fig. 1).

Clearly, for such precision to be useful, *accuracy* is critical (Fig. 1) (Deschout et al., 2014). In the context of localization microscopy, accuracy is very sensitive to proper choice of theoretical PSF, and the latter depends on experimental conditions. Thus, if a fluorophore is bound to a molecule of interest with a single, flexible linker and can change its orientation unhindered by its environment, its rapid thermal motion will cause it to image as an isotropic superposition of itself. Consequently, its theoretical PSF is very well approximated with a 2D Gaussian plus a constant "background" (Mortensen et al., 2010). This approximation is very popular in single-molecule localization studies. When it is justified, localization is accurate (Churchman et al., 2005; Pertsinidis et al., 2010), and even sub-nanometer accuracy has been demonstrated with advanced feedback-controlled instruments (Pertsinidis et al., 2010).

A fluorophore bound with a single, flexible linker may, however, be constrained in its thermal motion by a crowded molecular environment (Burghardt, 2011; Burghardt, Josephson, & Ajtai, 2011) or by interactions with neighboring proteins or nucleotide residues through hydrophobic or stacking effects (Iqbal et al., 2008; Pertsinidis et al., 2010). The orientation of a fluorophore also may be fixed deliberately to a molecule of interest (Aguet, Geissbühler, Märki, Lasser, & Unser, 2009; Forkey, Quinlan, Shaw, Corrie, & Goldman, 2003; Mortensen et al., 2010; Reifenberger et al., 2009; Toprak et al., 2006). In such cases, the PSF is typically anisotropic, and if it is approximated by a 2D Gaussian, localization accuracy is compromised: systematic errors may amount to tens of nanometers (Enderlein, Toprak, & Selvin, 2006; Stallinga & Rieger, 2010), irrespective of precision, which may still be nanometers. Thus, the true position of the fluorophore can be tens of nanometers away from the precisely estimated position!

Such anisotropic intensity distributions provide information about the fluorophore's orientation, however. Fluorophores with fixed orientation have been analyzed to extract fluorophore orientations, e.g., by using deliberately defocused images (Aguet et al., 2009; Reifenberger et al., 2009; Toprak et al., 2006), polarization-based measurements (Forkey et al., 2003), or engineered PSFs (Backlund et al., 2012). In some cases, fluorophore positions were also extracted (Aguet et al., 2009; Backlund et al., 2012; Lew & Moerner, 2014; Mortensen et al., 2010; Reifenberger et al., 2009; Stallinga & Rieger, 2012; Toprak et al., 2006). Unfortunately, in some instances (Reifenberger et al., 2009; Toprak et al., 2006), positions were inaccurate because a 2D Gaussian was fitted to focused images of fixed fluorophores to extract locations (Enderlein et al., 2006; Stallinga & Rieger, 2010). Importantly, none of these reports of methods nor our own (Mortensen et al., 2010) provided direct, rigorous verification of accuracy (with precision) for estimates of orientation and location of probes, e.g., in the form of an experimental demonstration that estimates of positions and orientations of fluorophores agree with a priori known results.

If optimally analyzed, such anisotropic intensity distributions may, however, provide positions of fluorescent probes with nanometer precision and accuracy and, simultaneously, from the same image, provide the orientations of the same probes with degree accuracy and precision. We demonstrate this later, and in that process we extend current experimental practice with fixed fluorophores to a two-color assay (Churchman et al., 2005). This allows us to determine separation and orientations simultaneously for two fluorophores in intramolecular proximity from a single spectrally separated image of them. This establishes the simultaneous use of two fixed fluorophores of different color as a structural tool for the study of molecular conformation by allowing measurements of orientation and position of one molecular domain with respect to another. The method works at short intramolecular distances that currently are studied mainly using single-molecule FRET (Roy, Hohng, & Ha, 2008), which is incapable of discriminating between intramolecular rotations and translations.

We verify experimentally that the method is (i) *accurate*, by demonstrating that its estimates of distances and relative orientations between two fixed fluorophores agree with the a priori known experimental reality, and (ii) about as *precise* as the information limit allows (Mortensen et al., 2010) (Fig. 1).

2. METHODS

2.1 dsDNA Internally Labeled with Cy3/Cy5 Dyes

We used two constructs of double-stranded DNA (dsDNA) to demonstrate the theoretical analysis and to experimentally establish two-color colocalization microscopy with fixed fluorophores as a structural tool. In our two constructs, Cy3 and Cy5 are covalently fused into the backbone

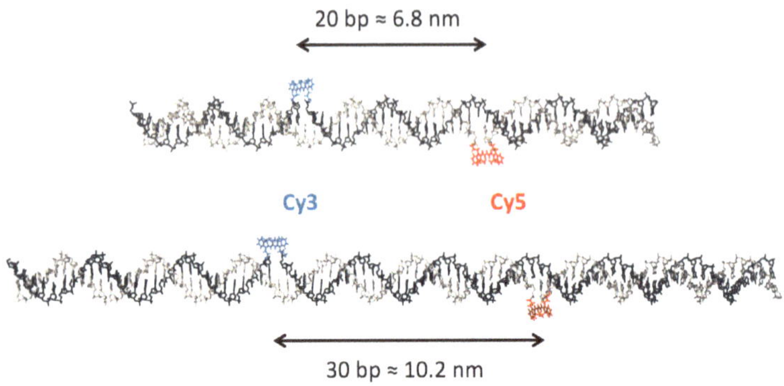

Fig. 2 Schematic illustration of the B-form dsDNA labeled internally with two fluorophores of different color. The Cy3 and Cy5 dyes are covalently attached to the backbone. We used two constructs: (i) In the first, the total length of each strand is 60 bp, and Cy3 and Cy5 label complementary strands 20 and 21 bp from each 5′ end, respectively, which makes the interdye separation 20 bp ($\approx$6.8 nm); (ii) In the other, the total length is 90 bp and the interdye separation is 30 bp ($\approx$10.2 nm). This integer multiple of 10 bp, one helix turn, ensures that the relative orientation between the fluorophores is equal in our two constructs. We modeled the B-form dsDNA as a simple double helix with a pitch of 3.4 nm distributed with 1.2 and 2.2 nm on the minor and major grooves, respectably. Simulations (Dolghih, 2009) have shown that the Cy3 and Cy5 may interact with the dsDNA located either in a minor or a major groove due to, presumably, hydrophobic effects. So we assumed that each fluorophore hinges with a radius of 1 nm and in a manner that keeps it parallel to the piece of helix backbone it replaces. This makes the relative angle between the two dipoles constant and equal to 74 degrees. We also assumed that the dsDNA was approximately parallel to the coverslip surface. Due to the possible minor/major groove interactions of the probes with the dsDNA, the distances between them may vary with $\pm$4 bp, corresponding to both probes interacting with the closest minor groove, respectively, both dyes interacting with the closest major groove. The distance between the probes changes only negligibly for combinations of major/minor groove interactions. *Figure adapted from Mortensen, K. I., Sung, J., Flyvbjerg, H., & Spudich, J. A. (2015). Optimized measurements of separations and angles between intra-molecular fluorescent markers.* Nature Communications, 6, *8621. http://doi.org/10.1038/ncomms9621.*

of each complementary strand with double attachments (Fig. 2). The constructs were purchased from Integrated DNA Technologies and were synthesized, hybridized, and HPLC purified before use. The orientation of each fluorophore is fairly rigidly fixed relatively to the dsDNA backbone (Ranjit, Gurunathan, & Levitus, 2009). The number of base pairs between the two dye molecules in a construct controls the distance and relative orientation between them. The total number of base pairs that a construct is made of gives its length. The lengths were, respectively, 60 and 90 bp with Cy3 and Cy5 located near the middle of complementary strands such that the two dyes are separated by, respectively, 20 or 30 bp (Fig. 2). We used a simple, geometrical model for the structure of the dsDNA-probe constructs (Fig. 2). The expected relative angle between the probes is identical in our two samples (Fig. 2), because they differ in probe separation by a single repeating unit of the dsDNA helix.

Note that the base pair sequences of the constructs were randomly generated with ratio AT:GC = 2:3, but a couple of base pairs near the Cy3 and Cy5 dye locations and at the ends of the dsDNA were replaced for more GC pairs to stabilize the fixation of the dyes (Mortensen et al., 2015). The dyes are embedded in the double helix structure without a base-paring partner, so stable base paring of the neighboring nucleotides is important for their stable fixation.

2.2 Dual-Color TIRF Microscopy Setup

The fluorescent probes attached to dsDNA were imaged using a conventional total internal reflection fluorescence (TIRF) microscope, with some modifications (Fig. 3). Specifically, we used an inverted TIRF microscope (Nikon, Eclipse TE2000-PSF) with an oil immersion TIRF objective lens (Nikon, Apo-TIRF 100 ×, NA 1.49) (Fig. 1). We used the built-in Perfect Focus System (PFS) to minimize possible focus drift and/or fluctuations during data collection. We adjusted the focal plane by using the offset lens in the PFS such that the diffraction-limited spots were tightly focused with minimum width. It was critical to firmly fix the focus since change of focus may affect the map between channels significantly. The microscope sample slide was firmly mounted on a sample holder on a *XYZ* motorized stage with a *Z*-axis piezo stage (ASI, PZ-2000). The *XYZ* motorized stage was mainly used to move the sample in the *XY* plane and to roughly adjust the focus by moving the sample along the *Z*-axis, while the *Z*-axis piezo stage was used for the finer adjustment of focus. The microscope system is enclosed in an acrylic box where the temperature is feedback stabilized at 18°C (InVivo Scientific). The stable temperature minimized possible drift caused by temperature fluctuations.

2.2.1 Excitation System

For excitation of all fluorescent probes, we used 561 and 643 nm CW DPSS lasers (Cobolt) whose powers were electronically controlled by acousto-optic tunable filters (AOTFs) (Gooch & Housego). The laser beams were spectrally cleaned by a multicolor excitation filter (Chroma, z488-491/561/640 ×), reflected on a multichannel dichroic mirror (Chroma,

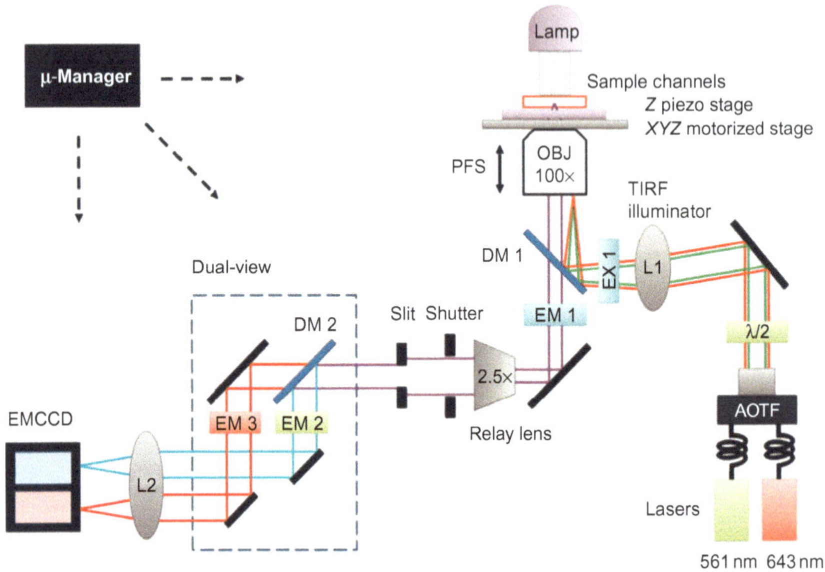

Fig. 3 Experimental setup. We used a commercial TIRF scope (Nikon, Eclipse TE2000-PSF) with an oil immersion TIRF objective lens (Nikon, Apo-TIRF 100×, NA 1.49). The Perfect Focus System (PFS) was used throughout to minimize the focus drift. The sample slide was firmly mounted on stages (*XYZ* motorized and *Z* piezo stage). 561 and 643 nm lasers were coupled and their power was modulated by the acousto-optic tunable filters (AOTFs). The laser beams were focused at the back-focal plane of the objective lens and the angle was controlled by the TIRF illuminator for TIRF excitation of the sample with low background noise. The fluorescence emission was filtered by an emission filter (EM1), expanded by a 2.5× relay lens for increased magnification, and split in the Dual-view (Cairn Research, Optosplit II) where the different emission spectra were split on a dichroic mirror (DM2) and each signal was further filtered by emission filters (EM2 and EM3). Finally, they were imaged on different areas of the EMCCD camera (Andor, iXON plus). Bright light from the lamp (Nikon, DiaLamp) was transmitted through the objective lens, split in the Dual-view, and imaged on the EMCCD camera whose high-contrast bright image was used to correct for emission path drift at the time of data analysis. See main text for details. *Figure adapted from Mortensen, K. I., Sung, J., Flyvbjerg, H., & Spudich, J. A. (2015). Optimized measurements of separations and angles between intra-molecular fluorescent markers.* Nature Communications, *6, 8621. http://doi.org/10.1038/ncomms9621.*

zt488-491/561/640rpc), and transmitted through the objective lens. We used a TIRF-illuminator to easily and accurately control the TIRF angle for low background noise. We also used a half-wave plate to make the excitation beams *p*-polarized, to excite fluorophores having excitation (and emission) dipoles in a broad range of orientations.

2.2.2 Emission System

The probes are separated by ~7 and ~10 nm, respectively, which are just a few percent of the diffraction limit (Fig. 2), and is even within the range of single-molecule FRET (Roy et al., 2008). Therefore, we spectrally separated the light emitted from Cy3 and Cy5 and imaged them on two different areas (Channels 1 and 2, respectively) of an EMCCD camera (Fig. 3). In fact, we did observed FRET between the dyes, especially at ~7 nm separation, when only the 561 nm laser was used for excitation, and hence only the donor (Cy3) was excited.

The emitted fluorescence was collected and collimated by the TIRF objective lens. The collimated light was transmitted through an emission filter (Chroma, zet488-491/561/640m), which allows transmission of light only for wavelengths in the range of 580–620 nm or >660 nm. The beam was directed toward the side-port of the microscope, followed by a C-mount 2.5 × relay lens (Nikon, MQD42120) to achieve a total 250 × magnification, which is sufficiently high to spatially resolve the anisotropic shape of each diffraction-limited Cy3/Cy5 spot. We used a Dual-view beam splitter (Cairn Research, Optosplit II) to spectrally split the Cy3 and Cy5 fluorescence, which makes adjustment of the emission paths easy. The Dual-view has a dichroic mirror (Chroma, 630 dcxr) that reflects the emission from Cy3, while transmitting the emission from Cy5. Those signals then passed through emission filters (Chroma, HQ610/75M and HQ580/60M for Cy3; HQ680/60M for Cy5). The separated paths were recombined and directed onto different areas on the EMCCD camera for imaging (Fig. 3). We carefully selected the filters to minimize any bleed-through of the Cy3 emission tail into the Cy5 channel, as it potentially could affect the measured PSFs from Cy5 dyes and thus compromise the analysis of these.

2.2.3 EMCCD Camera

To detect the dim light from single fluorophores, we used a highly sensitive EMCCD camera (Andor, iXon plus), which has a 1004 × 1002 pixel array consisting of 8-μm square pixels. The 250 × magnification of the microscope resulted in an effective pixel size of 8 μm/250 = 32 nm. We used a calibration bar with 2 μm-separated stripes, to confirm this. We used a gain setup (Pre-Gain: ×1, EM-Gain: ×400) that allowed sufficient amplification of the dim light signals from the single fluorophores.

2.2.4 *μ-Manager*

We used μ-manager (version 1.4.16) as a main program to control most of the components in the setup, including microscope, motorized and piezo stages, PFS, EMCCD camera, lasers, AOTF, and shutters. The μ-manager is a convenient tool to collect data in an easy and a consistent manner. Automated data collection was useful for high-throughput sampling.

2.2.5 *Poly-L-Lysine-Coated Coverslips*

The dsDNA molecules were immobilized nonspecifically and electrostatically to poly-L-Lysine (PLL)-coated microscope coverslips. The latter were prepared from glass coverslips (Fisher Scientific, #1, 22 × 22 mm^2) that were rinsed with Milli-Q machine-filtered water, dried with filtered nitrogen gas, and incubated in a UV ozone cleaner (Jelight, UVO Cleaner 42) for 30 min, which removed organic fluorescent junk from the surface. The cleaned coverslips were incubated in 0.01% PLL solution (EMS, 19320-A) for 5 min and dried overnight on coverslip staining outfits (Thomas Scientific, 8542E40). The condition of the dried PLL coverslip usually remained good for a few weeks. We carefully tested the surface cleanliness by imaging background fluorescence without dyes.

2.2.6 *Sample Flow Cell Chamber*

Three flow cell chambers, respectively, for the dsDNA (20 and 30 bp) and the 100-nm TetraSpeck fluorescent beads (Invitrogen, T7279) used as fiducial markers to map between channels, were made using a PLL-coated coverslip, a microscope slide (Gold Seal Micro Slides), and double-sided tape (Scotch, 3 M). First, 10 μL of 10 p*M* dsDNA 20 bp/30 bp were flowed into the first/third sample chambers, and 10 μL of 10 p*M* TetraSpeck beads was flowed into the second sample chamber. Second, the samples were incubated for 2 min and washed with 50 μL of the buffer to eliminate any weakly attached molecules or beads. Finally, 20 μL of imaging buffer was flowed through the chambers, after which they were sealed with vacuum grease to prevent drying. See Mortensen et al. (2015), for detailed information about buffers.

2.3 Calibration of the EMCCD Camera

Calibration of the EMCCD camera is necessary in order to convert its output signal to its equivalent number of photons. This is required to (i) calculate the precision with which one may localize a probe and (ii) to specify the full EMCCD output distribution when using maximum-likelihood estimation (MLE) (Mortensen et al., 2010). We first determined the constant offset in the camera's output signal from a dark image obtained with the shutter closed and found it to be 300. Next, we used two areas of constant photonic background from the two channels, respectively, recorded during dsDNA data recording. For each area, the distributions of pixel outputs were used to calibrate the camera. We used the full EMCCD output signal distribution (Mortensen et al., 2010; Ulbrich & Isacoff, 2007), which was fitted to the output pixel values, to determine the remaining values of the EMCCD parameters, namely, the gain, the expected input photon number, and the width of the Gaussian readout-noise distribution (Mortensen et al., 2015). We note that in cases where dedicated calibration measurements have not been performed, and areas of constant photonic background do not exist, a new method for *post festum*-calibration of an EMCCD camera from its images recently became available (Mortensen & Flyvbjerg, 2016).

The electron multiplication process in the EMCCD camera effectively results in variances that are inflated by an exact factor of two compared to pure shot noise. This is often referred to as excess noise (Hynecek & Nishiwaki, 2003; Mortensen & Flyvbjerg, 2016; Mortensen et al., 2010; Robbins & Hadwen, 2003; Ulbrich & Isacoff, 2007).

2.4 Considerations on Sample Stage Drift

In all our experiments, we are interested only in differences between positions measured in each of the two channels of the experiment. Such differences are unaffected by translational sample stage drift, as positions in both channels change by the same amount. Here, we assume that the drift is sufficiently slow to have only a negligible effect during each frame, such that motion blur (Vestergaard, Blainey, & Flyvbjerg, 2014; Vestergaard, Pedersen, Mortensen, & Flyvbjerg, 2015) may be neglected and therefore that each PSF is unaffected by the drift. We verify this assumption in the analysis of fluorescent spots. Later, we consider also possible drift in the optical components in the emission path, which does have the potential to change measured distances between probes in the two channels.

2.5 Calibration of the Map Between Color-Separated Channels

In order to calculate distances between positions in two different images of the same field of view, one must establish a map between the two images (Churchman et al., 2005; Pertsinidis et al., 2010). This map must convert coordinates between the images (color channels) with an accuracy and a precision that both add little error to the distances we wish to resolve (Fig. 2). In our case, we need a map with resolution in the single-nanometer range.

To establish this map between channels, one needs fiducial markers, i.e., isolated objects that image in both channels. As fiducial markers we used 100-nm TetraSpeck fluorescent beads, which are coated with four different dyes. They were diluted in buffer (Mortensen et al., 2015) for a 10 p*M* final concentration, vortexed, and sonicated for 1 min before they were used. They were nonspecifically attached to the PLL-coated coverslip surface at random positions. Only the orange and dark-red dyes of these versatile beads were excited by the lasers in our setup. Those two dyes imaged into the two different channels in the emission system (Fig. 3).

Ideally, a map between channels would simply be a constant vector, independent of the position in the field of view. In practice, however, such maps vary with position in field of view. Slow variation with position is expected due to chromatic aberrations. It might be modeled phenomenologically with a "global" map, a mathematical function covering the whole field of view. We did not do that because on top of the slow dependence on positions, we found a fast dependence with lower amplitude, a speckle pattern of aberrations that cannot be modeled with a global map. Instead we must model local maps phenomenologically. To this end we used a dense array of fiducial markers. This array may be obtained by translating single fiducial markers in the field of view (Churchman et al., 2005; Pertsinidis et al., 2010), or simpler, as here, using many fiducial markers imaged simultaneously at random positions. We used 93 time-lapse movies of fluorescent beads, each movie consisting of 90 frames.

2.5.1 Localization of Fiducial Marker Positions

The approximate positions of beads in both channels were determined automatically as local maxima above a threshold and then localized with nanometer precision using unweighted least-squares method with a 2D-Gaussian PSF plus a constant "background" PSF, which results in the so-called Gaussian Mask Estimator (GME) (Mortensen et al., 2010; Thompson et al., 2002) (Fig. 4).

2.5.2 Selection of Fiducial Markers for Analysis

To avoid analysis of beads that had been imaged too closely, we rejected any bead with center within 24 pixels of any other bead's center in the same image, checking both images. This conservative acceptance criterion qualified 7324 spots for further analysis. To eliminate the effect of translational drift of the sample stage, discussed earlier, we considered the vector

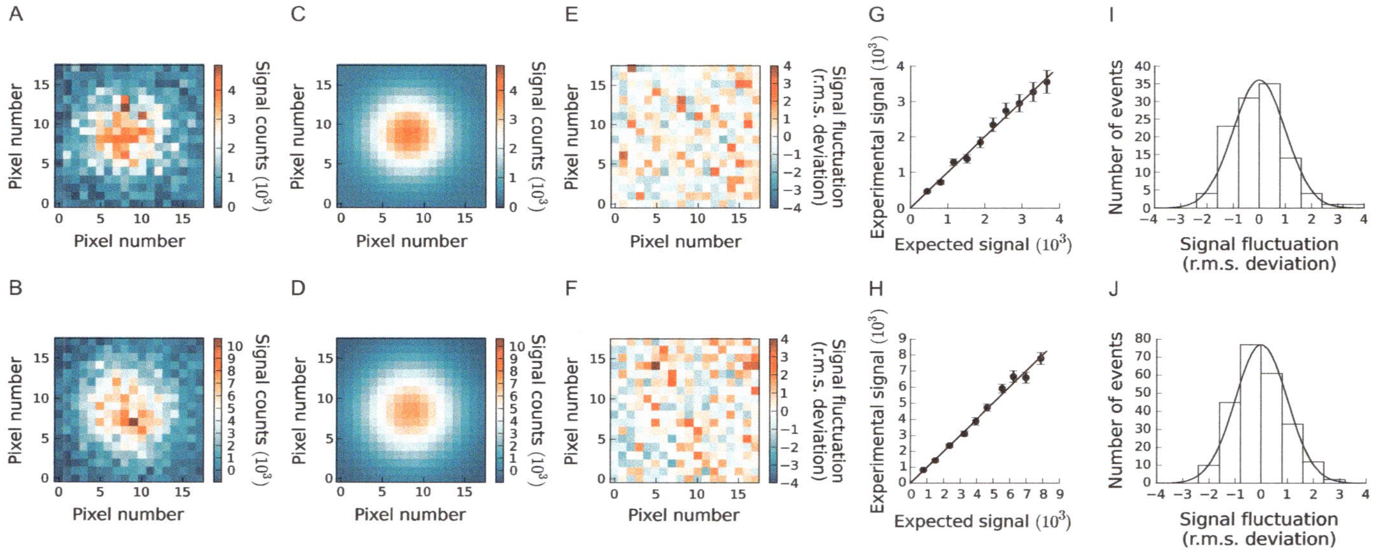

Fig. 4 Localization of fluorescent beads used as fiducial markers in order to map between the two channels in the experiment. (A and B) TIRF microscopy images of, respectively, orange (A) and dark red (B) emissions of the same multicolored fluorescent bead. (C and D) Theoretical images with parameter values obtained speedily with GME, i.e., least squares fitting with equal statistical weights for all pixel-signals in the fit, in conjunction with a 2D Gaussian-plus-a-constant PSF (Mortensen et al., 2010; Thompson et al., 2002). (E and F) Rescaled residuals. Each pixel shows the difference between measured and expected pixel signal values in units of its theoretical r.m.s. deviation as given by the *color bar*. The theoretical r.m.s. deviation was obtained using Eq. 6 in (Mortensen et al., 2010) with the parameters fitted in (C) and (D). (G and H) Measured signal values compared to expected values. We binned the expected signals and their associated pixels with experimental values according to the expected values. For each bin, the mean experimental signal is plotted against the expected signal with an *error bar* indicating the theoretical s.e.m. Data points shown should scatter about the straight line through the origin with unit slope, each in a Gaussian manner with s.d. equal to the shown *error bar*. (I and J) Histograms of rescaled residuals. For sufficiently large expected pixel signals, theory predicts a normal distribution with zero mean and unit s.d. (*solid line*) for the fluctuations in (E) as well as in (**F**).

difference between each bead's locations in the two images. These vector separations are unknown a priori and fluctuate through a movie, so we calculated their time-averages and standard deviations. To eliminate loosely fixed beads and artifacts from the automatic detection scheme, we conservatively accepted a bead for further analysis only if the components of its vector separation between channels fluctuate with normal distributions around their mean values and that with s.d. given theoretically by the number of photons in the diffraction limited images of the bead (Mortensen et al., 2010) ($p > 0.015$, Pearson's chi-squared test for distributions) (Figs. 1 and 5).

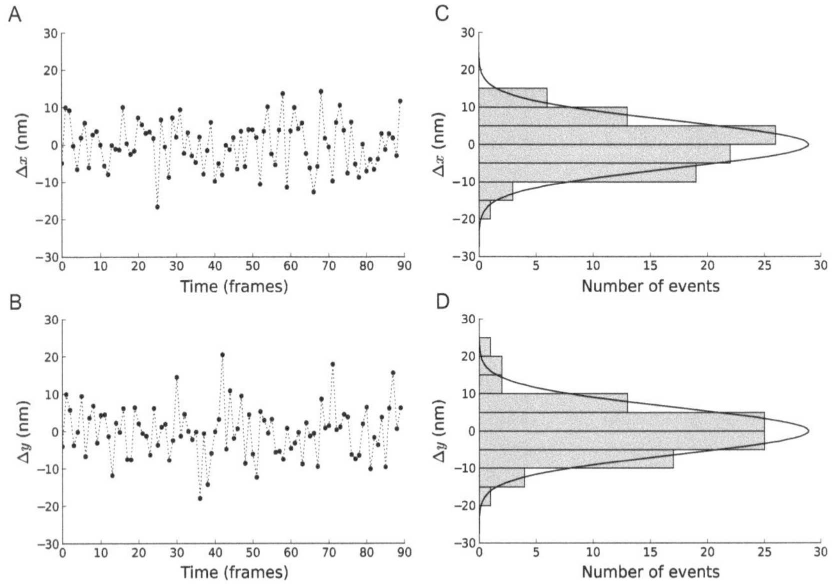

Fig. 5 Time series of fluctuations in the vector distance in the image plane between two spectrally separated images of the same fluorescent bead. (A) Using GME, we localized the center of the image of a single fluorescent bead in both channels for each frame in a time-lapse movie consisting of 90 frames. For each frame, we computed the separation between such images along the *x*-axis and subtracted its time-averaged value. The resulting residuals are shown here (*connected black dots*). (B) Same as (A) for the separation along the *y*-axis. (C) Histogram of the residuals from (A) overlaid with a normal distribution centered on the origin and with s.d. equal to the theoretical s.d. for the estimator of separations (*solid line*). The distribution of residuals of measured separations is consistent with the theoretical expectation, the Gaussian distribution shown, with $p = 0.30$. This *p*-value was found using Pearson's chi-squared test for the distribution. (D) Same as (C) for the residuals from (B). Here, we found consistency with $p = 0.73$. We used a bead as fiducial marker only if the fluctuations in the vector separation between its dark red and orange images were consistent with photon shot noise with $p > 0.015$ for both coordinates.

This cutoff was chosen to maximize the number of beads included in the analysis, while avoiding inclusion of any obviously spurious beads. The latter group stood out, because they contributed with a sub-cutoff peak in the otherwise uniform p-value distribution. We also disregarded beads with estimated mean widths that differed from the population average by more than three standard deviations. This eliminated beads that had imaged with overlapping PSFs. These filters left us with 5622 beads, each labeled with multiple fluorophores of multiple colors, qualified as fiducial markers on which a map can be based.

2.5.3 Precision of Distances Between Color-Separated Locations of Fiducial Markers

Slow changes (drift) of any origin in the separations between color-separated images of the same beads would reduce the precision with which these distances can be determined and thus the quality of our map. To check for such drift, we calculated the power spectra of each time series of vector separations. In the absence of drift and other sources of autocorrelations in time, the time series of vector separations between color-separated images of the same bead is a constant vector plus an independent white noise on each coordinate, due to localization errors caused by photon shot noise. Consequently, the expected value for the power spectra of the two coordinates of the vector separations is independent of frequency for nonzero frequencies, it is constant and equal to the variance of the estimated separations. By virtue of the central limit theorem, the experimental power spectral values consequently are exponentially distributed with expected value the variance of the estimated separations (Fig. 6). We found that the power spectra were consistent with the hypothesis of negligible autocovariance, because the p-values (Pearson's chi-squared test for distributions) for this hypothesis found across the population of beads were approximately uniformly distributed in the interval $[0,1]$.

Also, we found no discernable correlations between the equal-time values of the two coordinates of the vector separation, because the p-values (Pearson's test for correlation) encountered in the population of beads were uniformly distributed in $[0,1]$. In this case, measurements are statistically independent, and one may calculate the error on the time-averaged vector separations as usual as s.e.m. = s.d./$\sqrt{N}$, where $N = 90$ is the number of frames in each time series. Typically, we found an s.d. of $\sim$6 nm, hence an s.e.m. of $\sim$0.6 nm (Fig. 6). Therefore, though GME is less precise than other methods (Mortensen et al., 2010), it is sufficiently precise for our

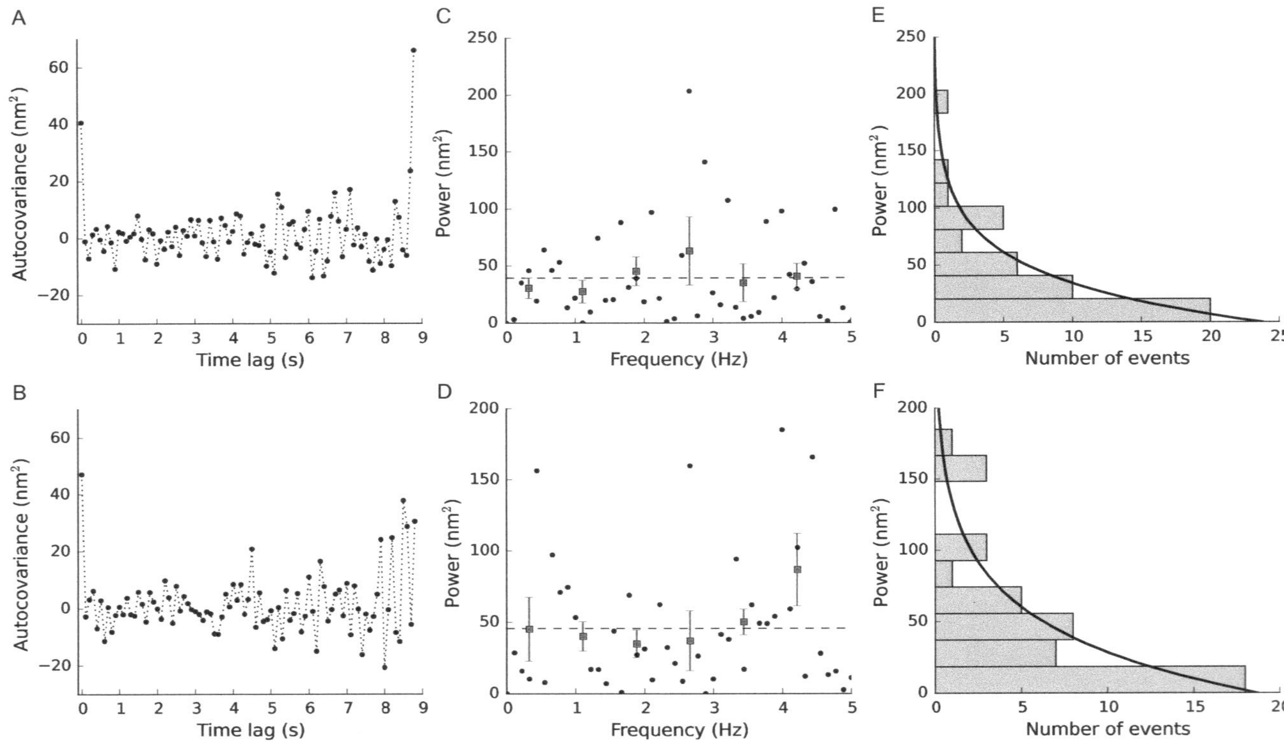

Fig. 6 See legend on opposite page.

purpose here and attractive because of its speed: we can establish a large number of fiducial markers with 6 Å precision with acceptable speed.

2.6 Properties of the Map Between Color-Separated Channels

The resulting map of fiducials between channels reveals smooth spatial variations across the imaging region (Fig. 7), but these smooth variations are superposed with a speckled pattern of variations with characteristic length-scale near the diffraction limit (Mortensen et al., 2015, supplementary fig. 7). This speckle pattern could be due to small geometrical imperfections in the optics; the mathematical shapes of lens designs and the actual physical shapes of lenses must differ on some length scale; and good economical engineering lets them differ on the largest length scale still irrelevant for the intended use of the optics. We use the optics beyond its intended use, when we use it for super-resolution, and hence we must encounter these controlled geometrical aberrations in our super-resolving map between color channels, i.e., as chromatic aberrations. For this reason, we had to abandon the idea of calibrating a single, global map valid at every position in the field of view (Mortensen et al., 2015).

Instead, to minimize errors due to these speckle effects in our map between channels, we mapped each coordinate locally. A given point in the field of view was mapped between channels, by linear least-squares fitting of two second-degree polynomials in the coordinates x and y—

Fig. 6 Assessment of possible fluctuations in the vector separation between a bead's images in the two channels. (A) From the residuals of separations between a bead's orange and dark red images along the x-axis, we calculated the autocovariance (*connected black dots*). Its value at zero lag is the variance of the residuals. (B) Same as (A), but for the separations along the y-axis. (C) The power spectral values of the residuals in (A) (*black dots*). Uncorrelated noise gives rise to a uniform (white) power spectrum with value equal to the variance of the signal (*dashed line*). Block averages of the power spectral values (*square points*) are consistent with such white noise. (D) Same as (C) for the input data from (B). (E) Histogram of power spectral values from (C). These values are independent and identically exponentially distributed with mean value equal to the variance of residues (*solid line*), if they originate in white noise with finite variance. The measured distribution is consistent with this theoretical expectation with $p=0.90$. This p-value was found using Pearson's chi-squared test for the distribution using 10 bins of equal integrated theoretical probability. Consequently, due to the absence of temporal correlations, the time-averaged separation has s.e.m. = s.d./$\sqrt{N}$. Here, $N=90$ is the number of frames in the movie and s.d. = 6.3 nm, so s.e.m. = 0.7 nm. Thus, our estimates for the vector difference between the two images of a bead have subnanometer stochastic errors. (F) Same as (E) for the power spectral values in (D). Here, we found $p=0.99$ and s.e.m. = 0.7 nm.

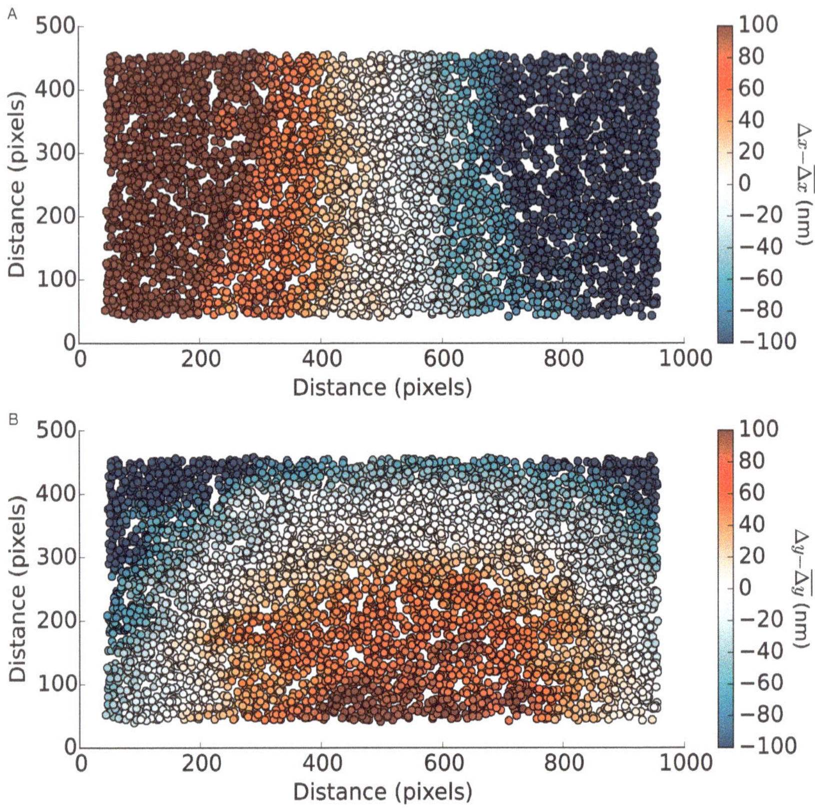

Fig. 7 Map of fiducials between channels. (A) We estimated the time-averaged positions of 5622 beads in Channel 1 as well as in Channel 2. For each bead, its average position in the field of view is plotted (*colored points*), and the *point's color* indicates the extent to which the difference in *x*-coordinate between the point's coordinates in the two channels differs from the population average of the same quantity. The effective pixel size was 32 nm. (B) Same as (A) for the difference in *y*-coordinates. Panels (A) and (B) show that the map between channels measured experimentally with fiducial beads is not a constant vector translation when measured with the precision we need. On top of the constant vector translation we see smooth spatial variations here, as expected from optics. But if this smooth variation is modeled and subtracted from the difference in *x*-coordinate between each point's coordinates in the two channel, a speckle pattern of vector-differences in coordinates remains with a characteristic length scale comparable to the diffraction limit (Mortensen et al., 2015, supplementary fig. 7). This pattern makes a global map impossible to calibrate. We calibrated the map locally instead. *Figure adapted from Mortensen, K. I., Sung, J., Flyvbjerg, H., & Spudich, J. A. (2015). Optimized measurements of separations and angles between intra-molecular fluorescent markers.* Nature Communications, 6, *8621. http://doi.org/10.1038/ncomms9621.*

one polynomial for each coordinate being mapped—to the mappings of all beads in a circular area centered on the point and with radius equal to 22 pixel widths (Fig. 8). Typically, such an area contained ~25 beads, which overdetermines each second-degree polynomial in the two coordinates with its six degrees of freedom. This excess of fiducial beads was necessary, because the physical centers of the two relevant dyes are offset relatively to each other in the individual bead (Fig. 8): the map determined by two such fitted polynomials does not map the fiducial beads it is based on, onto their actual images. This agrees with unpublished observations of other groups (personal communication). The vector difference between a bead's mapped coordinates and its actual coordinates are essentially random: this error shows no spatial correlations. This is presumably because the two relevant distributions of dyes have different locations within any given bead. Combined with the random orientation of the beads on the surface, this explains our observed random mapping errors. These errors add to the shot noise errors on bead positions to an error with s.d. ~5 nm (Fig. 8). By using an excess of beads in a map, their randomly oriented offsets of color centers cause errors with random signs, i.e., they almost cancel each other. We propagated these errors to calculate the error of the mapped position and found it to be ~2 nm (Fig. 8). This mapping error should be included each time one assesses distances between positions measured in each of the two channels. We do that everywhere later. The accuracy of the map is verified implicitly by consistency in the measured distances between the probes on the dsDNA. Had the beads been imaged significantly out of focus, one would expect the map between channels to change with the focus due to aberrations. Here, however, all variation in distance measurements below is accounted for by shot noise, which leaves no room for systematic errors.

2.7 Correction for Drift in the Optical Components of the Emission Path

During experiments, we occasionally observed drift in both imaging windows with respect to the EMCCD pixel array (Fig. 9). This drift was likely due to drift in the macroscopic springs that control the mirrors in the Dual-view and thus the alignment of each channel onto the EMCCD camera. For the objective of measuring distances between probes based on position measurements in two different emission channels, slow drift in the optical components that determine where those channels image on the camera

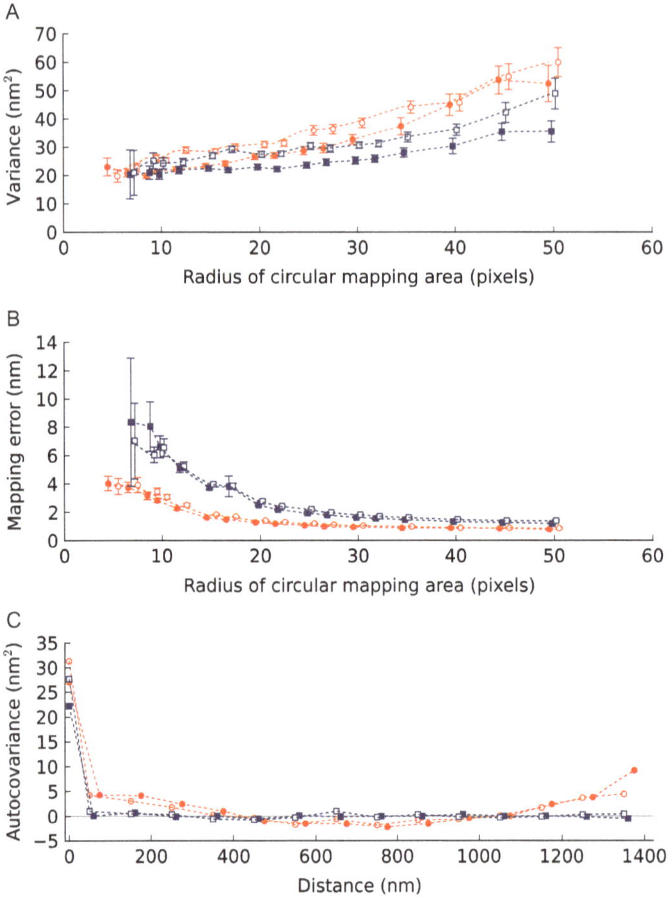

Fig. 8 Properties of the local map calibration using fluorescent beads as fiducial markers. (A) Ability of locally calibrated maps to capture the spatial variation in the global map. We used a locally determined low-degree polynomial to map each coordinate of a position of interest (POI) to its corresponding position in the other channel. We used a circular region around the POI and calibrated the map using all accepted beads in this area as fiducials. The polynomial coefficients of the local map were determined with a linear least-squares fit. We divided the imaging region (Fig. 7) into mapping areas of different sizes. For each size of mapping area, we determined, respectively, a first-degree (*circles*) and second-degree (*squares*) polynomial in the *x*- (*filled symbols*) and *y*-coordinate (*open symbols*). We calculated the variance of the residuals between the mapped and the measured positions of the beads in the target channel, where we accounted for the fit of the map and the number of beads included in the fit. Because the number of beads in an area exceeds the degrees of freedom in the polynomial, the quality of a fitted map is quantified by this variance, and this value vanishes only when all bead coordinates from one channel are mapped perfectly onto their measured coordinates in the other channel. Smaller mapping areas lead to lower variance, because the

(Pertsinidis et al., 2010) would be detrimental for the accuracy of our distance measurements, because our map between channels would be rendered unreliable, as it was recorded at a time different from the time it was used.

To detect and correct for this, we first measured the imaging positions of the emission paths by recording bright light from a lamp (Nikon, DiaLamp) passing through the imaging windows of the two channels (Fig. 3). This amplified the contrast to the background at the windows' edges (Fig. 9). We measured the positions of these edges by fitting a theoretical light intensity distribution for diffraction by a rectangular window to the measured distributions (Fig. 9). The model is a rectangular aperture with diffraction at its edges. This diffraction causes a smooth transition in intensity from bright light inside each rectangle to dark background outside it. This intensity profile we modeled with an error function, which results from using a Gaussian

---

polynomial can better approximate the actual spatial variation in the map in a small area than in a larger area. However, below radii of ~20 pixels for the second-degree polynomial, the residual variance is constant, indicating a contribution to the variance, which is independent of the area size, and most likely is due to the two dyes being distributed differently inside our fiducial beads. Notice that the second-degree polynomial consistently outperforms the first-degree polynomial for radii larger than ~10 pixels. The effective pixel size was 32 nm. (B) Mapping error for the POI. For each map in (A), we calculated the mapping error for the POI by propagating the variances for each bead as estimated in (A). To this end, we assumed that the residuals of the beads in a map are spatially uncorrelated. This is the case only when a map captures essentially all spatial variation, ensured by using a sufficiently small mapping area (the plateau in (A)). On the other hand, the mapping error decreases with the number of beads included in the map. Therefore, the optimal mapping area balances these two effects. We used a radius of 22 pixels with the second-degree polynomial, which amounts to a mapping error of ~2 nm. We proceed to show that this area is sufficiently small to ensure uncorrelated residuals. (C) Autocovariance in space of the mapping residuals for a circular mapping area of radius of 22 pixels. To validate that the map captures essentially all spatial variation, we calculated the autocovariance function of the residuals by binning beads by their Euclidean distance to each other. We bias-corrected the autocovariance to take into account the polynomial fitting and the number of beads contributing to a map. Then we averaged the autocovariance over all maps. The autocovariance of the first-degree polynomial (*connected circles*) shows positive correlation for short distances, which demonstrates that the first-degree polynomial is incapable of capturing all spatial variation. The autocovariance of the second-degree polynomial (*connected squares*) is essentially equal to a delta function, hence shows no spatial correlations. This demonstrates the validity of the calculated mapping error in (B) for this size of mapping area using a second-degree polynomial. *Figure adapted from Mortensen, K. I., Sung, J., Flyvbjerg, H., & Spudich, J. A. (2015). Optimized measurements of separations and angles between intra-molecular fluorescent markers.* Nature Communications, 6, *8621. http://doi.org/10.1038/ncomms9621.*

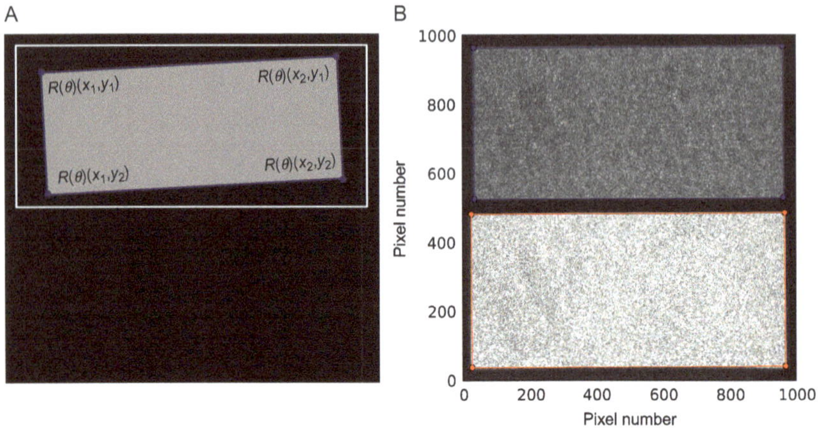

Fig. 9 Detection and correction of drift in the optical components in the emission path of the microscope. To quantify such drift with nanometer precision, we shone bright light through the microscope. This revealed the imaging windows clearly for both channels. (A) We measured the position and orientation of each of these imaging windows by fitting its light distribution with a *rectangular area* in which diffraction at the edges is accounted for. To this end, we fitted the tilted *rectangle's position coordinates* of its edges (x_1, y_1, x_2, y_2) and its rotation (θ) with respect to the camera pixel array (*white rectangle*), along with the intensity inside and outside the imaging window. The standard rotation matrix in two dimensions $R(\theta)$ rotates the rectangle into position. Here, rotation with respect to the *white reference rectangle* is exaggerated for clarity. (B) Fits of the imaging windows for the two channels (*upper and lower rectangles*) for a single image are shown. We repeated this for 10 consecutive images. Then we compared the fitted parameters for such measurements done immediately after the collection of dsDNA data and the map calibration. Typically, we found no significant change in the rotation of the imaging windows (Student's *t*-test), between these time points. The edge coordinate differences between two experiments were computed for each of the four position coordinates. In all cases, the change in distances for the *right and left edges*, respectively, *top and bottom edges*, were consistent with each other (Student's *t*-test), which demonstrates that the entire window moves without changing its shape. Consequently, we averaged the distances moved by the *right and left edges*, respectively, *top and bottom edges*, to quantify the drift of the imaging windows along each of the two axes. Typically, such distances were estimated with an s.e.m. of ~0.5 nm. We corrected for drift, if the estimated edge distances exceeded twice their s.e.m. In this way, we detected no significant drift in the emission path between the time points of 20 bp dsDNA data collection and the mapping calibration. Between the time points of 30 bp dsDNA data collection and the mapping calibration, however, we detected drift magnitudes of 1.9 ± 0.6 and 1.6 ± 0.6 nm, respectively, for the *x*- and *y*-coordinates of Channel 1 and 1.3 ± 0.3 and 1.3 ± 0.3 nm, respectively, for the *x*- and *y*-coordinates of Channel 2. *Figure adapted from Mortensen, K. I., Sung, J., Flyvbjerg, H., & Spudich, J. A. (2015). Optimized measurements of separations and angles between intra-molecular fluorescent markers.* Nature Communications, 6, *8621. http://doi.org/10.1038/ncomms9621.*

approximation for the point-spread function. We parameterized this model rectangle with a common constant background, total intensity, rotation, width, and four location parameters (Fig. 9). Typically, the shape and orientation of the rectangle did not change between the time of measurements of dsDNA data and corresponding calibrations of the map, but occasionally we detected translational drift. If this drift was significant (Fig. 9), we corrected for it. In practice, the position of a dye in one channel was translated to account for the drift experienced by its imaging window w.r.t. the position of that imaging window at the time of the map calibration measurements. Then the corrected dye position was mapped onto the other channel. The position of the other dye was simply translated to account for drift in that channel. This procedure determined edges and allowed correction for drift in the emission path, both with nanometer precision (Fig. 9).

2.8 Analysis of Images of Single, Fixed Fluorophores

A fluorophore emits photons via a dipole transition, and if the dipole is fixed in space (Fig. 2), the diffraction-limited image of the fluorophore (Fig. 10) depends radically on the polar and azimuthal angles of that emission dipole's orientation in space and on the objective's distance from focus (Aguet et al., 2009; Böhmer & Enderlein, 2003; Enderlein et al., 2006; Mortensen et al., 2010; Reifenberger et al., 2009; Toprak et al., 2006). This dependence complicates traditional analysis of data (Enderlein et al., 2006), because the resulting spot no longer is rotationally symmetric. However, the asymmetry of the image may be turned into an advantage, because it allows extraction of the orientation of the fluorophore simultaneously with its position in the plane. To achieve this, one needs an analytical expression for the PSF as a function of its position and orientation parameters. With that, values for these parameters can be extracted directly from an image of a fluorophore using MLE. MLE is known to be the optimal statistical procedure, when the number of photons in an imaged PSF is large. When this is the case, MLE produces estimates that are accurate (i.e., unbiased) and scatter around this accurate value with the maximum precision possible for an unbiased estimator, according to information theory (Fig. 1) (Rao, 1973). Thus, any other estimator can do only worse.

2.8.1 Theoretical PSF

With this strategy in mind, we modeled the fluorophore emission as a dipole transition to obtain an analytical expression for the theoretical PSF.

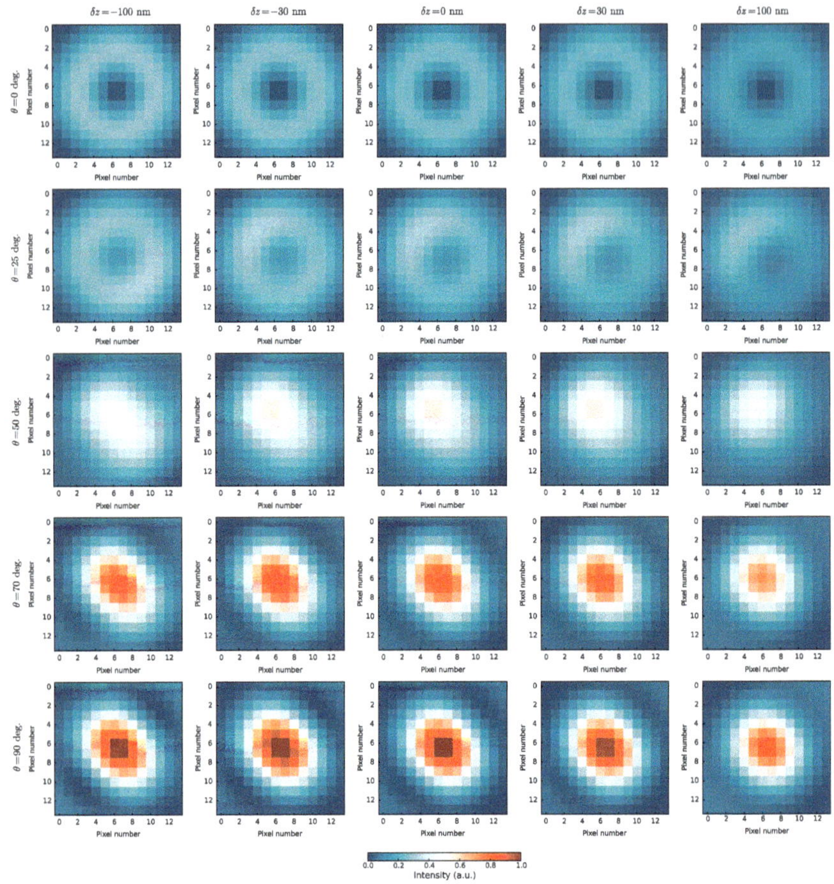

Fig. 10 Point spread functions for fluorophores with fixed orientations imaged near the focus. The *top left panel* shows the theoretical intensity distribution from an image of a fluorophore that is fixed in orientation during imaging with azimuthal and polar angles of 225 and 0 degrees, respectively, and imaged at a distance of −100 nm from the focal plane. The fluorophore's position coincides with the center of the pixel array. We used an emission wavelength of 570 and 32 nm square pixels throughout, while the remaining parameters are set identical to the ones used in the analysis of the experimental dsDNA data. Other panels show the intensity distribution for fluorophores oriented at different polar angles (*rows*) and imaged at different focal depths (*columns*). The intensity scale is identical in all panels and given by the *color bar* at the *bottom*. *Figure adapted from Mortensen, K. I., Sung, J., Flyvbjerg, H., & Spudich, J. A. (2015). Optimized measurements of separations and angles between intra-molecular fluorescent markers.* Nature Communications, 6, *8621. http://doi.org/10.1038/ncomms9621.*

A complete vector description of the emission was then applied to the refraction at the relevant interfaces in the microscope and diffraction by the objective. Assuming perfect imaging—i.e., loss-free light propagation and negligible optical aberrations—we obtained the theoretical PSF (Mortensen et al., 2010). The PSF's dependence on focus was included in its analytical description by taking into account the appropriate phase changes (Böhmer & Enderlein, 2003) in the electric and magnetic field components in its calculation. For each probe, we used the average emission wavelength incident on the camera after the various emission filters and dichroics in the emission path were accounted for (Fig. 3). We used 580 and 670 nm for Cy3 and Cy5, respectively. Note that the polarization of the excitation beam only affects the brightness but not the shape of the PSF, which is solely determined by the orientation of its emission dipole.

2.8.2 Maximum-Likelihood Estimation

We here use the theoretical PSF with MLE to extract simultaneously the position and the orientation of a molecule, directly from its nearly focused image, using a slightly modified version of the procedure described previously (Mortensen et al., 2010).

In that process, we also determine the total number of source photons emitted by the probe, a constant background level, and now additionally the objective's distance from its focus (Fig. 10). Although the latter does not directly carry information about the molecule, its determination is paramount for accurate estimates of the molecule's position and orientation (Mortensen et al., 2015). Using this methodology, we found that the effective focal depth in the two channels maybe differed slightly; specifically, we found that the Cy3 and the Cy5 fluorophores, respectively, yielded defocuses of -30 ± 40 nm and -70 ± 25 nm (mean $\pm$ s.d.). It is conceivable that these effective focal depths could be corrected for by using appropriate lenses positioned in the emission channels, although we have not verified this experimentally.

2.8.3 Critical Speedup of the Method

We used an analytical approximation to the PSF of a fixed fluorophore. This approximation was sufficiently accurate and precise for the purpose of estimating location and orientation (Mortensen et al., 2015). For each wavelength and for each value of the defocus in increments of 1 nm in a range of ± 250 nm around the focus, we obtained an accurate analytical approximation to the full PSF by proceeding as previously (Mortensen et al., 2010). This tabulation of this analytical approximation allowed us to fit the value of the instantaneous defocus along with the other parameters in each fit.

The MLE was done with available software (Mortensen et al., 2015), which also yielded the theoretical covariance matrix for the estimates. The present method for simultaneous estimation of location and orientation of an isolated dipole we refer to as diPOLE for brevity. Thus diPOLE uses MLE based on the true theoretical PSF of an isolated fixed dipole that is more or less in focus, but with the theoretical PSF well approximated analytically to achieve a dramatic and necessary speedup of computation with essentially no cost in the form of degraded accuracy or precision.

2.9 Two-Color Localization Microscopy with Fixed Fluorophores

The fluorescent light from each dsDNA molecule was spectrally separated.

Each channel's image (Fig. 11) was fitted individually using diPOLE. This yielded the fluorophores' locations in the sample plane and their spatial orientations.

2.9.1 Verification of the Theoretical PSF

As discussed earlier, MLE is optimal when used in conjunction with an accurate theoretical PSF. As often is the case with the sharpest tools, high efficiency comes at a high risk. To minimize this risk, a crucial step in the analysis verifies that the applied theoretical model for the PSF (in its accurate analytical approximation) is in accurate agreement with the experimentally measured intensity distributions of fixed fluorophores. We compared the distributions in a number of ways (Fig. 11). First, we directly compared measured PSFs pixel by pixel to each pixel's expected value (Fig. 11). Discrepancies between measured and expected pixel values were rescaled with their theoretical r.m.s. deviation as calculated from the theoretical covariance matrix and assuming photon shot noise to be the only cause of discrepancy. Second, we binned pixels with similar expected values for the light intensity they register. We calculated the mean value for the expected signals of all pixels in a given bin. Against this value, we plotted the mean value and s.e.m. of the actually recorded values for all pixels in the same bin (Fig. 11). Repeating this for the full range of expected pixel values, we expect the resulting points to fall on a straight line through the origin with unit slope for accurate theoretical descriptions of the diffraction limited

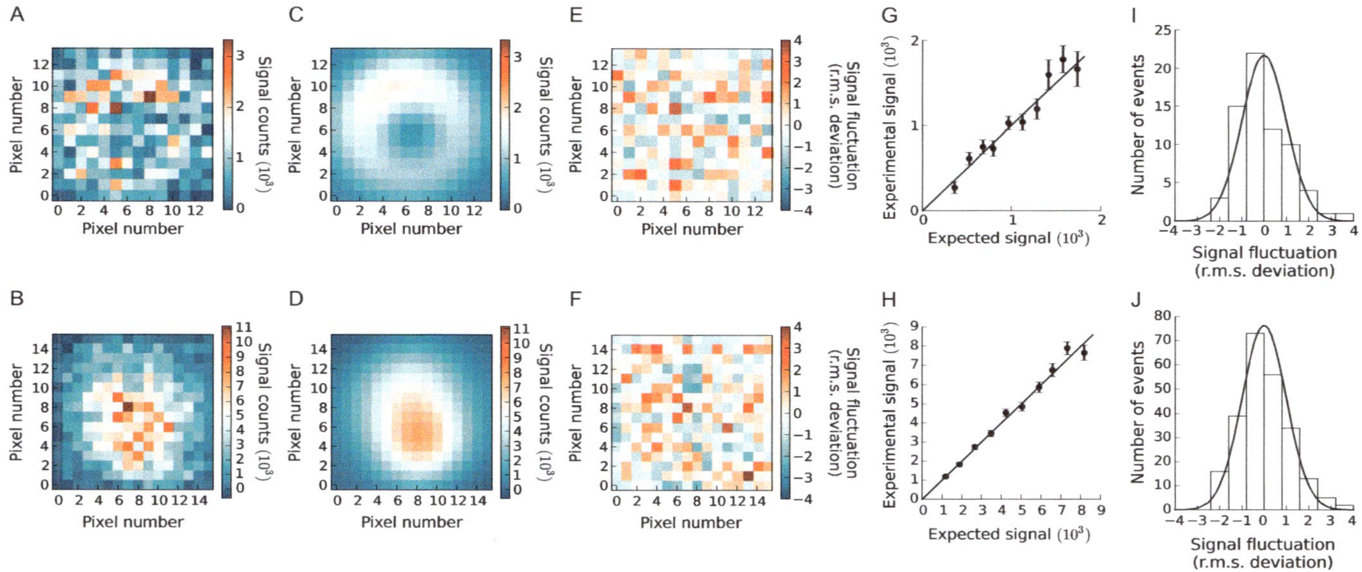

Fig. 11 Point spread functions for Cy3 and Cy5 fluorophores internally labeling complementary strands on the same dsDNA molecule (Fig. 2) at 20 bp separation. "Signal counts" denotes the raw output signal from the EMCCD camera's pixels, not to be confused with photon counts. (A and B) TIRF microscopy images of Cy3 and Cy5 fluorophores, respectively. (C and D) Theoretical images with parameter values obtained by applying diPOLE to fixed probe images in (A) and (B), respectively, and using the dyes' average emission wavelengths of 580 and 670 nm. (E and F) Rescaled residuals. Each pixel shows the difference between measured and expected pixel signal values in units of its theoretical r.m.s. deviation as given by the *color bar*. (G and H) Measured signal values compared to expected values. We binned the expected signals and associated pixels with a bin if the expected signal fell in it. For each bin, the mean experimental signal is plotted against the expected signal with an *error bar* indicating the theoretical s.e.m. Data points shown should scatter about the straight line through the origin with unit slope, each in a Gaussian manner with s.d. equal to the shown error bar. (I and J) Histograms of rescaled residuals. For sufficiently large expected pixel signals, theory predicts the standard normal distribution (*solid line*) for the fluctuations in (E) and (F), respectively.

patterns. This is clearly our case here. Finally, we assessed the distribution of the rescaled signal fluctuations (Fig. 11). For sufficiently large signals (often satisfied everywhere because of background fluorescence), such rescaled fluctuations should scatter as a standard normal distribution, for an accurate theoretical PSF (Fig. 11). The experimental intensity distributions (Fig. 11) and the corresponding theoretical distributions (Fig. 11) are in near-perfect agreement, showing that the theoretical PSF in its analytical approximation is appropriate for the present experimental practice.

2.9.2 Possible Slack in Fluorophore Fixation

Recently, dipoles with not entirely fixed orientations have been studied for the effects on localization of their slack in orientation (Stallinga, 2015; Stallinga & Rieger, 2012). Here, however, we demonstrated near-perfect agreement between the theoretical PSF with fully fixed dipole orientation and the measured intensity distributions (Fig. 11). While this agreement was compelling, it alone did not guarantee that accurate estimates resulted from applying that PSF to experimental data. We therefore demonstrated (see below) that it did, using two differently colored fluorophores fixed within the well-known structure of dsDNA. Thus, the PSF described earlier models reality sufficiently well to guarantee robust estimates of orientations and positions of fixed fluorophores.

3. RESULTS

3.1 Measurements of Intramolecular Distances

When using a sparse distribution of molecules, selection of single molecules labeled with single dyes that had not already been photobleached is easy due to their colocalization in the images. For analysis, one should discard fluorophores that were not fixed in orientation during imaging, assessed as the reproducibility of the estimated azimuthal and polar angles of the probes.

On this basis, we selected dsDNA molecules for analysis. From each frame in a time-lapse movie of the probes, we simultaneously estimated each probe's position coordinates in the xy-plane (Fig. 12) and angles of dipole orientation (Fig. 12). From these we calculated the probe separation in the xy-plane (Fig. 12), using the calibrated local map, and the relative angle between them (Fig. 12). These measures allow us to compare intramolecular distances and orientations among randomly oriented molecules. The vector

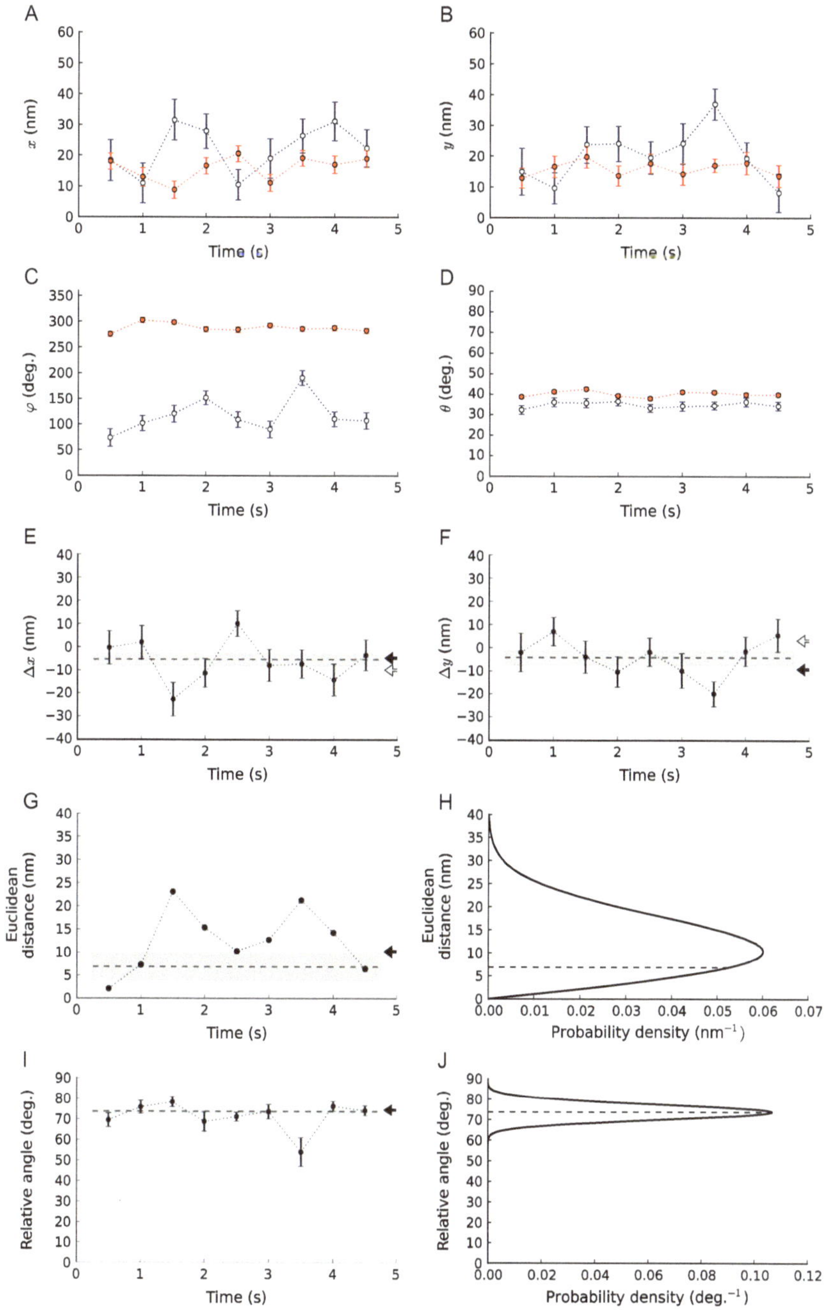

Fig. 12 See legend on opposite page.

difference between the position coordinates of the two probes in a pair (Fig. 12) was unaffected by sample drift during a movie. Each vector difference scattered normally around its mean values with an s.d. that agreed with the minimal value allowed by the information limit given theoretically by the number of photons in probe images, demonstrating that the method is as precise as possible (Fig. 1). We calculated the 2D Euclidean distance between probes in a pair from their time-averaged vector differences (Fig. 12). For these distance measurements, we calculated the weighted averages of the vector separation. As weights we used the inverse of the theoretical covariance matrix, assuming shot noise was the only source of variation.

The distance agrees with the value expected from a simple model of the dsDNA structure (Fig. 2). It is important to average vector differences before

Fig. 12 Distance and angle between a pair of fluorophores fixed in one DNA molecule, estimated using diPOLE. (A) Time series at 2 Hz of repeatedly estimated position coordinates (x) of a pair of fixed Cy3 (*open circles*) and Cy5 (*filled circles*) fluorophores, attached internally to the same dsDNA molecule at 30 bp separation (Fig. 2). For comparison, coordinates were mapped between image channels as described in the text. *Error bars* assume theoretical shot noise as only source of error. (B) Same as (A) for position coordinate y. (C) Same as (A) for the azimuthal angle (φ). (D) Same as (A) for the polar angle (θ). (E) Difference in x-coordinates (*black circles*) from (A). We calculated the weighted mean value (*dashed line*) and the theoretical s.e.m. (*gray region*). The data are consistent with a normal distribution centered on this mean value. The *arrows* indicate two predictions for the x-coordinate difference based alone on the estimated angles in (C and D) and the dsDNA structure (Fig. 14). (F) Same as (E) for the difference in y-coordinates. (G) Euclidean distance between the fluorophores, estimated frame-by-frame (*black circles*). The *black arrow* indicates the expected value from the dsDNA structure. We estimated the Euclidean distance between the probes (*dashed line*) as the Euclidean length of the movie-average of the frame-by-frame vector difference between the probes shown in (E and F). *Gray region* shows the theoretical s.d. of this estimate. (H) Non-Gaussian distribution (Churchman, Flyvbjerg, & Spudich, 2006) (*solid line*) of frame-by-frame Euclidean distances, resulting from probe separation (*dashed line*) being comparable to statistical errors on its estimates in (G). Note that this distribution peaks well above its true distance. (I) Relative angle between the emission dipoles of the probes. Its expected value based on the dsDNA structure is indicated by the *black arrow*. We calculated the weighted mean value (*dashed line*) with theoretical s.e.m. (*gray region*). (J) Distribution of relative angles (*solid line*). In general, it is a non-Gaussian distribution (Mortensen et al., 2015), but here it is indistinguishable from a Gaussian centered on the estimate from (I). Consequently, the relative angle is estimated as the movie average over estimates in individual frames shown in (**I**). *Figure adapted from Mortensen, K. I., Sung, J., Flyvbjerg, H., & Spudich, J. A. (2015). Optimized measurements of separations and angles between intra-molecular fluorescent markers.* Nature Communications, 6, *8621. http://doi.org/10.1038/ncomms9621.*

calculating the Euclidean distance. If the Euclidean distance between probes in a pair is calculated frame-by-frame in a movie and then averaged, this average overestimates the true Euclidean distance (Churchman et al., 2006) (Fig. 12). Also, had the analysis falsely assumed that the dyes were imaged exactly in focus, the resulting distance estimate would have been significantly compromised. In that case, each coordinate of an estimated position would be subject to a systematic error that depends on the focus mismatch and the polar angle of the probe. Such systematic errors may exceed 10 nm for a position coordinate per probe for intermediate polar angles and typical values of the objective's distance from the focus (Mortensen et al., 2015).

3.2 Measurements of Intramolecular Relative Orientations

To estimate the angle between emission dipoles, we calculated the scalar product between their vectors (Fig. 12). Like the angle between the dipoles, this scalar product is independent of the orientation of the molecule to which the probes are fixed. Consequently, the angle between probe pairs is the same for all DNA molecules, apart from possible molecule-to-molecule variation. The noise on these angle estimates is so small compared to the expected value for the angle that the appropriate non-Gaussian distribution (Mortensen et al., 2015) of relative angles is well approximated by a Gaussian (Fig. 12I and J). Thus, the weighted average of the individual relative angles is an unbiased estimate for the mean relative angle and we used that. As weights we used the theoretical variances calculated from the theoretical covariance matrix, assuming shot noise as only source of variation.

From these angles, we calculated the relative angle between probes in a pair (Fig. 12), which is independent of all drift and of pitch, yaw, and roll of the dsDNA molecule. The relative angles scatter tightly around their mean value (Fig. 12), but typically do so with an s.d. that is slightly larger than the one expected from photon shot noise alone. This slack compromised neither the accuracy of the estimate nor the Gaussian approximation of the distribution of the relative angles (Fig. 12J) (Mortensen et al., 2015). Justified by the latter, we calculated the mean relative angle between probes and its s.e.m. from the Gaussian distributed relative angles. In the single molecule analyzed in Fig. 12 with probes separated by 30 bp as an example, diPOLE applied to the fluorescent probes yields a Euclidean distance and relative angle between them that is in fine agreement with a simple model of the dsDNA structure (Figs. 2 and 12).

3.3 Performance of the Method

For the demonstration of accuracy and precision of the method, distances and relative angles were measured over time for 10 molecules with 20 bp separations and 10 molecules with 30 bp separations (Fig. 13). We calculated the average separation for each sample and found 8 ± 1 nm (mean $\pm$ theoretical s.e.m.) for probes separated by 20 bp and 12 ± 1 nm (mean $\pm$ theoretical s.e.m.) for probes separated by 30 bp.

3.3.1 Accurate and Precise Measurements of Intramolecular Separations

The distance estimates for each group of molecules scatter about their respective mean separations (Fig. 13) with an s.d. that is consistent with the information limit (Fig. 1). The population means agree with the values expected from a simple model of the internally labeled dsDNA (Fig. 2), i.e., with expected separations of 6.8 nm (20 bp) and 10.2 nm (30 bp). The latter values are subject to small uncertainties due to possible interactions of the dyes with the major or minor grooves of the dsDNA (Fig. 12) (Dolghih, 2009). This demonstrates that diPOLE applied to a single pair of probes provides an accurate and precise estimate of their separation at the nanometer level and that diPOLE applied to a small number of probe pairs discriminates a 10 bp difference.

3.3.2 Accurate and Precise Measurements of Intramolecular Relative Orientations

The expected relative angle between the probes is identical in our two samples (Figs. 2 and 13), because they differ in probe separation by a single repeating unit of the dsDNA helix. Each molecule yields a mean relative angle, which is determined so precisely (Figs. 11I and 13B) that discernable systematic variation between molecules is revealed (Fig. 13). This is likely due to dye–surface and/or dye–DNA interactions. An interaction resulting in an angstrom change of one end of a fluorophore may give rise to a $\sim$5 degree change in relative orientation. This easily explains the excess scatter of the mean relative orientations in Fig. 13B. For the population, we found a mean value of 71 ± 2 degrees (mean $\pm$ experimental s.e.m.), where this s.e.m. predominantly reflects the systematic variation in the results likely due to surface interaction. This result is consistent with the expected relative angle of 74 degrees based on the simple dsDNA model structure (Fig. 2).

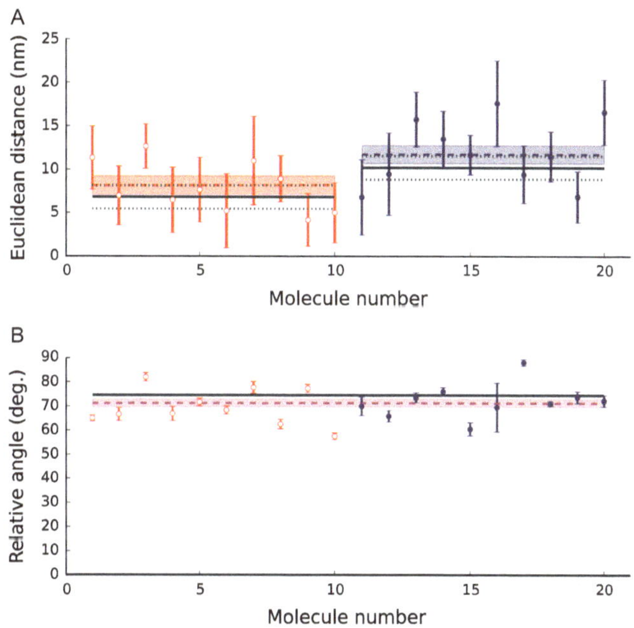

Fig. 13 Demonstration of the accuracy and precision of diPOLE. (A) Euclidean distance estimates between probes at 20 bp (*open circles*) and 30 bp (*filled circles*) separation, respectively. Error bars are theoretical, assuming shot noise as only source of error. The expected Euclidean separation based on the dsDNA structure (*black solid lines*) is 6.8 and 10.2 nm, for 20 and 30 bp separation, respectively, with an uncertainty (*dotted lines*) due to possible minor and major groove interactions between fluorophores and dsDNA (Fig. 2). We find weighted mean separations of, respectively, 8 ± 1 and 12 ± 1 nm (mean $\pm$ theoretical s.e.m.) (*dashed lines with colored area*), which are in perfect agreement with the expected values (*black solid lines*) and in tantalizing agreement with the values expected if both probes are stuck in their extremal positions in the minor groove (*dotted lines* coinciding with *dashed lines*). (B) Same as (A) for relative angles. Error bars are experimentally calculated s.e.m. The expected relative angle between the probes (*black solid line*) is 74 degrees in both samples because they differ in probe separation by a full repeat of the dsDNA. There is no discernable difference between the relative angles measured in the two samples. We found a mean orientation of 71 ± 2 degrees (mean $\pm$ experimental s.e.m.) (*dashed line* with *shaded region*), which agrees with the expected value. The individual angle estimates scatter more than photon shot noise alone predicts. We therefore report here the experimental s.e.m. for realism. *Figure adapted from Mortensen, K. I., Sung, J., Flyvbjerg, H., & Spudich, J. A. (2015). Optimized measurements of separations and angles between intra-molecular fluorescent markers.* Nature Communications, 6, *8621. http://doi.org/10.1038/ncomms9621.*

3.4 Unique Measurements of the 3D Orientation of Doubly Labeled dsDNA

The method's accuracy and precision are sufficiently good to allow determination of 3D orientation of the dsDNA molecule: the estimated fluorophore orientations alone specify two possible 3D orientations of the assumed model dsDNA structure (Fig. 14), where the degeneracy is due to the fact that the emission dipoles are not directional. To determine the possible spatial orientations of the dsDNA construct, we required that the tangent vectors of the dsDNA structure agreed with the estimated

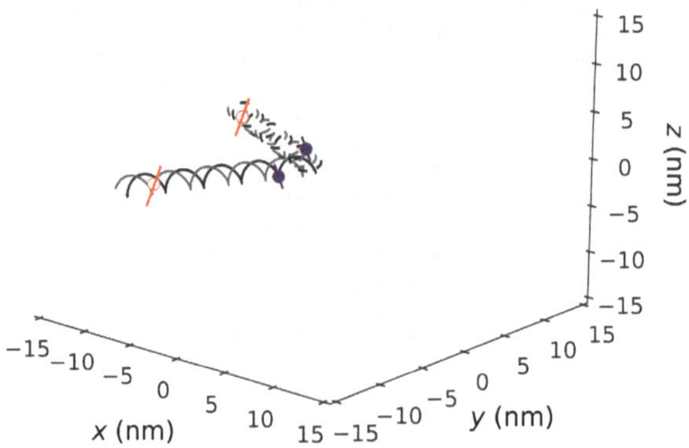

Fig. 14 Given the orientations in space of the two emission dipoles, we still do not know the orientation in space of our dsDNA construct: two are possible. An emission dipole moment (Cy3, *filled circles with line segments*; Cy5, *open circles with line segments*) determines a direction in space only modulo π, i.e., a direction and its opposite direction are the same thing when describing a dipole's orientation. Consequently, knowledge of the polar and azimuthal angles (modulo π) for both probes is not sufficient to determine the orientation in space of the dsDNA construct they are part of. Thus, two different orientations of this construct have the same probe angles, as illustrated here with the probe angles estimated for the molecule discussed in Fig. 12. Here, the two strands of each dsDNA are shown in *black* and *gray*, respectively, and the two orientations of the DNA are shown with *full* and *dashed lines*, respectively. (The probes are separated by 30 bp, and only 50 bp of the full construct is shown to avoid clutter.) These two different orientations of the construct have different angles with the optical axis, however, so the vector differences in position between the probes projected onto the *xy*-plane are different. We compare these to the measured vector differences (Fig. 12E and F), and this comparison uniquely favors one solution (*black arrows* in Fig. 12E and F; dsDNA with *full lines* here) over the other (*white arrows* in Fig. 12E and F; dsDNA with *dashed lines* here). *Figure adapted from Mortensen, K. I., Sung, J., Flyvbjerg, H., & Spudich, J. A. (2015). Optimized measurements of separations and angles between intra-molecular fluorescent markers.* Nature Communications, *6, 8621. http://doi.org/10.1038/ncomms9621.*

orientations of the emission dipoles of Cy3 and Cy5, and then we solved numerically for the pitch, yaw, and roll angles of the dsDNA (Fig. 14). Each of the two possible 3D orientations predicts a 2D vector difference for the fluorophore positions in the image plane (Fig. 12E and F). Comparison of the measured vector differences to the predicted values from the two candidate orientations singles out one orientation (black arrows, Figs. 12E and F, and 14) over the other (white arrows, Figs. 12E and F, and 14). Thus, diPOLE extracts the full 3D orientation, i.e., pitch, yaw, and roll, of the dsDNA molecule.

4. DISCUSSION

We have demonstrated that distances measured with diPOLE are optimally precise (Fig. 1): the scatter of estimates of the mean probe distance obtained with diPOLE (Fig. 13) is fully explained by photon shot noise, excess noise (Hynecek & Nishiwaki, 2003; Mortensen et al., 2010; Robbins & Hadwen, 2003; Ulbrich & Isacoff, 2007), and the finite number of beads included in the calibration of the map between the channels. Specifically, we found precisions of mean distance estimates for individual molecules of ~2 to 5 nm. Multiple paths to increased precision exist: using additional, or brighter, images of a fluorophore pair would increase this precision. Doing that would also reduce the standard error of the population-averaged distances below their current values of ~1 nm. So would inclusion of additional molecules in the analysis. Similarly, note that improving the ~2 nm precision of each map between channels is possible simply by inclusion of additional beads in map calibrations (i.e., without need for more advanced fiducial markers, such as dual-color fluorescently labeled dsDNA strands). In each of these cases, precision would improve as one over the square root of the relevant statistical quantity, i.e., the number of photons, images, molecules, or beads. The current resolution, however, is sufficient to establish both nanometer accuracy and precision of distances measured within an individual molecule with diPOLE. When such distances are averaged over a small number of molecules, a 10 bp difference, between 20 and 30 bp, is resolved, we have shown.

On the other hand, diPOLE yields mean estimates of the relative angle between the probes (Fig. 13) that are so precise that a molecule-to-molecule variation is revealed, which, however, is on the order of a few degrees and easily explained by, e.g., surface interaction.

At 2 Hz, we recorded approximately 1000–20,000 photons per image, depending on the fluorophore's orientation with respect to the excitation light field, its position in the field of view, and its possible function as a FRET donor or acceptor. This suggests that the method will have a time resolution that is an order of magnitude better using similar excitation laser intensity, if one considers only the brightest probes. Increased excitation laser power will also improve time resolution, albeit at the expense of faster photobleaching. In the near future, brighter dyes and new oxygen-scavenging systems that extend fluorescent life-times are likely to improve the temporal resolution of our method as both are pursued in academia and commercially for their potential applications in many fields, including single-molecule fluorescence and live-cell imaging.

One bottleneck that currently is limiting diPOLE from becoming generally applicable to all systems is the biochemical challenge to fix the orientation of dye with respect to the molecule of interest. However, with rapid progress in molecular biology and protein engineering, the future offers increasing opportunities to solve the fixation issue.

5. CONCLUSION AND PERSPECTIVES

Here, we have described the methodology required to employ a fluorescence-based single-molecule tool, diPOLE, for the study of molecular conformations. We have provided evidence for its performance and shown that it is precise and accurate to within a few nanometers and a few degrees, when applied to fixed fluorescent probes attached to a single molecule. We have demonstrated that it works also in the intramolecular range, and it thus complements smFRET there, by adding the ability to discriminate between translational and rotational conformational changes to the single-molecule toolbox.

The experimental verification demonstrated in this chapter makes fixed fluorescent probes a reliable tool in single- and dual-color assays for simultaneous probing of positions and orientations directly from images. With a single-color assay, macromolecular dynamics that involves translational and/or rotational motion can be monitored. A two-color assay enables dynamic studies of changes in relative orientation and position, e.g., originating from intra- or inter-macromolecular interactions, between discrete molecular states that are sufficiently rigid to ensure fixation of the spatial orientation of the fluorescent probes. All that is now possible with ease and confidence and a resolution that was inaccessible till now in single-molecule studies.

The results presented here open up for further method developments: (i) The probe-dsDNA construct itself (Fig. 2) may be useful for investigating the protein–DNA interaction, as an advanced dual-color probe. One might, e.g., study conformations of dsDNA in nucleosomes as a function of different posttranslational modifications, thus improving our understanding of chromatin regulation at the single-molecule level. (ii) While diPOLE and smFRET are two complementary techniques, a hybrid might provide more information about probe separation in 3D as well as probe orientations.

REFERENCES

Aguet, F., Geissbühler, S., Märki, I., Lasser, T., & Unser, M. (2009). Super-resolution orientation estimation and localization of fluorescent dipoles using 3-D steerable filters. *Optics Express*, *17*(8), 6829–6848. http://doi.org/10.1364/OE.17.006829.

Backlund, M. P., Lew, M. D., Backer, A. S., Sahl, S. J., Grover, G., Agrawal, A., … Moerner, W. E. (2012). Simultaneous, accurate measurement of the 3D position and orientation of single molecules. *Proceedings of the National Academy of Sciences of the United States of America*, *109*(47), 19087–19092. http://doi.org/10.1073/pnas.1216687109.

Betzig, E., Patterson, G. H., Sougrat, R., Lindwasser, O. W., Olenych, S., Bonifacino, J. S., … Hess, H. F. (2006). Imaging intracellular fluorescent proteins at nanometer resolution. *Science (New York, N.Y.)*, *313*(5793), 1642–1645. http://doi.org/10.1126/science.1127344.

Bobroff, N. (1986). Position measurement with a resolution and noise-limited instrument. *Review of Scientific Instruments*, *57*(6), 1152–1157. http://doi.org/10.1063/1.1138619.

Böhmer, M., & Enderlein, J. (2003). Orientation imaging of single molecules by wide-field epifluorescence microscopy. *Journal of the Optical Society of America B*, *20*(3), 554–559. http://doi.org/10.1364/JOSAB.20.000554.

Born, M., & Wolf, E. (1999). *Principles of optics* (2nd ed.). New York: Cambridge University Press.

Burghardt, T. P. (2011). Single molecule fluorescence image patterns linked to dipole orientation and axial position: Application to myosin cross-bridges in muscle fibers. *PLoS One*, *6*(2), e16772. http://doi.org/10.1371/journal.pone.0016772.

Burghardt, T. P., Josephson, M. P., & Ajtai, K. (2011). Single myosin cross-bridge orientation in cardiac papillary muscle detects lever-arm shear strain in transduction. *Biochemistry*, *50*(36), 7809–7821. http://doi.org/10.1021/bi2008992.

Churchman, L. S., Flyvbjerg, H., & Spudich, J. A. (2006). A non-Gaussian distribution quantifies distances measured with fluorescence localization techniques. *Biophysical Journal*, *90*(2), 668–671. Retrieved from, http://www.pubmedcentral.nih.gov/articlerender.fcgi?artid=1367071&tool=pmcentrez&rendertype=abstract.

Churchman, L. S., Okten, Z., Rock, R. S., Dawson, J. F., & Spudich, J. A. (2005). Single molecule high-resolution colocalization of Cy3 and Cy5 attached to macromolecules measures intramolecular distances through time. *Proceedings of the National Academy of Sciences of the United States of America*, *102*(5), 1419–1423. http://doi.org/10.1073/pnas.0409487102.

Deschout, H., Cella Zanacchi, F., Mlodzianoski, M., Diaspro, A., Bewersdorf, J., Hess, S. T., & Braeckmans, K. (2014). Precisely and accurately localizing single emitters in fluorescence microscopy. *Nature Methods*, *11*(3), 253–266. http://doi.org/10.1038/nmeth.2843.

Dolghih, E. (2009). *Theoretical studies of dye-labeled DNA systems with applications to fluorescence resonance energy transfer (Doctoral dissertation)*. University of Florida.

Enderlein, J., Toprak, E., & Selvin, P. R. (2006). Polarization effect on position accuracy of fluorophore localization. *Optics Express*, *14*(18), 8111–8120. Retrieved from, http://www.ncbi.nlm.nih.gov/pubmed/19529183.

Forkey, J. N., Quinlan, M. E., Shaw, M. A., Corrie, J. E. T., & Goldman, Y. E. (2003). Three-dimensional structural dynamics of myosin V by single-molecule fluorescence polarization. *Nature*, *422*(6930), 399–404. http://doi.org/10.1038/nature01529.

Hynecek, J., & Nishiwaki, T. (2003). Excess noise and other important characteristics of low light level imaging using charge multiplying CCDs. *IEEE Transactions on Electron Devices*, *50*(1), 239–245.

Iqbal, A., Arslan, S., Okumus, B., Wilson, T. J., Giraud, G., Norman, D. G., … Lilley, D. M. J. (2008). Orientation dependence in fluorescent energy transfer between Cy3 and Cy5 terminally attached to double-stranded nucleic acids. *Proceedings of the National Academy of Sciences of the United States of America*, *105*(32), 11176–11181. http://doi.org/10.1073/pnas.0801707105.

Lew, M. D., & Moerner, W. E. (2014). Azimuthal polarization filtering for accurate, precise, and robust single-molecule localization microscopy. *Nano Letters*, *14*, 6407–6413.

Lippincott-Schwartz, J., & Manley, S. (2009). Putting super-resolution fluorescence microscopy to work. *Nature Methods*, *6*(1), 21–23. http://doi.org/10.1038/nmeth.f.233.

Mortensen, K. I., Churchman, L. S., Spudich, J. A., & Flyvbjerg, H. (2010). Optimized localization analysis for single-molecule tracking and super-resolution microscopy. *Nature Methods*, 7(5), 377–381.

Mortensen, K. I., & Flyvbjerg, H. (2016). "Calibration-on-the-spot": How to calibrate an EMCCD camera from its images. *Scientific Reports*, *6*, 28680. http://doi.org/10.1038/srep28680.

Mortensen, K. I., Sung, J., Flyvbjerg, H., & Spudich, J. A. (2015). Optimized measurements of separations and angles between intra-molecular fluorescent markers. *Nature Communications*, *6*, 8621. http://doi.org/10.1038/ncomms9621.

Ober, R. J., Ram, S., & Ward, E. S. (2004). Localization accuracy in single-molecule microscopy. *Biophysical Journal*, *86*(2), 1185–1200. http://doi.org/10.1016/S0006-3495(04)74193-4.

Pertsinidis, A., Zhang, Y., & Chu, S. (2010). Subnanometre single-molecule localization, registration and distance measurements. *Nature*, *466*(7306), 647–651. http://doi.org/10.1038/nature09163.

Ranjit, S., Gurunathan, K., & Levitus, M. (2009). Photophysics of backbone fluorescent DNA modifications: Reducing uncertainties in FRET. *The Journal of Physical Chemistry*, *113*(22), 7861–7866. http://doi.org/10.1021/jp810842u.

Rao, C. R. (1973). *Linear statistical inference and its applications*. New York, NY: Wiley.

Reifenberger, J. G., Toprak, E., Kim, H., Safer, D., Sweeney, H. L., & Selvin, P. R. (2009). Myosin VI undergoes a 180 degrees power stroke implying an uncoupling of the front lever arm. *Proceedings of the National Academy of Sciences of the United States of America*, *106*(43), 18255–18260. http://doi.org/10.1073/pnas.0900005106.

Robbins, M. S., & Hadwen, B. J. (2003). The noise performance of electron multiplying charge-coupled devices. *IEEE Transactions on Electron Devices*, *50*(5), 1227–1232.

Roy, R., Hohng, S., & Ha, T. (2008). A practical guide to single-molecule FRET. *Nature Methods*, *5*(6), 507–516. http://doi.org/10.1038/NMETH.1208.

Rust, M. J., Bates, M., & Zhuang, X. W. (2006). Sub-diffraction-limit imaging by stochastic optical reconstruction microscopy (STORM). *Nature Methods*, *3*(10), 793–795. http://doi.org/Doi 10.1038/Nmeth929.

Stallinga, S. (2015). Effect of rotational diffusion in an orientational potential well on the point spread function of electric dipole emitters. *Journal of the Optical Society of America. A, Optics, image science, and vision*, *32*(2), 213–223.

Stallinga, S., & Rieger, B. (2010). Accuracy of the gaussian point spread function model in 2D localization microscopy. *Optics Express*, *18*(24), 24461–24476. Retrieved from, http://www.ncbi.nlm.nih.gov/pubmed/21164793.

Stallinga, S., & Rieger, B. (2012). Position and orientation estimation of fixed dipole emitters using an effective Hermite point spread function model. *Optics Express*, *20*(6), 5896–5921. http://doi.org/10.1364/OE.20.005896.

Thompson, R. E., Larson, D. R., & Webb, W. W. (2002). Precise nanometer localization analysis for individual fluorescent probes. *Biophysical Journal*, *82*(5), 2775–2783.

Toprak, E., Enderlein, J., Syed, S., McKinney, S. A., Petschek, R. G., Ha, T., … Selvin, P. R. (2006). Defocused orientation and position imaging (DOPI) of myosin V. *Proceedings of the National Academy of Sciences of the United States of America*, *103*(17), 6495–6499. http://doi.org/10.1073/pnas.0507134103.

Ulbrich, M. H., & Isacoff, E. Y. (2007). Subunit counting in membrane-bound proteins. *Nature Methods*, *4*(4), 319–321. http://doi.org/10.1038/NMETH1024.

Vestergaard, C. L., Blainey, P. C., & Flyvbjerg, H. (2014). Optimal estimation of diffusion coefficients from single-particle trajectories. *Physical Review E*, *89*(2), 022726. http://doi.org/10.1103/PhysRevE.89.022726.

Vestergaard, C. L., Pedersen, J. N., Mortensen, K. I., & Flyvbjerg, H. (2015). Estimation of motility parameters from trajectory data. *The European Physical Journal Special Topics*, *224*(7), 1151–1168. http://doi.org/10.1140/epjst/e2015-02452-5.

Yildiz, A., Forkey, J. N., McKinney, S. A., Ha, T., Goldman, Y. E., & Selvin, P. R. (2003). Myosin V walks hand-over-hand: Single fluorophore imaging with 1.5-nm localization. *Science (New York, N.Y.)*, *300*(5628), 2061–2065. http://doi.org/10.1126/science.1084398.

CHAPTER SEVEN

Precisely and Accurately Inferring Single-Molecule Rate Constants

C.D. Kinz-Thompson, N.A. Bailey, R.L. Gonzalez Jr.[1]
Columbia University, New York, NY, United States
[1]Corresponding author: e-mail address: rlg2118@columbia.edu

Contents

Abstract

The kinetics of biomolecular systems can be quantified by calculating the stochastic rate constants that govern the biomolecular state vs time trajectories (i.e., state trajectories) of individual biomolecules. To do so, the experimental signal vs time trajectories (i.e., signal trajectories) obtained from observing individual biomolecules are often idealized to generate state trajectories by methods such as thresholding or hidden Markov modeling. Here, we discuss approaches for idealizing signal trajectories and calculating stochastic rate constants from the resulting state trajectories. Importantly, we provide an analysis of how the finite length of signal trajectories restricts the precision of these approaches and demonstrate how Bayesian inference-based versions of these approaches allow rigorous determination of this precision. Similarly, we provide an analysis of how the finite lengths and limited time resolutions of signal trajectories restrict the accuracy of these approaches, and describe methods that, by accounting for the effects of the finite length and limited time resolution of signal trajectories, substantially improve this accuracy. Collectively, therefore, the methods we consider here enable a

Methods in Enzymology, Volume 581
ISSN 0076-6879
http://dx.doi.org/10.1016/bs.mie.2016.08.021
© 2016 Elsevier Inc.
All rights reserved.

rigorous assessment of the precision, and a significant enhancement of the accuracy, with which stochastic rate constants can be calculated from single-molecule signal trajectories.

1. INTRODUCTION

In single-molecule, kinetic studies of biomolecular systems, experimental data consisting of a signal originating from an individual biomolecule are collected as a function of time (Tinoco & Gonzalez, 2011). This signal is, or can be converted into, a proxy for the underlying biomolecular state of the system. For instance, the intramolecular fluorescence resonance energy transfer (FRET) efficiency (E_{FRET}) that is measured between two fluorophore-labeled structural elements of an individual biomolecule in a single-molecule FRET (smFRET) experiment depends on the distance between the two structural elements and can therefore be converted into a proxy for the conformational state of the biomolecule (Roy, Hohng, & Ha, 2008). Similarly, the distance that is measured between two optically trapped microbeads that are tethered to each other by an individual biomolecule in a single-molecule force spectroscopy experiment is a proxy for the conformational state of the biomolecule (Greenleaf, Woodside, & Block, 2007; Moffitt, Chemla, Smith, & Bustamante, 2008). Investigating the kinetics of biomolecular systems using such single-molecule approaches eliminates the ensemble averaging that is inherent to bulk approaches. Thus, these approaches can reveal transient and/or rare kinetic events that are typically obscured by ensemble averaging, but that can often be critically important for elucidating biological mechanisms. In order to take full advantage of the unique and powerful mechanistic information provided by single-molecule experiments, however, the observed signals must be sensitive enough to unambiguously resolve the biomolecular states that are sampled during the experiment.

To obtain relevant kinetic information about a biomolecular system from such single-molecule experiments, the inherently noisy, experimental signal vs time trajectories (i.e., signal trajectories) obtained from observing individual biomolecules are typically transformed, or idealized, into biomolecular state vs time trajectories (i.e., state trajectories). This idealization process is not trivial, as limitations in signal and temporal resolution can easily obscure the relevant biomolecular states. Under the most favorable conditions, a researcher can sometimes manually select the signal data point where the biomolecule transitions to a new state. Unfortunately, this process is subjective and time consuming, and often the data are not sufficiently resolved to use this approach. A second method involves manually setting a signal threshold that, once crossed by the experimental signal, indicates a transition to a new state but this approach is still subjective and difficult to implement when more than two biomolecular states are present. A third, more rigorous and widely adopted method uses hidden Markov models (HMMs) to transform the inherently noisy signal trajectories into state trajectories by estimating the underlying, "hidden" state responsible for producing the signal during each measurement period in a signal trajectory (Colquhoun & Hawkes, 1977, 1981). An advantage of using HMMs for this transformation is that they can manage many states simultaneously and that methods have been developed to select the correct number of states present in the trajectory (Bronson, Fei, Hofman, Gonzalez, & Wiggins, 2009; Bronson, Hofman, Fei, Gonzalez, & Wiggins, 2010; van de Meent, Bronson, Wiggins, & Gonzalez, 2014; van de Meent, Bronson, Wood, Gonzalez, & Wiggins, 2013). Regardless of the method that is used to idealize a signal trajectory into the corresponding state trajectory, the state trajectories can then be used to calculate stochastic rate constants and obtain kinetic information about the observed biomolecular system.

Herein, we begin by comparing the deterministic rate constants that are obtained from ensemble kinetic studies with the stochastic rate constants that are obtained from single-molecule kinetic studies as a means for introducing the conceptual framework that is typically used to analyze and interpret single-molecule kinetic data. We then clarify the basis of several approaches for calculating stochastic rate constants from single-molecule state trajectories. We go on to describe how the finite lengths of signal trajectories restrict the precision of these approaches and demonstrate how Bayesian inference-based versions of these approaches provide a natural method to account for the precision of the calculated stochastic rate constants. We then end by addressing how the finite lengths and limited time resolutions of signal trajectories restrict the accuracy of these approaches, and describing methods to correct for the effects of the finite length and limited time resolution of the signal trajectories in order to increase the accuracy of these approaches. The methods we examine here for assessing the precision and improving the accuracy of the approaches that are currently used to calculate stochastic rate constants from single-molecule data greatly improve the analysis and interpretation of single-molecule kinetic experiments.

2. SINGLE-MOLECULES AND STOCHASTIC RATE CONSTANTS

In bulk kinetic experiments, the large number of molecules present in an ensemble yields well-defined, ensemble-averaged approaches to equilibrium that mask the individual behaviors of the underlying molecules (McQuarrie, 1963). Thus, these approaches to equilibrium are traditionally described as time-dependent changes in the concentrations of reactants, reaction intermediates, and/or products that are modeled using phenomenological, differential rate equations (Van Kampen, 2007). Notably, bulk reaction kinetics and the rate equations that are used to model them are: (i) continuous in that individual molecules are not observed to undergo reactions, but rather the reaction is observed and described in terms of changes in concentrations, and (ii) deterministic in that an initial set of concentrations determines the subsequent values of the concentrations. By fitting changes in the concentrations of reactants, reaction intermediates, and/or products as they approach their equilibrium concentrations to these deterministic rate equations, one can obtain the deterministic rate constants that characterize the kinetics of the bulk system (Zhou, 2010).

In contrast with bulk reaction kinetics, however, single-molecule reaction kinetics are: (i) discrete in that individual molecules are observed to undergo reactions and (ii) stochastic in that, even at equilibrium, reactions occur at random times that are often, but not always, independent of previous conditions. These differences between bulk and single-molecule reaction kinetics make it inappropriate to use the deterministic rate equations used to describe bulk reaction kinetics to account for the stochastic reactions that are observed at the single-molecule level (McQuarrie, 1963). Therefore, in order to describe single-molecule reaction kinetics, stochastic approaches like the chemical master equation and the stochastic simulation algorithm were developed to model the time evolution of discrete reactions in which the behavior of individual molecules could be observed (Gillespie, 1976, 1977, 2007; McQuarrie, 1967; Van Kampen, 2007; Zwanzig, 2001). These stochastic methods aim to quantify the kinetics of the molecular system by modeling the occurrence of individual reactions with probability distributions that are governed by stochastic rate constants, as opposed to modeling changes in concentrations with differential equations that are governed by deterministic rate constants. In order to quantify the kinetics of biomolecular systems observed in signal trajectories recorded using single-molecule biophysical techniques, therefore, we must adopt such a stochastic approach.

Consider the reaction coordinate of a biological process, such as protein folding or ligand binding. Due to the multiplicity of interactions present in biomolecular systems, the forward and reverse reactions along this reaction coordinate can often be considered as separate, elementary reactions that occur randomly and independently of the history of the system (i.e., in what states the biomolecule has been in and for how long) (Colquhoun & Hawkes, 1995; Zwanzig, 1997). Such stochastic reactions are called Markovian when the probability of a reaction occurring (i.e., a transition between states) depends only upon the current state of system; when these probabilities are time dependent, or, rather, depend upon the previous state(s) of the system, the reaction is called non-Markovian. As a result of the constant transition probability of Markovian reactions, the lengths of time that a biomolecule spends in a particular state before a transition occurs, called the dwell times, t, are distributed according to an exponential distribution of the form

$$p(t_i|\, k_i) = k_i e^{-k_i t_i}, \tag{1}$$

where $p(t_i|\, k_i)$ is the probability density function (PDF) of a dwell time in the ith state lasting length t_i given the stochastic rate constant k_i, where $k_i = \sum_{j \neq i} k_{ij}$. Here, k_i is the net stochastic rate constant out of the ith state, and k_{ij} are the stochastic rate constants governing the individual Markovian reactions out of the ith state. For instance, if there are multiple, parallel, Markovian reactions out of the ith state, the net stochastic rate constant that describes the length of time spent in the ith state, k_i, will be the sum of the individual stochastic rate constants governing each of the parallel reactions, k_{ij}. Effectively, the dwell times in the ith state, t_i, even if they are sorted into only those that transition to the jth state, will be distributed according to this net stochastic rate constant, k_i. It follows then that, regardless of the final state, the average dwell time spent in the ith state, $\langle t_i \rangle$, is the reciprocal of this net stochastic rate constant, k_i. Finally, while it is not possible to distinguish among the collection of stochastic rate constants, k_{ij}, that describe the individual Markovian reactions exiting the ith state by analyzing the observed dwell times spent in the ith state, t_i, the number of times that an individual molecule enters a particular jth state will depend upon the stochastic rate constant k_{ij} and can therefore be used to quantify k_{ij}.

Interestingly, the ergodic hypothesis asserts that the dwell time PDF for an individual molecule observed for a very long amount of time is equivalent to the dwell time PDF comprised of many identical, individual molecules, each observed for very short periods of time (Van Kampen, 2007). Thus, because many experimental factors, such as the photobleaching of fluorophores, limit the length of time that an individual biomolecule can be continuously observed, the latter approach of observing many individual biomolecules for very short periods of time is often taken. Regardless of which approach is taken, Onsager's regression hypothesis (Onsager, 1931; Zwanzig, 2001) asserts that this "microscopic" dwell time PDF of an individual molecule is equivalent to the "macroscopic" relaxation to equilibrium of an ensemble of molecules described by traditional chemical kinetics. Therefore, when monitoring the reaction of one biomolecule, or of multiple, identical, individual biomolecules, the observed single-molecule reaction kinetics are equivalent to those that would be measured in bulk, if it were possible to observe them despite the ensemble averaging—this is especially significant for situations where the biomolecular population or event of interest is too rarely sampled to observe using a bulk, ensemble-averaged signal.

Before describing how to quantify the single-molecule stochastic rate constants k_i and k_{ij} described earlier, we must note the several complications that have already arisen. First, the exponential dwell time PDF described earlier assumes that time is continuous, but single-molecule signal trajectories are comprised of a sequence of discrete measurements that are spaced by, at minimum, the acquisition period of the measurement during which the signal was time averaged to acquire a single data point. Errors can therefore be introduced into these stochastic rate constant calculations if the discretized state trajectories misrepresent the temporal behavior of the molecule(s) as it samples state space (i.e., the finite set of states available to it). Second, these stochastic rate constant calculations require several assumptions about the observed single-molecule data, including that a sufficient number of events were observed to accurately represent the ensemble average, that there are no subpopulations present in the sample, and that the system is at equilibrium and will not change over time, resulting in non-Markovian behavior. These assumptions are inherently difficult to confirm due to the small amounts of information present in a state trajectory from an individual molecule.

3. CALCULATING STOCHASTIC RATE CONSTANTS FROM SIGNAL TRAJECTORIES

3.1 Approaches to Calculating Stochastic Rate Constants

As mentioned earlier, stochastic rate constants govern the Markovian nature with which a single-molecule samples state space during a reaction. The dwell time, t, in a particular state is governed by the sum of all of the stochastic rate constants exiting that state, while the number of transitions between particular states depends upon the particular stochastic rate constant describing that reaction. Later, we discuss how stochastic rate constants for Markovian reactions can be quantified by considering the distribution of dwell times, or the probability of transitioning between particular states. Before describing these methods, however, we will briefly discuss how the properties of state trajectories that facilitate stochastic rate constant calculations can be quantified such that they can be easily incorporated into the various stochastic rate constant calculation methods.

The state trajectories described earlier are each composed of a series of sequential, discretized data points, where each data point indicates the state occupied, during a measurement period of length τ, by the single molecule corresponding to the signal trajectory being analyzed; it is worth noting that this state was inferred from a time-averaged signal collected during the measurement period τ. From these sequential data points that comprise a state trajectory, we can obtain a dwell time list, $\boldsymbol{n}_{ij}$, where each entry is the number of contiguous measurement periods, τ's, that the single molecule is observed to spend in a state, i, before transitioning to a second state, j. This is a discretized list of the dwell times in state i, t_i, that transition to state j, and it has the form: $\boldsymbol{n}_{ij} = [5, 13, 12, 7, \ldots]$. Additionally, we can construct a counting matrix, $\boldsymbol{M}$, for each state trajectory where the matrix elements, $\boldsymbol{M}_{ij}$, represent the number of times that the state trajectory began in state i at measurement n (i.e., at time $t = 0$) and ended in state j at measurement $n + 1$ (i.e., at time $t = \tau$). $\boldsymbol{M}$ is related to $\boldsymbol{n}_{ij}$ such that the off-diagonal elements of $\boldsymbol{M}$, $\boldsymbol{M}_{ij}$, are the number of entries in the corresponding $\boldsymbol{n}_{ij}$, and the on-diagonal elements of $\boldsymbol{M}$, $\boldsymbol{M}_{ii}$, are

$$\boldsymbol{M}_{ii} = \sum\nolimits_{j \neq i} \left(\left(\Sigma \boldsymbol{n}_{ij} \right) - \boldsymbol{M}_{ij} \right), \tag{2}$$

where $\sum \boldsymbol{n}_{ij}$ is the sum of the entries in $\boldsymbol{n}_{ij}$. $\boldsymbol{M}$ may be row normalized, such that each element in a row (i.e., with the same i) is divided by the sum of that row to yield the transition matrix, $\boldsymbol{P}$. The off-diagonal elements of the transition matrix $\boldsymbol{P}$, $\boldsymbol{P}_{ij}$, give the frequency that an individual molecule in state i has transitioned to state j at the next measurement period. Later, we detail several methods to explain how the stochastic rate constants that characterize kinetic processes may be obtained from the calculated dwell time list, $\boldsymbol{n}_{ij}$, counting matrix, $\boldsymbol{M}$, or transition matrix, $\boldsymbol{P}$.

3.1.1 Dwell Time Distribution Analysis

One method to calculate stochastic rate constants from a state trajectory is by analyzing the distribution of observed dwell times. A state trajectory can be thought of as a sequence of discrete measurements that report on whether a transition has occurred between two measurements. These "transition trials" are reminiscent of a series of repeated Bernoulli trials from probability theory (Resnick, 1992), which are events where the outcome is either a success with probability p, or a failure with probability $1-p$. In this analogy, a successful Bernoulli trial would be when the single-molecule transitions from state i at measurement period n to state j at measurement period $n+1$, whereas a failed Bernoulli trial would be when, instead, the single molecule remains in state i at measurement period $n+1$.

The number of repeated, failed trials before a success (i.e., a transition) occurs is distributed according to the geometric distribution probability mass function (PMF) (Resnick, 1992),

$$P(n|p) = p(1-p)^n, \tag{3}$$

where n is the number of failed trials and p is the probability of a success. Therefore, the PMF of the number of measurement periods until a transition occurs in a Markovian state trajectory can be modeled using the geometric distribution. From the geometric distribution, we expect that the mean number of successive measurement periods in state i, $\langle n_i \rangle$, until a transition out of state i occurs is

$$\langle n_i \rangle = \frac{1-P_i}{P_i}, \tag{4}$$

where P_i is the probability of a successful transition out of state i to any other state, j. Given a particular state trajectory in a Markovian system, an estimate of the mean number of measurement periods before a transition out of state i occurs, $\langle n_i \rangle$, would then allow the probability of a successful transition out of state i to be calculated by solving this equation. The maximum-likelihood estimate of $\langle n_i \rangle$ is the total number of measurement periods observed to be in state i divided by the total number of transitions out of state i,

$$\langle n_i \rangle = \frac{\sum_j (\Sigma n_{ij})}{\sum_{j \neq i} M_{ij}}, \tag{5}$$

where $\sum n_{ij}$ is the sum of all entries in n_{ij} and the M_{ij} are the total number of observed transitions from state i to state j. Solving this equation yields the probability of a successful transition,

$$P_i = \frac{\sum_{j \neq i} M_{ij}}{\sum_{j \neq i} M_{ij} + \sum_j (\Sigma n_{ij})}. \tag{6}$$

As mentioned in the previous section, the dwell times that a single molecule will spend in a particular state before transitioning to a different state, t's, in a Markovian system are distributed according an exponential distribution. Therefore, for such a Markovian system, the probability that a transition out of state i occurs within a measurement period, τ, in a signal trajectory is the integral of the exponential distribution PDF from $t=0$ to $t=\tau$, the measurement period, which is

$$P_i = \int_0^{\tau} k_{ij} \cdot e^{-k_i \cdot t} \cdot dt = \left(1 - e^{-k_i \cdot \tau}\right). \tag{7}$$

This equation implies that a stochastic rate constant can be calculated as

$$k_i = \frac{-\ln(1 - P_i)}{\tau}, \tag{8}$$

if the transition probability, P_i, can be quantified as described earlier. Notably, the stochastic rate constant obtained by considering the dwell times in a particular state will be a sum of multiple stochastic rate constants, except in cases when there is only one state to transition to (e.g., two-state systems). Analyzing only the dwell times that a single molecule spends in state i before transitioning to a particular state j still yields the same sum of the stochastic rate constants, and not the associated k_{ij}. The major advantage of analyzing the distribution of dwell times, however, is that deviations from Markovian behavior can be observed as nongeometric distribution and then this non-Markovian behavior can be analyzed.

Interestingly, a careful consideration of these equations reveals a limitation in the application of this dwell time distribution analysis method, which is the fact that the geometric distribution requires the state trajectories to have discrete dwell times that last $\{0, 1, 2, \ldots\}$ measurement periods, τ, before a transition occurs. Unfortunately, in a state trajectory, dwell times, t, of zero measurement periods, τ, are never included in the dwell time lists, n_{ij}, because a dwell time must be at least one measurement period, τ, long for it to be associated with a particular state. The result is an undercounting of $\boldsymbol{M}$ due to the exclusion of all zero measurement period-long dwell times ($n = 0$, or, equivalently, $t < \tau$), and a subsequent miscalculation of P_i. This undercounting is exacerbated by the fact that, from the geometric distribution, the highest probability dwell times are the zero measurement period-long dwell times ($n = 0$). As a result, stochastic rate constants calculated using the dwell time distribution analysis method are misestimates and, more specifically, underestimates of the true stochastic rate constant. Nonetheless, this underestimate can easily be accounted for by conditioning the geometric distribution PMF such that only dwell times that are greater than zero measurement periods, τ, in length are considered ($n > 0$, or, equivalently, $t > \tau$).

Here, we will condition the geometric distribution PMF so that it only considers $n > 0$, and denote these discrete dwell time lengths with $n^\star \in \{1, 2, \ldots\}$ to maintain clarity. From the law of conditional probability, we note that

$$
\begin{aligned}
P(n^\star|p, n > 0) &= \frac{P(n|p, n = 0 \cap n > 0)}{P(n|p, n = 0)}, \\
P(n^\star|p) &= \frac{p(1-p)^{n^\star}}{1 - \left(1 - (1-p)^{0+1}\right)}, \\
P(n^\star|p) = p(1-p)^{n^\star - 1} &= \frac{1}{1-p} \cdot P(n|p).
\end{aligned}
\tag{9}
$$

Therefore, the geometric distribution PMF conditioned upon all dwell times being greater than zero measurement periods in length is equivalent to the regular geometric distribution PMF divided by $1 - p$. Because $P(n^\star|\, p)$ is proportional to $P(n|p)$ in a manner that does not depend upon n, the expectation values of $p(n^\star|p)$ (e.g., the mean) are also proportional to those of $P(n|p)$ in the same manner due. Therefore,

$$
\langle n^\star \rangle = \frac{1}{1-p} \langle n \rangle. \tag{10}
$$

We can then follow the same derivation of P_i above in Eq. (6), but substitute this expression for $\langle n^{\star}\rangle$ in place of $\langle n\rangle$. This yields,

$$\langle n_i^{\star}\rangle = \frac{\sum_j \Sigma n_{ij}}{\sum_{j\neq i} M_{ij}} = \frac{1}{1-P_i}\cdot\frac{1-P_i}{P_i}, \text{ and therefore}$$

$$P_i = \frac{\sum_{j\neq i} M_{ij}}{\sum_j \Sigma n_{ij}}. \tag{11}$$

Interestingly, this is the identical result for the transition probability P_{ij} that is obtained with the transition probability expansion analysis described in the following section.

For further insight into this expression, consider that, from the Poisson distribution, the expected value for the number of transitions out of state i is $\langle M_i\rangle = k_i \cdot T_i$, where T_i is the total time spent in state i. Then, from Eq. (11), we find that

$$P_i = \frac{\sum_{j\neq i} M_{ij}}{\sum_j \Sigma n_{ij}} \approx \frac{T_i \sum_j k_{ij}}{T_i/\tau} = \sum_j k_{ij}\tau = k_i\tau. \tag{12}$$

Note that the expression for the P_i that is calculated here is different then in Eq. (7). From the Taylor series

$$e^x = 1 + \frac{x^1}{1!} + \frac{x^2}{2!} + \ldots, \tag{13}$$

we see that Eq. (12) is the Taylor series expansion of the transition probability given by Eq. (7), but truncated after the first-order term. Notably, since this expression in Eq. (12) is conditioned upon only the observation of dwell times, t's, that are greater than zero measurement periods, τ's, in length, this conditioned, dwell time distribution analysis is insensitive to the types of missed dwells in state i that are less than one measurement period, τ, long. As we will show further below, however, it is sensitive to other types of missed events.

Finally, it is worth noting that stochastic rate constants for a particular reaction pathway out of state i, k_{ij}, can be calculated from k_i by equating the splitting probability, p_{ij}^{split}, and the observed branching ratios as

$$p_{ij}^{\text{split}} = \frac{k_{ij}}{\sum_j k_{ij}},$$

$$\frac{M_{ij}}{\sum_{j \neq i} M_{ij}} \approx \frac{k_{ij}}{k_i}, \text{ and therefore}$$

$$k_{ij} = \frac{M_{ij}}{\sum_{j \neq i} M_{ij}} \cdot k_i. \tag{14}$$

Since we will not discuss this approach in the section on Bayesian inference further below, we note here that the calculation of the k_{ij} described earlier can be recast with a Bayesian inference approach by utilizing a Dirichlet distribution as the conjugate prior and a multinomial distribution as the likelihood function (vide infra). Regardless, while this dwell time distribution analysis approach to calculating individual stochastic rate constants is quite effective, and it has the benefit of allowing the dwell times to be checked for non-Markovian behavior that would render the calculated stochastic rate constants much less meaningful, a more straightforward method to calculate the stochastic rate constants for each parallel reaction pathway of an individual molecule is to analyze the transition probabilities for each pathway.

3.1.2 Transition Probability Expansion Analysis

Another method for calculating stochastic rate constants is to consider the observed frequency with which a single-molecule transitions from one state to another. For the discrete state trajectories considered here, this is equivalent to determining whether the single molecule in state i during a measurement period, n, is in state j during the subsequent measurement period, $n+1$. Since these data consist of multiple Bernoulli trials of whether or not the transition has occurred, the probability of a particular transition can be modeled with the binomial distribution. The binomial distribution is appropriate for modeling the number of successful trials (i.e., transitions from state i to state j, M_{ij}) from a certain number of performed trials (i.e., the number of times the single molecule was in state i in the state trajectory, $\sum n_{ij}$) that can each succeed with a fixed probability (i.e., P_{ij}), and is written as

$$P(m|n,p) = \binom{n}{m} p^m (1-p)^{n-m}, \tag{15}$$

where m is the number of successful trials, n is the total number of trials, and p is the probability of a successful trial. From the mean of the binomial distribution, $\langle m \rangle = np$, we will take frequentist approach to statistics and substitute

$$M_{ij} \approx \left(\sum\nolimits_{j\neq i} (\Sigma n_{ij})\right) P_{ij} = \left(\sum_j M_{ij}\right) P_{ij}. \tag{16}$$

Here, we have equated P_{ij} with the observed frequency of the transitions from state i to state j. However, in an experiment, only a finite number of transitions from state i to state j are observed; as such, the equality will only be approximate. Regardless, according to the central limit theorem, as the number of measurements increase, M_{ij} should approach the mean value dictated by the binomial distribution; thus, barring a small number of measurements (e.g., less than ~100 measurements), we might reasonably estimate that

$$P_{ij} = \frac{M_{ij}}{\sum_j M_{ij}}, \tag{17}$$

and from this expression, estimate k_{ij} using Eq. (8).

Now, we will consider the accuracy of calculating a stochastic rate constant in this manner. Interestingly, given a particular amount of time spent in state i in a state trajectory, T_i, the Poisson distribution indicates that

$$\langle M_{ij} \rangle = k_{ij} \cdot T_i \approx k_{ij} \cdot \left(\sum\nolimits_j M_{ij}\tau\right), \tag{18}$$

where the substitution for T_i is generally accurate, excepting the types of missed events which we will discuss further below. With this in mind, by substituting Eq. (18) into Eq. (17), we find that

$$P_{ij} = \frac{k_{ij} \cdot \sum_j M_{ij} \cdot \tau}{\sum_j M_{ij}} = k_{ij} \cdot \tau. \tag{19}$$

Therefore, rather than being corrected to a Taylor series expansion of the transition probability truncated at the first-order term, as was the case in the dwell time distribution analysis approach described in the previous section, this method of calculating transition probabilities is inherently a Taylor series expansion of the transition probability truncated at the first-order term. Regardless, the transition probability expansion analysis approach described here and the dwell time distribution analysis approach described earlier are therefore equivalent methods of calculating stochastic rate constants, which are accurate only when $k_{ij}\tau$ is small (i.e., much less than one) and the higher-order terms of the Taylor series expansion are therefore negligible.

When $k_{ij}\tau$ is large (i.e., approaching and greater than one), however, the probability of experimentally recording measurements where more than one state is occupied during a measurement period becomes substantially high. Neither the process of idealizing a signal trajectory into a state trajectory nor performing the first-order expansion of the Taylor series is well justified in such a situation. Regardless, before discussing the precision associated with calculating stochastic rate constants from individual molecules, we would like to note here that the transition probability expansion analysis approach described in this section has the added benefit of being insensitive to missed dwells, as will be discussed further below. Finally, as will also be discussed further below, this type of analysis approach is analogous to using the transition matrix from an HMM for P_{ij}.

3.2 Methods for Calculating Stochastic Rate Constants

3.2.1 Manual Idealization of Signal Trajectories

In order to calculate stochastic rate constants using either the dwell time distribution- or transition probability expansion analysis methods described earlier, a signal trajectory must first be idealized into a state trajectory. This state trajectory can then be quantified as described earlier to obtain the parameters necessary to calculate stochastic rate constants. One approach to idealizing a signal trajectory is to identify the states that are sampled by the signal trajectory, as well as the measurement periods during which transitions between the states take place, by visual inspection (e.g., as in Ha et al., 1999; Zhuang et al., 2000, 2002). Even in cases where the experimental signals corresponding to the various states are well separated and the signal trajectory has an excellent signal-to-noise ratio, however, it is still difficult and time consuming to locate the exact measurement period during which a transition occurs. In cases where the signals are insufficiently separated and/or the signal trajectory has a poor signal-to-noise ratio, therefore, this method can become quite subjective, such that different researchers, who will generally have slightly different criteria for what constitutes a state or a transition, can produce different state trajectories from the same signal trajectory, and thus different stochastic rate constants.

A more robust approach is to systematically employ a user-defined signal threshold such that transitions from one state to another state can be pinpointed by identifying the measurement periods in a signal trajectory during which the signal crosses the threshold (e.g., as in Blanchard, Gonzalez, Kim, Chu, & Puglisi, 2004; Blanchard, Kim, Gonzalez, Puglisi, & Chu, 2004; Gonzalez, Chu, & Puglisi, 2007; Lee, Blanchard, Kim, Puglisi, & Chu, 2007). Typically, thresholds are defined by generating a histogram of all of the signal values that are sampled throughout the entire signal trajectory, and subsequently identifying signal boundaries (i.e., thresholds) for each state that minimize overlap of the signal values corresponding to neighboring states. When more than two states are present, different thresholds can be used to define each state so as to allow for more flexibility when dealing with multiple states; however, it can be difficult to unambiguously specify these thresholds. Unless the signals corresponding to the various states are well separated and the signal-to-noise ratio of the signal trajectory is exceptional, there is often significant overlap between the signal values corresponding to neighboring states. As a result, natural fluctuations in the signal due to noise can result in spurious transitions that cross the threshold. This will result in the misidentification of transitions in the state trajectory, which can propagate into a misestimation of the stochastic rate constants. One approach to guard against the effects of these spurious transitions, as well as to dispel concern about the subjectivity of a user-defined signal threshold, is to repeat the process of idealizing the signal trajectory and calculating the stochastic rate constants using several, slightly different values for the user-defined signal thresholds (e.g., favoring one state, favoring the other state, exactly between, etc.), and demonstrating the robustness of the calculated stochastic rate constants to the choice of threshold (e.g., as in Gonzalez et al., 2007; Lee et al., 2007).

3.2.2 Hidden Markov Models

HMMs are a popular method to analyze signal vs time trajectories obtained from biophysics experiments (Andrec, Levy, & Talaga, 2003; Bronson et al., 2009; Chung, Moore, Xia, Premkumar, & Gage, 1990; McKinney, Joo, & Ha, 2006; Qin, Auerbach, & Sachs, 2000; van de Meent et al., 2014)—detailed descriptions can be found elsewhere (Bishop, 2006; Colquhoun & Hawkes, 1995). Briefly, in an HMM, the time-averaged signal recorded during each measurement period, τ, in a signal trajectory is assumed to be representative of some "hidden" state (i.e., the state trajectory). The underlying, hidden state trajectory, which is not directly observed, is then assumed to behave as a Markovian process that is governed according to transition probabilities. As discussed earlier, the transition probabilities of a single molecule in a Markovian system are related to stochastic rate constants governing the biomolecular system. With an HMM, the probability that a signal originates from a particular hidden state is calculated while considering the hidden state of the previous time period in order to explicitly account for the transition probability. Notably, in an HMM, the values of the signal that are observed when a single molecule is in a particular hidden state are typically assumed to be distributed according to a normal distribution PDF (i.e., the observed signals will be a Gaussian mixture model). Using this approach, one "estimates" an HMM that describes the signal in terms of a discrete number of states, and that provides, as parameters, the signal emission probabilities of each state as well as the transition probabilities as a transition probability matrix, $\boldsymbol{P}$, from each state.

With the optimal estimate of the HMM describing a signal trajectory, two different methods can be used to calculate stochastic rate constants. In the first method, the idealized, state trajectory can be obtained from the HMM and then quantified as described for use with the dwell time distribution, or transition probability expansion analysis approaches. This idealized, state trajectory is obtained by applying the Viterbi algorithm to the HMM in order to generate the Viterbi path (Viterbi, 1967). The Viterbi path, which gives the idealized state trajectory directly, is the most likely sequence of hidden states that not only would yield the observed signal values given the optimal signal emission probabilities, but that would most likely have arisen from the optimal transition probabilities. As such, it is important to note that, by using an HMM to idealize a signal trajectory, the resulting idealized state trajectory and emission- and transition probabilities have been forced to be as Markovian as possible. Therefore, if there is any non-Markovian behavior present in the biomolecular system under investigation, it will be masked and made to appear Markovian. To avoid this shortcoming of HMMs, the idealized state trajectory can be generated using a different approach, such as thresholding.

The second method for calculating stochastic rate constants from the optimal HMM estimate involves directly using the transition probabilities obtained from the HMM. While, on its surface, this method seems to bypass the use of idealized, state trajectories, the process of estimating the optimal HMM that describes the data inherently involves estimating the hidden states that generated the signal trajectory and therefore involves the use of idealized, state trajectories. From an HMM, individual stochastic rate constants can be calculated using Eq. (8) and the transition probability matrix, which is analogous to that calculated from an idealized, state trajectory. This approach is equivalent to transition probability expansion analysis. As with calculating stochastic rate constants from the Viterbi path, it must be noted that this second HMM method also enforces Markovian behavior.

Finally, we note that in the smFRET literature alone, there are several software packages available for HMM-based analysis of E_{FRET} trajectories. Of these packages, there are two types of approaches to estimating the optimal HMM that describes the data: maximum-likelihood approaches (e.g., QuB (Qin, Auerbach, & Sachs, 1997), HaMMy (McKinney et al., 2006), and SMART (Greenfeld, Pavlichin, Mabuchi, & Herschlag, 2012)) and Bayesian approaches (e.g., vbFRET (Bronson et al., 2009; Bronson et al., 2010) and ebFRET (van de Meent et al., 2014, 2013)). There are many benefits to using Bayesian HMMs over maximum-likelihood HMMs. First, unlike Bayesian HMMs, maximum-likelihood HMMs are fundamentally ill-posed mathematical problems—essentially, individual states can "collapse" onto single data points, which yields a singularity with infinite likeliness that is not at a reasonable HMM estimate. Second, as we will discuss in the next section, Bayesian approaches naturally incorporate the precision with which a certain amount of data can determine the parameters of the HMM by learning the probability distribution of the transition probabilities instead of finding one set of transition probabilities. In addition to providing the precision, this allows one to combine the results from multiple, individual molecules, and simultaneously learn consensus, stochastic rate constants from an ensemble of single molecules. Third, while maximum-likelihood approaches can result in HMMs that are significantly overfit and that consequently overestimate the number of hidden states present in a signal trajectory, Bayesian approaches are inherently able to select the correct number of hidden states present in a signal trajectory. For example, with maximum-likelihood HMMs, a better HMM estimate of the signal trajectory is obtained simply by adding additional hidden states; in the extreme case, there would be one hidden state for each data point. Although the HMM in this extreme case would fit the data perfectly, it would not be very meaningful, nor would it be a useful model for predicting the future behavior of the system. While the use of heuristic approaches such as the Bayesian and Akaike Information Criteria (BIC and AIC, respectively) has been proposed to help select the correct number of states in maximum-likelihood HMMs, these are approximations to true Bayesian approaches that are valid only under certain conditions and that, in practice, we find do not work well for the HMM-based analysis of smFRET data. Additionally, Bayesian HMMs have been shown to be more accurate than maximum-likelihood HMMs for the analysis of signal trajectories where the dwell times, t's, in the hidden states are transient relative to the measurement period, τ (Bronson et al., 2009). Finally, there is effectively no added computational cost between the maximum-likelihood and Bayesian approaches to HMMs, as both implement the same algorithms to calculate the probabilities associated with the HMM (e.g., the forward–backward algorithm), so speed is not a concern. Given the benefits of the Bayesian approach over the maximum-likelihood approach for HMMs, we recommend using Bayesian HMMs when analyzing signal trajectories from single-molecule biophysical experiments.

4. PRECISION OF CALCULATED RATE CONSTANTS

4.1 Using Bayesian Inference to Quantify Precision

The finite length of a signal trajectory ensures that only a finite number of randomly distributed dwell times and transitions will be observed during the duration of the signal trajectory. The fact that only a finite number of dwell times and transitions are observed in a signal trajectory limits the precision with which a stochastic rate constant can be calculated from that signal trajectory. With the observation of more dwell times and transitions, this precision will increase, and eventually the value of the calculated stochastic rate constant will converge to the value of the "true" stochastic rate constant. Here, we demonstrate how to rigorously quantify this precision, and therefore the amount of information contained in a single signal trajectory, through the use of Bayesian inference.

One simplistic attempt to account for variability in the number of dwell times and transitions that are observed is to report the statistical uncertainty in the calculated stochastic rate constant in the context of "bootstrapping" of the data (Efron, 1979). Bootstrapping is an attempt to simulate the data of future experiments from a set of observed data. From the analysis of bootstrapped, "future" data, any variation in subsequently calculated properties can be attributed to the uncertainty present in the original dataset. For example, when calculating stochastic rate constants from a state trajectory as described earlier, the bootstrapping process would involve creating a resampled dataset, $\boldsymbol{n}_{ij}'$, by randomly sampling from $\boldsymbol{n}_{ij}$ with replacement such that, after each sample is drawn, the sampled data point is placed back into the population before the next sample is drawn. The new, bootstrapped transition probability, P_{ij}', can then be calculated from $\boldsymbol{n}_{ij}'$, and this yields new, bootstrapped stochastic rate constants, k_{ij}'. The bootstrapping process is then repeated several times, and the reported stochastic rate constant k_{ij} is given as the mean of the set of bootstrapped k_{ij}', with the uncertainty of the reported k_{ij} given as the standard deviation of the set of bootstrapped k_{ij}'. It is important to note, however, that bootstrapping inherently assumes that the collected data accurately represent the characteristics of an infinitely large amount of data. Consequently, bootstrapping artificially inflates the dataset in a way that perpetuates any misrepresentations of an infinitely large amount of data that are present in the actual dataset. The smaller the collected dataset is, the more likely it is to misrepresent this infinitely large amount of data. In practice, bootstrapping single-molecule results, where there are often only several hundreds of individual molecules in a dataset, perpetuates these misrepresentations and leads to inaccurate rate constants, all the while not providing a reasonable estimate of the statistical error present in the calculation.

Consider the following, extreme, hypothetical calculation where only one transition with a one measurement period-long dwell time (i.e., $\boldsymbol{n}_{ij} = [1]$) has been observed in one signal trajectory. Using the conditioned dwell time distribution- or transition probability expansion analysis approaches, we find that, in this case, P_{ij} is equal to 1.0 and that all of the bootstrapped P_{ij}' are also equal to 1.0. Thus, in this case, there is no uncertainty in the calculation of the transition probability, or, consequently, in the stochastic rate constant, and this stochastic rate constant is infinitely large. Nonetheless, we know intuitively that the stochastic rate constant is probably not infinity and that there must be some uncertainty in this calculation, even though it employs only one measurement. The uncertainty lies in the fact that the one transition we have observed simply cannot be representative of the stochastic rate constant governing an entire ensemble or even an individual molecule. Likewise, we should suspect that P_{ij} is probably a poor estimate of the true transition probability. It is easy to imagine that after recording a few more measurements from that hypothetical single molecule, we might calculate a different value of P_{ij}, and that the extra data would give us a better sense of the uncertainty in P_{ij}. This extreme example illustrates how the analyses of the stochastic rate constant calculations described earlier are insufficient by themselves, even when supplemented by bootstrapping. Fortunately, in contrast to these intrinsic shortcomings, Bayesian inference provides a statistically rigorous manner with which to encode our intuition that the number of observations should change our knowledge about P_{ij}, and systematically address the uncertainty in the calculation of stochastic rate constants.

Bayesian inference is a statistical method grounded in the Bayesian approach to probability (see Sivia & Skilling, 2006 for a pedagogical introduction). In Bayesian inference approaches, the parameters of a model that has been developed to describe experimentally observed data are treated as probability distributions that reflect the consistency of the particular parameter values with the data. These probability distributions can then be updated if new data is acquired so as to be consistent with the new, as well as any previous, data. This approach is analogous to the way that a scientific hypothesis is tested and then updated with each new laboratory experiment (Sivia & Skilling, 2006). In the context of quantifying a state trajectory, Bayesian inference allows us to formulate a hypothesis about the underlying stochastic rate constants of a system (i.e., the probability of certain stochastic rate constants producing the observed state trajectory) and then to update that hypothesis as each transition, or lack thereof, is observed in the state trajectory. In this way, we can use Bayesian inference to describe the probability distribution of a stochastic rate constant as each measurement period in a signal trajectory is analyzed.

The foundation of Bayesian inference is Bayes' rule, which can be written mathematically as

$$P(\Theta|\, D) \propto P(D|\, \Theta) \cdot P(\Theta), \tag{20}$$

where Θ represents the parameters of the model and D represents the data values. The first, second, and third terms are referred to as the "posterior," the "likelihood," and the "prior," respectively. Bayes' rule can be expressed verbally as: the probability of the model's parameter values after observing the data is proportional to the product of (i) the probability of observing the data given those particular parameter values and (ii) the initial probability of those parameters. More succinctly, the posterior probability is proportional to the product of the likelihood and the prior probability.

With a model for experimental data (i.e., expressions for the likelihood and the prior probability distribution), we can calculate the posterior probability distribution and learn about the distribution of parameter values that are consistent with the experimental data. Unfortunately, for some models, these calculations can be analytically and numerically difficult, making their practical use relatively intractable. However, there are certain conditions that significantly simplify these calculations. Specifically, certain pairs of likelihood functions and prior distributions are complementary in that they yield posterior distributions that are of the same algebraic form as the prior distribution. In such a case, the prior is called the conjugate prior for that particular likelihood function. The benefit of using a conjugate prior with its corresponding likelihood function is that simple updating rules can be applied to the parameters of the conjugate prior probability distribution to yield the resulting posterior probability distribution. These calculations typically amount to the addition of certain experimental values. As such, the use of conjugate priors and likelihood functions circumvents the computationally expensive need to calculate the posterior probability distribution for every possible point in the entire probability space.

Here we describe how to employ Bayesian inference using conjugate priors and likelihood functions in both the dwell time distribution- and the transition probability expansion analysis approaches described earlier for calculating stochastic rate constants from state trajectories in a manner that is extremely tractable, and easy to employ.

4.2 Bayesian Dwell Time Distribution Analysis

To perform Bayesian inference upon the exponentially distributed dwell times that a single molecule will spend in a particular state in a state trajectory, we must first identify the likelihood function and its conjugate prior probability distribution that will serve as a model of the observed data. As described earlier, the number of consecutive, discrete measurement periods, n, that such a single molecule will spend in a particular state is distributed according to the geometric distribution PMF. Therefore, in this model, the geometric distribution PMF is the likelihood function for observing some number of sequential measurements in state i before transitioning to state j, and this depends only upon one parameter: the transition probability out of state i, P_i. Mathematically, the geometric distribution PMF is constructed such that the conjugate prior for this likelihood function is the beta distribution PDF,

$$p(P|\alpha,\beta) = P^{\alpha-1}(1-P)^{\beta-1}/B(\alpha,\beta), \tag{21}$$

where $B(x,y)$ is the beta function of x and y.

The beta distribution PDF is often used to describe the probability of a probability, P (in this case, of a successful transition out of state i, P_i), because, much like a probability, the PDF is defined continuously between 0 and 1 (Bishop, 2006). Additionally, the beta distribution PDF is a function of only two parameters, α and β, which have intuitive interpretations relating to probabilities. Notably, when $\alpha = \beta = 1$, the beta distribution is flat, as all values of P have equal probabilities. In this case, the beta distribution mathematically expresses a lack of knowledge about P in a similar manner as the equal, a priori probability assumption of statistical mechanics (Van Kampen, 2007). Along these lines, larger values of α and/or β yield more defined and peaked distributions, which expresses the increased knowledge about P. As we will discuss later, the process of performing Bayesian inference amounts to modifying the initial values of α and β in a data-dependent manner to yield a posterior, beta distribution PDF with updated values of α and β. In this sense, Bayesian inference mathematically encodes a method to express the incremented knowledge that originates from new information.

By using the geometric distribution PMF as a likelihood function, and the beta distribution PDF as its conjugate prior, we can now calculate the posterior probability distribution of the transition probability, P_i, from a state trajectory. We begin by assuming that all transition probabilities are initially equally probable. Therefore, the prior probability distribution is a beta distribution PDF with $\alpha = \beta = 1$. The posterior probability distribution will be another beta distribution PDF where α and β are interpreted as α_0 plus the number of successful transitions, and β_0 plus the number of unsuccessful transitions, respectively, where the subscript 0 refers to the prior probability distribution. Thus, for the transitions out of state i in a state trajectory, the posterior probability distribution is a beta distribution with $\alpha = 1 + \sum_{j \neq i} M_{ij}$ and $\beta = 1 + \sum_j \left(\sum n_{ij}\right)$. Therefore, since the mean of the beta distribution is $\alpha/(\alpha + \beta)$, the mean transition probability out of state i after having observed the state trajectory is

$$\langle P_i \rangle = \frac{1 + \sum_{j \neq i} M_{ij}}{1 + \sum_{j \neq i} M_{ij} + 1 + \sum_j \left(\Sigma n_{ij}\right)}. \tag{22}$$

This mean value of the transition probability converges to the maximum-likelihood estimate of P_i given in the previous section when $\alpha \gg 1$ and $\beta \gg 1$. Note that the maximum-likelihood estimate of P_i is equivalent to the mode of the beta distribution PDF, which is $(\alpha - 1)/(\alpha + \beta - 2)$.

The benefit of this Bayesian inference approach is that the posterior probability distribution of P_i not only provides a mean value but also speaks to the uncertainty inherent in P_i due to limited number of dwell times observed in state i. This uncertainty can be expressed in the form of a credible interval. A credible interval, which is similar to the frequentist idea of a confidence interval, is the range in which a certain percentage of the probability density of the PDF resides; typically one uses a 95% credible interval as this is similar to $\pm 2\sigma$ for a normal distribution, but this choice is arbitrary. The upper and lower boundaries of the credible interval can be found through the inverse of the cumulative distribution function of the beta distribution. Many standard computational programs come with a function to do this, which is sometimes called the "inverse function of the regularized incomplete beta function," $I_x(\alpha, \beta)$, where α and β are the posterior probability distribution parameters and x is the fraction of the boundary (e.g., 0.025 for 2.5%). For instance, in Matlab this function is called *betaincinv*.

Finally, let us consider the application of this Bayesian approach to observed data from a state trajectory where the length of a dwell time must be at least one measurement period in length ($n \geq 1$) in order to be associated with a particular state, as discussed earlier. Previously, we conditioned the geometric distribution PMF to only consider dwell times of at least one measurement period in length to address this problem. Now we must adapt our Bayesian inference approach to allow for this conditioning. Due to the linearity of this conditioning, and since the total likelihood function is the product of the likelihood function from each individual data point, the conditioned posterior probability distribution contains an extra term of $\left(\frac{1}{1-P}\right)^{\left(\sum_{j\neq i} M_{ij}\right)}$. This is equivalent to setting $\beta' = \beta - \sum_{j\neq i} M_{ij}$, where β' is the parameter used in the beta distribution for the posterior probability distribution, and β is the parameter calculated above. Using α and β' as the parameters for a beta distribution PDF, the posterior probability distribution of the transition probability, P_i, can be accurately and precisely quantified as a function of each successive, observed dwell time, even though dwell times of zero length are missed in the state trajectory (Fig. 1). With a sufficient number of measurements, this approach yields the same mean transition probability as the maximum-likelihood estimate of the transition probability expansion analysis, thereby rendering this approach insensitive to some types of missed events that we will discuss further below.

To be concrete, we will use this Bayesian dwell time distribution analysis approach to analyze the extreme, hypothetical case of the single observed transition introduced in the previous section. The posterior probability distribution would be a beta distribution with $\alpha = (1+1) = 2$, and

$\beta' = (1 + 1 - 1) = 1$. This yields $\langle P_i \rangle = 0.66$, with a lower bound of $P_i = 0.16$, and an upper bound of $P_i = 0.99$ for the 95% credible interval. Notably, the mean value of the transition probability calculated using the Bayesian dwell time distribution analysis approach is not infinitely large, as was the estimate of P_i using the maximum-likelihood approach as described earlier, and, by noting that the credible interval is consistent with a wide range of transition probabilities, this method inherently accounts for the large uncertainty in the transition probability that we intuitively expect (Fig. 1).

The transition probabilities calculated with this approach can also be transformed into the stochastic rate constants with Eq. (8). Therefore, this Bayesian inference-based method also provides an intuitive, explicit expression for how the uncertainty in the stochastic rate constants, k_i, diminishes with additional observations. One interesting case is that when no

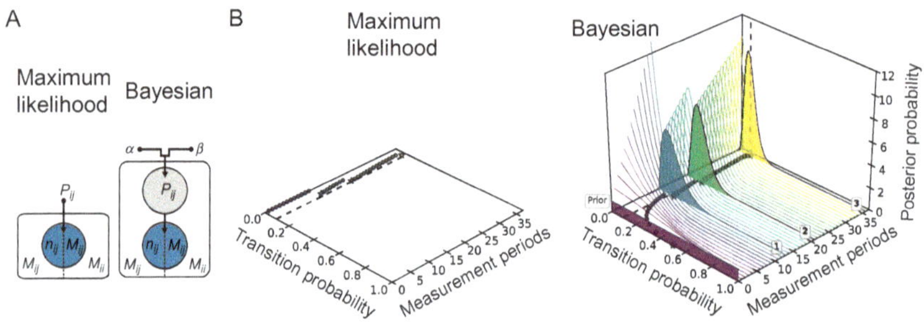

Fig. 1 Maximum-likelihood vs Bayesian approaches to calculating transition probabilities. (A) Graphical models of maximum-likelihood (ML) (*left*), and Bayesian (*right*)-based methods for calculating the transition probability from state *i* to state *j*, P_{ij}. Each model is divided in half to give the dwell time distribution analysis (*left*) or the transition probability expansion analysis (*right*). *Blue circles* represent the observed variables, *gray circles* represent hidden variables, and *black dots* represent fixed parameters. The Bayesian model expands upon the ML model by using a probability distribution to describe P_{ij}. (B) The calculations of P_{ij} from three dwell times using ML (*left*)- and Bayesian (*right*)-based approaches are plotted as a function of increasing measurement periods (i.e., observations in a state trajectory). The true transition probability is shown with a *dashed line*. Both the ML value and the mean of the posterior probability distribution value of P_{ij} calculated with dwell time distribution analysis (*stars*) and transition probability expansion analysis (*circles*) are shown. Additionally, for the Bayesian approach, the posterior probability distributions are plotted for dwell time distribution analysis (*filled curves*) and for transition probability expansion analysis (*thin curves*). The prior probability distribution and the numbers of the dwell times are denoted in *boxes*. Notably, the Bayesian-based approach yields nonzero transition probabilities and also provides the uncertainty in P_{ij} in the form of a probability distribution.

measurements have been made, the posterior distribution of the rate constants is equivalent to the prior distribution; all rate constants from 0 to ∞ are therefore equally probable. Thus, this analysis method is a very objective approach to analyzing transition probabilities from discrete state trajectories, and it is one that intrinsically encodes a statistically rigorous approach to the precision of such calculations.

4.3 Bayesian Transition Probability Expansion Analysis

We can also extend the transition probability expansion analysis approach to account for the precision of these calculations in a statistically robust manner with the application of Bayesian inference. Since the probability of undergoing a transition from state i to state j during a measurement period, n, was modeled with the binomial distribution, the binomial distribution will be the likelihood function used to perform Bayesian inference. The binomial distribution depends upon a single parameter: the probability of a success, P, which, in this case, is the transition probability P_{ij}. Mathematically, the conjugate prior to the binomial distribution is also the beta distribution, which is consistent with the interpretation of the beta distribution as describing the probability of a probability. Without any foreknowledge of the transition probability or, equivalently, the stochastic rate constant, we will use a flat, uninformative prior of $\alpha_0 = \beta_0 = 1$. From this prior probability distribution, the resulting posterior probability distribution for P_{ij} is a beta distribution with $\alpha = 1 + M_{ij}$, and $\beta = 1 + \left(\sum_j M_{ij}\right) - M_{ij}$. Interestingly, while this posterior probability distribution can be quantified for each observed transition trial, it is equivalent to the posterior probability distribution calculated using Bayesian dwell time distribution analysis once all of the transition trials that comprise a particular dwell time have been analyzed (Fig. 1). For the extreme example of a state trajectory with one transition from a one measurement period-long dwell time ($\boldsymbol{n}_{ij} = [1]$) the posterior probability distribution would then be $\alpha = (1 + 1) = 2$, and $\beta = (1 + 1 - 1) = 1$. The mean and the credible interval for the beta distribution can then be calculated as described earlier, as can the stochastic rate constants related to these transition probabilities.

Interestingly, a more encompassing, Bayesian approach to inferring transition probabilities is obtained by considering all of the parallel reaction pathways out of state i at once. In this case, the multivariate generalization of the binomial distribution, which is called the multinomial distribution, is more appropriate for the likelihood function, as it models the probability of a

Bernoulli trial where there are different types of successes—although only one type of success is chosen at a time. The conjugate prior to the multinomial distribution is the Dirichlet distribution,

$$p(\boldsymbol{x}|\,\boldsymbol{\alpha}) = \frac{1}{\boldsymbol{B}(\boldsymbol{\alpha})}\prod\nolimits_{i=1}^{K} x_i^{\alpha_i - 1}, \tag{23}$$

where bold characters denote a vector and $\boldsymbol{B}(\boldsymbol{x})$ is the multinomial beta function of $\boldsymbol{x}$. Unsurprisingly, the Dirichlet distribution is the multivariate generalization of the beta distribution; in fact, in the case of only one type of success (i.e., in one dimension), they are equivalent. Analogously, we will use a flat, uninformative prior of $\alpha_{ij} = 1$, such that each jth element of $\boldsymbol{\alpha}$ is unity. As a result, the posterior probability distribution is $\alpha_{ij} = 1 + M_{ij}$. In order to analyze the transition probability of an individual reaction pathway out of state i from this posterior probability distribution of the transition probabilities for all the possible transitions, we can marginalize the posterior Dirichlet distribution. The result is that the posterior probability distribution for one of the reaction pathways is a beta distribution with $\alpha = \alpha_{ij} = 1 + M_{ij}$, and $\beta = \beta_{ij} = \left(\sum\nolimits_j \alpha_{ij}\right) - \alpha_{ij} = \left(\sum\nolimits_j 1\right) + \left(\sum\nolimits_j M_{ij}\right) - \left(1 + M_{ij}\right)$. This is equivalent to the binomial result for a two-state system given at the start of this section. Regardless, the most notable aspect of this treatment is that the mean posterior probability distribution is equivalent to the transition probability matrix that is calculated using an HMM. Notably, the Bayesian-based HMMs go even further and utilize Dirichlet distributions such as this one to describe the posterior probability distributions of the transition probabilities (Bronson et al., 2009, 2010; van de Meent et al., 2014, 2013). As such, both this Bayesian transition probability expansion analysis approach and the Bayesian-based HMMs are able to describe the precision associated with the transition probabilities calculated from a finite number of transitions by calculating a credible interval from the marginalized distribution as described earlier.

Importantly, unlike maximum-likelihood methods, the Bayesian inference-based approach to transition probability expansion analysis enables the statistically robust analysis of trajectories where there are not only zero transitions to a particular state, but also when there are no transitions at all during a state trajectory. In these cases, the on-diagonal elements of $\boldsymbol{M}$, M_{ii}, will reflect the measurements from the state trajectory that were assigned to state i, even though the final state was unclear. In doing so, the prior probability distribution accounts for the numerical instability that would otherwise yield infinitely precise estimates of stochastic rate constants that are zero when using the maximum-likelihood approach.

5. ACCURACY OF CALCULATED STOCHASTIC RATE CONSTANTS

5.1 Characterizing Missed Events

While discretized, idealized state trajectories can be used to analyze the single-molecule reactions, many factors complicate the quantification of these state trajectories and limit the amount of information that can be extracted from them. For instance, if the underlying single-molecule reaction is faster than the time resolution (i.e., the integration time of each measurement) of the experimental technique used to record the signal trajectories from which the state trajectories originate, then there is a risk that excursions to states with dwell times, t, that are significantly shorter than the measurement period, τ, will be missed. The consequence of this type of situation is that the idealized, discretized state trajectory will contain missing transitions, misclassified transitions, and missed dwells such that it is no longer a reasonable representation of the underlying single-molecule reaction. As a rule of thumb, the effects of missed events in a state trajectory begin to become pronounced when, for a stochastic rate constant, k, the condition $k\tau > 0.1$ is true (i.e., k is greater than about 1/10th of the acquisition rate). This is because, for a Markovian reaction, the exponential distribution dictates that when $k\tau = 0.1$ about 10% of the dwell times will be shorter than the measurement period, τ. This percentage increases as the stochastic rate constant increases, leading to a substantial number of missed events. In the sections that follow, we discuss how missing such events when transforming signal trajectories into state trajectories complicates the process of analyzing single-molecule kinetic data using state trajectories (Fig. 2), and then discuss how one might correct for these effects in order to ensure the accuracy of analyzing single-molecule kinetic data using state trajectories.

5.1.1 Finite Length of Signal Trajectories

Many factors limit the length of the signal trajectories that can be collected from individual biomolecules using single-molecule kinetic techniques. Superficially, the patience of the experimenter and the practical data storage limitations of computers restrict this length. Practically, the stability of the biomolecular system can limit the length of an experiment; for instance,

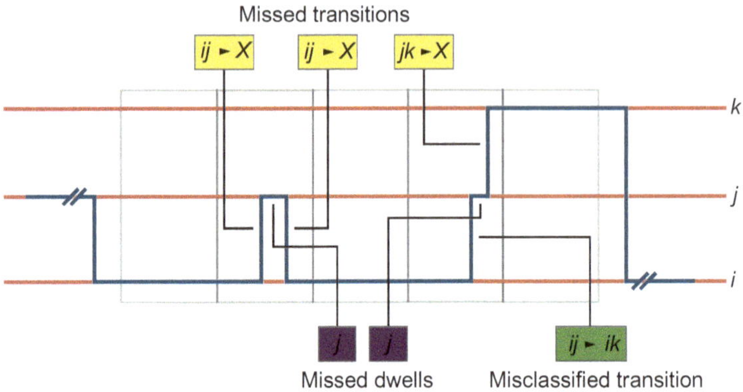

Fig. 2 Types of missed events. An example of a single-molecule's path through state space is shown in *blue*, and it transitions between three states (*i*, *j*, and *k*) shown in *red*. Measurement periods over which the experimental signal is time averaged are shown as alternating *white* and *gray boxes*. Missed transitions are shown in *yellow*, misclassified transitions are shown in *green*, and missed dwells are shown *purple*.

many enzymes become inactive after a certain time spent at room temperature under in vitro conditions, or, depending on the acid-base properties of the reactants and products of the reaction being investigated, the buffering capacity of a buffer might saturate. More commonly, however, is the fact that the signal corresponding to an individual molecule can simply be lost, for instance, by photobleaching of a fluorophore, or by dissociation of a tether, and such an event terminates the signal trajectory. Regardless of the cause, signal trajectories are finite in length and do not extend infinitely. Thus, considering the ergodic hypothesis, the data from a single molecule will consequently not contain enough information to completely characterize a system. In an extreme case, one can imagine a state trajectory where no transitions occur before signal loss. Such a situation places a clear limitation on the precision with which the dynamics of the single-molecule system can be quantified. This consideration applies to all state trajectories generated during the analysis of single-molecule kinetic data, because all of these trajectories will have a finite length.

5.1.2 Missed Transitions

Consider a single molecule that dwells in a particular state, i, for some length of time, t. Eventually, the single molecule will transition to a new state, j. If the dwell time, t, in j is shorter than the measurement period, τ, there is a chance that the single molecule might transition back to i during the measurement period (Fig. 2, yellow boxes). This is more likely to occur with increasingly fast rate constants for the transition from j to i. In such a case, neither the initial transition from i to j nor the subsequent transition from j to i would be registered in the state trajectory. Instead, the single molecule would appear to have remained in i throughout this measurement period, n, —not having transitioned to the new state; this event is called a missed transition, and they affect the $\boldsymbol{n}_{ij}$, and thus $\boldsymbol{M}$. The direct consequence of the missed transition is that the number of transitions from i to j, M_{ij}, would be underestimated, ultimately resulting in an underestimation of k_{ij}. Additionally, as a result of the missed transition, the initial dwell time in i would be overestimated, because it would be the combined length of the initial dwell time and the following dwell time in i, consequently resulting in an overestimation of M_{ii}, and, ultimately, an underestimation of k_{ij}. Similarly, in this example, the transition back from j to i is also missed, resulting in an underestimation of M_{ji}, and therefore an underestimation of k_{ji}.

5.1.3 Misclassified Transitions

A related occurrence is that of misclassified transitions, rather than of missed transitions. In this case, a single molecule beginning in state i could transition to state j, where it dwells for a period of time, t, that is less than the measurement period, τ. Instead of transitioning back from j to i, as in the example above, however, the single molecule could transition to a third, distinct state, k. In this case, the initial dwell time in i can approximately be correctly measured from the state trajectory, but the transition from i to j will be misclassified as a transition from i to k, and the transition from j to k will be entirely missed (Fig. 2, green box). As a result of this misclassification, M_{ij} will be underestimated, while M_{jk} will be overestimated. These misestimations result in an underestimation of k_{ij}, and an overestimation of k_{jk}. Moreover, in cases where j is an obligatory intermediate in the transition from i to k, such misclassified transitions could lead to an incorrect kinetic model in which the fact that j is an obligatory intermediate is not deduced and, instead, direct transitions from i to k are erroneously concluded to occur.

5.1.4 Missed Dwells

In the example of the missed transition from state i to state j given above, we described a dwell time, t, in state j that was shorter than the measurement period, τ. This transient dwell time, which resulted in the missed transition, is called a missed dwell because it is so short that the time spent in j was not registered in the state trajectory (Fig. 2, purple boxes). While the missed dwell is closely related to the missed transitions (it is causal), it and its effects are conceptually distinct from a missed transition. The missed dwell in j yields an underestimation of M_{jj}, and, consequently, an overestimation of k_{jx}, where x stands for any state accessible from j. However, it also can provide drastic overestimates of the entries in $\boldsymbol{n}_{ix}$, which, as we show later, can seemingly distort otherwise normal Markovian behavior.

5.2 Correcting Rate Constants for the Finite Length of Signal Trajectories

Biomolecular systems may undergo very long-lived dwell times, t, relative to the finite length of a signal trajectory. For example, in an smFRET experiment, signal loss due to fluorophore photobleaching can occur before a transition occurs. In such a case, the entire state trajectory is typically discarded and is not included in any subsequent dwell time distribution analysis. This is because the arbitrary experimental end-time of the signal trajectory truncates the last and only dwell time, and it is therefore unclear to which $\boldsymbol{n}_{ij}$ such a dwell time belongs. As a result, such long-lived dwell times are typically unclassified, and systematically excluded from further analyses, which can result in a misestimated counting matrix, $\boldsymbol{M}$, but also, it reduces the amount of data in $\boldsymbol{M}$ to a point where any subsequent calculation of a stochastic rate constant will be extremely imprecise.

Fortunately, there is a straightforward correction that can be employed to correct for this loss of the excluded data, which relies on a control experiment. By including the unclassified dwell times in the ith state into M_{ii}, the counting matrix is augmented to account for the effect of not having observed a transition during the finite length of the signal trajectory. This is true if the finite length of the trajectory is due to stochastic causes (e.g., photobleaching, or dissociation of a tether) or deterministic causes (e.g., prematurely terminated data collection) (Wang, Caban, & Gonzalez, 2015). Notably, the uncertainty in the transition probabilities quantified by the Bayesian inference approaches introduced in the previous sections accounts for the unobserved transitions. One complicating factor, however, is that any resulting stochastic rate constant calculated from this counting matrix will be the sum of the parallel reaction pathways of both the reaction under consideration, as well as the stochastic causes of signal termination. Mathematically, this can be expressed as

$$k_{ij}^{\mathrm{obs}} = k_{ij} + k_{st}, \tag{24}$$

where k_{ij}^{obs} is the observed stochastic rate constant from states i to j calculated from the augmented counting matrix, and k_{st} is the stochastic rate constant governing the stochastic termination of the signal trajectory. Fortunately, k_{st} can be measured using a control experiment performed at the single-molecule level or at the ensemble level (e.g., by measuring the rate of photobleaching or of dissociation of a tether. Therefore, the true stochastic rate constant in the absence of these signal-terminating processes, k_{ij}, can be calculated using Eq. (24). Finally, we note that this correction can easily be extended to address additional considerations, such as inactive subpopulations, as it simply entails modifying the on-diagonal elements of the counting matrix, $\boldsymbol{M}$, to account for otherwise ignored contributions.

5.3 Correcting Stochastic Rate Constants for Missed Dwells and Transitions

One well-characterized method to correct for the effects of missed dwells and missed transitions upon the calculation of stochastic rate constants is through the augmentation of the kinetic mechanism with "virtual states" (Crouzy & Sigworth, 1990). This method originated in the field of single-molecule conductance measurements on ion channels, where researchers such as Colquhoun and Hawkes pioneered the use of HMMs to analyze the stochastic kinetics of individual ion channel opening and closing events (Colquhoun & Hawkes, 1995). The general approach of this method to correct stochastic rate constants is to consider the expected number of missed dwells in a particular state. These expected, missed dwells are then classified into virtual states, which then account for any missed dwells without artificially contaminating the dwells that were actually observed. While this method was developed in Crouzy and Sigworth (1990), and reviewed several times since (Colquhoun & Hawkes, 1995; Stigler & Rief, 2012), we briefly explore it here for completeness.

Assume that there is some "cutoff time," τ_c, for which a dwell time shorter than τ_c would become a missed dwell in a state trajectory. Interestingly, τ_c is related to the distinction in signal between two states in a signal trajectory, more than to a particular dwell time. For instance, if one is assigning states in a state trajectory based upon the crossing of a threshold, then τ_c is the amount of time in a state that yields a time-averaged signal that crosses that threshold. Along these lines, τ_c is also related to the noise and other particulars of the recording equipment used in the experiment. Unfortunately, it remains an open question as to how to determine τ_c exactly (Crouzy & Sigworth, 1990; Stigler & Rief, 2012). For example, consider the asynchronicity of the stochastic transitions between states relative to the start of a measurement period in a signal trajectory. For an arbitrary dwell time of length $t=\tau$, the measurement period length, a single molecule will, at least, occupy the state for one-half of a measurement period, and, at most, for all of a measurement period; the exact amount depends upon the exact times when the transition occurred, and when the measurement began. Regardless, given an evenly spaced threshold, both of these observed dwell times of length τ would time average the signal past the threshold—either during the measurement period, n, where the transition occurred or during the neighboring one, $n+1$. However, given several dwell times of the exact same length $\tau/2<t<\tau$, only some of these dwell times would pass the threshold and be detected; the success of these detections would depend only upon the stochastic time of the transition relative to the beginning and end-time of the measurement period. Therefore, any static value of τ_c stochastically excludes only some, but not all, of the dwell times of these lengths. Regardless, τ_c should hypothetically be between 0 and τ.

To perform the stochastic rate constant correction, consider a single-molecule experiment on a reversible, two-state system, $1 \rightleftharpoons 2$, with forward and reverse stochastic rate constants of k_{12} and k_{21}, and where measurements are made with a measurement time period, τ. For instance, this reaction could be a conformational change, ligand binding event, or folding process between states 1 and 2 of a biomolecular system. In this case, the forward reaction occurs from state $i=1$ only to state $j=2$, while the reverse reaction occurs from state $i=2$ only to state $j=1$. For a particular observed dwell time in state 1, the following dwell time in state 2 can either be a missed dwell or an observed dwell if it is of length $t<\tau_c$, or $t>\tau_c$, respectively. Since each missed dwell can induce a missed transition in a state vs time trajectory, this criterion also allows us to split the true number of transitions in a state trajectory into those that are observed transitions, and those that are missed transitions. Furthermore, by considering the mean of the Poisson distribution, an equivalent statement can be made for the stochastic rate constants; therefore, the true stochastic rate constants can be partitioned as

$$k_{\text{true}} = k_{\text{observed}} + k_{\text{missed}}. \tag{25}$$

From Eq. (25), we can calculate a corrected stochastic rate constant, $k_{\text{corrected}}$, in place of k_{true}, by utilizing a virtual state to account for the contribution for k_{missed}. Since we know the dwell times assigned to the virtual state are those that were missed, this expression can be written as

$$k_{\text{corrected}} = k_{\text{observed}} + f_{\text{missed}} \cdot k_{\text{corrected}}, \tag{26}$$

where f_{missed} is the fraction of the total transitions that are missed transitions. Because a missed dwell in the subsequent state causes a missed transition in the state of interest, for a Markovian system, f_{missed} in state 1 is the fraction of dwell times in state 2 that are less than τ_c, which is $f_{\text{missed}} = 1 - e^{-k_{21}\tau_c}$. An equivalent expression can be written for the f_{missed} in state 2. Therefore, by substituting this expression into Eq. (26), we find

$$k_{12,\text{corrected}} = k_{12,\text{observed}} \cdot e^{k_{21,\text{corrected}}\tau_c},$$

and

$$k_{21,\text{corrected}} = k_{21,\text{observed}} \cdot e^{k_{12,\text{corrected}}\tau_c}. \tag{27}$$

This resultant set of coupled equations is nonlinear, so the solution to the corrected stochastic rate constants can be calculated numerically by minimizing the sum of squares of these equations (Stigler & Rief, 2012). Without the correction, the observed stochastic rate constants for a two-state system begin to become inaccurate when the stochastic rate constants become faster than one-tenth of the acquisition rate, τ^{-1}. This correction increases the region over which stochastic rate constants can be accurately calculated, such that the corrected stochastic rate constants are now ~90% accurate when they approach the acquisition rate; the inaccuracy is partially due to an unclear choice of τ_c, and also that the correction assumes a well-quantified k_{observed}, which may not be the case, especially given any misclassified transitions. Additionally, there are sets of true stochastic rate constants that do not provide a solution to these equations, and those that do unfortunately have two solutions—one with faster stochastic rate constants and one with slower stochastic rate constants—so, it can be challenging to pick the proper solution (Crouzy & Sigworth, 1990).

5.3.1 Seemingly Non-Markovian Behavior Induced by Missed Dwells

While we have described how to partially account for missed dwells and missed transitions when calculating stochastic rate constants from state trajectories, the assumptions used to both calculate the observed stochastic rate constants and correct the observed stochastic rate constants rely on the system being Markovian. Experimentally, many single-molecule systems seem to exhibit non-Markovian behavior (Austin, Beeson, Eisenstein, Frauenfelder, & Gunsalus, 1975; English et al., 2006), and this is typically assessed, if at all, by checking to see whether the discrete dwell times observed in a particular state are distributed according to the geometric distribution PMF (Markovian) or not (non-Markovian). Again, all of the methods described earlier that directly address stochastic rate constants assume Markovian behavior and should not be applied in the case of non-Markovian behavior. Additionally, it is worth noting that model selection for HMMs depends upon this assumption as well (Bronson et al., 2009, 2010; van de Meent et al., 2014, 2013). With these limitations in mind, here we demonstrate that one particularly detrimental consequence of missed dwells in an otherwise Markovian state trajectory is the introduction of seemingly non-Markovian behavior.

To demonstrate the introduction of seemingly non-Markovian behavior into a Markovian system during the analysis of state trajectories, consider a single-molecule kinetic experiment that is performed on a reversible, two-state, Markovian system, $1 \rightleftharpoons 2$, with forward and reverse rate constants k_{12} and k_{21}, respectively. As before, in this case the forward reaction occurs from state $i=1$ only to state $j=2$, while the reverse reaction occurs from state $i=2$ only to state $j=1$. If even one of these stochastic rate constants is relatively fast compared to the acquisition rate, there will be many missed dwells for that state. To be concrete, one such system might be where $k_{12}=0.5\ \mathrm{s}^{-1}$, $k_{21}=10.0\ \mathrm{s}^{-1}$, and $\tau=0.1$ s; here k_{21} is equal to the acquisition rate, while k_{12} is 20 times slower, and we therefore expect that there will be many missed dwells in state 2. The subsequent missed events can be readily observed in a signal trajectory (Fig. 3).

After idealizing this signal trajectory into a state trajectory, perhaps by using a threshold, the observed length of each dwell time is used to calculate the stochastic rate constants. While the observed length of a dwell time in the state trajectory depends upon the true length of the dwell time in question, it also depends upon the true lengths of previous and subsequent dwell times. This is evident by considering the effect that a missed dwell has upon the state trajectory. Consider a dwell time in state 1 that is longer than the measurement period, τ, which is followed by a dwell time in state 2 that is shorter than the measurement period, τ (Fig. 2). Since this short dwell time in the transiently occupied state 2 will be missed, the previous and subsequent dwell times in state 1 are compounded together to create an overly long, observed dwell time in the state trajectory. These compounded dwell times can be composites of two, three, four, or higher integer numbers of dwells in state 1; where the exact number is one more than the number of missed events in state 2. This dwell time compounding phenomenon also occurs for the dwell times in state 2, when the subsequent dwell time in state 1

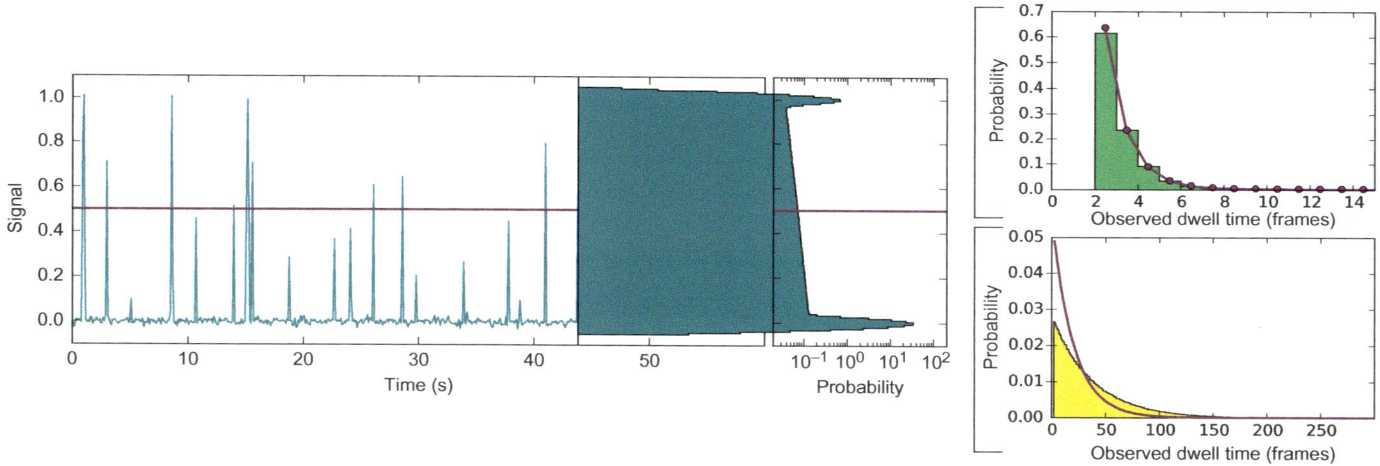

Fig. 3 Seemingly non-Markovian behavior from a Markovian, two-state system. (*Left*) Plot of the first 60 s of a signal trajectory from a simulated two-state, Markovian system. A state trajectory for this system was simulated for 2.5×10^6 s, and then the corresponding signal with signal means of 0 and 1 for states 1 and 2, respectively, was time averaged for each measurement period to create a signal trajectory. A negligible amount of Gaussian noise was added for visibility. The *purple line* denotes the threshold used to idealize the data back into a state trajectory for analysis. Many dwells in the upper state are so transient that they result in missed dwells and missed transitions. (*Right*) Histograms of the observed dwell times for state 1 (*top*) and state 2 (*bottom*). The *purple curves* are geometric distribution PMFs conditioned upon dwell times greater than one measurement period, which were calculated using the exact stochastic rate constants that were used for the simulation. Deviations are the apparent non-Markovian behavior. Contributions from observed dwell times comprised of compounded dwell times are shown as *dashed curves*.

is too short; nonetheless, in this example, there are rarely any missed dwells in state 1 because k_{12} is so slow.

Observed dwell times that are actually several compounded dwell times introduce seemingly non-Markovian behavior into the state trajectory. This is apparent when inspecting the distribution of the lengths of the observed dwells in the state trajectory (Fig. 3, left). If the system is Markovian, these discrete dwell times should be distributed according to the geometric distribution PMF as described earlier. However, it is clear that the geometric distribution PMF does not adequately describe the distribution of these observed dwell times in this example, especially for the dwell times in state 1 (Fig. 3, right). Despite the Markovian behavior used to simulate this two-state system, the fast stochastic rate constant k_{21} yields a dwell time distribution with behavior that is seemingly non-Markovian. As such, it is important to recognize that, in this particular situation, analysis methods that assume Markovian behavior would be deemed inappropriate for analyzing this data. Beyond the problem of attempting to correct for missed events, in order to accurately calculate stochastic rate constants from single-molecule kinetic data recorded on systems governed by such fast stochastic rate constants, a new approach must be developed that would effectively enable "temporal superresolution" of the data collected from any single-molecule kinetic technique. Recently, we have developed a Bayesian inference-based method to do this that we call Bayesian inference for the analysis of sub-temporal resolution data (BIASD) (manuscript in preparation).

6. CONCLUSIONS

While we highlighted several well-established methods for calculating stochastic rate constants from state trajectories, the reporting of the precision associated with the resultant stochastic rate constants has often been underappreciated. Here, we have shown that a common Bayesian inference-based framework is able to provide the uncertainty associated with the analysis of every data point in a statistically robust manner; thus, not only does it provide an intuitive method to integrate concerns about precision into stochastic rate constant calculations, but it helps to maximize the efficiency of the experiment by enabling the analysis of the entirety of the data. In addition, we categorized the types of missed events that often appear after idealizing signal trajectories into state trajectories and discussed the consequences of these missed events as well as some of the methods that can be implemented to account for them and improve the accuracy of stochastic rate constant calculations. Perhaps in the future, more detailed, statistical descriptions of the underlying molecular dynamics present in the signal trajectories, such as that offered by BIASD, will be developed to further overcome these limitations.

ACKNOWLEDGMENTS

This work was supported by two NIH-NIGMS grants (R01 GM084288 and R01 GM119386) and a Camille Dreyfus Teacher-Scholar Award to R.L.G. C.D.K.-T. was supported by the Department of Energy Office of Science Graduate Fellowship Program (DOE SCGF), made possible in part by the American Recovery and Reinvestment Act of 2009, administered by ORISE-ORAU under contract number DEAC05-06OR23100, and by Columbia University's NIH Training Program in Molecular Biophysics (T32-GM008281). N.A.B. was supported by the Department of Defense (DoD) through the National Defense Science & Engineering Graduate Fellowship (NDSEG) Program (32 CFR 168a) administered by the American Society for Engineering Education (FA9550-11-C-0028).

REFERENCES

Andrec, M., Levy, R. M., & Talaga, D. S. (2003). Direct determination of kinetic rates from single-molecule photon arrival trajectories using hidden Markov Models. *The Journal of Physical Chemistry. A*, *107*, 7454–7464.

Austin, R. H., Beeson, K. W., Eisenstein, L., Frauenfelder, H., & Gunsalus, I. C. (1975). Dynamics of ligand binding to myoglobin. *Biochemistry*, *14*, 5355–5373.

Bishop, C. M. (2006). *Pattern recognition and machine learning*. New York: Springer.

Blanchard, S. C., Gonzalez, R. L., Kim, H. D., Chu, S., & Puglisi, J. D. (2004). tRNA selection and kinetic proofreading in translation. *Nature Structural and Molecular Biology*, *11*, 1008–1014.

Blanchard, S. C., Kim, H. D., Gonzalez, R. L., Puglisi, J. D., & Chu, S. (2004). tRNA dynamics on the ribosome during translation. *Proceedings of the National Academy of Sciences of the United States of America*, *101*, 12893–12898.

Bronson, J. E., Fei, J., Hofman, J. M., Gonzalez, R. L., & Wiggins, C. H. (2009). Learning rates and states from biophysical time series: A Bayesian approach to model selection and single-molecule FRET data. *Biophysical Journal*, *97*, 3196–3205.

Bronson, J. E., Hofman, J. M., Fei, J., Gonzalez, R. L., & Wiggins, C. H. (2010). Graphical models for inferring single molecule dynamics. *BMC Bioinformatics*, *11*, S2.

Chung, S. H., Moore, J. B., Xia, L. G., Premkumar, L. S., & Gage, P. W. (1990). Characterization of single channel currents using digital signal processing techniques based on hidden Markov models. *Philosophical Transactions of the Royal Society of London. Series B, Biological Sciences*, *329*, 265–285.

Colquhoun, D., & Hawkes, A. G. (1977). Relaxation and fluctuations of membrane currents that flow through drug-operated channels. *Proceedings of the Royal Society of London B: Biological Sciences*, *199*, 231–262.

Colquhoun, D., & Hawkes, A. G. (1981). On the stochastic properties of single ion channels. *Proceedings of the Royal Society of London. Series B, Biological Sciences*, *211*, 205–235.

Colquhoun, D., & Hawkes, A. G. (1995). The principles of the stochastic interpretation of ion-channel mechanisms. In *Single-channel recording* (pp. 397–482). USA: Springer.

Crouzy, S. C., & Sigworth, F. J. (1990). Yet another approach to the dwell-time omission problem of single-channel analysis. *Biophysical Journal*, *58*, 731–743.

Efron, B. (1979). Bootstrap methods: Another look at the jackknife. *The Annals of Statistics*, *7*, 1–26.
English, B. P., Min, W., van Oijen, A. M., Lee, K. T., Luo, G., Sun, H., et al. (2006). Ever-fluctuating single enzyme molecules: Michaelis-Menten equation revisited. *Nature Chemical Biology*, *2*, 87–94.
Gillespie, D. T. (1976). A general method for numerically simulating the stochastic time evolution of coupled chemical reactions. *Journal of Computational Physics*, *22*, 403–434.
Gillespie, D. T. (1977). Exact stochastic simulation of coupled chemical reactions. *The Journal of Physical Chemistry*, *81*, 2340–2361.
Gillespie, D. T. (2007). Stochastic simulation of chemical kinetics. *Annual Review of Physical Chemistry*, *58*, 35–55.
Gonzalez, R. L., Chu, S., & Puglisi, J. D. (2007). Thiostrepton inhibition of tRNA delivery to the ribosome. *RNA*, *13*, 2091–2097.
Greenfeld, M., Pavlichin, D. S., Mabuchi, H., & Herschlag, D. (2012). Single molecule analysis research tool (SMART): An integrated approach for analyzing single molecule data. *PloS One*, *7*, e30024.
Greenleaf, W. J., Woodside, M. T., & Block, S. M. (2007). High-resolution, single-molecule measurements of biomolecular motion. *Annual Review of Biophysics and Biomolecular Structure*, *36*, 171–190.
Ha, T., Zhuang, X. W., Kim, H. D., Orr, J. W., Williamson, J., & Chu, S. (1999). Ligand-induced conformational changes observed in single RNA molecules. *Proceedings of the National Academy of Sciences of the United States of America*, *96*, 9077–9082.
Lee, T.-H., Blanchard, S. C., Kim, H. D., Puglisi, J. D., & Chu, S. (2007). The role of fluctuations in tRNA selection by the ribosome. *Proceedings of the National Academy of Sciences of the United States of America*, *104*, 13661–13665.
McKinney, S. A., Joo, C., & Ha, T. (2006). Analysis of single-molecule FRET trajectories using hidden Markov modeling. *Biophysical Journal*, *91*, 1941–1951.
McQuarrie, D. A. (1963). Kinetics of small systems. I. *The Journal of Chemical Physics*, *38*, 433.
McQuarrie, D. A. (1967). Stochastic approach to chemical kinetics. *Journal of Applied Probability*, *4*, 413–478.
Moffitt, J., Chemla, Y., Smith, S., & Bustamante, C. (2008). Recent advances in optical tweezers. *Annual Review of Biochemistry*, *77*, 205–228.
Onsager, L. (1931). Reciprocal relations in irreversible processes. I. *Physics Review*, *37*, 405–426.
Qin, F., Auerbach, A., & Sachs, F. (1997). Maximum likelihood estimation of aggregated Markov processes. *Proceedings of the Biological Sciences*, *264*, 375–383.
Qin, F., Auerbach, A., & Sachs, F. (2000). A direct optimization approach to hidden Markov modeling for single channel kinetics. *Biophysical Journal*, *79*, 1915–1927.
Resnick, S. I. (1992). *Adventures in stochastic processes*. Basel: Birkhauser Verlag.
Roy, R., Hohng, S., & Ha, T. (2008). A practical guide to single-molecule FRET. *Nature Methods*, *5*, 507–516.
Sivia, D. S., & Skilling, J. (2006). *Data analysis: A Bayesian tutorial*. Oxford: Oxford University Press.
Stigler, J., & Rief, M. (2012). Hidden Markov analysis of trajectories in single-molecule experiments and the effects of missed events. *ChemPhysChem*, *13*, 1079–1086.
Tinoco, I., & Gonzalez, R. L. (2011). Biological mechanisms, one molecule at a time. *Genes & Development*, *25*, 1205–1231.
van de Meent, J.-W., Bronson, J. E., Wiggins, C. H., & Gonzalez, R. L. (2014). Empirical bayes methods enable advanced population-level analyses of single-molecule FRET experiments. *Biophysical Journal*, *106*, 1327–1337.
van de Meent, J.-W., Bronson, J. E., Wood, F., Gonzalez, R. L., Jr., & Wiggins, C. H. (2013). Hierarchically-coupled hidden Markov models for learning kinetic rates from single-molecule data. In *Proceedings of the 30th international conference on machine learning*.

Van Kampen, N. G. (2007). *Stochastic processes in physics and chemistry*. Amsterdam: Elsevier.

Viterbi, A. J. (1967). Error bounds for convolutional codes and an asymptotically optimum decoding algorithm. *IEEE Transactions on Information Theory*, *13*, 260–269.

Wang, J., Caban, K., & Gonzalez, R. L. (2015). Ribosomal initiation complex-driven changes in the stability and dynamics of initiation factor 2 regulate the fidelity of translation initiation. *Journal of Molecular Biology*, *427*, 1819–1834.

Zhou, H.-X. (2010). Rate theories for biologists. *Quarterly Reviews of Biophysics*, *43*, 219–293.

Zhuang, X., Bartley, L. E., Babcock, H. P., Russell, R., Ha, T., Herschlag, D., et al. (2000). A single-molecule study of RNA catalysis and folding. *Science*, *288*, 2048–2051.

Zhuang, X., Kim, H., Pereira, M. J., Babcock, H. P., Walter, N. G., Chu, S., et al. (2002). Correlating structural dynamics and function in single ribozyme molecules. *Science*, *296*, 1473–1476.

Zwanzig, R. (1997). Two-state models of protein folding kinetics. *Proceedings of the National Academy of Sciences of the United States of America*, *94*, 148–150.

Zwanzig, R. (2001). *Nonequilibrium statistical mechanics*. Oxford: Oxford University Press.

CHAPTER EIGHT

Quantification of Functional Dynamics of Membrane Proteins Reconstituted in Nanodiscs Membranes by Single Turnover Functional Readout

M.E. Moses*, P. Hedegård†,1, N.S. Hatzakis*,1
*Department of Chemistry & Nanoscience Center, University of Copenhagen, Copenhagen, Denmark
†Niels Bohr Institute & Nanoscience Center, University of Copenhagen, Copenhagen, Denmark
[1]Corresponding authors: e-mail address: Hedegard@nbi.ku.dk; hatzakis@nano.ku.dk

Contents

Abstract

Single-molecule measurements are emerging as a powerful tool to study the individual behavior of biomolecules, revolutionizing our understanding of biological processes. Their ability to measure the distribution of behaviors, instead of the average behavior, allows the direct observation and quantification of the activity, abundance, and lifetime of multiple states and transient intermediates in the energy landscape that are typically averaged out in nonsynchronized ensemble measurements. Studying the function of membrane proteins at the single-molecule level remains a formidable challenge, and to date there is limited number of available functional assays. In this chapter, we describe in detail our recently developed methodology to reconstitute membrane proteins such as the integral membrane protein cytochrome P450 oxidoreductase on membrane systems such as Nanodiscs and study their functional dynamics by recordings at the fundamental resolution of individual catalytic turnovers using prefluorescent substrate analogues. We initially describe the methodology for reconstitution, surface

Methods in Enzymology, Volume 581
ISSN 0076-6879
http://dx.doi.org/10.1016/bs.mie.2016.08.026
© 2016 Elsevier Inc.
All rights reserved.

immobilization, and data acquisition of individual enzyme catalytic turnovers. We then explain in detail the statistical analysis, with an emphasis on the model development, the potential pitfalls for correctly identifying the abundance, lifetime, and likelihood of sampling protein functional states. This methodology may enable studies of functional dynamics and their role in biology for a spectrum of membrane proteins.

ABBREVIATIONS

APD avalanche photodiode
BSA bovine serum albumin
DTT dithiothreitol
FAD flavin adenine dinucleotide
FCS fluorescent correlation spectroscopy
FMN flavin mononucleotide
MSP membrane scaffold protein
ND nanodisc
PLL-PEG poly-(L-lysine) grafted with poly(ethyleneglycol)
POR cytochrome P450 oxidoreductase
SM single molecule

1. INTRODUCTION AND OUTLINE

Single-molecule (SM) studies over the last decade have provided an unprecedented wealth of dynamic information, revolutionizing our understanding of protein dynamics and behavior in ways unattainable by conventional bulk measurements (Banerjee & Deniz, 2014; Cornish & Ha, 2007; Hatzakis, 2014; Joo, Balci, Ishitsuka, Buranachai, & Ha, 2008; Jorgensen & Hatzakis, 2013; Michalet, Weiss, & Jager, 2006; Moerner, 2007; Schuler & Eaton, 2008; Xie, Choi, Li, Lee, & Lia, 2008). SM studies offer the direct observation of heterogeneities rather than the average biophysical property from which they can be inferred (Choi, Cai, Frieda, & Xie, 2008; Guydosh & Block, 2009; Lu, Xun, & Xie, 1998; Mukhtasimova, Lee, Wang, & Sine, 2009; Yang et al., 2003; Zhao et al., 2011). Being free of the averaging effect of nonsynchronized ensemble measurements, SM studies offer the direct observation of previously unknown transient intermediates in the energy landscape as well as the existence of functional dynamics and the abundance and lifetimes of enzyme states and their role in regulation (Hatzakis, 2014).

The two principal approaches (Cornish & Ha, 2007; Walter, Huang, Manzo, & Sobhy, 2008) for directly observing individual molecules may be broadly categorized in the ones involving micromechanical manipulations and fluorescence, while recent advances achieve their combination (Hohng, Lee, Lee, & Jo, 2014). Micromechanical manipulation involves optical and magnetic tweezers (Greenleaf, Woodside, & Block, 2007; Oddershede, 2012) as well as atomic force microscopy (Dufrene, 2008; Engel & Muller, 2000; Fisher, Marszalek, & Fernandez, 2000; Muller & Dufrene, 2008; Muller, Helenius, Alsteens, & Dufrene, 2009), while fluorescent measurements involve polarization lifetime, particle tracking (Shih et al., 2013; Veigel & Schmidt, 2011; Yang, 2010), FRET (Akyuz et al., 2015; Cornish & Ha, 2007; Ha, Kozlov, & Lohman, 2012; Honda et al., 2014; Ning, Fei, & Gonzalez, 2014; Uphoff, Reyes-Lamothe, Garza de Leon, Sherratt, & Kapanidis, 2013), and intensity measurement (Cornish & Ha, 2007; Engelkamp et al., 2006; Roeffaers et al., 2007; Smiley & Hammes, 2006; Walter et al., 2008).

Studies at the fundamental limit of individual product formation cycles by us (Engelkamp et al., 2006; Hatzakis et al., 2006, 2012; Kuznetsova et al., 2008; Laursen et al., 2014) and others (Brender et al., 2005; Edman, Foldes-Papp, Wennmalm, & Rigler, 1999; English et al., 2006; Goldsmith et al., 2011; Shi et al., 2004; Velonia et al., 2005) identified time-dependent activity fluctuation for an individual enzyme and the existence of more than one functional state (dynamic disorder) (Brender et al., 2005; Edman et al., 1999; Engelkamp et al., 2006; English et al., 2006; Goldsmith et al., 2011; Hatzakis et al., 2006, 2012; Kuznetsova et al., 2008; Laursen et al., 2014; Shi et al., 2004; Velonia et al., 2005) as well as heterogeneities in activity both in between seemingly identical enzymes of the population (static disorder) (Brender et al., 2005; Craig, Arriaga, Wong, Lu, & Dovichi, 1996; Edman et al., 1999; Engelkamp et al., 2006; English et al., 2006; Goldsmith et al., 2011; Hatzakis et al., 2006, 2012; Kuznetsova et al., 2008; Laursen et al., 2014; Shi et al., 2004; Velonia et al., 2005; Xue & Yeung, 1995). These functional states may originate from enzymes sampling the free energy landscape and the existence of conformational states not optimized for catalysis. The accurate quantification of dynamic disorder and the number of functional states involve treating long trajectories of enzyme events, correlation of intensity changes ($C_I(\tau)$), the probability density of the waiting time between consecutive events ($p_w(t)$), and the correlation of waiting times (C_n). Correctly evaluating noisy trajectories has been attempted over the last years; we (Hatzakis et al., 2012) and others (Terentyeva et al., 2012; Terentyeva, Hofkens, Komatsuzaki, Blank, & Li, 2013) have developed methodologies of quantifying the number of functional states underlying time-dependent activity fluctuations. Time-dependent activity fluctuations until recently were primarily observed by high temporal resolution parked beam studies resolving individual turnover events. Recent studies with lower temporal resolution directly observed activity fluctuations to extend to multiple seconds and minutes timescales (Iversen et al., 2014; Li et al., 2015; Veshaguri et al., 2016). Here, we will focus our analysis on data at the individual turnover rate.

To date, SM studies have primarily been done on soluble proteins with few studies on membrane-spanning proteins (Hatzakis et al., 2012; Iversen et al., 2014; Li et al., 2015; Veshaguri et al., 2016; Wadsater et al., 2012) mainly due to difficulties in handling them. The limited SM studies on membrane-spanning proteins are primarily performed for proteins solubilized in detergents. Solubilizing membrane proteins with detergent may significantly alter their stability dynamics and function (Bayburt & Sligar, 2010; Borch & Hamann, 2009; Cruz & Edmondson, 2007; Gohon & Popot, 2003; Laursen et al., 2014; Mi et al., 2008), seen in multiple recent studies on structurally diverse membrane-spanning protein ranging from single-helix membrane-spanning cytochrome P450 oxidoreductase (POR) (Laursen et al., 2014) to ABC (Zoghbi, Cooper, & Altenberg, 2016) and glutamate transporter (Akyuz, Altman, Blanchard, & Boudker, 2013) to mention a few. The lipid composition of membranes dictates their physical properties (e.g., hydrophobic thickness, membrane viscosity, surface charge, interfacial curvature) which in turn affects the function of many membrane proteins (Alami, Dalal, Lelj-Garolla, Sligar, & Duong, 2007; Bayburt & Sligar, 2010; Borch & Hamann, 2009; Cruz & Edmondson, 2007; Lee, 2004; Shaw, Pureza, Sligar, & Morrissey, 2007). Hence to gain information about the intrinsic function of a membrane protein, it is beneficial to mimic the native lipid environment, rather than use detergent micelles.

This chapter consists of two parts: The first part involves the methodology for acquisition of SM functional dynamics for membrane-spanning proteins and the second the evaluation of functional dynamics by statistical analysis. We will initially discuss the experimental methodology to reconstitute in membrane systems the membrane-spanning enzymes, their immobilization on passivated surfaces, and how to record functional dynamics of individual enzymes. As a model system, we will use the membrane-spanning enzyme P450 oxidoreductase reconstituted in membrane patches, nanodiscs (NDs). Consequently, we will discuss how to analyze functional data acquired by parked beam measurements in a reliable manner and how to quantify functional dynamics and the occupancy as well as the abundance of functional states. Special focus will be on the potential pitfalls of analyzing waiting time distribution and ways to overcome them. Lastly, we will focus on how to extend our analysis to multimeric enzyme systems and how to study the activity and possible allosteric coupling of (N) functional states of each of the monomers.

2. LIPID MEMBRANE SYSTEMS FOR SM FUNCTIONAL STUDIES

SM characterization of membrane proteins requires methods to maintain the proteins in a functionally active and water-soluble form. These have been obtained by encapsulating the hydrophobic parts of the protein in detergent micelles, or incorporating the protein in a native-like environment such as liposomes, supported lipid bilayers, and recently NDs. For many SM measurements, it is further advantageous to obtain individual (or a controlled number of) proteins immobilized and spatially separated with a fixed orientation.

Membranes hold the grave advantage that their lipid composition may be fine-tuned to mimic the native environment of membrane proteins. Supported lipid bilayers allow membrane proteins to freely defuse in the plane of the bilayer which is imperative when tracking translatory protein movement (Gudmand et al., 2010; Pinaud, Clarke, Sittner, & Dahan, 2010; Veigel & Schmidt, 2011). The continuous observation of single protein function (if uncorrelated to the diffusion) in real time entails compartmentalization of the bilayer (e.g., by lithographically defined barriers) (Groves, Ulman, & Boxer, 1997; Iversen et al., 2014). Liposomes, on the other hand, are ideal systems to confine spatially individual enzymes in a biocompatible environment and study their behavior (Akyuz et al., 2015; Hatzakis et al., 2012; Li et al., 2015; Veshaguri et al., 2016). They offer (a) the exterior part of the bilayer to study the behavior of peripherally associated proteins, (b) their membrane to reconstitute membrane proteins and study their function and dynamics (Fruh, IJzerman, & Siegal, 2011; Jorgensen et al., 2013; Veshaguri et al., 2013), and (c) lastly, their hollow low permeability interior for directly observing the function of membrane transporter (Akyuz et al., 2015; Li et al., 2015; Veshaguri et al., 2016). Membrane protein orientation, loading efficiency (Li et al., 2015; Mathiasen et al., 2014; Veshaguri et al., 2016), as well as liposome size distributions (Christensen, Bolinger, Hatzakis, Mortensen, & Stamou, 2012; Hatzakis et al., 2009; Larsen, Hatzakis, & Stamou, 2011) and their effect on protein function should be considered when performing studies on liposomes.

NDs have proven to be an alternative tool for SM characterization of membrane proteins by providing a water-soluble, stable, and monodisperse reconstitution medium with narrow and tuneable size distribution and controllable lipid composition (Alami et al., 2007; Bayburt, Grinkova, & Sligar, 2002; Bayburt & Sligar, 2010; Borch & Hamann, 2009; Denisov, Grinkova, Lazarides, & Sligar, 2004; Laursen et al., 2014; Wadsater et al., 2012). NDs do not have hollow interior, eliminating their possibility to study membrane transporter function. Immobilizing protein containing NDs fixes the orientation of the protein perpendicular to the surface plane, eliminating orientation limitations. Furthermore, the number of loaded proteins per ND can be controlled by adjusting the protein to NDs ratio during the preparation. This stoichiometric control is crucial when studying membrane proteins at the SM level (Laursen et al., 2014; Wadsater et al., 2012).

2.1 ND Membrane Patches for Membrane-Spanning Protein Immobilization

The term ND (first introduced by Bayburt et al. (2002)) refers to a small (9.8–12.9 nm in diameter) discoidal phospholipid bilayer encircled by genetically modified membrane scaffold proteins (MSPs). The amphiphilic nature of the MSPs allows them to wrap around the nonpolar edges of the discoidal phospholipid bilayer, thereby assembling small patches of water-soluble bilayers. Consequently, the size of the NDs can be controlled by choosing MSPs of specific lengths. Changing the lipid composition of the NDs not only allow mimicking the native environment of a specific protein; introducing labeled lipids also allows labeling of the NDs with a low number of chromophores, which in turn minimizes cross talk. Labeling of NDs, containing single-membrane proteins, has further advantage of facilitating immobilization and dye-labeling of individual proteins without adhering tags and chromophores directly to the protein structure which may alter its intrinsic function.

2.2 Theory of ND Preparation

NDs are formed from a mixture of buffered MSP and detergent/phospholipid micelles upon removal of detergent by addition of porous polystyrene beads (see Fig. 1). The stoichiometry of MSP and lipids is crucial for the formation of a homogeneous monodisperse solution of NDs, and depends on the length of the MSP and the lipid composition. For example, the optimal lipid to MSP ratio for NDs formed of MSP1 and dipalmitoyl phosphatidylcholine is 200 and yields 171 $\pm$ 6 lipids per disc, while the optimum MSP to lipid ratio of NDs formed of 1-palmitoyl-2-oleoyl-phosphatidylcholine is 150 and yields 129 $\pm$ 9 lipids per disc (Bayburt et al., 2002). A greater than optimum lipid to MSP ratio will yield populations of larger lipoprotein particles, while a lower ratio will yield lipid-poor particles (Bayburt et al., 2002).

Incorporating the target membrane proteins in the NDs is done by adding detergent-solubilized protein to the MSP/detergent/phospholipid mixture prior to detergent removal. Insertion of proteins in the NDs will displace a number of lipids depending on the structural properties of the target proteins, hence entails further adjustments of the lipid to MSP ratio. Finding the optimal stoichiometry of lipids and MSP for a specific target protein entails empirical optimization by measuring the size distribution of assembled particles at different lipid ratios by size-exclusion chromatography (SEC) (Bayburt et al., 2002; Bayburt & Sligar, 2010).

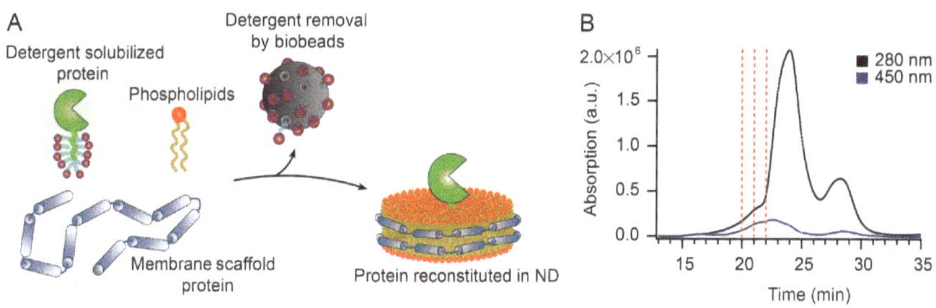

Fig. 1 Reconstitution of membrane proteins in nanodiscs. (A) Schematic illustration of ND reconstitution. Removing detergent from a mixture of MSPs, phospholipids, and detergent-solubilized membrane proteins results in NDs formation loaded with membrane proteins. (B) Size-exclusion chromatograph of cytochrome P450 oxidoreductase in NDs. *Gray color* represents aromatic amino acid residues (absorbance at 280 nm), i.e., detection of both MSP and POR. *Blue color* represents flavin cofactors of POR (absorbance at 450 nm). *Dotted red lines* highlight elusion fractions (21 and 22 min) of high loading efficiency.

Choosing MSP will not only dictate the size of assembled NDs but also affect parameters such as loading efficiency and stability. Using MSP1E3D1, instead of the smaller MSP1D1, to reconstitute the membrane-associated enzyme POR hindered coelution of oligomers, presumably due to less POR–MSP interactions (Bertram et al., 2015; Laursen et al., 2014). The self-assembly of NDs is initiated by detergent removal, either by dialysis or by porous polystyrene beads. It is in general favorable to use a detergent with high critical micelle concentration and small micelle size in order to separate efficiently the detergent molecules from MSP and target integral membrane proteins (Bayburt et al., 2002). Here, cholate is the detergent used for ND preparation.

2.3 Case Study of POR

As a model system, we will use the membrane-associated enzyme POR. POR is found in all kingdoms and is situated in the membrane of the endoplasmic reticulum where it activates all cytosolic P450 enzymes (50 in humans) (Pandey & Fluck, 2013). Structure, function, and pathology of POR are extensively reviewed elsewhere by Pandey and Fluck (2013). Aberrations in POR function result in metabolic disorders with phenotypic spectrum ranging from hormone deficiency to skeletal dysplasia (Fluck, Mullis, & Pandey, 2009; Fluck, Nicolo, & Pandey, 2007; N. Huang, Agrawal, Giacomini, & Miller, 2008; N. W. Huang et al., 2005; Nicolo, Fluck, Mullis, & Pandey, 2010; Xia et al., 2011).

2.3.1 Reconstitution of POR in NDs

For reconstitution of POR in NDs, MSP1D1E3 was used as MSP, and a mixture of 1,2-dilauroyl-*sn*-glycero-3-phospho-(1′-rac-glycerol) (25%), 1,2-dilauroyl-*sn*-glycero-3-phosphocholine (69.8%), *N*-(biotinoyl)-1,2-dihexadecanoyl-*sn*-glycero-3-phosphoethanolamine (DHPE-biotin) (4%), and DiO (1.2%) was used to mimic the lipid composition of the native membrane, and to label individual NDs with biotin and DiO. This lipid/MSP combination forms NDs of ~ 13 nm in diameter, containing ~ 240 lipids and 2 MSPs; hence, each ND contains, on average, 9.6 biotin labels and 2.9 DiO molecules.

Ideally, each ND should contain a single dye molecule to minimize cross talk in the FCS/microscopy measurements. However, from Poisson statistics, we know that an average of 1 dye per ND would result in a significant number of NDs having no dye leaving them undetectable in the microscope. We found an average number of 3 labeled lipids per ND as optimum to allow imaging by microscope but having limited cross talk.

High loading efficiency of protein in NDs is generally desirable; however, for SM studies, it is imperative to avoid reconstitution of more than a single protein per ND, and this is achieved by using a low ratio (1:10) of POR per ND in the preparation. A simple simulation, using Poisson statistics, provides the probability of reconstituting 0–4 proteins per ND, depending on the protein per ND ratio (Laursen et al., 2014). The simulation confirms that at 0.1 POR per ND, close to 100% of loaded NDs will contain only a single protein. In practice, this was measured by SM bleaching step analysis, for which both proteins and NDs were labeled and colocalized by total internal reflection microscopy (TIRF) (see Fig. 2). This allows us to quantify the number of bleaching steps and consequently the number of Cy5 labels per ND (see Fig. 2A). The result showed that ~ 81% of the loaded NDs contained a single labeled protein, ~ 16% and ~3% displayed double and triple bleaching steps, respectively, which may be due to ND containing more than one protein or single proteins labeled with more than one dye (see Fig. 2).

SEC allows higher yield of loaded NDs, enhancing the loading efficiency from 10% (in the preparation mixture), to ~80% loaded NDs in the final sample. Loaded NDs will have a slightly larger Stoke radius compared to unloaded NDs, hence will elute faster than unloaded NDs (see Fig. 1B); this is utilized to optimize the final loading efficiency. The faster elution of

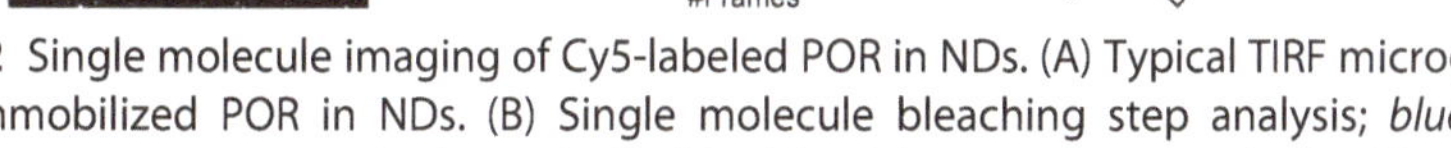

Fig. 2 Single molecule imaging of Cy5-labeled POR in NDs. (A) Typical TIRF micrograph of immobilized POR in NDs. (B) Single molecule bleaching step analysis; *blue* and *red* curves represent single and double bleaching step, respectively. Bleaching step analysis of 576 colocalized POR in NDs revealed that 81% of NDs contained a single-labeled POR. Double (16%) and triple (3%) bleaching steps may arise from NDs loaded with more than one POR or a single POR with more than one label. *Source: Figure based on data from Laursen, T., Singha, A., Rantzau, N., Tutkus, M., Borch, J., Hedegård, P., et al. (2014). Single molecule activity measurements of cytochrome P450 oxidoreductase reveal the existence of two discrete functional states.* ACS Chemical Biology, 9 *(3), 630–634.*

loaded NDs is observed in Fig. 1B as a shoulder in the 280 nm absorption spectrum, two fractions of high loading efficiency (21 and 22 min) are marked in the figure by dashed red lines.

2.3.2 Protocol for ND Reconstitution of POR

Prior to reconstitution, POR enzymes were expressed, purified, and stored in detergent (cholate) at −80°C. POR lipids (solubilized in cholate) and MSP aliquots are thawed in ice to minimize temperature stress.

2.3.2.1 ND Preparation

1. Appropriate amounts of MSPs, enzymes (POR), and the desired lipids are mixed (e.g., for 0.1 POR per ND: 440 μL of 0.153 nmol/μL MSP1D1E3 + 161 μL of 50 nmol/μL lipid mixture + 599 μL of 0.006 nmol/μL POR).
2. Dithiothreitol (DTT) is added to the mixture to a final concentration of 5 m*M*.
3. In the case of POR reconstitution, the cofactors flavin adenine dinucleotide (FAD) and flavin mononucleotide (FMN) are added to respective final concentrations of 100 μ*M*, to ensure reconstitution of functional POR enzymes (i.e., POR may lose its FMN and FAD cofactors during reconstitution, and thus become dysfunctional).
4. The mixture is incubated for 1 h on ice on a shaking platform. (Meanwhile one might prepare the HPLC for SEC, see step 1 in Section 2.3.2.2.)
5. Cholate is removed by addition of biobeads (Bio-Rad SM-2). The biobeads are kept on ice before ∼1 g is added to the ND mixture. The mixture is incubated for 1 h on ice on a shaking board.
6. To further remove detergent, ∼3 g of biobeads are added to the mixture and incubated for 3 h on ice on a shaking board.
7. After incubation, the mixture is centrifuged briefly and the supernatant is removed from the biobeads and centrifuged for 10 min (20, 000 × *g*, 4°C). The supernatant is transferred to new vials and is ready to be purified by SEC.

Note that in a typical experiment, precipitation should not be observed in the centrifuge step. In case, however, some precipitation or a pellet is observed, an indication of protein aggregation and failed protein reconstitution. This is not the case with POR but may be observed for less robust membrane proteins.

2.3.2.2 Purification of ND Reconstituted POR by SEC

HPLC settings: 0.5 mL/min, pressure < 15 bar.

1. HPLC is prepared by flushing the system with water for 20 min. The column (Superdex 200 30/100 GL) is flushed with 2× bed volume of water and 1× bed volume of ND buffer.
2. A sample of assembled NDs is run through the HPLC and 5 fractions are collected in the interval 20–25 min. The fractions are stored at 4°C while evaluating absorption at 280 nm (aromatic amino acids) and 450 nm (the flavin cofactors of POR) (see Fig. 1B).
3. From evaluation of the two measured absorptions, desired fraction are pooled and stored at − 80°C.

2.3.3 Surface Preparation/Immobilization

Immobilization of ND has been done in various ways by incorporating tags (such as 6× His and FLAG) to the MSP and introducing these to a surface coated with the respective antibodies (Borch, Torta, Sligar, & Roepstorff, 2008). Here, we utilize the strong binding affinity of neutrAvidin and biotin to immobilize NDs (Laursen et al., 2014; Wadsater et al., 2012) (see Fig. 3A). NDs are labeled with biotin by introducing a fraction of biotin-labeled lipids to the MSP/detergent/lipid/protein mixture. Surface functionalization

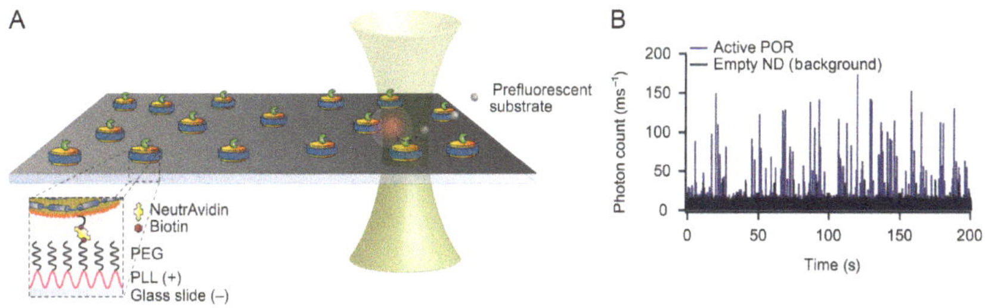

Fig. 3 Parked beam functional assay at the fundamental limit of individual catalytic cycles. (A) Individual enzymes, reconstituted in NDs, are immobilized on a glass cover slide by biotin-neutrAvidin adhesion. The glass slide (partially negatively charged by plasma cleaning) is functionalized by PLL-PEG/PLL-PEG-biotin, allowing neutrAvidin to tether biotin-labeled NDs to the surface. FCS allows continuous observation of individual enzymatic turnover of prefluorescent substrate by individual enzymes. (B) Typical time trace of enzymatic product (resorufin) formation by a single POR enzyme reconstituted in ND (*blue*). Control measurement on an empty ND shows the background (*gray*). *Source: Figure based on data from Laursen, T., Singha, A., Rantzau, N., Tutkus, M., Borch, J., Hedegård, P., et al. (2014). Single molecule activity measurements of cytochrome P450 oxidoreductase reveal the existence of two discrete functional states.* ACS Chemical Biology, 9 *(3), 630–634.*

is done with either bovine serum albumin (BSA)/BSA-biotin or poly-(L-lysine) grafted with poly-(ethyleneglycol)(PLL-PEG)/PLL-PEG-biotin; in both cases, neutrAvidin is introduced to the surface to adhere the biotin-labeled NDs to the biotin coated surface.

2.3.3.1 Protocol

BSA/BSA-biotin functionalization of glass cover slides does not require plasma cleaning, hence step 2 can be omitted.

1. Glass cover slides (type and size) are cleaned prior to surface functionalization.
   - **(a)** Glass slides are sonicated for 10 min in a solution of HellmaneX 2% vol., and washed five times with MilliQ water. This step is done three times.
   - **(b)** Glass slides are sonicated for 10 min in MilliQ water, and washed five times with MilliQ water. This step is done three times.
   - **(c)** Finally, the glass slides are sonicated once in methanol, the solvent is discarded, and the glass slides are stored in new methanol until use.
2. Glass slides are plasma cleaned to remove small organic impurities from the surface, and to probe the surface with negative charge (this facilitates adhesion of the positively charged lysine molecules of the PLL-PEG). Prior to plasma cleaning, the glass slides are dried under a flow of nitrogen gas. For our setup (Harrick Plasma PDC-32G/PlasmaFlo), we clean the slides in N_2 gas, at low pressure (< 300 mTorr) and high radio frequency power (18 W) for > 2 min.
3. Individual glass slides are fixed in a customized teflon/aluminum microscope chamber and filled with a solution of either PLL-PEG/PLL-PEG-biotin (100:1) or BSA/BSA-biotin (20:1) in 15 m*M* HEPES buffer pH 5.6, and incubated for 30 min.
4. Excess BSA/BSA-biotin and PLL-PEG/PLL-PEG-biotin are removed by exchanging the chamber volume five times with HEPES buffer.
5. The HEPES buffer is exchanged by one volume of 0.1 g/L neutrAvidin (in 15 m*M* HEPES buffer pH 5.6), and allowed to incubate for a few minutes before excess neutrAvidin is removed by exchanging the chamber volume five times with the desired protein/ND buffer. (The functionalized surfaces can be stored at 5°C for at least 3 days before use.)
6. NDs are introduced to the surface by adding one volume of low concentration ($\sim$ 1 n*M*) biotin-labeled NDs to the chamber. Excess NDs are removed by exchanging the chamber volume 10 times with ND buffer. (In our case, this approach will yield a surface density of $\sim$ 400 NDs per

field of view (83 μm ×83 μm); we advise optimizing this final step to reach a desired surface density.).

2.3.4 Single-Turnover Activity Readouts

Direct observation of dynamic protein function in real time is achieved by continuous detection of enzymatic turnover of prefluorescent substrate (i.e., a nonfluorescent substrate that becomes highly fluorescent upon enzymatic reaction) by individual immobilized proteins (Hatzakis, 2014; Hatzakis et al., 2012; Jorgensen & Hatzakis, 2013; Laursen et al., 2014). Using avalanche photodiodes (APDs) provides the essential sensitivity and time resolution for detecting single fluorophores produced in the enzymatic reaction. The continuous replenishment of prefluorescent substrate practically eliminates bleaching issues and allows prolonged observation of individual products events. The main limitations of this methodology arise from the limited number of available prefluorescent substrates, the potential depletion of substrate, and increase of background signal for very long observation times. The latter can be easily overcome by using flow cells that allow constant injection of substrate and removal of product. The major advantage is the direct and continuous observation of enzymatic turnover events for several minutes or even hours, allowing statistically robust analysis of functional dynamics. Importantly, they offer recordings of individual product formation events, an important observable compared to the structural dynamics that are anticipated to underlie it.

Things to consider when establishing SM assays with prefluorescent substrate include: (a) Are the fluorescent properties of the product significantly different than the substrate so as to detect individual turnover events? (b) Does the on-time of the product facilitate SM detection (i.e., does the fluorescent product molecule stay long enough in the detection volume to produce signal detectable above the background)? (c) Do product molecules bind nonspecifically in the detection volume (e.g., to the surface or the NDs), thereby giving rise to detection of "false" activity?

2.3.5 Parked Beam Microscope Setup

Activity of single POR was measured by observing turnover of prefluorescent substrate (resazurin) with fluorescent correlation spectroscopy (FCS) (Laursen et al., 2014). The measurements were carried out on a Leica TCS SP5 inverted confocal microscope with an Hex PL APO CS × 100 (NA 1.46) oil immersion objective. Photon arrivals were detected by APDs. NDs were labeled with DiO and excited by a 488-nm laser line, resorufin (the fluorescent product) was excited with a 514-laser line, and

POR was labeled with Cy5 and excited with a 633-laser line. The fluorescent signal was split by a 625-nm beam splitter. Signal from NDs (DiO) and POR (Cy5) was filtered through ET585/40 and HQ675/50 emission filters, respectively.

Individual enzymes were selected by evaluating colocalized signal from DiO (ND) and Cy5 (POR). Subsequently, a single-enzyme loaded ND is placed under the parked beam of the confocal microscope. Substrate turnover was measured sequentially for a total number of no less than 50 selected enzymes. Typical trajectory length was 200 s. Due to the constant replenishment of prefluorescent substrate, bleaching does not limit the measurements. The trajectory length is mainly limited by enzyme denaturation due to radiation of high-intensity laser.

3. STATISTICAL ANALYSIS OF TRAJECTORIES OF INDIVIDUAL TURNOVER CYCLES

In this section, we will discuss how to analyze data of individual turnover cycles. Each enzymatic turnover provides a short burst of light (on time) before the product diffuses away from the confocal volume. The most frequently used technique to analyze these data involves their initial binning to photons per time resulting in the stochastic trajectory of events like the ones shown in Fig. 3. A threshold is consequently applied to separate the short high-intensity on-times from the background. The time between consecutive catalytic cycles, the waiting time (τ), is consequently measured, and its probability density ($p_w(t)$) as well as waiting time autocorrelation may be used to quantify functional dynamics. To date, however, both the bin size and the threshold position are often chosen by a trial-and-error approach as there is no unique way to determine them. Such arbitrary selection was recently found to affect the data significantly and possibly lead to artifactual conclusions (Terentyeva et al., 2012, 2013). The same studies outlined the inherent limitation of employing waiting time histograms to quantify the number of functional states, especially if the number of exponentials needed for the fit is more than three.

To extract information reliably from the stochastic trajectory of event like the one shown in Fig. 3, we recently proposed to initially employ the enzyme's contribution to the intensity autocorrelation $C_I(\tau)$ to quantify the number of functional states (Hatzakis et al., 2012). $C_I(\tau)$ is independent of (a) binning as it is calculated from the photon arrival times and (b) the applied threshold, both of which may cause artifactual results for noisy data (Terentyeva et al., 2012). By employing this simplified approach, we recently showed that the functional dynamics of most, though not all, enzymes were correctly described by sampling two functional states.

In this section, we will first discuss how to use $C_I(\tau)$ as an unbiased first tool to quantify the number of functional states and how to employ the amplitude and the decay of $C_I(\tau)$ to extract information of their interconversion rates. We will consequently focus on how to extend our analysis to multimeric enzyme systems where each of their monomers exhibits (N) functional states and define conditions to differentiate allosterically coupled from independent functional dynamics

3.1 Model

We shall be considering a model which describes a protein with n states. Each of these states has different activities described by rate constants k_1, …, k_n. This means that a product formation will take place in a short time interval Δt after a time t with probability $k_1P(A_{1t})\Delta t + \cdots k_nP(A_{nt})\Delta t$. Here, $P(A_{it})$ is the probability that the molecule is in fact in state i at time t. We shall further assume that a recovery is very short in the timescales we are considering, and that the protein returns to the state i immediately after the product formation has taken place. The numerical value of the rates k_i will depend on the concentration of substrate molecules. At low concentrations, there will be a linear dependence: $k_i = \alpha_i N_s$, where N_s is the substrate concentration. At higher concentration, the value of the k_i will saturate. The exact concentration where the crossover between linear and saturated concentration dependence occur will depend on the system.

The protein can switch between the various states. These switches are described by a set of rate constants k_{ij}. We have that the probability that a switch from state i to state j is taken place in a short time interval Δt is given by $k_{ji}\Delta t$. The switching rates are expected to have no, or only a very weak, dependence on substrate concentration.

Each product formation event is associated with a burst of photons, and we are interested in calculating the intensity of photons $I(t)$ as a function of time. This function will typically be a sequence of spikes coming at random times. It can described mathematically as follows

$$I(t) = I_0 \sum_n \delta(t - t_n). \tag{1}$$

Here, I_0 is the intensity of photons at a single event, while $\delta(t - t_n)$ is a narrow function with a very short width centered around the time t_n of the nth event. We will calculate the average intensity $\langle I(t) \rangle$ and the intensity correlation function $\langle I(t + \tau)I(t) \rangle$. Further, we will calculate the probability density, $p_w(t)$, that the waiting time between two consecutive events will last a time t, and finally, the waiting time correlation function $\langle \tau_n \tau_1 \rangle$. Here, τ_1 is the waiting time between two events labeled 0 and 1, and τ_n is the waiting time between events $n - 1$ and n. These quantities are the most accessible experimentally. The waiting times can be a bit tricky, since photons from a background source having nothing to do with the system can interfere and break a waiting time period for the system into two. The problem of background photons is not so severe when it comes to the simple intensity measurements. If we assume that the background is not correlated with the system, then the measured intensity correlation function can be written

$$\langle I(t+\tau)I(t) \rangle = \langle I_s(t+\tau)I_s(t) \rangle + \langle I_b(t+\tau)I_b(t) \rangle, \tag{2}$$

where the first term is associated with the system, while the second term is coming from the background. If further the background signal is essentially white noise, i.e., has a very short lived correlations, then it is easy to separate the system signal from the background signal.

To get started, we first consider the probability of finding the protein in state i at time t. It is denoted $Q_i(t)$. It is useful to organize all these functions as a vector. Define

$$Q(t) = \begin{pmatrix} Q_1(t) \\ \vdots \\ Q_n(t) \end{pmatrix}. \tag{3}$$

We can also organize the various rate constants in two matrices K and M_0:

$$K = \begin{pmatrix} k_1 & \cdots & 0 \\ \vdots & \ddots & \vdots \\ 0 & \cdots & k_n \end{pmatrix}, \quad M_0 = \begin{pmatrix} \sum_i k_{i1} & \cdots & -k_{1n} \\ \vdots & \ddots & \vdots \\ -k_{n1} & \cdots & \sum_i k_{in} \end{pmatrix}. \tag{4}$$

K is a diagonal matrix with all the activity rates in the diagonal, while the switching rates between states are organized in M_0. Note that the sum of elements in each column of M_0 is 0. This fact can also be written mathematically using the vector $e = (1, \ldots, 1)$. We have simply

$$eM_0 = 0. \tag{5}$$

With these definitions, we can write the rate equations satisfied by the Q probabilities in the following compact form

$$\frac{dQ}{dt} = -M_0 Q. \tag{6}$$

Given (5), we know that 0 is an eigenvalue of the matrix M_0 (with e being a left eigenvector). Further, it is known that all other eigenvalues of M_0 are positive. We will denote the right eigenvector associated with the eigenvalue 0 by Q_0, so that $M_0 Q_0 = 0$. This is vector of probabilities if the sum of all elements is 1, i.e., if $eQ_0 = 1$. We will so normalize the vector Q_0. The solution to (6) is given by

$$Q(t) = e^{-M_0 t} Q(0). \tag{7}$$

We see that in the limit $t \to \infty$, $Q(t) \to Q_0$.

So far, we have not considered the activity events. Let us assume that an event took place at time 0. We shall then consider the probability that an event will take place in a short time interval immediately after time t, i.e., the probability that we have been waiting precisely a time t for the next event. Using the two symbols e and K, we can write this probability as

$$p_w(t)\Delta t = eKP(t)w\Delta t. \tag{8}$$

Here, $P(t)$ is an $n \times n$ matrix of probabilities $P_{ij}(t)$, which denotes the probability that the protein is in state i at time t given that it was in state j at time 0 and that *no* event took place in the time between 0 and t. The vector w is a vector of probabilities, w_j, that an event took place in state j at time 0. This is given by

$$w = \frac{KQ_0}{eKQ_0}. \tag{9}$$

Here, we use Q_0 since we have no knowledge of the state of the protein at time 0. We note that the sum of these probabilities is 1, i.e., we have $ew = 1$.

The probabilities $P(t)$ satisfy the rate equations

$$\frac{dP}{dt} = -MP(t), \tag{10}$$

where the matrix $M = M_0 + K$. Due to the presence of K, the eigenvalues of M are all positive and probability will "leak" out as a function of time, because it is increasingly improbable that the system has *not* had an event as time goes.

The solution to Eq. (10) is

$$P(t) = e^{-Mt}, \tag{11}$$

since at time 0, we have $P_{ij}(0) = \delta_{ij}$, i.e., $P(0)$ is the unit $n \times n$ matrix.

Having now established the waiting time distribution, $p_w(t)$, we can calculate the moments of this distribution:

$$\langle \tau^m \rangle = \int_0^\infty \tau^m p_w(\tau) d\tau. \tag{12}$$

Using the solution, we arrive at the result

$$\langle \tau^m \rangle = m! eK \frac{1}{M^{m+1}} w. \tag{13}$$

In order to calculate these, it is useful to note that

$$KQ_0 = MQ_0, \tag{14}$$

since $M_0 Q_0 = 0$. We get

$$\langle \tau^0 \rangle = 1, \quad \langle \tau \rangle = (eKQ_0)^{-1}, \quad \langle \tau^2 \rangle = 2\frac{eK^{-1}Q_0}{eKQ_0}. \tag{15}$$

The second result is particularly interesting, since it can give us the average number of events in 1 s, denoted A (for activity):

$$A = \frac{1}{\langle \tau \rangle} = eKQ_0 = \sum_{i=1}^{n} k_i Q_{0i}, \tag{16}$$

i.e., it is the weighted average of the rate constants of the different states with the weights being the probability of finding the protein in the states, a quite intuitive result. We also note that the activity is directly proportional to the rates k_i and hence proportional to the substrate concentration N_s at low concentrations and saturating at higher concentrations.

3.1.1 Intensity

As noted, the actual intensity is a sequence of events taking place at random times t_n. Let us assume that an event took place immediately after time 0. The probability density that a definite sequence took place is given by

$$p(t_1, \ldots, t_n) = eKP(t_n - t_{n-1}) \cdots P(t_1 - 0)w \tag{17}$$

So, the probability average of the intensity is given by

$$\langle I(t) \rangle = I_0 \sum_{n=0}^{\infty} \int_0^{\infty} d\tau_1 \cdots \int_0^{\infty} d\tau_n \delta(t - t_n) eKP(\tau_n) \cdots KP(\tau_0) w \tag{18}$$

For times close to 0, the system still remembers that an event took place at time 0. But for later times, $\langle I(t) \rangle$ will tend to a constant which is the quantity we are interested in. This can be recovered by the following mathematical procedure:

$$\langle I(t) \rangle \to s \int_0^{\infty} e^{-st} \langle I(t) \rangle dt, \quad \text{with } s \to 0. \tag{19}$$

With

$$\int_0^{\infty} e^{-st} P(t) dt = \frac{1}{s + M}, \tag{20}$$

it is straightforward mathematics to show that

$$\langle I(t) \rangle \to I_0 se \frac{1}{1 - K \dfrac{1}{s + M}} w = I_0 se(s + M) \frac{1}{s + M_0} w. \tag{21}$$

Note that we have $s + M_0$ in the denominator. This means that the function is singular at values of s equal to the eigenvalues of M_0. One of these eigenvalues is precisely 0, so the limit $s \to 0$ is easy to evaluate using the left and right eigenvectors belonging to the 0 eigenvalue:

$$\langle I(t) \rangle \to I_0 eKQ_0 = I_0 A. \tag{22}$$

Using the same technique, we can find the intensity correlation function, or rather the Laplace transform of that function

$$LI(s) = \int_0^{\infty} e^{-s\tau} \langle I(t + \tau) I(t) \rangle d\tau. \tag{23}$$

It becomes

$$LI(s) = I_0^2 e(s + M) \frac{1}{s + M_0} KQ_0. \tag{24}$$

For s close to 0, the function is dominated by the 0 eigenvalues of M_0 and we get

$$LI(s \approx 0) = I_0^2 \frac{A^2}{s}. \tag{25}$$

As noted above, $LI(s)$ has singularities at the eigenvalues of M_0. By using the eigenvectors of M_0 denoted by Q_m for right eigenvectors and Q_m^i for the left eigenvectors and the associated eigenvalues, we can do the inverse Laplace transform and get

$$\langle I(t+\tau)I(t)\rangle - \langle I(t)\rangle^2 = I_0^2 \sum_{m=1}^{n-1} (e(-\lambda_m + M)Q_m)\left(Q_m^i K Q_0\right) e^{-\lambda_m \tau}. \tag{26}$$

We see that the correlation function is a multiexponential function with $n - 1$ different decays.

3.1.2 Waiting Times

Consecutive waiting times will be correlated since the state of the protein just before an event, i.e., at the end of one waiting time, is the same as the state of the protein right after the event, i.e., at the start of a new waiting time. Using the techniques derived above, we can calculate the correlation of waiting time number 1 and waiting time number n in the sequence of waiting times. For $n > 1$, we have

$$\begin{aligned}
\langle \tau_n \tau_1 \rangle &= \int_0^\infty d\tau_n \cdots \int_0^\infty d\tau_n e K \tau_n P(\tau_n) \cdots K \tau_1 P(\tau_1) w \\
&= eK \frac{1}{M^2} \left(K \frac{1}{M}\right)^{n-2} K \frac{1}{M^2} w \\
&= e \frac{1}{K} \left(K \frac{1}{M}\right)^n \frac{Q_0}{A},
\end{aligned} \tag{27}$$

while for $n = 1$, we have

$$\langle \tau^2 \rangle = 2e \frac{1}{M} \frac{Q_0}{A}. \tag{28}$$

It is interesting to study the matrix $K\frac{1}{M}$. The eigenvalue equation is $K\frac{1}{M}v = \lambda v$. Multiplying this equation with e results in $ev = \lambda ev$. This in turn means that either $\lambda = 1$ or $ev = 0$. It is easy to see that the vector w is in fact an eigenvector with eigenvalue $\lambda = 1$. We next write the vector $Q_0 = w + Q_1$, where the vector Q_1 is a linear combination of the eigenvectors not being w, i.e., which satisfy $eQ_1 = 0$. Using this in the expression for $\langle \tau_n \tau_1 \rangle$, we get

$$
\begin{aligned}
\langle \tau_n \tau_1 \rangle &= e\frac{1}{K}\frac{w}{A} + e\frac{1}{K}\left(K\frac{1}{M}\right)^n \frac{Q_1}{A} \\
&= \frac{1}{A^2} + e\frac{1}{K}\left(K\frac{1}{M}\right)^n \frac{Q_0 - w}{A}.
\end{aligned}
\tag{29}
$$

Note that the matrix $K\frac{1}{M}$ also can be written $(1+M_0K^{-1})^{-1}$. It is, hence, the eigenvalues of the matrix M_0K^{-1} which controls the decay of the waiting time correlations.

We can now summarize our findings. We have considered three functions of the data: waiting time distribution, intensity correlations, and waiting time correlations. Each are multiexponential functions. In the following table, their properties are summarized.

Quantity	**Number of Exp**	**Matrix**	**Function**
Waiting time	n	$M_0 + K$	$e^{-\lambda t}$
Intensity correlation	$n - 1$	M_0	$e^{-\lambda t}$
Waiting time correlation	$n - 1$	M_0K^{-1}	$(1+\lambda)^{-n}$

3.1.3 Example: Two-State Model

In this section, we will consider the special case of a two-state model. The two rate matrices are

$$
K = \begin{pmatrix} k_1 & 0 \\ 0 & k_2 \end{pmatrix}, \quad M_0 = \begin{pmatrix} k_{21} & -k_{12} \\ -k_{21} & k_{12} \end{pmatrix}. \tag{30}
$$

The eigenvalues of M_0 are 0 and $k_{12} + k_{21}$. The eigenvector Q_0 and the vector w are given by

$$
Q_0 = \frac{1}{k_{12} + k_{21}} \begin{pmatrix} k_{12} \\ k_{21} \end{pmatrix}, \quad w = \frac{1}{k_1k_{12} + k_2k_{21}} \begin{pmatrix} k_1k_{12} \\ k_2k_{21} \end{pmatrix}, \tag{31}
$$

and the activity

$$
A = \frac{k_1k_{12} + k_2k_{21}}{k_{12} + k_{21}}. \tag{32}
$$

The waiting time distribution function is a sum of two exponential decays.

The intensity correlation function is a single exponential decay. Using the general formula, we get

$$\langle I(t+\tau)I(t)\rangle - \langle I(t)\rangle^2 = \frac{(k_1-k_2)^2 k_{12}k_{21}}{(k_{12}+k_{21})^2} e^{-(k_{12}+k_{21})t}. \tag{33}$$

The factor in front of the exponential is the variance of the intensity, i.e., a measure of much the intensity fluctuates: $\langle (t)^2 \rangle - \langle I(t) \rangle^2$. The variations themselves scale with the square root of this factor. We see that this will be proportional to $|k_2 - k_1|$ and hence will be proportional to the substrate concentration N_s.

In order to find the waiting time correlations, we need the eigenvalues of the matrix M_0K^{-1}. They are 0 and $\frac{k_{12}}{k_2} + \frac{k_{21}}{k_1}$. With this, we get for $n > 1$

$$\langle \tau_n \tau_1 \rangle - \langle \tau \rangle^2 = \frac{\frac{k_{12}k_{21}}{k_1 k_2}\left(\frac{k_1 - k_2}{k_1 k_{12} + k_2 k_{21}}\right)^2}{\left(1 + \frac{k_{12}}{k_2} + \frac{k_{21}}{k_1}\right)^n}. \tag{34}$$

Here, we see that since k_1 and k_2 are proportional to the substrate concentration (below saturation), the correlation function will have a higher value—proportional to N_s^2—and decay slower with n for larger values of N_s.

As an example, we consider POR. Here, we have recorded light emitted from single product formation (plus background) and determined a set of waiting times. The four graphs in Fig. 4 show the data.

We see that the intensity correlation function is very close to being a single exponential decay, suggesting that the protein can be modeled by a simple two-state model. A fit gives a decay rate of $\gamma_I = 0.081\ \text{ms}^{-1}$. For the waiting time distribution, we then expect a double exponential function. A fit to such a function gives two decay rates $\gamma_{w1} = 0.57\ \text{ms}^{-1}$ and $\gamma_{w2} = 0.026\ \text{ms}^{-1}$. Finally, the waiting time correlation function is expected to be single exponential function of the type $A/(1+\gamma_{wc})^n$, and this is indeed the case. The best fit gives a value $\gamma_{wc} = 0.171$. We have four pieces of independent information, and by equating the measured rates with the theoretical, we can obtain values for the model's reaction rates. They become: $k_1 = 0.025\ \text{ms}^{-1}$, $k_2 = 0.492\ \text{ms}^{-1}$, $k_{12} = 0.0801\ \text{ms}^{-1}$, and $k_{21} = 2 \times 10^{-4}\ \text{ms}^{-1}$. Note that to extract the activity and interconversion rates of the functional states accurately with error bars of less than 10% typically requires more than 10^3 turnover cycles.

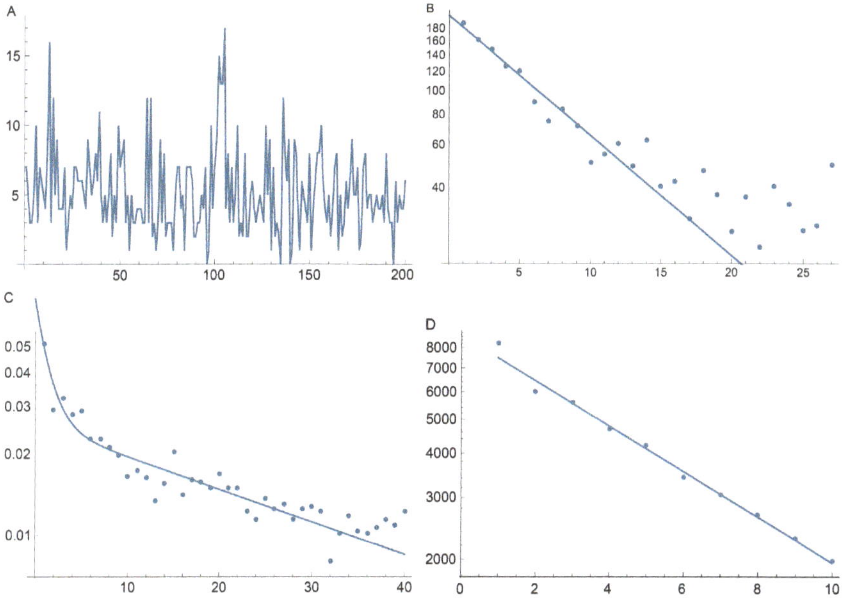

Fig. 4 *Upper left*: The first 200 ms of the raw intensity data. *Upper right*: Intensity correlation function. *Lower left*: Waiting time distribution. *Lower right*: Waiting time correlation function.

To check for consistency, we can now compare the theoretical activity, which with the above model parameters gives $0.027\ \text{ms}^{-1}$, while the directly measured value is $0.028\ \text{ms}^{-1}$. All seems consistent.

Being able to extract all four rates accurately allows the quantification of the probability of sampling the highly active state P_{act} ($P_{act} = k_{21}/(k_{21} + k_{12})$) (Hatzakis et al., 2012; Laursen et al., 2014). P_{act}, as well as k_1 and k_2, values may vary between otherwise identical enzymes of the same experiment, a term ascribed as static heterogeneities. As an example for POR, we recorded static heterogeneities P_{act} varied from 0.3% to 5% (Laursen et al., 2014), while less significant for lipase from Candida Antarctica (0.2–0.45) (Hatzakis et al., 2012). To overcome high static heterogeneities requires statistically robust analysis and acquisition of a high number of long activity traces.

3.1.4 Multimeric Systems

Often one is dealing with multimeric proteins consisting of multiple identical copies of the same protein, like dimers, trimers, and quadromers. In the experiment, one is collecting light from all subsystems. This will create problems, since the observed waiting times are *not* the waiting times of the individual systems, since an event in one subsystem can be followed by one in another subsystem. If we can assume that the subsystems are behaving individually, then the intensity correlation function is a simple sum of intensity correlation functions for each subsystems.

The way to handle such problems is to consider a combination of systems as one single system. Assume as an example a combination of two 2-state models, and denote the two states in each system A and B. The combined system now has four states denoted AA, AB, BA, and BB. The combined system will have K and M_0 matrices of rate constants which have the form

$$K = \begin{pmatrix} k_1 + k_1 & 0 & 0 & 0 \\ 0 & k_1 + k_2 & 0 & 0 \\ 0 & 0 & k_2 + k_1 & 0 \\ 0 & 0 & 0 & k_2 + k_2 \end{pmatrix}, \tag{35}$$

$$M_0 = \begin{pmatrix} k_{21} + k_{21} & -k_{12} & -k_{12} & 0 \\ -k_{21} & k_{12} + k_{21} & 0 & -k_{12} \\ -k_{21} & 0 & k_{12} + k_{21} & -k_{12} \\ 0 & -k_{21} & -k_{21} & k_{12} + k_{12} \end{pmatrix} \tag{36}$$

Using these and the general formulas, one can then proceed with the analysis. One can even introduce interactions between the two systems. If the activity of the monomers are correlated (cooperative or anticooperative), e.g., the switching rate of one system depends on the state of the other system, then the appropriate rate constants should be changed.

If no such interactions are relevant, then it is easy to prove that the eigenvalues of the combined matrices M_0 and $M_0 + K$ can be obtained from the eigenvalues of the individual systems. The new eigenvalues are simply all additive combinations of eigenvalues of the individual systems: $\lambda = \lambda_i + \lambda_j$. Hence, it is fairly easy to construct the waiting time distribution and the intensity correlations. The intensity correlation function is particularly simple, since it is guaranteed to be the sum of the intensity correlation functions for each subsystem. There is no such simple relation between the eigenvalues of M_0K^{-1} and those of the individual systems. Therefore, the waiting time correlations require more involved calculations.

3.1.5 Data Analysis

In this section, we mention a few points that are important when performing the actual data analysis. The most direct data set is probably the intensity as a sequence of intensities closely spaced in time: I_1, I_2, I_3, ... The average intensity is of course no problem, while the intensity correlation function is more tricky. The best and most efficient tool is based on the mathematical Wiener–Khintchine theorem. Consider a function of time $a(t)$ (it could be the detected light intensity) which is recorded in a total time T. Denote by $a(\omega)$ the Fourier transform of this function. Consider now the absolute square of that function $C(\omega) = |a(\omega)|^2$. The theorem states that this is the Fourier transform of the average of the intensity correlation function, i.e., of

$$C(\tau) = \frac{1}{T}\int_0^T a(t+\tau)a(t)dt. \tag{37}$$

So, the procedure is to Fourier transform the recorded signal. Take the absolute square of the result and finally, back Fourier transform of that square. Most computer data packages have built-in efficient Fourier transformation capabilities.

3.1.6 Binning

As a last point, we should like to warn against improper conclusions based on binned data. One would be tempted to bin, e.g., activity data, and plot the average activity of a certain enzyme by collecting all data in time windows with a much longer duration than that of the raw data. This might result in apparent slow activity fluctuations that may be misinterpreted as originating from the enzyme moving through different states on a timescale of multiple seconds. We recently showed (Hatzakis et al., 2012) that this may not necessarily be the correct interpretation and that activity fluctuations are natural statistical fluctuations of simple two-state models.

4. SUMMARY

In this chapter, we discussed how membrane systems like NDs may be used to reconstitute and spatially confine membrane proteins for SM fluorescence measurements with a focus on resolving their function at the individual turnover level. NDs are a powerful alternative to liposomes, especially for orientation-sensitive SM functional approaches. We then explained in detail the statistical analysis required to identify protein behavior faithfully from trajectories of individual turnovers, emphasizing on the model development, the potential pitfalls for correctly identifying the abundance, lifetime, and likelihood of sampling protein functional states. We showed that intensity autocorrelation is a reliable tool to identify the abundance and consequently how to utilize the waiting time distributions and autocorrelation to extract lifetime. This methodology may enable studies of functional dynamics and their role in biology for spectrum of membrane protein.

ACKNOWLEDGMENTS

The research leading to these results has received funding from the NSH, the Villum Foundation "Young Investigator Program", and the Center for Synthetic Biology "bioSYNergy" supported by the UCPH Excellence Program for Interdisciplinary Research

REFERENCES

Akyuz, N., Altman, R. B., Blanchard, S. C., & Boudker, O. (2013). Transport dynamics in a glutamate transporter homologue. *Nature*, *502*(7469), 114.

Akyuz, N., Georgieva, E. R., Zhou, Z., Stolzenberg, S., Cuendet, M. A., Khelashvili, G., et al. (2015). Transport domain unlocking sets the uptake rate of an aspartate transporter. *Nature*, *518*(7537), 68.

Alami, M., Dalal, K., Lelj-Garolla, B., Sligar, S. G., & Duong, F. (2007). Nanodiscs unravel the interaction between the SecYEG channel and its cytosolic partner secA. *EMBO Journal*, *26*(8), 1995–2004.

Banerjee, P. R., & Deniz, A. A. (2014). Shedding light on protein folding landscapes by single-molecule fluorescence. *Chemical Society Reviews*, *43*(4), 1172–1188.

Bayburt, T. H., Grinkova, Y. V., & Sligar, S. G. (2002). Self-assembly of discoidal phospholipid bilayer nanoparticles with membrane scaffold proteins. *Nano Letters*, *2*(8), 853–856.

Bayburt, T. H., & Sligar, S. G. (2010). Membrane protein assembly into nanodiscs. *FEBS Letters*, *584*(9), 1721–1727.

Bertram, N., Laursen, T., Barker, R., Bavishi, K., Moller, B. L., & Cardenas, M. (2015). Nanodisc films for membrane protein studies by neutron reflection: Effect of the protein scaffold choice. *Langmuir*, *31*(30), 8386–8391.

Borch, J., & Hamann, T. (2009). The nanodisc: A novel tool for membrane protein studies. *Biological Chemistry*, *390*(8), 805–814.

Borch, J., Torta, F., Sligar, S. G., & Roepstorff, P. (2008). Nanodiscs for immobilization of lipid bilayers and membrane receptors: Kinetic analysis of cholera toxin binding to a glycolipid receptor. *Analytical Chemistry*, *80*(16), 6245–6252.

Brender, J. R., Dertouzos, J., Ballou, D. P., Massey, V., Palfey, B. A., Entsch, B., et al. (2005). Conformational dynamics of the isoalloxazine in substrate-free p-hydroxybenzoate hydroxylase: Single-molecule studies. *Journal of the American Chemical Society*, *127*(51), 18171–18178.

Choi, P. J., Cai, L., Frieda, K., & Xie, S. (2008). A stochastic single-molecule event triggers phenotype switching of a bacterial cell. *Science*, *322*(5900), 442–446.

Christensen, S. M., Bolinger, P. Y., Hatzakis, N. S., Mortensen, M. W., & Stamou, D. (2012). Mixing subattolitre volumes in a quantitative and highly parallel manner with soft matter nanofluidics. *Nature Nanotechnology*, 7(1), 51–55.

Cornish, P. V., & Ha, T. (2007). A survey of single-molecule techniques in chemical biology. *ACS Chemical Biology*, *2*(1), 53–61.

Craig, D. B., Arriaga, E. A., Wong, J. C. Y., Lu, H., & Dovichi, N. J. (1996). Studies on single alkaline phosphatase molecules: Reaction rate and activation energy of a reaction catalyzed by a single molecule and the effect of thermal denaturation—The death of an enzyme. *Journal of the American Chemical Society, 118*(22), 5245–5253.

Cruz, F., & Edmondson, D. E. (2007). Kinetic properties of recombinant MAO-A on incorporation into phospholipid nanodisks. *Journal of Neural Transmission, 114*(6), 699–702.

Denisov, I. G., Grinkova, Y. V., Lazarides, A. A., & Sligar, S. G. (2004). Directed self-assembly of monodisperse phospholipid bilayer nanodiscs with controlled size. *Journal of the American Chemical Society, 126*(11), 3477–3487.

Dufrene, Y. F. (2008). Towards nanomicrobiology using atomic force microscopy. *Nature Reviews Microbiology, 6*, 674–680.

Edman, L., Foldes-Papp, Z., Wennmalm, S., & Rigler, R. (1999). The fluctuating enzyme: A single molecule approach. *Chemical Physics, 247*(1), 11–22.

Engel, A., & Muller, D. J. (2000). Observing single biomolecules at work with the atomic force microscope. *Nature Structural Biology, 7*, 715–718.

Engelkamp, H., Hatzakis, N. S., Hofkens, J., De Schryver, F. C., Nolte, R. J. M., & Rowan, A. E. (2006). *Do enzymes sleep and work? Chemical Communications, (9), 935–940.* .

English, B. P., Min, W., van Oijen, A. M., Lee, K. T., Luo, G., Sun, H., et al. (2006). Ever-fluctuating single enzyme molecules: Michaelis-Menten equation revisited. *Nature Chemical Biology, 2*(2), 87–94.

Fisher, T. E., Marszalek, P. E., & Fernandez, J. M. (2000). Stretching single molecules into novel conformations using the atomic force microscope. *Nature Structural Biology, 7*, 719–724.

Fluck, C. E., Mullis, P. E., & Pandey, A. V. (2009). Modeling of human P450 oxidoreductase structure by in silico mutagenesis and MD simulation. *Molecular and Cellular Endocrinology, 313*(1–2), 17–22.

Fluck, C. E., Nicolo, C., & Pandey, A. V. (2007). Clinical, structural and functional implications of mutations and polymorphisms in human NADPH P450 oxidoreductase. *Fundamental & Clinical Pharmacology, 21*(4), 399–410.

Fruh, V., IJzerman, A. P., & Siegal, G. (2011). How to catch a membrane protein in action: A review of functional membrane protein immobilization strategies and their applications. *Chemical Reviews, 111*(2), 640–656.

Gohon, Y., & Popot, J. L. (2003). Membrane protein-surfactant complexes. *Current Opinion in Colloid & Interface Science, 8*(1), 15–22.

Goldsmith, R. H., Tabares, L. C., Kostrz, D., Dennison, C., Aartsma, T. J., Canters, G. W., et al. (2011). Redox cycling and kinetic analysis of single molecules of solution-phase nitrite reductase. *Proceedings of the National Academy of Sciences of the United States of America, 108*(42), 17269–17274.

Greenleaf, W. J., Woodside, M. T., & Block, S. M. (2007). High-resolution, single-molecule measurements of biomolecular motion. *Annual Review of Biophysics and Biomolecular Structure, 36*, 171–190.

Groves, J. T., Ulman, N., & Boxer, S. G. (1997). Micropatterning fluid lipid bilayers on solid supports. *Science, 275*(5300), 651–653.

Gudmand, M., Rocha, S., Hatzakis, N. S., Peneva, K., Mullen, K., Stamou, D., et al. (2010). Influence of lipid heterogeneity and phase behavior on phospholipase A(2) action at the single molecule level. *Biophysical Journal, 98*(9), 1873–1882.

Guydosh, N. R., & Block, S. M. (2009). Direct observation of the binding state of the kinesin head to the microtubule. *Nature, 461*(7260), 125–129.

Ha, T., Kozlov, A. G., & Lohman, T. M. (2012). Single-molecule views of protein movement on single-stranded DNA. *Annual Review of Biophysics, 41*, 295–319.

Hatzakis, N. S. (2014). Single molecule insights on conformational selection and induced fit mechanism. *Biophysical Chemistry*, *186*, 46–54.

Hatzakis, N. S., Bhatia, V. K., Larsen, J., Madsen, K. L., Bolinger, P. Y., Kunding, A. H., et al. (2009). How curved membranes recruit amphipathic helices and protein anchoring motifs. *Nature Chemical Biology*, *5*(11), 835–841.

Hatzakis, N. S., Engelkamp, H., Velonia, K., Hofkens, J., Christianen, P. C. M., Svendsen, A., et al. (2006). Synthesis and single enzyme activity of a clicked lipase-BSA hetero-dimer. *Chemical Communications*, (19), 2012–2014.

Hatzakis, N. S., Wei, L., Jorgensen, S. K., Kunding, A. H., Bolinger, P. Y., Ehrlich, N., et al. (2012). Single enzyme studies reveal the existence of discrete functional states for monomeric enzymes and how they are "selected" upon allosteric regulation. *Journal of the American Chemical Society*, *134*(22), 9296–9302.

Hohng, S., Lee, S., Lee, J., & Jo, M. H. (2014). Maximizing information content of single-molecule FRET experiments: Multi-color FRET and FRET combined with force or torque. *Chemical Society Reviews*, *43*, 1007–1013.

Honda, M., Okuno, Y., Hengel, S. R., Martín-López, J. V., Cook, C. P., Amunugama, R., et al. (2014). Mismatch repair protein hMSH2-hMSH6 recognizes mismatches and forms sliding clamps within a D-loop recombination intermediate. *Proceedings of the National Academy of Sciences of the United States of America*, *111*, E316–E325.

Huang, N., Agrawal, V., Giacomini, K. M., & Miller, W. L. (2008). Genetics of P450 oxidoreductase: Sequence variation in 842 individuals of four ethnicities and activities of 15 missense mutations. *Proceedings of the National Academy of Sciences of the United States of America*, *105*(5), 1733–1738.

Huang, N. W., Pandey, A. V., Agrawal, V., Reardon, W., Lapunzina, P. D., Mowat, D., et al. (2005). Diversity and function of mutations in p450 oxidoreductase in patients with Antley-Bixler syndrome and disordered steroidogenesis. *American Journal of Human Genetics*, *76*(5), 729–749.

Iversen, L., Tu, H.-L., Lin, W.-C., Christensen, S. M., Abel, S. M., Iwig, J., et al. (2014). Ras activation by SOS: Allosteric regulation by altered fluctuation dynamics. *Science*, *345*(6192), 50–54.

Joo, C., Balci, H., Ishitsuka, Y., Buranachai, C., & Ha, T. (2008). Advances in single-molecule fluorescence methods for molecular biology. *Annual Review of Biochemistry*, 77, 51–76.

Jorgensen, S. K., & Hatzakis, N. S. (2013). Insights in enzyme functional dynamics and activity regulation by single molecule studies. *Biophysical Reviews and Letters*, *8*(3–4), 137–160.

Jorgensen, S. K., McMillan, D., Krzeminski, L., Stamou, D., Jeuken, L. J. C., & Hatzakis, N. S. (2013). Single proton pump activity measurements on single vesicles for a quinol heme-copper oxidase. *Biophysical Journal*, *104*(2), 277A–278A.

Kuznetsova, S., Zauner, G., Aartsma, T. J., Engelkamp, H., Hatzakis, N., Rowan, A. E., et al. (2008). The enzyme mechanism of nitrite reductase studied at single-molecule level. *Proceedings of the National Academy of Sciences of the United States of America*, *105*(9), 3250–3255.

Larsen, J., Hatzakis, N. S., & Stamou, D. (2011). Observation of inhomogeneity in the lipid composition of individual nanoscale liposomes. *Journal of the American Chemical Society*, *133*(28), 10685–10687.

Laursen, T., Singha, A., Rantzau, N., Tutkus, M., Borch, J., Hedegård, P., et al. (2014). Single molecule activity measurements of cytochrome P450 oxidoreductase reveal the existence of two discrete functional states. *ACS Chemical Biology*, *9*(3), 630–634.

Lee, A. G. (2004). How lipids affect the activities of integral membrane proteins. *Biochimica et Biophysica Acta, Biomembranes*, *1666*(1–2), 62–87.

Li, M., Jorgensen, S. K., McMillan, D. G. G., Krzeminski, L., Daskalakis, N. N., Partanen, R. H., et al. (2015). Single enzyme experiments reveal a long-lifetime proton leak state in a heme-copper oxidase. *Journal of the American Chemical Society*, *137*(51), 16055–16063.

Lu, H. P., Xun, L. Y., & Xie, X. S. (1998). Single-molecule enzymatic dynamics. *Science*, *282*(5395), 1877–1882.

Mathiasen, S., Christensen, S. M., Fung, J. J., Rasmussen, S. G. F., Fay, J. F., Jorgensen, S. K., et al. (2014). Nanoscale high-content analysis using compositional heterogeneities of single proteoliposomes. *Nature Methods*, *11*(9), 931–934.

Mi, L. Z., Grey, M. J., Nishida, N., Walz, T., Lu, C. F., & Springer, T. A. (2008). Functional and structural stability of the epidermal growth factor receptor in detergent micelles and phospholipid nanodiscs. *Biochemistry*, *47*(39), 10314–10323.

Michalet, X., Weiss, S., & Jager, M. (2006). Single-molecule fluorescence studies of protein folding and conformational dynamics. *Chemical Reviews*, *106*(5), 1785–1813.

Moerner, W. E. (2007). New directions in single-molecule imaging and analysis. *Proceedings of the National Academy of Sciences of the United States of America*, *104*(31), 12596–12602.

Mukhtasimova, N., Lee, W. Y., Wang, H. L., & Sine, S. M. (2009). Detection and trapping of intermediate states priming nicotinic receptor channel opening. *Nature*, *459*(7245), 451–454.

Muller, D. J., & Dufrene, Y. F. (2008). Atomic force microscopy as a multifunctional molecular toolbox in nanobiotechnology. *Nature Nanotechnology*, *3*, 261–269.

Muller, D. J., Helenius, J., Alsteens, D., & Dufrene, Y. F. (2009). Force probing surfaces of living cells to molecular resolution. *Nature Chemical Biology*, *5*, 383–390.

Nicolo, C., Fluck, C. E., Mullis, P. E., & Pandey, A. V. (2010). Restoration of mutant cytochrome P450 reductase activity by external flavin. *Molecular and Cellular Endocrinology*, *321*(2), 245–252.

Ning, W., Fei, J. Y., & Gonzalez, R. L. (2014). The ribosome uses cooperative conformational changes to maximize and regulate the efficiency of translation. *Proceedings of the National Academy of Sciences of the United States of America*, *111*, 12073–12078.

Oddershede, L. B. (2012). Force probing of individual molecules inside the living cell is now a reality. *Nature Chemical Biology*, *8*, 879–886.

Pandey, A. V., & Fluck, C. E. (2013). NADPH P450 oxidoreductase: Structure, function, and pathology of diseases. *Pharmacology and Therapeutics*, *138*(2), 229–254.

Pinaud, F., Clarke, S., Sittner, A., & Dahan, M. (2010). Probing cellular events, one quantum dot at a time. *Nature Methods*, 7(4), 275–285.

Roeffaers, M. B. J., De Cremer, G., Uji-i, H., Muls, B., Sels, B. F., Jacobs, P. A., et al. (2007). Single-molecule fluorescence spectroscopy in (bio)catalysis. *Proceedings of the National Academy of Sciences of the United States of America*, *104*, 12603–12609.

Schuler, B., & Eaton, W. A. (2008). Protein folding studied by single-molecule FRET. *Current Opinion in Structural Biology*, *18*(1), 16–26.

Shaw, A. W., Pureza, V. S., Sligar, S. G., & Morrissey, J. H. (2007). The local phospholipid environment modulates the activation of blood clotting. *Journal of Biological Chemistry*, *282*(9), 6556–6563.

Shi, J., Palfey, B. A., Dertouzos, J., Jensen, K. F., Gafni, A., & Steel, D. (2004). Multiple states of the Tyr318Leu mutant of dihydroorotate dehydrogenase revealed by single-molecule kinetics. *Journal of the American Chemical Society*, *126*(22), 6914–6922.

Shih, S. M., Engel, B. D., Kocabas, F., Bilyard, T., Gennerich, A., Marshall, W. F., et al. (2013). Intraflagellar transport drives flagellar surface motility. *eLife*, *2*, e00744.

Smiley, R. D., & Hammes, G. G. (2006). Single molecule studies of enzyme mechanisms. *Chemical Reviews*, *106*, 3080–3094.

Terentyeva, T. G., Engelkamp, H., Rowan, A. E., Komatsuzaki, T., Hofkens, J., Li, C. B., et al. (2012). Dynamic disorder in single-enzyme experiments: Facts and artifacts. *ACS Nano*, *6*(1), 346–354.

Terentyeva, T. G., Hofkens, J., Komatsuzaki, T., Blank, K., & Li, C. B. (2013). Time-resolved single molecule fluorescence spectroscopy of an alpha-chymotrypsin catalyzed reaction. *Journal of Physical Chemistry B*, *117*(5), 1252–1260.

Uphoff, S., Reyes-Lamothe, R., Garza de Leon, F., Sherratt, D. J., & Kapanidis, A. N. (2013). Single-molecule DNA repair in live bacteria. *Proceedings of the National Academy of Sciences of the United States of America*, *110*, 8063–8068.

Veigel, C., & Schmidt, C. F. (2011). Moving into the cell: Single-molecule studies of molecular motors in complex environments. Nature Reviews. *Molecular Cell Biology*, *12*(3), 163–176.

Velonia, K., Flomenbom, O., Loos, D., Masuo, S., Cotlet, M., Engelborghs, Y., et al. (2005). Single-enzyme kinetics of CALB-catalyzed hydrolysis. *Angewandte Chemie, International Edition*, *44*(4), 560–564.

Veshaguri, S., Christensen, S. M., Kemmer, G. C., Ghale, G., Møller, M. P., Lohr, C., et al. (2016). Direct observation of proton pumping by a eukaryotic P-type ATPase. *Science*, *351*(6280), 1469–1473.

Veshaguri, S., Tutkus, M., Lundgaard, C. V., Tonnesen, A., Mathiasen, S., Jorgensen, S. K., et al. (2013). Developing an assay to probe activation and conformational dynamics of beta 2-adrenergic receptor on single molecule level. *Biophysical Journal*, *104*(2), 61A–61A.

Wadsater, M., Laursen, T., Singha, A., Hatzakis, N. S., Stamou, D., Barker, R., et al. (2012). Monitoring shifts in the conformation equilibrium of the membrane protein cytochrome P450 reductase (POR) in nanodiscs. *The Journal of Biological Chemistry*, *287*(41), 34596–34603.

Walter, N. G., Huang, C. Y., Manzo, A. J., & Sobhy, M. A. (2008). Do-it-yourself guide: How to use the modern single-molecule toolkit. *Nature Methods*, *5*, 475–489.

Xia, C., Panda, S. P., Marohnic, C. C., Martasek, P., Masters, B. S., & Kim, J. J. (2011). Structural basis for human NADPH-cytochrome P450 oxidoreductase deficiency. *Proceedings of the National Academy of Sciences of the United States of America*, *108*(33), 13486–13491.

Xie, X. S., Choi, P. J., Li, G. W., Lee, N. K., & Lia, G. (2008). Single-molecule approach to molecular biology in living bacterial cells. *Annual Review of Biophysics*, *37*, 417–444.

Xue, Q. F., & Yeung, E. S. (1995). Differences in the chemical reactivity of individual molecules of an enzyme. *Nature*, *373*(6516), 681–683.

Yang, H. (2010). Progress in single-molecule spectroscopy in cells. *Current Opinion in Chemical Biology*, *14*, 3–9.

Yang, H., Luo, G. B., Karnchanaphanurach, P., Louie, T. M., Rech, I., Cova, S., et al. (2003). Protein conformational dynamics probed by single-molecule electron transfer. *Science*, *302*(5643), 262–266.

Zhao, Y. F., Terry, D. S., Shi, L., Quick, M., Weinstein, H., Blanchard, S. C., et al. (2011). Substrate-modulated gating dynamics in a Na(+)-coupled neurotransmitter transporter homologue. *Nature*, *474*(7349), 109–113.

Zoghbi, M. E., Cooper, R. S., & Altenberg, G. A. (2016). The lipid bilayer modulates the structure and function of an ATP-binding cassette exporter. *The Journal of Biological Chemistry*, *291*, 4453–4461.

CHAPTER NINE

Putting Humpty–Dumpty Together: Clustering the Functional Dynamics of Single Biomolecular Machines Such as the Spliceosome

C.E. Rohlman*, M.R. Blanco[†,1], N.G. Walter[†,2]
*Albion College, Albion, MI, United States
[†]Single Molecule Analysis Group and Center for RNA Biomedicine, University of Michigan, Ann Arbor, MI, United States
[2]Corresponding author: e-mail address: nwalter@umich.edu

Contents

Abstract

The spliceosome is a biomolecular machine that, in all eukaryotes, accomplishes site-specific splicing of introns from precursor messenger RNAs (pre-mRNAs) with high fidelity. Operating at the nanometer scale, where inertia and friction have lost the dominant role they play in the macroscopic realm, the spliceosome is highly dynamic and assembles its active site around each pre-mRNA anew. To understand the structural dynamics underlying the molecular motors, clocks, and ratchets that achieve functional accuracy in the yeast spliceosome (a long-standing model system), we have developed single-molecule fluorescence resonance energy transfer (smFRET) approaches that report changes in intra- and intermolecular interactions in real time. Building on our work using

[1] Current address: California Institute of Technology, 1200 East California Boulevard, Pasadena, California 91125, USA.

Methods in Enzymology, Volume 581
ISSN 0076-6879
http://dx.doi.org/10.1016/bs.mie.2016.08.022
© 2016 Elsevier Inc.
All rights reserved.

hidden Markov models (HMMs) to extract kinetic and conformational state information from smFRET time trajectories, we recognized that HMM analysis of individual state transitions as independent stochastic events is insufficient for a biomolecular machine as complex as the spliceosome. In this chapter, we elaborate on the recently developed smFRET-based Single-Molecule Cluster Analysis (SiMCAn) that dissects the intricate conformational dynamics of a pre-mRNA through the splicing cycle in a model-free fashion. By leveraging hierarchical clustering techniques developed for Bioinformatics, SiMCAn efficiently analyzes large datasets to first identify common molecular behaviors. Through a second level of clustering based on the abundance of dynamic behaviors exhibited by defined functional intermediates that have been stalled by biochemical or genetic tools, SiMCAn then efficiently assigns pre-mRNA FRET states and transitions to specific splicing complexes, with the potential to find heretofore undescribed conformations. SiMCAn thus arises as a general tool to analyze dynamic cellular machines more broadly.

1. INTRODUCTION

In all eukaryotes, precursor messenger RNA (pre-mRNA) splicing, i.e., the removal of introns and ligation of flanking exons, is a crucial step in the maturation of an mRNA. In higher eukaryotes such as humans, alternative splicing allows for cell- and tissue specific, tightly regulated expression of diverse protein isoforms from a single gene (de Klerk & t Hoen, 2015; Nilsen & Graveley, 2010; Papasaikas & Valcarcel, 2016). A vast majority (~94%) of all human genes contain introns with an average size of 1000–2000 nucleotides (nt), approximately 10 times the size of the protein-coding exons (Barash et al., 2010; Braunschweig, Gueroussov, Plocik, Graveley, & Blencowe, 2013; Calarco, Zhen, & Blencowe, 2011; Chen & Manley, 2009; de Klerk & t Hoen, 2015; Kornblihtt et al., 2013; Lander et al., 2001; Papasaikas & Valcarcel, 2016; Rino & Carmo-Fonseca, 2009). At multimegadalton, perhaps the largest and most elaborate macromolecular machine of the cell (Chen & Moore, 2014; Matera & Wang, 2014; Nilsen, 2003; Papasaikas & Valcarcel, 2016), the spliceosome is responsible for most cellular splicing processes. It must reliably identify intron–exon boundaries, precisely ligate adjacent exons without inadvertent skipping, and appropriately regulate alternative splicing to meet the biological requirements of each cell independent of its environment (Kornblihtt et al., 2013; Matera & Wang, 2014; Naftelberg, Schor, Ast, & Kornblihtt, 2015; Papasaikas & Valcarcel, 2016; Wang & Cooper, 2007). Intron excision with single-nucleotide precision is thus no easy task. Not surprisingly, up to 50% of all mutations leading to human disease are thought to originate from splicing defects (Cartegni, Chew, & Krainer, 2002; Kornblihtt et al., 2013;

Li et al., 2016; Matera & Wang, 2014; Naftelberg et al., 2015; Poulos, Batra, Charizanis, & Swanson, 2011; Scotti & Swanson, 2016; Wang & Cooper, 2007; Zhang & Manley, 2013).

The spliceosome lacks a preformed catalytic core. Rather, for each splicing cycle a spliceosome is—just like the proverbial Humpty–Dumpty—put together anew on an intron-containing pre-mRNA substrate in a stepwise manner that requires both binding and release of small nuclear ribonucleoproteins (snRNPs), each formed by protein cofactors binding to a single small nuclear RNA (snRNA) (Fig. 1; Matera & Wang, 2014; Papasaikas & Valcarcel, 2016; Wahl, Will, & Luhrmann, 2009; Will & Luhrmann, 2011). Notably, splicing entails two successive phosphoryl transfer reactions that are isoenergetic and have been shown to also be reversible, at least under a certain set of ionic conditions (Smith & Konarska, 2008; Tseng & Cheng, 2008). Nonetheless, spliceosome assembly requires at least eight RNA-dependent ATPases of the so-called DExD/H-box subfamily (Jarmoskaite & Russell, 2014; Koodathingal & Staley, 2013; Staley & Guthrie, 1998). These enzymes, which catalyze successive rearrangements of RNA and protein components, are believed to enhance the fidelity of splicing by acting as proofreading clocks that allow the spliceosome to discard mutant substrates at multiple steps in the assembly and catalytic pathway (Burgess & Guthrie, 1993; Couto, Tamm, Parker, & Guthrie, 1987; De, Schmitzova, & Pena, 2016; Horowitz, 2011; Jarmoskaite & Russell, 2014; Koodathingal, Novak, Piccirilli, & Staley, 2010; Koodathingal & Staley, 2013; Mayas, Maita, Semlow, & Staley, 2010; Mayas, Maita, & Staley, 2006; Xu & Query, 2007).

For the budding yeast *Saccharomyces cerevisiae*, a vast array of tools for genetic and biochemical manipulation are available, making this simple, unicellular eukaryote an ideal model organism to study many biological processes. Much has been learned about the dynamic yeast spliceosome since its discovery (Fig. 1; Chen & Moore, 2014; Matera & Wang, 2014; Papasaikas & Valcarcel, 2016; Wahl et al., 2009; Will & Luhrmann, 2011). Assembly is initiated by the ATP-independent formation of a commitment complex in which the U1 snRNP interacts with the 5′ splice site (5′SS), while BBP and Mud2 interact with the branchpoint (BP) sequence. If the BP is mutated (UACUAAC → UACUACC), this complex cannot form and spliceosome assembly is blocked. With a wild-type (WT) precursor, the commitment complex is converted to the prespliceosome by the ATP-dependent binding of U2 to the BP, followed by association of the U4/U6.U5 triple snRNP and Prp19 complexes. Before the first

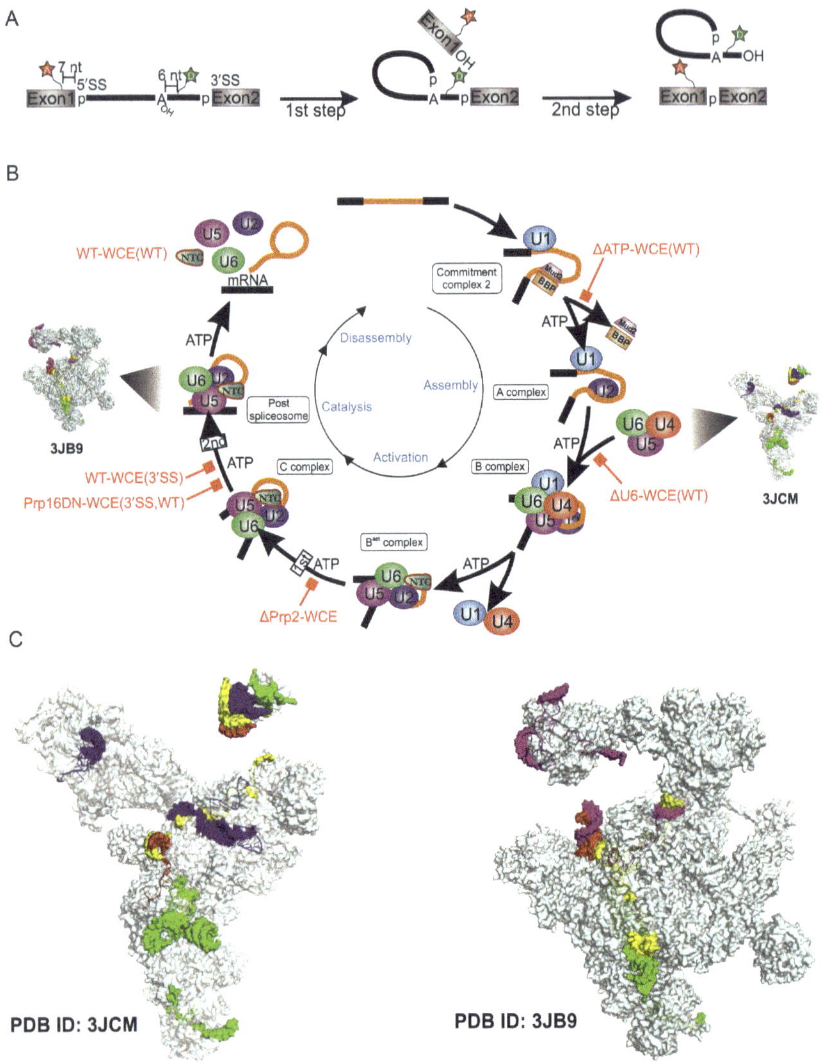

Fig. 1 The pre-mRNA splicing cycle. (A) The Ubc4 pre-mRNA substrate used to monitor pre-mRNA dynamics by smFRET contains Cy5 and Cy3 fluorophores seven nucleotides upstream of the 5′SS and six nucleotides downstream of the BP, respectively. (B) The spliceosome assembly and catalysis pathway is thought to progress in a stepwise manner requiring ATP at several steps of assembly. The biochemical and genetic stalls referred to in this chapter are indicated by *red blocks*. (C) Molecular models of spliceosomal intermediates (with more currently emerging). Solvent accessible surface map of spliceosomal intermediates based on reported cryo-EM structures (PDB IDs 3JCM and 3JB9). RNA components are highlighted in *red* for substrate or product RNA (pre-mRNA or lariat), *blue* for U4 snRNA, *yellow* for U6 snRNA, *green* for U5 snRNA, and *purple* for U2 snRNA. Images were generated in PyMol (DeLano, 2002).

transesterification can occur, U1 and U4 snRNPs must be released from the assembled spliceosome to form the activated B^{act} complex, again in ATP-dependent rearrangements. The DExD/H-box helicase Prp2 then converts this complex to the catalytically active B^{act} complex by destabilizing the U2 snRNP protein complexes (SF3a and SF3b) such that the BP adenosine becomes available for a nucleophilic attack on the distal 5′SS (Fig. 1B). This first chemical step, which results in formation of the looped lariat intermediate and free 5′ exon, can occur even in the presence of a mutation at the 3′ splice site (3′SS mutant UAG → UAC) known to block the second transesterification. After the Prp16- and Prp22-dependent second chemical step of splicing, the spliceosome undergoes ATP-dependent disassembly from the lariat intron and mature mRNA (Fig. 1A and B).

Recent advances in X-ray crystallography and particularly cryogenic electron microscopy (cryo-EM) have started to reveal near-atomic resolution structures of three select spliceosomal complexes from budding (Nguyen et al., 2015; Nguyen, Galej, Fica, et al., 2016; Wan et al., 2016) and fission yeast (Hang, Wan, Yan, & Shi, 2015; Yan et al., 2015), as well as from human (Agafonov et al., 2016; Pomeranz Krummel, Oubridge, Leung, Li, & Nagai, 2009), with the hope that more gaps around the splicing cycle may soon be filled (Fig. 1 and Table 1; Nguyen, Galej, Fica, et al., 2016). Despite three decades of study leading to these insights, however, until recently there was still precious little known about the timing and coordination of the multiple compositional and conformational rearrangements, which together constitute its enzymatic mechanism and endow the spliceosome with high efficiency and fidelity. This gap in knowledge was in large part due to, first, a paucity of biophysical tools suited to

Table 1 Currently Available High-Resolution Cryo-EM Yeast Spliceosome Structures

PDB ID	Resolution (Å)	Components	Organism	Publication
5GAN	3.7	30 proteins + U4/U6 and U5 snRNAs	*S. cerevisiae*	Nguyen, Galej, Bai, et al. (2016)
3JCM	3.8	U5 snRNA, U4/U6 snRNA, 30 proteins Prp8, Snu114, pre-mRNA	*S. cerevisiae*	Wan et al. (2016)
3JB9	3.6	37 Proteins, three snRNAs, 1 RNA lariat	*S. pombe*	Yan et al. (2015)

dissect such a complex, highly dynamic and heterogeneous machine in a holistic manner. Second, splicing is still studied in a crude yeast whole-cell extract (WCE) in vitro, where multistep reversible processes are obscured by asynchronous progression along the pathway. Third, splicing in WCE—developed in 1985 (Brody & Abelson, 1985)—is generally inefficient, leading to only a fraction of molecules undergoing one or both steps of splicing (Fig. 2). In addition, while RNA (–protein) conformational rearrangements often occur at subsecond timescales, it is known that pre-mRNAs are spliced in vitro only over 15–30 min (Das et al., 2006), requiring techniques for in vitro studies that span a wide range of timescales.

Single-molecule fluorescence resonance energy transfer (smFRET) presents a solution uniquely suited to these challenges and can provide dynamic information to animate the emerging static structures, a critically important step toward understanding the splicing mechanism. More specifically, we have developed prism-based total internal reflection fluorescence (P-TIRF) microscopy-based smFRET assays that monitor the distances between site-specifically placed donor and acceptor fluorophores (Figs. 1A, 2, and 3). Alternatively, a wide-field objective-type total internal reflection fluorescence microscope can be used (Roy, Hohng, & Ha, 2008). Analysis of the resulting time trajectories using smFRET histograms, hidden Markov models (HMMs), and various transition density plots (TDPs) has begun to dissect pre-mRNA conformational changes during splicing (Abelson et al., 2010; Blanco, Johnson-Buck, & Walter, 2012; Blanco et al., 2015; Blanco & Walter, 2010; Johnson-Buck, Blanco, & Walter, 2012; Kahlscheuer, Widom, & Walter, 2015; Krishnan et al., 2013; Semlow, Blanco, Walter, & Staley, 2016; Widom, Dhakal, Heinicke, & Walter, 2014). In particular, we identified a 135 nucleotide efficiently spliced yeast pre-mRNA, *UBC4*, in which donor and acceptor were placed either in the exons adjacent to the 5′SS and 3′SS or in the exon and intron adjacent to the 5′SS and BP, giving us unique windows into substrate dynamics. A strength of single-molecule approaches is that the macromolecular machines in the sample need not be synchronized, allowing the observation of both equilibrium and nonequilibrium processes (Walter, Huang, Manzo, & Sobhy, 2012). This is especially important when trying to understand highly reversible conformational steps, which are required in biological systems for proofreading functions and to facilitate pathway bifurcations as encountered in, for example, alternative splicing. However, the sheer number of states and interstate transitions became a challenge early on (Abelson et al., 2010), prompting us to develop smFRET techniques that employ

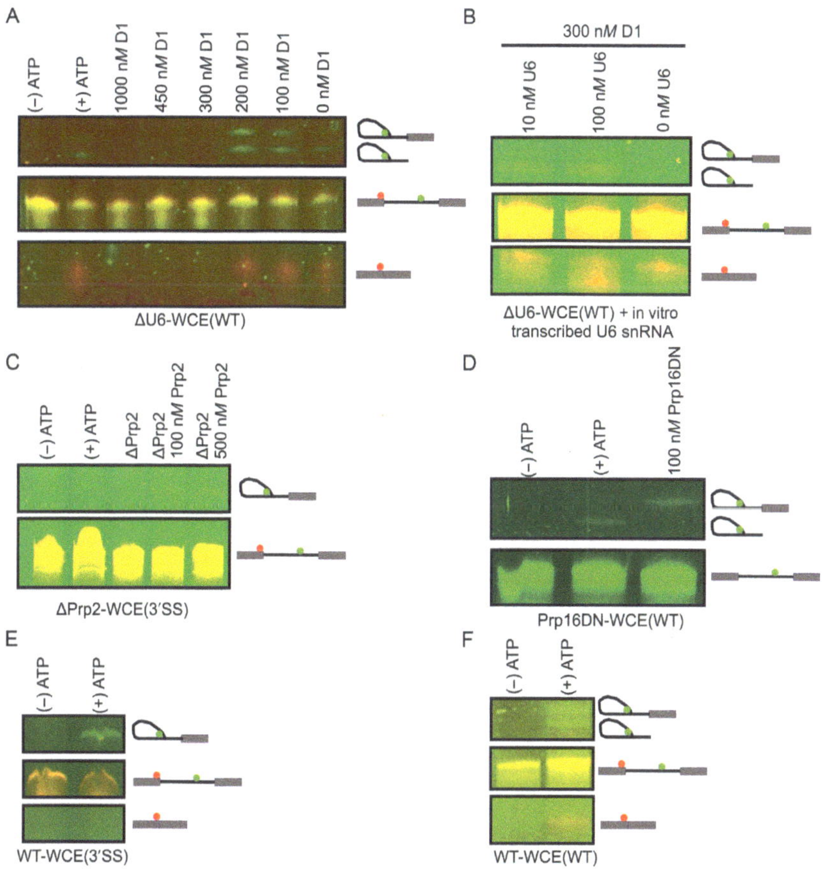

Fig. 2 Confirmation of blockage and reconstitution of splicing by in vitro splicing assays. Denaturing polyacrylamide gels were scanned with a variable mode Typhoon imager. The intron and intron-lariat products are observed in the Cy3 scan (*green*), and the mature mRNA product is visualized in the Cy5 scan (*red*). (A) The optimized concentration of oligodeoxynucleotide D1 required to deplete U6 snRNA (300 n*M*) was determined by titrating increasing amounts of D1 into the in vitro splicing assay. (B) Using the previously determined optimal concentration of D1 (300 n*M*, A), extract viability was confirmed through reconstitution with in vitro transcribed U6 snRNA. (C) Incubation of *prp2-1 cef1-TAP* yeast whole-cell extract at 37°C for 40 min completely blocks splicing activity (ΔPrp2 lane). Addition of recombinant Prp2p to the extract results in reconstitution of splicing, as expected. (D) Addition of recombinant dominant mutant Prp16DN to yeast extract stalls splicing after the first chemical step. (E) Incubation of 3′SS mutant substrate with WCE stalls splicing after the first step, while incubation with a WT substrate (F) results in efficient progression through both steps of splicing. *Reproduced with permission from Blanco, M. R., Martin, J. S., Kahlscheuer, M. L., Krishnan, R., Abelson, J., et al. (2015). Single molecule cluster analysis dissects splicing pathway conformational dynamics.* Nature Methods, 12, *1077–1084.*

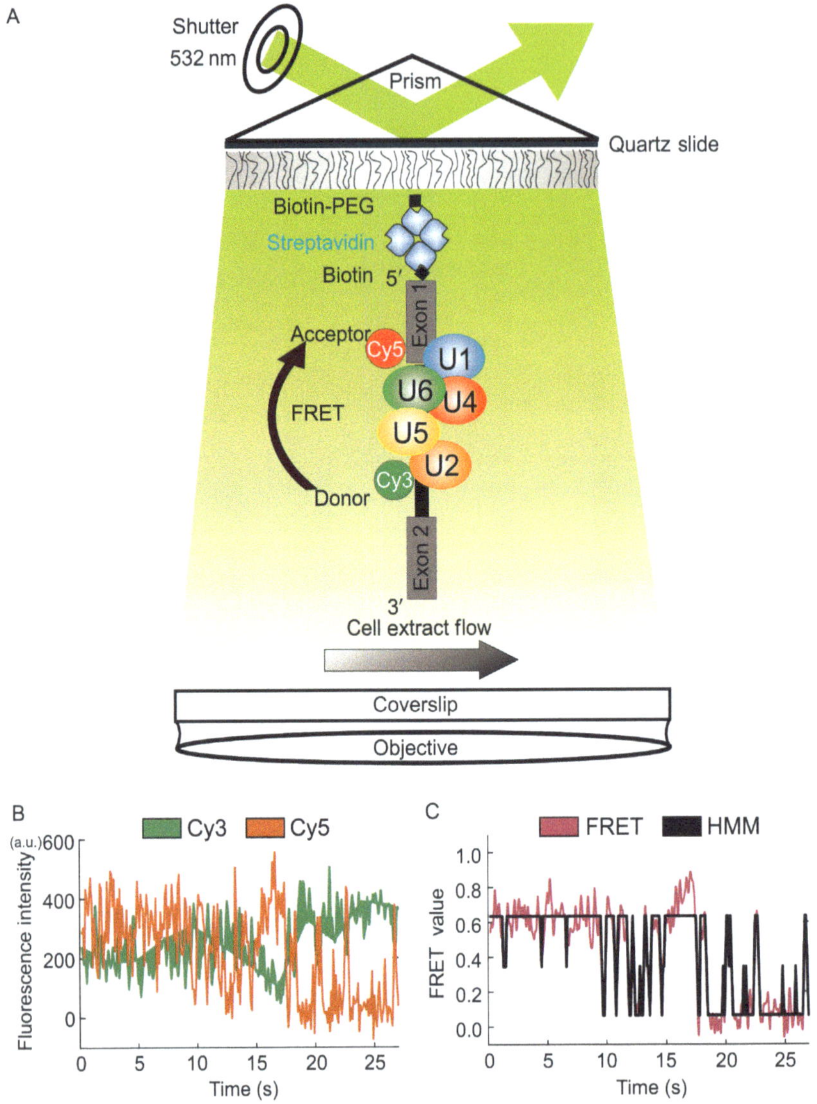

Fig. 3 Typical acquisition of smFRET data. (A) Prism-based TIRF microscope with microfluidic sample chamber containing the 5′ biotinylated Ubc4 pre-mRNA substrate incubated with various yeast splicing extracts (Abelson et al., 2010; Blanco et al., 2015). (B) Raw single-molecule time trace showing the anticorrelated donor (*green*) and acceptor (*red*) intensities. (C) The corresponding FRET trace (*purple*) and the HMM trace as assigned by vbFRET (*black*). *Reproduced in part with permission from Blanco, M. R., Martin, J. S., Kahlscheuer, M. L., Krishnan, R., Abelson, J., et al. (2015). Single molecule cluster analysis dissects splicing pathway conformational dynamics.* Nature Methods, 12, *1077–1084.*

either biochemical enrichment of specific complexes via glycerol gradient centrifugation and immunoprecipitation (termed Single-Molecule Pull-Down FRET, or SiMPull-FRET; Kahlscheuer et al., 2015; Krishnan et al., 2013; Semlow et al., 2016) or bioinformatic clustering for "computational purification" termed Single-Molecule Cluster Analysis, SiMCAn (Blanco et al., 2015) to pinpoint the structural dynamics of specific splicing intermediates. Our approach of focusing on the pre-mRNA substrate with SiMPull-FRET and SiMCAn is particularly timely given the emerging cryo-EM structures of relevant spliceosomal complexes typically lack at least part of the substrate (Fig. 1C and Table 1) and can thus be uniquely complemented and animated by substrate data.

SiMCAn translates time-resolved smFRET data into a temporal sequence of conformational changes that single pre-mRNA molecules adopt throughout splicing. In its initial application, SiMCAn discovered a previously undescribed low-FRET conformation adopted late in splicing by a 3′SS mutant (Blanco et al., 2015). Conversely, the growing number of high-resolution structures of spliceosomal components and complexes will aid in the refinement of SiMCAn-derived conformational models, allowing us to test the hypothesis that specific sequences of structural changes facilitate splicing. As an ultimate goal, SiMCAn has the potential to help delineate the mechanism of pre-mRNA splicing by dissecting—at single-molecule and single-nucleotide resolution—the complete "movie" of conformational changes throughout a full splicing cycle (Fig. 1), or any other biomolecular process. MATLAB scripts to perform all SiMCAn analyses are available as Supplementary Software with our original publication (Blanco et al., 2015).

2. EXPERIMENTAL METHODS AND DATA ANALYSIS

Data from smFRET studies have directly established that populations of single molecules often exhibit variant behaviors that include dominant, rare, long-lived, or transient species, as well as singular and parallel (i.e., alternate) pathways. Observations of large populations of molecules provide a statistical sample that mirrors ensemble measurements, while also being sensitive enough to describe the component subpopulations. The strengths of smFRET methods have established them as a useful probe of RNA molecular catalysis and dynamics, yet their application to more complex and multistep systems faces additional hurdles. Molecular systems that display two or three prominent FRET states are often accurately identified using either thresholding or HMM approaches (Blanco & Walter, 2010).

However, the analysis of multistate smFRET data may be subject to user judgment and bin size selection bias. With a higher number of states, the analysis becomes increasingly unreliable. Spliceosomally catalyzed pre-mRNA splicing in particular is composed of a multitude of conformational states and substates that rapidly interconvert. Analysis of smFRET trajectories from pre-mRNA splicing in vitro required new approaches to ensure unbiased sampling and accurate modeling of the interactions and rearrangements that take place between the RNA, protein, and pre-mRNA components of the spliceosomal machinery. Understanding the subtlety of these dynamics required comparison of smFRET trajectories featuring different conformational states and kinetic properties from across a range of experimental conditions. The complexity of pre-mRNA splicing precluded fitting all data with a single HMM model, which would impose a constraining single, preordained kinetic model on all molecules and conditions.

More generally, the results from kinetic measurements at the single-molecule level are fundamentally different from classical ensemble-averaged data due to their statistical nature (Moffitt, Chemla, & Bustamante, 2010). When we began addressing the challenge of mining the increasingly complex data obtained from spliceosomal complexes for commonalities, we realized that a simple alignment of time trajectories did not yield satisfactory results since the dwell time in any given state is part of an exponential probability distribution. That is, single molecules with both very long and very short dwell times in a given state are found regardless of the underlying rate constants, which only modulates the probability of their observation (and short dwell times are always the most likely). We therefore returned to using HMMs, aiming to overcome their underlying limitations. In particular, a Markov transition from state A to state B is assumed to take place with total state memory loss, a feature that is fundamentally not the case for a biological process wherein a specific functional intermediate is a product of all preceding steps and itself funnels the biomolecular machine into a subset of accessible subsequent steps. Put differently, all of biology is inherently dependent on energy being used to bias the direction of molecular interactions and transformations, thus overcoming the dominance of complete randomness dictated by the Second Law of Thermodynamics. In the case of biomolecular machines such as the spliceosome that operate at the nanoscale—where inertia and friction have lost their preponderant role exerted at the macroscale—intrinsic, random thermal motions are biased directionally by helicases that, as molecular motors, utilize the energy from ATP hydrolysis to achieve biased Brownian ratcheting (Kahlscheuer et al., 2015;

Krishnan et al., 2013). To discern this nonrandomness in our smFRET data, we turned to a hybrid method wherein HMMs are used to capture and summarize the molecular FRET states of single molecules and the transitions between them in a FRET similarity matrix (FSM), followed by "alignment" of the single-molecule FSMs to identify clusters of common molecular behaviors (Blanco et al., 2015). A second level of clustering based on the abundance of dynamic behaviors when splicing is stalled at defined intermediate steps through biochemical and genetic tools then allowed us to assign pre-mRNA FRET states and transitions to specific splicing complex intermediates. In the following, we elaborate on relevant considerations for this type of analysis.

2.1 Acquisition and Initial Analysis of smFRET Data

Clustering analysis has been used in a variety of fields to help group large datasets (e.g., gene expression profiles) based on a similarity metric that distinguishes similar from disparate behaviors (D'haeseleer, 2005). Computational clustering provides the necessary analytical power to tease out individual subpopulations from a larger single-molecule dataset. We here discuss the application of SiMCAn (Blanco et al., 2015) a hierarchical clustering tool to extract the consensus behaviors of pre-mRNAs throughout the assembly and catalytic steps of the yeast spliceosome. The technique is generalizable for other datasets with more or less complicated dynamics whose analysis by traditional smFRET tools is insufficient, too error prone or too cumbersome to extract the full scale of information available. The algorithms to perform all SiMCAn analyses are available as MATLAB scripts (The MathWorks, Inc., Natick, MA) as part of the Supplementary Software of our original publication (Blanco et al., 2015).

In smFRET, a donor fluorophore (often Cy3) is excited with typically a 532-nm laser, with the resulting emission of both donor and acceptor fluorophore (often Cy5) recorded on a P-TIRF microscope setup (Fig. 3A). Importantly, SiMCAn is truly agnostic to the nature and origin of the data. Molecules selected for further analysis by SiMCAn should fulfill certain criteria, however. The molecule's fluorescence time trajectory should last at minimum, ~30 data points (or, more generally, at least three times the mean dwell time of the observed FRET states) before photobleaching of the donor fluorophore (Cy3), show anticorrelated changes in donor (Cy3) and acceptor (Cy5) intensity, undergo single-step photobleaching, and contain active acceptor (Cy5) fluorophore throughout, as evidenced upon direct 635-nm laser excitation at the end of each data acquisition (Blanco et al., 2015). After extracting the signal, FRET values are calculated by dividing the cross talk corrected intensity of the acceptor emission by the total emission from both donor and acceptor (Fig. 3B and C). Trajectories are truncated to just before the first fluorophore photobleaches. To reliably identify genuine FRET states and their transition kinetics in inherently noisy datasets, each selected smFRET trace is then fitted with an HMM, which traces the most probable sequence of FRET states and derives their dwell times, using vbFRET (Bronson, Fei, Hofman, Gonzalez, & Wiggins, 2009) in the MATLAB environment with no assumptions about the values or distributions (Fig. 3C). In principle, any idealized trajectories from HMM-fitted data (generated using vbFRET, HaMMy, QuB, etc.) (Blanco & Walter, 2010) subsequently can be fed into SiMCAn. Multiple models with differing number of states should initially be used to determine the underlying FRET states; typically, the entire dataset for each condition is analyzed by the iterative application of the Viterbi and BaumWelch algorithms to generate idealized trajectories (Blanco & Walter, 2010). The number of states assumed in the idealization can be varied from 2 to 11 and the corresponding fits evaluated using the Bayesian information criterion (BIC). The number of states that result in the best BIC score should be used in the subsequent analysis. Using the BIC penalizes models with extraneous states that do not result in a significant improvement in the LogLikelihood, allowing one to select the most appropriate model by balancing goodness of fit and model parsimony. Many of the currently available software packages for HMM fitting of smFRET trajectories have this procedure incorporated into their analysis.

For pre-mRNA splicing, we binned each HMM assigned FRET state into the closest of 10 evenly spaced FRET values (0.05–0.95, increment of 0.10), chosen in relation to the error of ±0.05 expected from a measurement of a standard FRET value in the range of 0 and 1 (Abelson et al., 2010; Blanco et al., 2015). This information is already much more fine-grained than a simple FRET occupancy histogram, which plots frame-by-frame FRET values for a set of molecules collected over the first 100 frames (10 s of real time) of all FRET trajectories in a given dataset, equivalent to typical ensemble FRET data (Fig. 4A; Blanco & Walter, 2010). In such a histogram, all information on the connectivity of specific states, their interconversion kinetics, and variance in FRET value is lost. A more appropriate way to capture the details of HMM data of many molecules is the use of TDPs, weighted by the number of times a particular FRET transition

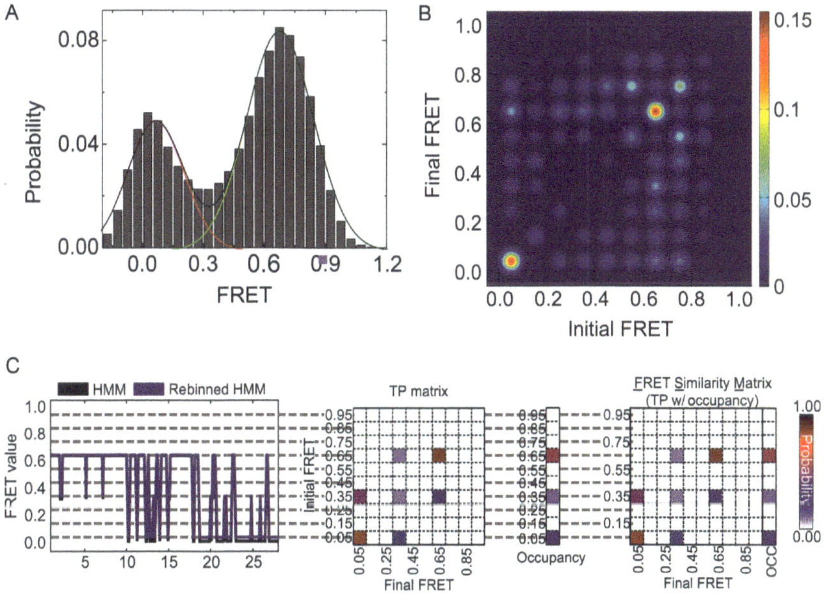

Fig. 4 Basic smFRET data analysis. (A) FRET probability distribution, summarizing all FRET values observed over the first 100 data points of 6079 molecules. (B) Transition occupancy density plot (TODP) for the data shown in (A), represented as a heat map of the fraction of molecules showing any particular transition. (C) Assigned FRET trace before (*black*) and after (*blue*) reassignment to the closest of 10 evenly spaced states (0.05–0.95, increments of 0.10, *gray dashed lines*). The transition probability (TP) matrix corresponds to the rebinned FRET trace in the *left panel* and represents the transition probability between each of the 10 binned FRET values. The FRET similarity matrix (FSM) combines both the TP matrix and the FRET state occupancies over the entire smFRET trace. *Reproduced in part with permission from Blanco, M. R., Martin, J. S., Kahlscheuer, M. L., Krishnan, R., Abelson, J., et al. (2015). Single molecule cluster analysis dissects splicing pathway conformational dynamics.* Nature Methods, 12, *1077–1084.*

occurs, or transition occupancy density plots (TODPs), weighted by the fraction of molecules displaying a particular transition (Figs. 4 and 5; Blanco & Walter, 2010). Of note, to eliminate possible effects of the high concentrations of proteins directly on the fluorophores, one may additionally evaluate the separated donor and acceptor signals by HMM and introduce a scoring function to select for anticorrelated donor and acceptor changes that reflect verified changes in FRET (Blanco & Walter, 2010). To better parse the complexity of pre-mRNA splicing, we decided to enrich specific splicing intermediates through seven biochemical and genetic stalls or combinations thereof (Fig. 2), as well as recorded smFRET data within

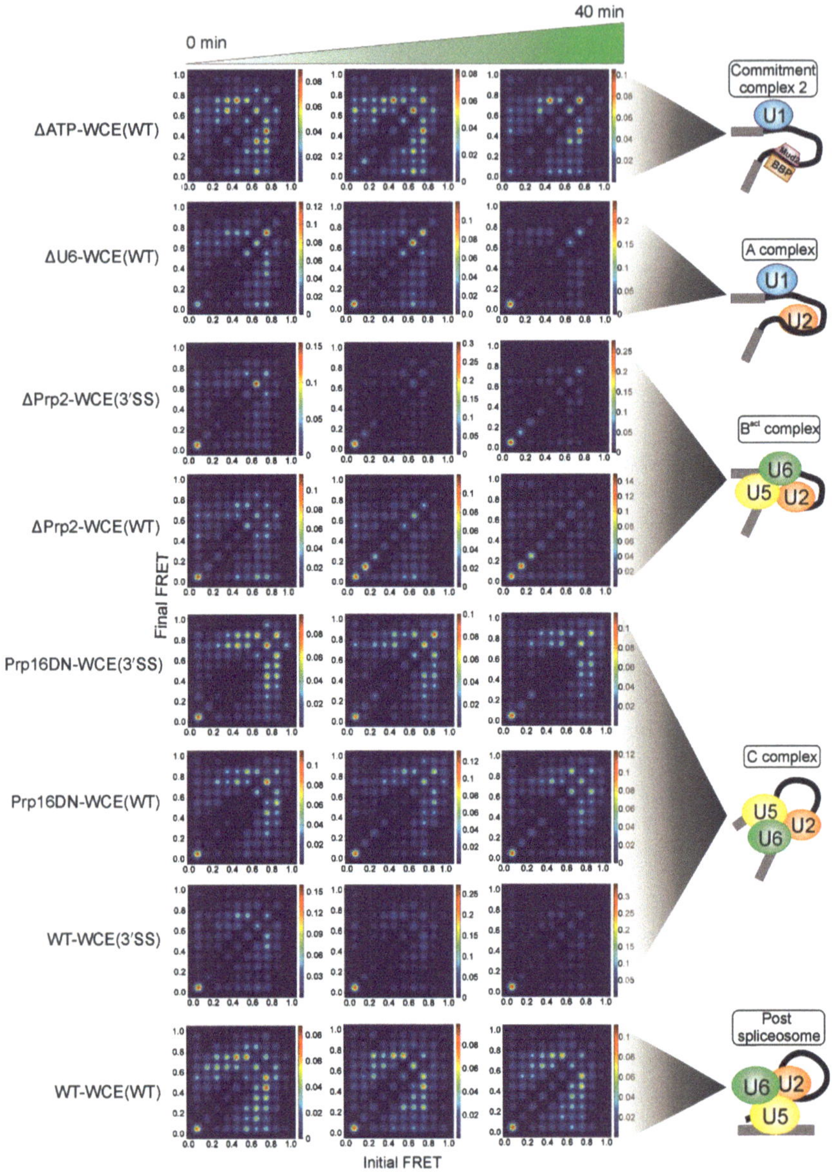

Fig. 5 Transition occupancy density plots (TODPs) from smFRET-monitored pre-mRNA splicing experiments. For each of the eight experimental conditions over the time course of the smFRET experiments, TODPs were used to depict the most probable transitions between an initial FRET state (*x*-axis) and a final FRET state (*y*-axis). *Reproduced with permission from Blanco, M. R., Martin, J. S., Kahlscheuer, M. L., Krishnan, R., Abelson, J., et al. (2015). Single molecule cluster analysis dissects splicing pathway conformational dynamics.* Nature Methods, 12, *1077–1084.*

three time windows of 0–8 min (early), 18–23 min (middle), and 33–40 min (late) after starting each assay. Including the absence of any stall, we collected a total of 10,680 time traces and corresponding HMMs (Blanco et al., 2015). The resulting TODPs show significant differences, but commonalities across conditions are challenging to discern by eye (Fig. 5), necessitating a more sophisticated analysis as offered by SiMCAn.

2.2 SiMCAn: Representing Single Molecules as FRET Similarity Matrices

In SiMCAn, each HMM of a single molecule is used to construct a symmetric transition probability (TP) matrix that delineates the molecule's FRET states as well as the probability per time unit for transitioning from one to any other state, including itself (Fig. 4C). The TP matrix thus summarizes for each molecule all conformational states it adopts and their interconversion kinetics, at a level of precision commensurate with the error of the underlying smFRET measurement, in our case as a 10 × 10 matrix (Blanco et al., 2015). Each TP matrix is then combined with an additional column describing the percent of time (before photobleaching) a given trace spends in each FRET state, resulting in the FSM (Fig. 4C). SiMCAn thus reduces each single-molecule trajectory to an easily comparable unit of information, the FSM, a data abstraction that is possible regardless of the number of FRET states or distinct molecular behaviors among the sampled species. The FSM allows the user to compare the multitude of FRET trajectories through hierarchical cluster analysis in an unbiased fashion, as described in the following section.

2.3 SiMCAn: The First Layer of Clustering

The FSM datasets collected for all experimental conditions are mixed, and then a pairwise distance between all possible trajectories is calculated using a Euclidean distance metric. In the case of pre-mRNA splicing, prior to SiMCAn clustering 4601 "static" molecules (no transitions with occupancy in only one state) were identified based on their FSMs among our total of 10,680 time traces and analyzed separately. Their features are simple so that they can be treated separately, and they are removed to avoid that they dominate clusters of other dynamic molecules with long lived, slowly interconverting FRET states. It cannot be distinguished whether these "static" molecules truly reside in only one state, adopt one long-lived state that happened to photobleach before transitioning into another state, or instead reside in two or more states so short lived that they time average into what only appears, at the available time resolution, as a single state with a mean FRET value.

The remainder, in our case the FSMs of 6079 dynamic molecules, are clustered using Ward's method for hierarchical clustering, in which the Euclidean distance between any pair of clusters r and s is given by Eq. (1):

$$d(r,s) = \sqrt{\frac{2n_r n_s}{n_r + n_s}} \left(\bar{x_r^2} - \bar{x_s^2}\right)^{1/2}, \tag{1}$$

where n is the number of elements and x is the centroids for the respective clusters r and s. This clustering algorithm is implemented natively in MATLAB. The goal is to group together molecules with short distances (high similarity) and connect them with other groups of molecules based on their average intercluster distance. This hierarchical clustering produces a large tree where each molecule is represented by a branch end point, or leaf (Fig. 6A). This tree can be cut (pruned) at various heights, leaving more or less clusters. A pruning threshold needs to be selected so that the intracluster distance (similarity within a cluster) is minimized and the intercluster distance (dissimilarity from other clusters) is maximized. The cutoff for this ideal number of clusters to represent a given dataset is best determined by Eq. (2), a k-means calculation as follows:

$$k = \frac{1}{B}\sum\left(\log\left(W_k^r\right) - \log(W_k)\right), \tag{2}$$

where B is the number of cluster sets and W_k^r is the average intercluster distance in a random set of k clusters, while W_k is the average intercluster distance of the actual clusters. In the case of pre-mRNA splicing, we pruned the tree to a height of 25 distinct clusters based on an inflection point encountered in the k-means plot over the number of clusters (Fig. 6B). That is, while the k-means used to evaluate the intercluster distances decreases monotonically as more clusters are chosen, at some point the number of clusters is so high that many similar FRET traces become separated. One therefore judiciously selects an inflection point in the k-means plot where an additional increase in the number of clusters has a smaller effect than the last increase had. The goal is that each final cluster represents a unique, distinguishable molecular behavior. For our splicing data, the resulting 25 dynamic clusters were named based on the first and second most occupied FRET states within the cluster (e.g., cluster 0.65–0.05 primarily occupies 0.65 and 0.05 FRET states) and were combined with the 10 static clusters, named for their sole FRET state (e.g., 0.05-S). Bootstrap analysis can be used to ask whether appropriate clusters capture all additional FSMs added to the

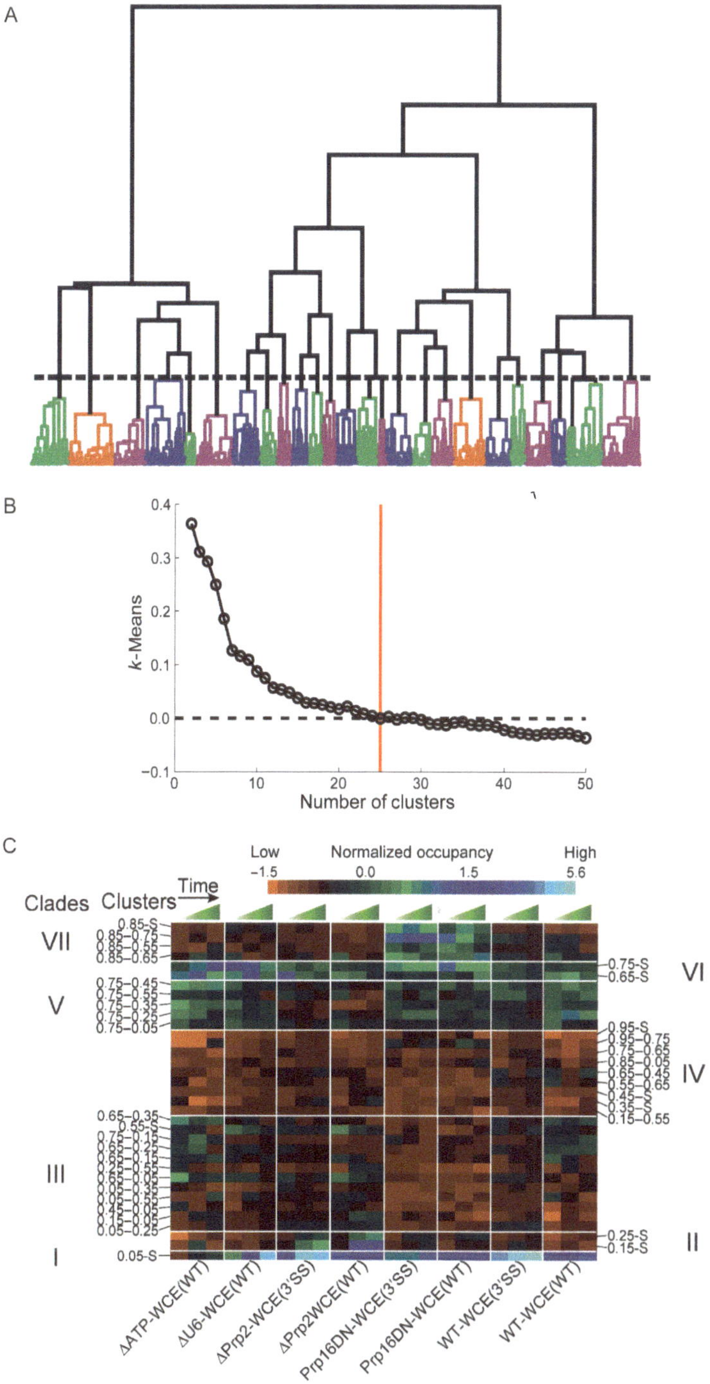

Fig. 6 See figure legend on next page.

dataset. This first layer of SiMCAn clustering is performed sufficiently unsupervised to generate a model-free tree to represent the types of molecular behaviors encountered in a given dataset objectively, without constraining the analysis with any a priori expectations.

2.4 SiMCAn: A Second Layer of Clustering Links the Data to Biology

In the case of pre-mRNA splicing, substrate mutations or extract modifications that block splicing at specific intermediates (Fig. 1B) can be used to limit the observed smFRET dynamics to steps prior to that block, thus enriching certain molecular behaviors and limiting others. This combination of a first, model-free round of clustering with a second round incorporating a priori knowledge from biochemically and genetically manipulated yeast extracts allowed further dissection of splicing in yeast. In particular, we sought to identify clusters whose occupancies are similarly enriched or depleted for a group of conditions, i.e., follow a similar pattern of high and low occupancies across these conditions, suggesting that they can be grouped into a "clade" (or cluster by a different name) of clusters (Fig. 6C). To this end, the occupancy within all 35 first-round clusters across

Fig. 6 SiMCAn-based cluster analysis. (A) Hierarchical clustering of 6079 dynamic single molecules. Varying the tree cutoff heights upon grouping the cluster occupancy among the eight experimental conditions leads to distinct numbers of (color-coded) clades of clusters. smFRET data were collected for each substrate and extract mutant and globally fit to a 10-state HMM. An FSM was then constructed for each molecule and clustered. Each colored branch represents a set of molecules that share common FRET transition probabilities. The *dashed line* indicates the threshold of 25 clusters used to describe the data. Molecules with only one state (static) were clustered separately and later incorporated as separate clusters. Hierarchical clustering was performed utilizing MATLAB routines by calculating the pairwise Euclidean distance of every single molecule and building an agglomerative hierarchical cluster tree utilizing the unweighted average distance to compute distances between clusters. (B) Iterative measurement of intercluster distances using a modified *k*-means algorithm utilized to determine the number of clusters that best describes the experimental data. (C) A heat-map representation of the clustering of clusters for the eight experimental conditions. Clusters are group based on profiles to identify "clades" of similar abundance patterns. The abundance profiles for each cluster were constructed and clustered to determine if groups of clusters were shifting in groups, similar to groups of genes with similar expression profiles. This procedure was performed using MATLAB's clustergram function. *Reproduced in part with permission from Blanco, M. R., Martin, J. S., Kahlscheuer, M. L., Krishnan, R., Abelson, J., et al. (2015). Single molecule cluster analysis dissects splicing pathway conformational dynamics.* Nature Methods, 12, *1077–1084.*

all conditions was used as a new similarity matrix to compute their distances, again based on a Euclidean distance measurement. Clades were generated by the same iterative *k*-means approach as before, with the aim to generate a reasonable group of clusters whose occupancy patterns across conditions are most alike (as measured by the Euclidean distance). For our splicing dataset, this second level of clustering yielded an optimal tree height of seven clades that was able to capture the changes in dynamic behavior as the pre-mRNA progresses through the splicing cycle. For example, conditions that stall WT substrate splicing at the early A complex (i.e., ΔU6-WCE(WT), Fig. 1B) show a time-dependent increase in clade I (Fig. 6C; Blanco et al., 2015). This stable low-FRET conformation suggests that the pre-mRNA substrate is partially denatured in the A complex compared to pre-mRNA in the absence of spliceosomal components (Abelson et al., 2010), keeping the first-step reaction sites at a distance. As another example, under the ΔPrp2-WCE(WT) and ΔPrp2-WCE(3′SS) conditions, which both should similarly enrich for the activated B^{act} spliceosome before the first step of splicing since the 3′SS mutation only affects the second step (Fig. 1A and B), SiMCAn recognizes a pair of dominant low-FRET clusters, 0.25-S and 0.15-S, that are appropriately grouped into clade II (Fig. 6C). Since these two clusters are so close in FRET value, they likely represent the same static low-FRET B^{act} conformation found independently upon biochemical purification of the B^{act} complex (Krishnan et al., 2013). Notably, SiMCAn was able to distinguish these clusters from the similarly static, but even lower FRET cluster 0.05-S of the A complex, which is not resolvable in an smFRET histogram (Fig. 4A).

In this way, the second clustering based on SiMCAn can reveal known or anticipated features of specific spliceosomal intermediates. Significantly, SiMCAn also identified heretofore undescribed differences in molecular behavior between the WT and 3′SS mutant substrates upon incubation with WT splicing extract, i.e., comparing conditions WT-WCE(WT) and WT-WCE(3′SS). In particular, both substrates are expected to progress through most of the splicing cycle until the second chemical step (Fig. 1B) and accordingly exhibited quite a similar set of pre-mRNA conformations (Fig. 6C). However, the 3′SS over time additionally adopts a static 0.05-S cluster, suggesting that the 5′SS and BP are far separated in comparison to their relative proximity in the splicing intermediate prior to this point in the cycle (Fig. 1B), represented by conditions Prp16DN-WCE(WT) and Prp16DN-WCE(3′SS) (Fig. 6C). This 0.05-S state is more predominant for the 3′SS mutant than the WT substrate, consistent with the appearance of a conformation in which the 5′SS and BP become greatly separated only after the first step of splicing when the mutated 3′SS is first detected (Blanco et al., 2015). This discovery of a proof-reading step, which allows for the selection of alternative 3′ splice sites if the canonical 3′SS is removed or damaged through mutation, was subsequently confirmed by smFRET probing of biochemically isolated 3′SS mutant complexes (Semlow et al., 2016), demonstrating the power of SiMCAn in discovery.

2.5 Generalized Application of SiMCAn to Datasets of Varying Complexity

Generally, experimental sampling ought to scale with the total number of distinguishable FRET states and kinetics in a dataset to ensure a reasonable level of convergence of SiMCAn, as with any other analysis approach. We developed SiMCAn specifically for dealing with very large datasets of high complexity, where traditional HMM analysis becomes inadequate and commonalities between single-molecule traces are challenging to identify. The availability of a dataset with >10,000 smFRET traces as described earlier for the pre-mRNA splicing cycle is, however, still not typical in the field. An example of a dataset more typically encountered in current single-molecule studies is that of the Prp2-mediated conformational transition immediately prior to the first step of splicing (Krishnan et al., 2013). In short, the immobilized B^{act} complex containing FRET-labeled pre-mRNA was monitored as it progresses through the B* to the C complex upon addition of recombinant proteins Prp2, Spp2, and Cwc25 (Fig. 7A). Notably, SiMCAn rapidly (within minutes) identified subpopulations of pre-mRNA molecules, recapitulating the subpopulations of molecules that had originally been identified manually by visual inspection (Fig. 7; Krishnan et al., 2013). More specifically, the HMM-fitted FRET traces under conditions designed to enrich for the B^{act}, B*, and C complexes, respectively, were converted into FSMs and analyzed together using SiMCAn. Again, maximizing the intercluster distances while minimizing the intracluster distances using SiMCAn revealed nine dynamic and four static clusters as most consistent with the data (Fig. 7B). As seen in bar graphs of the fraction of molecules present in each cluster under each of the three experimental conditions (Fig. 7C), we found that a cluster of static low-FRET (0.3-S) molecules dominates the B^{act} conditions, consistent with our previous observation of such a dominant state in the B^{act} spliceosome (Blanco et al., 2015; Krishnan et al., 2013). By contrast, a static high-FRET cluster (0.7-S) is most

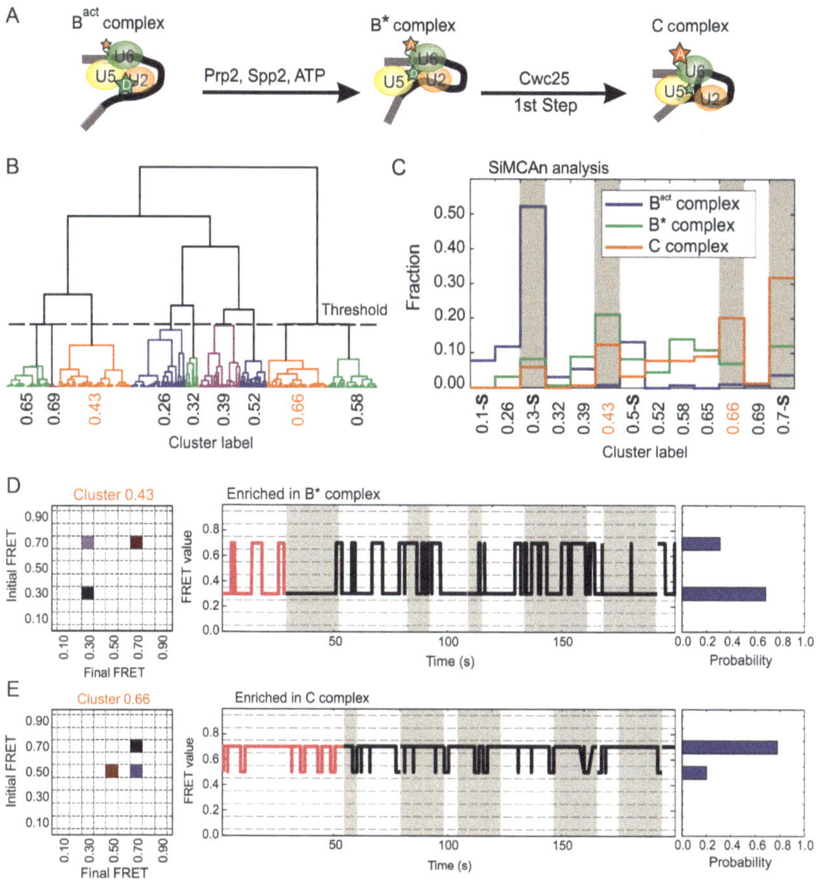

Fig. 7 Performance of SiMCAn on a previously analyzed smaller dataset describing the transition from the purified B^{act} to the C complex (Krishnan et al., 2013). (A) Protein requirements for the transition from the B^{act} complex through B* to the C complex. (B) Hierarchical tree based on hierarchical clustering analysis of the dynamic molecules refits with FRET states of 0.1, 0.3, 0.5, and 0.7 (Krishnan et al., 2013). Static molecules were identified and analyzed by SiMCAn separately. (C) Cluster occupancy bar graph showing the fraction of molecules from each experimental condition that occupy the nine dynamic and four static clusters found using SiMCAn. Dynamic clusters were labeled by the weighted average FRET value of the molecules within the cluster (e.g., 0.2563), while static clusters are labeled by the single state they describe (e.g., 0.1-S). *Gray bars* highlight the most populated clusters occupied by each of the complexes. (D and E) Dynamic clusters enriched in the B* (D, cluster 0.4267) and C (E, cluster 0.6478) complexes. Each representative for the B* and C complex shows the TP matrix of the cluster (*left*), the closest (*magenta*) and several random (*black*) traces from the cluster (*middle*), and the probability of FRET states within the cluster (*right*). *Reproduced with permission from Blanco, M. R., Martin, J. S., Kahlscheuer, M. L., Krishnan, R., Abelson, J., et al. (2015). Single molecule cluster analysis dissects splicing pathway conformational dynamics.* Nature Methods, 12, *1077–1084.*

abundant in the C complex (Fig. 7C), again consistent with manual sorting (Krishnan et al., 2013). Finally, SiMCAn also identified two dynamic clusters increasingly populated under B* (cluster 0.43, green) and C (cluster 0.66, red) complex conditions (Fig. 7C). Cluster 0.43 contains molecules with a short-lived high-FRET state most abundant under B* conditions, whereas cluster 0.66 contains molecules with a longer-lived high-FRET state that is enriched once Cwc25 has been added to favor the C complex (Fig. 7C), together matching our previous manual analysis (Krishnan et al., 2013). These results demonstrate that, when applied to a more typical experimental dataset, SiMCAn is able to segregate the data efficiently based on FRET states and differences in state-to-state interconversion kinetics in a way that derives biologically meaningful results, saves time and avoids potential user bias.

3. CONCLUSIONS AND OUTLOOK

The complex dynamic behavior of the many individual molecules we observed undergoing the pre-mRNA splicing cycle of budding yeast in vitro required the development of a novel approach to smFRET analysis. In recent years, hidden Markov modeling has vastly improved the amount of information one is able to extract from smFRET data. It is a powerful tool for the unbiased extraction of FRET states and kinetics from single-molecule trajectories. To leverage the TP matrix resulting from hidden Markov modeling, we created an FSM that can be utilized in hierarchical clustering algorithms. In turn, hierarchical clustering techniques allow us to group molecules with similar behaviors across conditions by applying the FSM as a measure of how similar any two molecules are. The FSM can help distinguish molecules with similar FRET states and rates of interconversion using SiMCAn, a tool available as supplemental MATLAB scripts associated with our original publication (Blanco et al., 2015). As a first application, we analyzed single-molecule data from pre-mRNAs labeled at the 5′SS and BP that were exposed to various yeast WCEs competent in splicing. To enrich for specific points along the splicing cycle, we introduced substrate mutations and depleted specific splicing factors. Blocking specific points along the splicing pathway allowed us to focus on dynamics that are relevant during the early and late stages of assembly, identifying both known and heretofore undescribed conformational states.

The ultimate goal in studies of any enzyme is to understand the link between the temporal sequence of conformational fluctuations and the chemical transformation(s) throughout a full catalytic cycle; in fact, it can be argued that the former begets the latter (Hammes-Schiffer & Benkovic, 2006). So far, this goal has been accomplished mostly for single-chain metabolic enzymes such as dihydrofolate reductase (Bhabha et al., 2011; Hammes-Schiffer & Benkovic, 2006) and adenylate kinase (Adk) (Henzler-Wildman et al., 2007). The overarching hypothesis underlying our work is that, similarly, a characteristic, helicase-directed sequence of stochastic conformational fluctuations leads to chemical catalysis in the more complex enzymatic machinery of the spliceosome. The power of SiMCAn-based clustering analysis lies in its ability to compare datasets efficiently and objectively by sorting single-molecule time traces with their stochastic idiosyncrasies into clusters of common FRET state and kinetic behaviors, and assessing the relative abundance of each cluster across any number of experimental conditions. SiMCAn thus emerges as a computational approach capable of translating smFRET data into temporal sequences, or "movies," of most common RNA conformational paths through a catalytic cycle that, combined with high-resolution cryo-EM and crystal structures, footprinting and other solution data will eventually make the spliceosome as well understood as single-chain enzymes are today.

Previous work has used clustering algorithms to identify distinct kinetic behaviors, but focused on small datasets with two or three FRET states and limited dynamics (Greenfeld, Pavlichin, Mabuchi, & Herschlag, 2012; Keller, Kobitski, Jaschke, Nienhaus, & Noe, 2014). SiMCAn performs well with such datasets, as demonstrated when applied to the B^{act} to B^{*} to C complex transition, but is adept at handling also more complex systems with higher numbers of states and complex kinetic networks under non-equilibrium conditions. That opens the door to more sophisticated experiments, for example, by labeling additional spliceosomal RNA and protein components. The number of potential targets for single-molecule experiments is great and facilitated by the wealth of genetic and biochemical tools available in the budding yeast splicing system (Hoskins et al., 2011). As more high-resolution structures of the spliceosome become available, they will provide further targets for specific interactions to monitor. The work presented here therefore presents a framework for further advances in what is sure to expand into the exciting and fruitful field of single-molecule splicing. Finally, SiMCAn stands ready to be applied to complex smFRET datasets that are sure to become more common as we expand the use of single-molecule tools to many more biomolecular machines.

ACKNOWLEDGMENTS

This work was supported by NIH Grant GM098023 to N.G.W. We thank Dr. Paul Lund for helpful insights.

REFERENCES

Abelson, J., Blanco, M., Ditzler, M. A., Fuller, F., Aravamudhan, P., et al. (2010). Conformational dynamics of single pre-mRNA molecules during in vitro splicing. *Nature Structural and Molecular Biology*, *17*, 504–512.

Agafonov, D. E., Kastner, B., Dybkov, O., Hofele, R. V., Liu, W. T., et al. (2016). Molecular architecture of the human U4/U6.U5 tri-snRNP. *Science*, *351*, 1416–1420.

Barash, Y., Calarco, J. A., Gao, W., Pan, Q., Wang, X., et al. (2010). Deciphering the splicing code. *Nature*, *465*, 53–59.

Bhabha, G., Lee, J., Ekiert, D. C., Gam, J., Wilson, I. A., et al. (2011). A dynamic knockout reveals that conformational fluctuations influence the chemical step of enzyme catalysis. *Science*, *332*, 234–238.

Blanco, M. R., Johnson-Buck, A. E., & Walter, N. G. (2012). Hidden Markov modeling. In G. C. K. Roberts (Ed.), *Encyclopedia of biophysics* (pp. 971–975). Springer.

Blanco, M. R., Martin, J. S., Kahlscheuer, M. L., Krishnan, R., Abelson, J., et al. (2015). Single molecule cluster analysis dissects splicing pathway conformational dynamics. *Nature Methods*, *12*, 1077–1084.

Blanco, M., & Walter, N. G. (2010). Analysis of complex single-molecule FRET time trajectories. *Methods in Enzymology*, *472*, 153–178.

Braunschweig, U., Gueroussov, S., Plocik, A. M., Graveley, B. R., & Blencowe, B. J. (2013). Dynamic integration of splicing within gene regulatory pathways. *Cell*, *152*, 1252–1269.

Brody, E., & Abelson, J. (1985). The "spliceosome": Yeast pre-messenger RNA associates with a 40S complex in a splicing-dependent reaction. *Science*, *228*, 963–967.

Bronson, J. E., Fei, J., Hofman, J. M., Gonzalez, R. L., & Wiggins, C. H. (2009). Learning rates and states from biophysical time series: A Bayesian approach to model selection and single-molecule FRET data. *Biophysical Journal*, *97*, 3196–3205.

Burgess, S. M., & Guthrie, C. (1993). A mechanism to enhance mRNA splicing fidelity: The RNA-dependent ATPase Prp16 governs usage of a discard pathway for aberrant lariat intermediates. *Cell*, *73*, 1377–1391.

Calarco, J. A., Zhen, M., & Blencowe, B. J. (2011). Networking in a global world: Establishing functional connections between neural splicing regulators and their target transcripts. *RNA*, *17*, 775–791.

Cartegni, L., Chew, S. L., & Krainer, A. R. (2002). Listening to silence and understanding nonsense: Exonic mutations that affect splicing. *Nature Reviews. Genetics*, *3*, 285–298.

Chen, M., & Manley, J. L. (2009). Mechanisms of alternative splicing regulation: Insights from molecular and genomics approaches. *Nature Reviews. Molecular Cell Biology*, *10*, 741–754.

Chen, W., & Moore, M. J. (2014). The spliceosome: Disorder and dynamics defined. *Current Opinion in Structural Biology*, *24*, 141–149.

Couto, J. R., Tamm, J., Parker, R., & Guthrie, C. (1987). A trans-acting suppressor restores splicing of a yeast intron with a branch point mutation. *Genes and Development*, *1*, 445–455.

Das, R., Dufu, K., Romney, B., Feldt, M., Elenko, M., et al. (2006). Functional coupling of RNAP II transcription to spliceosome assembly. *Genes and Development*, *20*, 1100–1109.

De, I., Schmitzova, J., & Pena, V. (2016). The organization and contribution of helicases to RNA splicing. *WIREs RNA*, *7*, 259–274.

de Klerk, E., & t Hoen, P. A. (2015). Alternative mRNA transcription, processing, and translation: Insights from RNA sequencing. *Trends in Genetics*, *31*, 128–139.

DeLano, W. L. (2002). *The PyMOL Molecular Graphics System*. Palo Alto, CA: DeLano Scientific. *http://www.pymol.org*.

D'haeseleer, P. (2005). How does gene expression clustering work? *Nature Biotechnology*, *23*, 1499–1501.

Greenfeld, M., Pavlichin, D. S., Mabuchi, H., & Herschlag, D. (2012). Single molecule analysis research tool (SMART): An integrated approach for analyzing single molecule data. *PLoS One*, *7*, e30024.

Hammes-Schiffer, S., & Benkovic, S. J. (2006). Relating protein motion to catalysis. *Annual Review of Biochemistry*, *75*, 519–541.

Hang, J., Wan, R., Yan, C., & Shi, Y. (2015). Structural basis of pre-mRNA splicing. *Science*, *349*, 1191–1198.

Henzler-Wildman, K. A., Thai, V., Lei, M., Ott, M., Wolf-Watz, M., et al. (2007). Intrinsic motions along an enzymatic reaction trajectory. *Nature*, *450*, 838–844.

Horowitz, D. S. (2011). The splice is right: Guarantors of fidelity in pre-mRNA splicing. *RNA*, *17*, 551–554.

Hoskins, A. A., Friedman, L. J., Gallagher, S. S., Crawford, D. J., Anderson, E. G., et al. (2011). Ordered and dynamic assembly of single spliceosomes. *Science*, *331*, 1289–1295.

Jarmoskaite, I., & Russell, R. (2014). RNA helicase proteins as chaperones and remodelers. *Annual Review of Biochemistry*, *83*, 697–725.

Johnson-Buck, A. E., Blanco, M. R., & Walter, N. G. (2012). Single-molecule fluorescence resonance energy transfer. In G. C. K. Roberts (Ed.), *Encyclopedia of Biophysics* (pp. 2329–2335). Springer.

Kahlscheuer, M. L., Widom, J., & Walter, N. G. (2015). Single-molecule pull-down FRET to dissect the mechanisms of biomolecular machines. *Methods in Enzymology*, *558*, 539–570.

Keller, B. G., Kobitski, A., Jaschke, A., Nienhaus, G. U., & Noe, F. (2014). Complex RNA folding kinetics revealed by single-molecule FRET and hidden Markov models. *Journal of the American Chemical Society*, *136*, 4534–4543.

Koodathingal, P., Novak, T., Piccirilli, J. A., & Staley, J. P. (2010). The DEAH box ATPases Prp16 and Prp43 cooperate to proofread 5′ splice site cleavage during pre-mRNA splicing. *Molecular Cell*, *39*, 385–395.

Koodathingal, P., & Staley, J. P. (2013). Splicing fidelity: DEAD/H-box ATPases as molecular clocks. *RNA Biology*, *10*, 1073–1079.

Kornblihtt, A. R., Schor, I. E., Allo, M., Dujardin, G., Petrillo, E., et al. (2013). Alternative splicing: A pivotal step between eukaryotic transcription and translation. *Nature Reviews. Molecular Cell Biology*, *14*, 153–165.

Krishnan, R., Blanco, M. R., Kahlscheuer, M. L., Abelson, J., Guthrie, C., et al. (2013). Biased Brownian ratcheting leads to pre-mRNA remodeling and capture prior to first-step splicing. *Nature Structural and Molecular Biology*, *20*, 1450–1457.

Lander, E. S., Linton, L. M., Birren, B., Nusbaum, C., Zody, M. C., et al. (2001). Initial sequencing and analysis of the human genome. *Nature*, *409*, 860–921.

Li, Y. I., van de Geijn, B., Raj, A., Knowles, D. A., Petti, A. A., et al. (2016). RNA splicing is a primary link between genetic variation and disease. *Science*, *352*, 600–604.

Matera, A. G., & Wang, Z. (2014). A day in the life of the spliceosome. *Nature Reviews. Molecular Cell Biology*, *15*, 108–121.

Mayas, R. M., Maita, H., Semlow, D. R., & Staley, J. P. (2010). Spliceosome discards intermediates via the DEAH box ATPase Prp43p. *Proceedings of the National Academy of Sciences of the United States of America*, *107*, 10020–10025.

Mayas, R. M., Maita, H., & Staley, J. P. (2006). Exon ligation is proofread by the DExD/H-box ATPase Prp22p. *Nature Structural and Molecular Biology*, *13*, 482–490.

Moffitt, J. R., Chemla, Y. R., & Bustamante, C. (2010). Methods in statistical kinetics. *Methods in Enzymology*, *475*, 221–257.

Naftelberg, S., Schor, I. E., Ast, G., & Kornblihtt, A. R. (2015). Regulation of alternative splicing through coupling with transcription and chromatin structure. *Annual Review of Biochemistry*, *84*, 165–198.
Nguyen, T. H., Galej, W. P., Bai, X. C., Oubridge, C., Newman, A. J., et al. (2016a). Cryo-EM structure of the yeast U4/U6.U5 tri-snRNP at 3.7 Å resolution. *Nature*, *530*, 298–302.
Nguyen, T. H., Galej, W. P., Bai, X. C., Savva, C. G., Newman, A. J., et al. (2015). The architecture of the spliceosomal U4/U6.U5 tri-snRNP. *Nature*, *523*, 47–52.
Nguyen, T. H., Galej, W. P., Fica, S. M., Lin, P. C., Newman, A. J., et al. (2016b). CryoEM structures of two spliceosomal complexes: Starter and dessert at the spliceosome feast. *Current Opinion in Structural Biology*, *36*, 48–57.
Nilsen, T. W. (2003). The spliceosome: The most complex macromolecular machine in the cell? *Bioessays*, *25*, 1147–1149.
Nilsen, T. W., & Graveley, B. R. (2010). Expansion of the eukaryotic proteome by alternative splicing. *Nature*, *463*, 457–463.
Papasaikas, P., & Valcarcel, J. (2016). The spliceosome: The ultimate RNA chaperone and sculptor. *Trends in Biochemical Sciences*, *41*, 33–45.
Pomeranz Krummel, D. A., Oubridge, C., Leung, A. K., Li, J., & Nagai, K. (2009). Crystal structure of human spliceosomal U1 snRNP at 5.5 Å resolution. *Nature*, *458*, 475–480.
Poulos, M. G., Batra, R., Charizanis, K., & Swanson, M. S. (2011). Developments in RNA splicing and disease. *Cold Spring Harbor Perspectives in Biology*, *3*, a000778.
Rino, J., & Carmo-Fonseca, M. (2009). The spliceosome: A self-organized macromolecular machine in the nucleus? *Trends in Cell Biology*, *19*, 375–384.
Roy, R., Hohng, S., & Ha, T. (2008). A practical guide to single-molecule FRET. *Nature Methods*, *5*, 507–516.
Scotti, M. M., & Swanson, M. S. (2016). RNA mis-splicing in disease. *Nature Reviews. Genetics*, *17*, 19–32.
Semlow, D. R., Blanco, M. R., Walter, N. G., & Staley, J. P. (2016). Spliceosomal DEAH-Box ATPases remodel pre-mRNA to activate alternative splice sites. *Cell*, *164*, 985–998.
Smith, D. J., & Konarska, M. M. (2008). Mechanistic insights from reversible splicing catalysis. *RNA*, *14*, 1975–1978.
Staley, J. P., & Guthrie, C. (1998). Mechanical devices of the spliceosome: Motors, clocks, springs, and things. *Cell*, *92*, 315–326.
Tseng, C. K., & Cheng, S. C. (2008). Both catalytic steps of nuclear pre-mRNA splicing are reversible. *Science*, *320*, 1782–1784.
Wahl, M. C., Will, C. L., & Luhrmann, R. (2009). The spliceosome: Design principles of a dynamic RNP machine. *Cell*, *136*, 701–718.
Walter, N. G., Huang, C., Manzo, A. J., & Sobhy, M. A. (2012). Do-it-yourself guide: How to use the modern single molecule toolkit. *Nature Methods*, *5*, 475–489.
Wan, R., Yan, C., Bai, R., Wang, L., Huang, M., et al. (2016). The 3.8 Å structure of the U4/U6.U5 tri-snRNP: Insights into spliceosome assembly and catalysis. *Science*, *351*, 466–475.
Wang, G. S., & Cooper, T. A. (2007). Splicing in disease: Disruption of the splicing code and the decoding machinery. *Nature Reviews. Genetics*, *8*, 749–761.
Widom, J. R., Dhakal, S., Heinicke, L. A., & Walter, N. G. (2014). Single-molecule tools for enzymology, structural biology, systems biology and nanotechnology: An update. *Archives of Toxicology*, *88*, 1965–1985.
Will, C. L., & Luhrmann, R. (2011). Spliceosome structure and function. *Cold Spring Harbor Perspectives in Biology*, *3*, pii: a003707.

Xu, Y. Z., & Query, C. C. (2007). Competition between the ATPase Prp5 and branch region-U2 snRNA pairing modulates the fidelity of spliceosome assembly. *Molecular Cell*, *28*, 838–849.

Yan, C., Hang, J., Wan, R., Huang, M., Wong, C. C., et al. (2015). Structure of a yeast spliceosome at 3.6-angstrom resolution. *Science*, *349*, 1182–1191.

Zhang, J., & Manley, J. L. (2013). Misregulation of pre-mRNA alternative splicing in cancer. *Cancer Discovery*, *3*, 1228–1237.

CHAPTER TEN

Single-Molecule FRET to Measure Conformational Dynamics of DNA Mismatch Repair Proteins

J.W. Gauer*, S. LeBlanc*, P. Hao†, R. Qiu†, B.C. Case‡, M. Sakato‡, M.M. Hingorani‡, D.A. Erie*,[1], K.R. Weninger†,[1]

*University of North Carolina at Chapel Hill, Chapel Hill, NC, United States
†North Carolina State University, Raleigh, NC, United States
‡Wesleyan University, Middletown, CT, United States
[1]Corresponding authors: e-mail address: derie@email.unc.edu; krwening@ncsu.edu

Contents

Abstract

Single-molecule FRET measurements have a unique sensitivity to protein conformational dynamics. The FRET signals can either be interpreted quantitatively to provide estimates of absolute distance in a molecule configuration or can be qualitatively interpreted as distinct states, from which quantitative kinetic schemes for conformational

Methods in Enzymology, Volume 581
ISSN 0076-6879
http://dx.doi.org/10.1016/bs.mie.2016.08.012
© 2016 Elsevier Inc.
All rights reserved.

transitions can be deduced. Here we describe methods utilizing single-molecule FRET to reveal the conformational dynamics of the proteins responsible for DNA mismatch repair. Experimental details about the proteins, DNA substrates, fluorescent labeling, and data analysis are included. The complementarity of single molecule and ensemble kinetic methods is discussed as well.

1. INTRODUCTION

1.1 Single-Molecule FRET to Measure Protein Conformation

Understanding how changes in protein conformation are linked to function is a key step in modeling molecular mechanisms for many enzymes. Among experimental methods that measure protein conformational changes, fluorescence resonance energy transfer (Förster resonance energy transfer, FRET) has proven to be of great use for sensitive detection of protein dynamics in physiologically relevant settings. FRET measurements are made by engineering macromolecules with two fluorophores attached at defined locations and measuring spectrally resolved fluorescence emission from these probes. Distance-dependent interaction between these fluorophores alters the overall emission spectrum, allowing the changes in signal to be interpreted as changes in molecular conformation. With proper corrections for systematic effects, these signals can be quantitatively related to nanometer dimensions, leading to the description of FRET as a "spectroscopic ruler" (Stryer & Haugland, 1967).

The power of FRET measurements to reveal dynamic protein conformational changes took a tremendous leap forward two decades ago when signals were successfully detected from a single molecule; this methodology is now called single-molecule FRET (smFRET) (Ha et al., 1996). Measuring FRET signals from single molecules avoids the averaging that dominates ensemble biochemical approaches. Single-molecule studies provide access to heterogeneous behaviors and the ability to monitor the dynamics of unsynchronizable populations of macromolecules, making it an ideal tool for investigating the role of conformational dynamics in enzyme function. Below, we describe single-molecule FRET assays that we have developed to study conformational dynamics of DNA mismatch repair (MMR) proteins.

1.2 Overview of Our DNA MMR Protein Experiments

Living organisms have a widely conserved system of proteins that locate and respond to DNA base–base mismatch and insertion errors made during replication, as well as damaged bases in DNA. The system, known as MMR, either repairs mismatches or leads to the activation of checkpoints that promote cell-cycle arrest or apoptosis in the case of damage (Bignami, Casorelli, & Karran, 2003; Fedier & Fink, 2004; Karran, Offman, & Bignami, 2003). MMR proteins play a fundamental role in genome stability and cellular homeostasis (Modrich & Lahue, 1996), and mutations in these proteins are associated with carcinogenesis (Heinen, 2016; Peltomaki, 2016).

MMR proteins MutS and MutL are responsible for detecting DNA mismatches and initiating the repair process. Both MutS and MutL homologues have DNA binding and ATPase activities that are essential for MMR in vivo (Iyer, Pluciennik, Burdett, & Modrich, 2006; Kunkel & Erie, 2005). Both proteins are homodimers in prokaryotes and heterodimers in eukaryotes. Here, our main goal is to describe smFRET methodology we have applied to our studies of the homodimeric MutS and MutL proteins from *Thermus aquaticus* (*Taq*). The mechanism of the *Taq* MMR system appears similar to that in most other species (with the notable exception of *Escherichia coli*), and therefore, we will simply call the proteins MutS and MutL in the following text.

The current consensus about key events occurring in MMR initiation is that MutS slides along double-stranded DNA searching for the rare mismatched or damaged base. It binds to mismatches (Fig. 1), at which point ATP induces conformational changes in MutS that can result in a mobile clamp state that moves along the DNA. This ATP and mismatch activated state of MutS has enhanced interactions with MutL, resulting in MutL stabilizing MutS at the mismatch (Groothuizen et al., 2015; Qiu et al., 2015; Schofield, Nayak, Scott, Du, & Hsieh, 2001). Interactions between the MutS-activated MutL complex and the replication processivity clamp (β-clamp in prokaryotes and PCNA in eukaryotes) can activate MutL to incise the newly replicated strand up to hundreds of bases from the mismatch both distally (preferential) and proximally to the mismatch (Kadyrov, Dzantiev, Constantin, & Modrich, 2006; Kadyrov et al., 2007; Pluciennik, Burdett, Lukianova, O'Donnell, & Modrich, 2009). MutS then stimulates exonucleases to begin degrading the newly synthesized strand from the MutL-created nick (or possibly from preexisting nicks/gaps in the strand). The resulting gap that typically extends past the mismatch is then

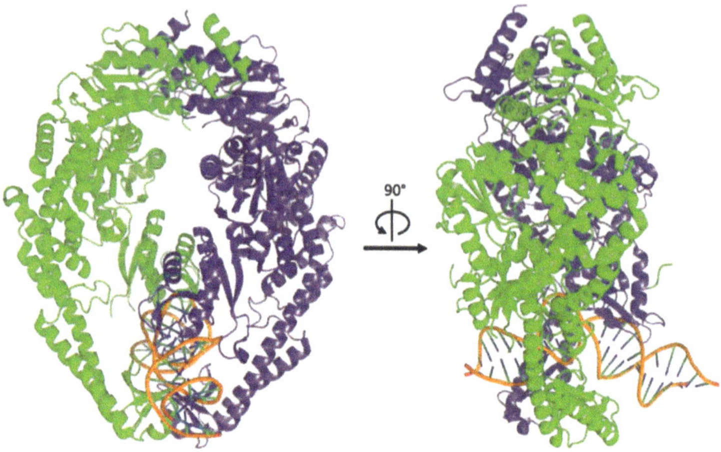

Fig. 1 *Front* view (*left*) and *side* view (*right*) of *Taq* MutS in complex with DNA containing a single thymine insertion (PDB ID: 1EWQ). The two subunits of the homodimer are colored *blue* and *green*, and the DNA is shown in *orange* (Obmolova et al., 2000).

resynthesized and remaining nicks are ligated to complete repair (Genschel & Modrich, 2003, 2009; Iyer et al., 2006; Kunkel & Erie, 2005, 2015; Liberti, Larrea, & Kunkel, 2013; Morita et al., 2010; Nick McElhinny, Kissling, & Kunkel, 2010; Pavlov, Mian, & Kunkel, 2003).

Initiation of MMR involves an ordered choreography of interactions between DNA, MutS, and MutL coupled with the ATPase reaction. Many studies using X-ray crystallography (Lamers et al., 2000; Natrajan et al., 2003; Obmolova, Ban, Hsieh, & Yang, 2000; Warren et al., 2007), AFM (Wang et al., 2003), and ensemble biochemistry (Fishel, 2015; Iyer et al., 2006; Kunkel & Erie, 2005) have suggested that these interactions involve substantial conformational changes in both the DNA and the proteins, but the dynamics of these events have been difficult to resolve. Our studies of these interactions at the single-molecule level have allowed us to characterize in detail large DNA bending transitions, conformational changes within the MutS mismatch binding domains that are dependent on ATP and the mismatch, as well as MutL interactions that alter these conformational trajectories. These observations, coupled with data from ensemble transient kinetics, have yielded a highly detailed view of the molecular mechanisms used by this enzyme system to identify, verify, and begin repairing DNA mismatch errors. The goal here is to describe these experimental smFRET approaches. The application of a broader range of single-molecule methods to study DNA MMR is reviewed elsewhere (Erie & Weninger, 2014). The studies described here are widely applicable to other biological systems in which the proteins can be purified and fluorescently tagged.

2. METHODS OF ACQUIRING FRET SIGNALS FROM *T. aquaticus* DNA MMR PROTEINS

For a FRET experiment, two fluorophores with distinct spectral characteristics must be attached to the molecule of interest (Fig. 2). The fluorophore with the shorter wavelength emission (called donor) is excited with an external light source. If the longer wavelength fluorophore (called acceptor) is further from the donor than the effective FRET range, then no FRET occurs and only donor emission is observed. FRET processes can excite the acceptor when it is within about 7–8 nm of the donor, which then decreases the donor emission and increases the acceptor emission. The

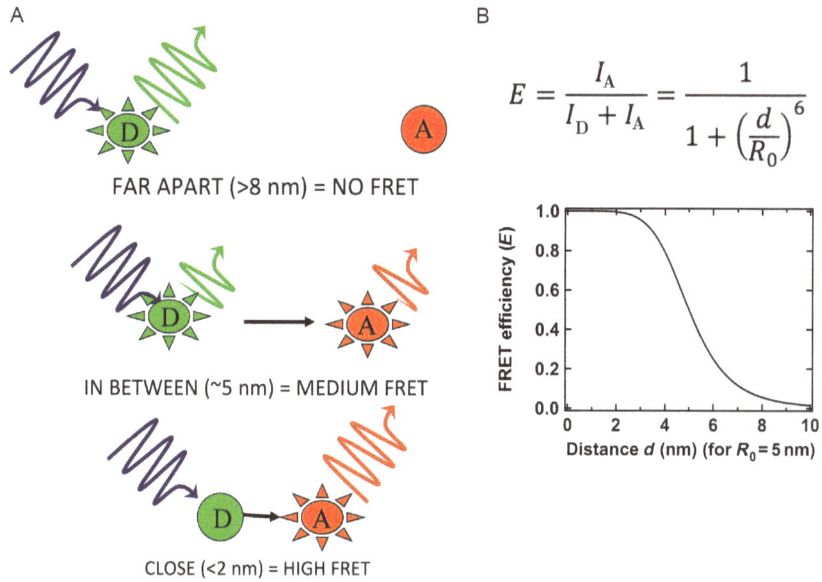

Fig. 2 Overview of the concept of FRET. (A) Schematic illustrating FRET exchange between the donor fluorophore (D, *green*) and the acceptor fluorophore (A, *red*). Donor fluorescence is excited by the external light source (*blue*), and as the distance between D and A decreases, emission from D decreases and emission from A increases. (B) The relation between FRET efficiency (E) and the separation of the fluorophores *d* is scaled by the R_0 parameter. Here $R_0 = 5$ nm is used.

FRET efficiency is defined as $E = I_A/(I_D + I_A)$, where I_D and I_A are the donor and acceptor emission intensities. FRET efficiency is a useful parameter because it is related to the distance d between the fluorophores by $E = 1/(1 + (d/R_0)^6)$, where R_0 (the Förster radius) sets the scale of the FRET effect. R_0 is a function of the properties of the fluorophores and typically ranges about 4–6 nm. The dependence on the sixth power of distance makes FRET very sensitive in the R_0 range (Fig. 2B).

Our experiments utilize total internal reflection fluorescence (TIRF) microscopy to observe MMR proteins interacting with DNA tethered to the surface of a flow cell. Methodological details of these experiments are described below.

2.1 Proteins

We work with MMR proteins from the thermophile *Taq*, which have proved to be stable in vitro and amenable to structure–function analysis by a wide variety of techniques, including smFRET. The *Taq* MutS monomer is 811 amino acids with a molecular weight ~90 kDa, and the *Taq* MutL monomer is 532 amino acids with a molecular weight ~59 kDa. MutS has one native cysteine that is mutated to alanine and, for the studies described below, methionine 88 is replaced with cysteine for labeling the DNA binding domain. Neither mutation alters DNA binding or ATPase rates (Sharma, Doucette, Biro, & Hingorani, 2013). *Taq* MutL has three native cysteines and none labeled when treated with maleimide dye; consequently, introduction of a surface-exposed cysteine is sufficient for site-specific labeling of MutL. Mutations were performed in the overexpression plasmid using the Quickchange II XL kit (Agilent). Note that *Taq* genes are GC rich, which makes PCR and sequencing difficult and less reliable. Multiple reactions are necessary to confirm successful manipulations.

2.1.1 Preparation of Taq MutS

Expression—Express MutS in pET3a plasmid in BL21-DE3 *E. coli* in LB media including 50 μg/mL Ampicillin at 37°C until OD 0.5 and then add IPTG to 0.1 m*M*. After expression for 3 h, collect cells by centrifugation.

Purification—Resuspend cells in lysis buffer (20 m*M* HEPES–NaOH pH 7.8, 100 m*M* NaCl, 1 m*M* EDTA, 0.1 m*M* DTT, 1 m*M* PMSF, protease inhibitor cocktail), sonicate to lyse, centrifuge at 40,000 rpm for 30 min at 4°C (Beckman MLA-80 Rotor), collect supernatant and heat to 65°C for 30 min (*Taq* MutS is thermostable), cool to room temperature, centrifuge at 8000 rpm for 20 min at 4°C (Eppendorf F34-6-38 Rotor), collect supernatant and add ammonium sulfate (36.6 g/100 mL supernatant, final ~60% saturation), agitate 4°C for 15 min, centrifuge 10,000 rpm for 20 min 4°C (Eppendorf F34-6-38 Rotor), resuspend pellet in Tris-50 buffer (20 m*M* Tris–HCl pH 7.8, 50 m*M* NaCl, 0.5 m*M* DTT), and dialyze against the same buffer with several changes. Load sample onto Hi-TRAP Sepharose Q column, wash with Tris-50 buffer, and then elute with an NaCl gradient to 1 *M* (MutS elutes around 150–300 m*M* NaCl). Dialyze the protein against Tris-50 buffer with several changes. Load onto a MonoQ column. Wash with Tris-50 buffer and then elute with an NaCl gradient to 1 *M* (MutS elutes around 200–300 m*M* NaCl). Dialyze in *Taq* storage buffer (20 m*M* HEPES–NaOH pH 7.8, 1 m*M* EDTA, 10% glycerol, 150 m*M* NaCl, 0.1 m*M* DTT) and snap freeze for storage at −80°C.

Labeling—To label MutS with maleimide fluorophores, exchange storage buffer with label buffer (20 m*M* phosphate, pH 7.2, 150 m*M* NaCl) using a Sephadex G-25 column (Nap5, PD10, GE Biosciences). Add TCEP at 10× concentration over free cysteines for 5 min at room temperature. Then add maleimide reactive dye at 10× concentration over free cysteines. Incubate 1 h to overnight. Remove unreacted dye using the Sephadex G-25 or G-50 desalting column or by dialysis against Tris-50 buffer.

2.1.2 Preparation of Taq MutL

Expression—Express MutL from pET17b plasmid in BL21-DE3 *E. coli* in LB media including 50 μg/mL Ampicillin at 37°C until OD 0.5, and then induce the expression with 0.5 m*M* IPTG. After expression for 3–4 h, collect cells by centrifugation.

Purification—Suspend cells in ice-cold lysis buffer (20 m*M* Tris–HCl pH 7.5, 0.5 *M* NaCl, 5 m*M* imidazole) augmented with protease inhibitors and PMSF. Sonicate on ice until cells are disrupted. Centrifuge at 35,000 rpm for 30 min (Beckman MLA-80 Rotor). Add supernatant to Ni-NTA agarose resin equilibrated in lysis buffer. Incubate 1 h at 4°C. Wash the Ni-NTA column in lysis buffer + 60 m*M* imidazole. Elute in lysis buffer + 300 m*M* imidazole (dimeric MutL elutes at ~300 m*M* imidazole). Dialyze against lysis buffer without imidazole overnight. Freezing before labeling with liquid nitrogen is acceptable.

Labeling—To label MutL with maleimide fluorophores, exchange buffer with label buffer (20 m*M* phosphate pH 7.2, 500 m*M* NaCl) using a P6 column (Bio-Rad Bio-Gel P-6DG). Note that *Taq* MutL binds nonspecifically to Sephadex G-25 matrix, but flows through P6 gel. Add TCEP at 10× concentration over free cysteines for 5 min at room temperature. Then add maleimide reactive dye at 10 × concentration over free cysteines. Incubate 1 h to overnight. Remove unreacted dye from MutL using the P6 desalting column in labeling buffer. Use immediately, freezing the labeled protein leads to aggregation for MutL.

2.2 Preparation of DNA Substrates

smFRET experiments have been performed with either 50 nucleotide (nt) or 550 nt dsDNA substrates. Both lengths of DNA include a biotin on one end to allow immobilization on a streptavidin-coated surface. The 550 nt dsDNA has a digoxin on the other end to allow end blocking by treatment with digoxigenin (dig) antibody. The 50 nt dsDNA substrate can be constructed from oligos ordered from commercial vendors. A 50 nt oligo with a 5′-biotin and a 3′-donor fluorophore is annealed to a 19 nt oligo with sequence complementary to the 3′-end of the 50 nt oligo (except a mismatched base at position 9). This 19 nt oligo has an acceptor fluorophore on its 3′-end that makes a FRET pair 19 bases apart with a mismatched base centered in between. A 31 nt oligo is annealed to fill the remaining ssDNA gap.

A more complex process is required to construct the 550 nt dsDNA substrate. PCR of a pUC-VSR plasmid template using oligos containing 5′-biotin and dig modifications is used to amplify a 550 nt section of the plasmid and introduce the biotin/dig moieties. A nicking endonuclease (Nt.BbvCI) is used to remove a central segment of one strand. Then an oligo is annealed to fill that gap. The oligo may include a mismatched base and fluorophores if desired. Ligase is used to seal the remaining nicks. A detailed protocol is given below:

Part 1—PCR and nicking

(1) PCR
  - **(a)** Prepare PCR reactions: 5 μL of 10 × polymerase buffer, 50 pmol F1-biotin primer, 50 pmol R2-primer*, 50 ng pUC-VSR plasmid, water to 49 μL, and 1 μL *Taq* polymerase (Invitrogen AccuPrime *Taq* DNA Polymerase) (*use different end-modified primers for different types of end-modified DNA; our studies include R2 (Normal) primer, LacO primer, and Dig primer).
  - **(b)** Run PCR: 95°C for 1 min; [95°C for 1 min—55°C for 33 s—69°C for 50 s]$_{\text{repeat } 32\times}$; 69°C for 5 min, 55°C for 5 min.
  - **(c)** Purify PCR product with PCR purification kit. Elute in 40 μL kit elution buffer. This is homo-duplex DNA.

(2) Nicking

(a) Prepare nicking reaction with 5 μL 10 × NEB buffer 4, 1 μg PCR product DNA, 45 μL water, 5 μL Nt.BbvCI (NEB, Cat# R0632S).

(b) Incubate 37°C for 16 h.

Part 2—Gapping, phosphorylation, annealing, and ligation

(1) Gapping:

a. Add 250 μL (5 × volume) of Buffer PBI (Qiagen) to each nicked sample

b. Incubate samples at 80°C for 20 min (heat inactivates any remaining Nt.BbvCI and also melts short nicked pieces away from 550 mer)

c. Immediately spin each reaction through a QIAquick spin column

d. Transfer to a new collection tube

e. Add 750 μL of buffer PE and spin column (containing ethanol)

f. Repeat spin in new collection tube to remove residual ethanol

g. Transfer to a new collection tube (with cap)

h. Add 30 μL of Buffer EB; allow to sit 1 min; spin to elute DNA

(2) Phosphorylation: (If your oligo is commercially phosphorylated, skip this step. The yield of fully ligated product will be reduced if the oligo is not ordered phosphorylated.)

a. Prepare your phosphorylation reactions with 20 pmol DNA, 2 μL 10 × kit reaction buffer, 0.2 μL ATP (100 μ*M* stock), 19 μL water, 1 μL PNK (T4 Polynucleotide Kinase, Fermentas, Cat# EK0031)

1. Incubate 37°C for 1 h and then 70°C for 5 min (denature enzyme). There is no need for purification at this stage as it will result in significant loss of DNA

(3) Annealing

a. Prepare annealing reactions. Mix gapped DNA with ~1.5 × mismatch-containing oligo. Note: excess oligo will result in multiple dyes on a single DNA molecule

b. Anneal at 70°C for 5 min and then ramp down temperature from 70°C to 16°C at 1°C per min (or let it cool down naturally to room temperature in heat block)

(4) Purify the annealed product using a PCR purification kit to remove extra oligo in the reaction. Elute in 45 μL elution buffer

(5) Ligation

a. Add 5 μL of *E. coli* DNA ligase buffer and 1.5 μL of *E. coli* DNA ligase to each reaction

b. Incubate at 16°C for 20 h

c. PCR purify the first ligation product (elute in 45 μL of elution buffer) and set up the second ligation reaction

d. Repeat three times

e. When the third ligation reaction is done, purify the final 550 nt DNA using a PCR purification kit

2.3 Considerations Related to Fluorescent Labeling

Several fluorescent tags are commercially available for labeling oligonucleotides (e.g., TAMRA, the Cy dyes, Alexa dyes). When choosing which specific fluorophores to use, it is important to select dyes with high extinction coefficients and quantum yields so that they can be easily detected at the single-molecule level. Also, the dyes' excitation and emission spectra should be compatible with the optical setup such that independent excitation and emission detection is possible for each dye. Donor–acceptor dye pairs commonly used in our smFRET experiments include Alexa 555-Alexa 647, Alexa 555-Cy5, TAMRA-Cy5, and Cy3-Cy5.

There are several unknown parameters that can complicate dye selection. Notably, the structures of several commercial fluorophores are proprietary; thus, any perturbations they may introduce into the DNA remain unknowable. Also, fluorophores that serve as effective reporters in bulk experiments may be prone to blinking or bleaching in single-molecule experiments, making them poor choices for conformational dynamics studies using smFRET. Finally, some dyes are prone to interactions with the DNA (e.g., stacking with the bases), which may change their fluorescence properties in ways that do not depend on DNA or MMR protein conformational changes.

Several chemical strategies exist to link fluorophores to DNA. Labeling the 5′- or 3′-ends of the DNA is one option; however, being limited to only the ends of the DNA may be problematic. While internally labeling DNA oligonucleotides allows flexibility in choosing labeling positions, there are fewer commercially available options for fluorophores. Two major types of internal labels are: (1) dyes attached covalently to thymine bases via a flexible linker extending from the major groove in the DNA, and (2) dyes (e.g., the Cy dyes) that can be incorporated directly into the DNA backbone. Note that these fluorophores are rigidly locked into a specific orientation relative to the DNA backbone as they are covalently attached at both ends, whereas fluorophores on a flexible linker have more conformational freedom. The Förster radius of a given FRET pair of fluorophores depends, in part, on the relative orientation of the transition dipoles of the two fluorophores (Lakowicz, 2006). Thus, limiting the conformational freedom of both fluorophores may lead to unpredictable changes in FRET due to changes in their relative orientations. In general, at least one freely rotating dye (i.e., those attached to flexible linkers) is recommended.

It is crucial that labeling positions are chosen beyond the binding footprint of the protein because the fluorescent properties are sensitive to changes in the local chemical environment. The crystal structure of *Taq* MutS shows protein–DNA contacts as far as 8 nucleotides from the mismatch (Fig. 1) (Obmolova et al., 2000); thus, fluorescent labels are placed outside this range.

2.4 TIRF Microscopy

Homemade flow cells constructed between a quartz microscope slide and a glass cover slip are built and samples are tethered to the surface through a biotin–streptavidin interaction. A quartz prism with index-matching oil at the contact with the quartz slide allows laser beam illumination to be directed to the sample above a microscope objective at an angle sufficiently steep that total internal reflection occurs at the quartz-buffer interface (Fig. 3A). The resultant evanescent field is used to excite fluorophores

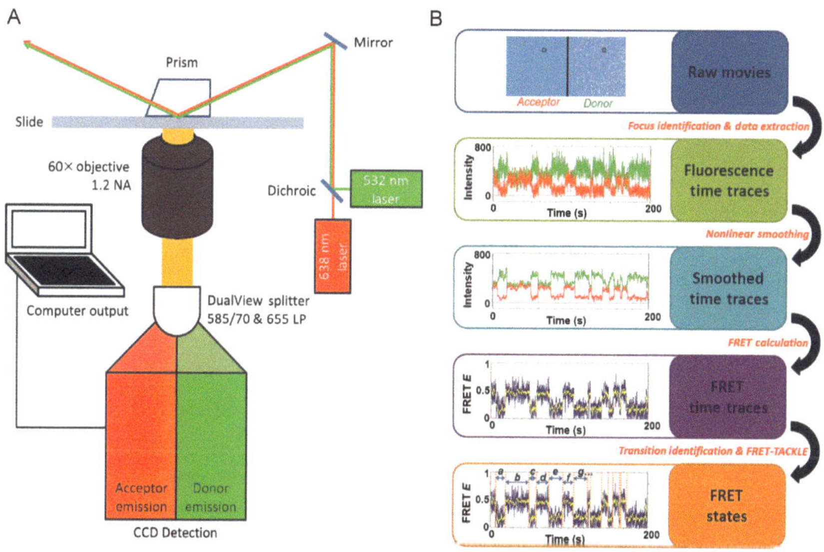

Fig. 3 Optical setup and data analysis pipeline. (A) An example of the total internal reflection fluorescence microscope. (B) A schematic of the data analysis pipeline. The specific trace is from experiments measuring DNA bending by MutS at a T-bulge mismatch by the method diagramed in Fig. 4A.

within a couple hundred nanometers of the interface. The illuminated area is imaged with a microscope objective onto an emCCD camera. A DualView image splitter (Photometrics) is used to split the image into two copies where each contains emission from either the donor or the acceptor spectral bands. These spectrally filtered images are relayed side by side onto the emCCD. Intensity vs time traces for spots identified to contain fluorophores are extracted from minute long movies. Detailed descriptions of flow-cell construction, sample immobilization, microscopy hardware, and extraction of single-molecule intensity vs time traces in donor and acceptor spectral emission bands are provided elsewhere (Choi, Weninger, & Bowen, 2012).

2.5 Overview of FRET Signals Reporting Different Conformational Aspects of DNA MMR Protein Complexes

We used smFRET to deduce conformations of several different aspects of the MutS:DNA complex (Fig. 4). Using a donor and acceptor attached to DNA (forming a FRET pair with the mismatched base halfway in between; Fig. 4A), MutS induced bending at the mismatched base can be measured (DeRocco, Anderson, Piehler, Erie, & Weninger, 2010; DeRocco, Sass, Qiu, Weninger, & Erie, 2014; Sass, Lanyi, Weninger, & Erie, 2010). When the DNA is straight, the fluorophores are further apart than when the DNA is kinked at the mismatch. We also labeled the DNA binding domain of dimeric MutS (domain I) with a mixture of donor and acceptor dyes (Fig. 4B). Single-molecule measurements allowed us to specifically select the fraction of dimers with one donor and one acceptor and measure FRET from this mixed-fluorophore population. smFRET experiments were designed to detect conformational transitions within the DNA binding domains of MutS in the absence or presence of DNA

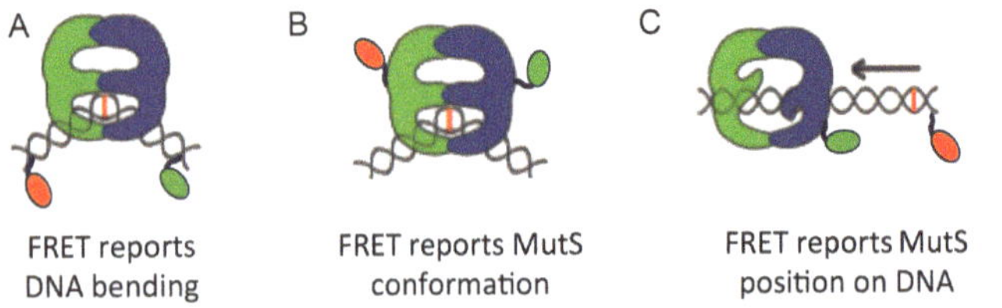

Fig. 4 Schematic of the types of FRET signals utilized in DNA MMR assays. FRET can report (A) DNA bending if both donor and acceptor are on the DNA flanking the mismatch, (B) MutS conformation if both donor and acceptor are on the protein, and (C) relative distance between a domain of MutS and a site on DNA if acceptor is on the DNA and donor is on MutS.

mismatches (Qiu et al., 2012). Finally, a third label combination, where a donor is on MutS and an acceptor is on DNA 9 nt away from a mismatched base (Fig. 4C), provided information on the position of MutS near a mismatch (Qiu et al., 2012, 2015). Labels in the N-terminal domain of dimeric MutL were also used to provide smFRET signals addressing the configuration of these domains as MutL interacts with MutS to initiate repair.

3. DATA ANALYSIS

3.1 Quantitative vs Qualitative Interpretation of FRET as Distance

Quantitative conversion of measured FRET ratios into the distance between the donor and acceptor fluorophores is possible. It is critical to address systematic issues related to the efficiency of the instrumentation for detection of fluorescence emission as well as the impact of the local molecular environment on mobility or quantum yields of the donor and the acceptor after they are attached to the DNA or protein molecules (McCann, Choi, Zheng, Weninger, & Bowen, 2010). smFRET is being developed as a quantitative tool for structural biology applications by making such systematic corrections to the FRET measurements (Brunger, Strop, Vrljic, Chu, & Weninger, 2011; Choi et al., 2010). In contrast, for our DNA MMR studies, we have not interpreted FRET as indicating a quantitative distance, but rather have focused on analysis of the kinetics of transitions among distinct states (Fig. 3B).

3.2 Edge Detection in Time Traces for Kinetics

Given proper surface coverage by a sample of labeled oligonucleotide and protein, a single movie collected as described above may contain fluorescence information for over 100 DNA molecules. While already a large amount of information, the amount of data to analyze multiplies with varying experimental conditions and replicate experiments, thus an automated computational analysis approach vital. Automated analyses, however, require verification to identify unexpected patterns and check for systematic errors. Below, a computational analysis pipeline is presented that systematically applies an analysis routine and outputs the results for a user interface.

3.2.1 Extracting the Fluorescence Time Traces of Individual DNA Molecules

Typical experiments produce movies made up of a series of approximately 1000 frames acquired at typically 10 Hz. As mentioned above, donor emission and acceptor emission from an area being imaged are recorded on side-by-side halves of the camera. By initially exciting only the acceptor dye, local maxima on the acceptor half of the image can be identified. These same locations mapped to the other half of the frame should locate the emissions from the donor dye. To determine the relative offset between the two halves of the image, broadband emission fluorescent beads fixed on a slide surface that appear in both halves of the image are imaged in separate calibration runs. These calibration images allow an experimentally determined offset to be determined, which is then applied to the experimental movies.

For each DNA molecule, the sum of the fluorescence intensities of a 3×3 pixel square centered on the local maxima pixel coordinates for the donor and acceptor dyes are determined. The local background intensity is subtracted, and those foci below a user set minimum threshold intensity are discarded. Foci too close to the camera edges or another focus are also discarded. For those foci that remain, the intensity is determined for each movie frame for both the donor and acceptor as the sum of the 4 highest pixels in the 3×3 square, ultimately resulting in a "time trace" of the donor and acceptor fluorescence emission intensities as a function of time (Fig. 3B, second row). Single-labeled donor or acceptor samples can be used to determine the fixed fraction of the emission that "leaks" into the unintended channel. This fixed fraction that is a function of the fluorophore emission spectra and the optical filters can be subtracted from the acquired data.

3.2.2 Smoothing the Donor and Acceptor Time Traces Using the Chung–Kennedy Filter

Fluorescence intensity time traces have substantial noise deriving primarily from the shot noise of detection. Several methods exist to smooth noisy data, such as box-car averaging, but most of these smoothing methods do not preserve sharp edges. To overcome this limitation, Chung and Kennedy developed a nonlinear smoothing algorithm designed specifically to smooth data while preserving edges (Chung & Kennedy, 1991; Haran, 2004). This algorithm smoothes data containing transitions (Fig. 5A) by first determining the averages for windows of data of various sizes on either side of a given data point (dubbed the "forward" and "backward" average windows, Fig. 5B). To preserve edges, "forward" and "backward" averages that contain transitions are given less weight in the overall average. The statistical weights assigned to the "forward" and "backward" averages are determined as follows: The standard deviations for windows of data on either side of the data point being smoothed (referred to as the "forward" and "backward"

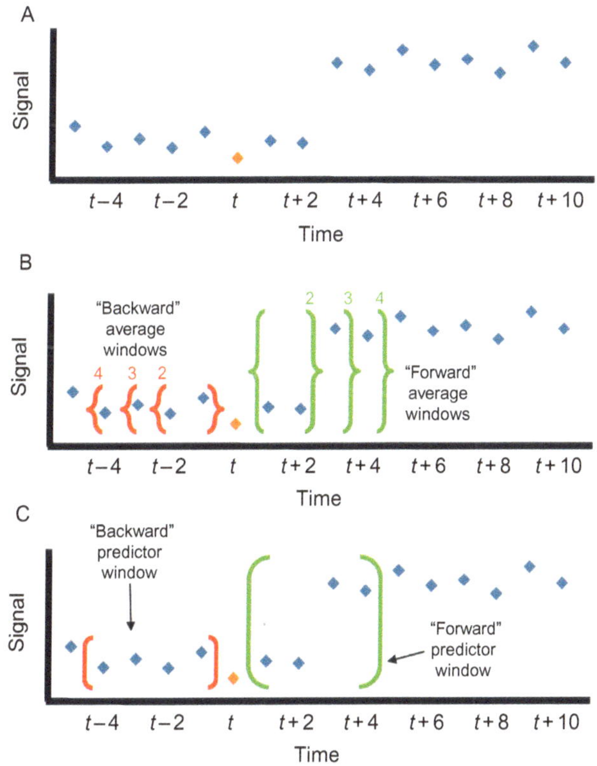

Fig. 5 Chung–Kennedy smoothing algorithm. (A) An example of data containing a transition between $t+2$ and $t+3$. (B) A schematic depicting three sizes of "forward" (*green*) and "backward" (*red*) average windows. (C) A schematic depicting four-point "forward" (*green*) and "backward" predictor windows.

predictor windows, Fig. 5C) are determined. The inverse of these standard deviations raised to a user-defined exponential term p is then used to calculate the statistical weights.

For example, consider a data point (at time t) with a transition occurring within the forward predictor window but not in the backward predictor window (Fig. 5A). The forward predictor window (Fig. 5C, green) would have a large standard deviation relative to the backward window (Fig. 5C, red). Thus, the averages from the forward average windows (Fig. 5B, green) would be assigned lower weight in the overall average, while the averages from backward average windows (Fig. 5B, red) would be assigned a larger statistical weight. This type of smoothing preserves transitions in the forward predictor window.

Importantly, in this approach the average windows and the predictor windows need not be the same size, nor is it necessary to use only one window size for each type of window. In fact, transitions that occur on different timescales can be preferentially preserved using windows of various sizes. Furthermore, the statistical weight term can be raised to the power of an empirically determined value (p) to exaggerate the different contributions of the forward and backward window averages. Thus, the input parameters of the average window size(s), predictor window size(s), and exponent term(s) can be defined by the user, and the optimal values in a given application can be empirically determined. Over filtering (i.e., introduction of false transitions) can be minimized by optimizing these input parameters.

When properly applied to the donor and acceptor time traces, the signal-to-noise ratio is significantly improved (Fig. 3B, third row), and transitions remain clear. Notably, in single-molecule FRET experiments, transitions in the donor and acceptor traces are expected to be anticorrelated (i.e., if the donor intensity increases, the acceptor intensity should simultaneously decrease, and vice versa). Because transitions in the donor and acceptor traces are expected to be simultaneous, the Chung–Kennedy smoothing algorithm can be improved by using the sum of the predictor window standard deviations for the both the donor and the acceptor to determine the statistical weights. As a result, simultaneous donor and acceptor intensity transitions will be more strongly preserved compared to uncorrelated donor and acceptor changes in intensity (Chung & Kennedy, 1991; Haran, 2004).

3.2.3 Screening Time Traces for Data Quality

Time traces of insufficient quality should be discarded from analysis. There are several reasons a time trace or portions of a time trace may not be worth analyzing. For example, the detected fluorescence emission intensities may be too low to reliably detect changes in intensity, or they may be too high to represent emission from only one molecule. More commonly, either the donor or the acceptor fluorophore (or both) may permanently or temporarily lose its fluorescence properties (i.e., bleach or blink) during data acquisition.

To identify unanalyzable regions of time traces, the fluorescence intensity of each data point can first be checked against user-defined minimum and maximum thresholds. To assess for bleaches or blinks in the remaining data, the average and standard deviation of a moving seven-point window can be used to determine a 95% confidence interval for each data point. If this interval includes zero, the data point can be considered as part of a bleaching or blinking event and discarded from analysis. Time traces containing both analyzable and unanalyzable regions may contain useful information (e.g., prior to photobleaching). These regions can then be further analyzed to determine their quality.

3.2.4 Calculating FRET and Identifying Transitions in the FRET Time Traces

Analyzable regions of the remaining donor and acceptor time traces can then be used to calculate single-molecule FRET efficiency using the equation $E = I_A/(I_D + I_A)$, where I_D and I_A represent the background and leakage corrected, fluorescence emission intensities of the donor and acceptor fluorophores, respectively (DeRocco et al., 2014; Lakowicz, 2006; Qiu et al., 2012, 2015; Sass et al., 2010). The resulting FRET time trace (Fig. 3B, fourth row) contains several pieces of information: (1) the FRET efficiency at each point in time; (2) the dwell time of a given FRET state provides kinetic information, such as the characteristic lifetimes associated with each state; and (3) the transitions provide insight into the preferred pathways of conformational changes. To extract this information, it is crucial to identify the time point at which the FRET time trace undergoes a transition.

Several methods exist to detect transitions in data (Blanco & Walter, 2010; Greenfeld et al., 2015; McKinney, Joo, & Ha, 2006; Shuang et al., 2014; van de Meent, Bronson, Wiggins, & Gonzalez, 2014). Applying a transition detection method at different levels of stringency can minimize misidentification of transitions, as transitions that withstand more stringent thresholds are more likely to be "real." Comparing the results of multiple methods can also help to overcome the limitations of using just one technique, as different detection methods will be better suited for detecting different types of changes (e.g., short- vs long-lived states) and will have different limitations. Edges detected by multiple methods at multiple stringency levels can then be assigned a higher confidence score. Presented here are two transition detection methods that can be independently applied at multiple thresholds to smFRET data.

Method 1: The Gaussian kernel method (Sass et al., 2010)—Mathematically, transitions in continuous functions can be identified by finding inflection points (i.e., maxima and minima in the first derivative of the time trace). Unfortunately, the FRET time traces are made up of discrete data points that have many apparent inflection points due to the significant noise in the signal. These issues can be circumvented by first convolving the FRET time traces with a Gaussian kernel of various widths and subsequently detecting inflection points in the convolved data. To ensure only "real" transitions are kept, a threshold can be incorporated. By changing the rigor of the threshold, the remaining "real" transitions can also be scored for confidence.

Method 2: The Chung-Kennedy method—The previously described smoothing algorithm developed by Chung and Kennedy uses increases in the predictor windows' standard deviations to calculate statistical weights (Chung & Kennedy, 1991; Haran, 2004). These increases in standard deviation can also be used to detect transitions in the FRET time traces as we describe here.

Consider a data point at time t with a transition occurring between time $t+2$ and $t+3$ (Fig. 6). Using a predictor window of four data points, the standard deviation of both the forward and backward predictor windows can be calculated at each value of t. (Fig. 6 depicts these calculations for $t-2$ to $t+7$.) In this example, the standard deviation of the forward predictor window at time t is at a local maximum. This maximum also occurs in the backward predictor window at time $t+5$ because both these windows contain the same range of data points. Accordingly, transitions can be detected by finding the midpoint between local maxima in the forward and backward predictor windows. To ensure only "real" transitions are kept, only the highest percentile local maxima are considered. By changing this percentile, the transitions can be scored for confidence. The most common source of false positives is bleaching and blinking events where either the donor or the acceptor fluorescence intensity is approximately zero. Regions of data previously identified as unanalyzable (i.e., bleaches and blinks) can be assigned a constant FRET not observed in the experiment, which eliminates their standard deviations and allows this transition detection method to function. Finally, a true change in FRET involves anticorrelated transitions in the donor and the acceptor signals (Haran, 2004). Including requirements that such anticorrelation be present above a threshold further aids confident identification of FRET transitions without falsely including bleaching and blinking events.

3.2.5 Alignment and Confirmation of Transitions in the FRET Time Traces

Typically, the timing of each transition identified by both methods is in agreement; however, the two described methods occasionally produce slightly disparate results. In these instances, any transitions occurring within 0.3 s can be empirically aligned taking the weighted average of the time detected by each method. The weights used in this average can be determined using the confidence scores provided by each method.

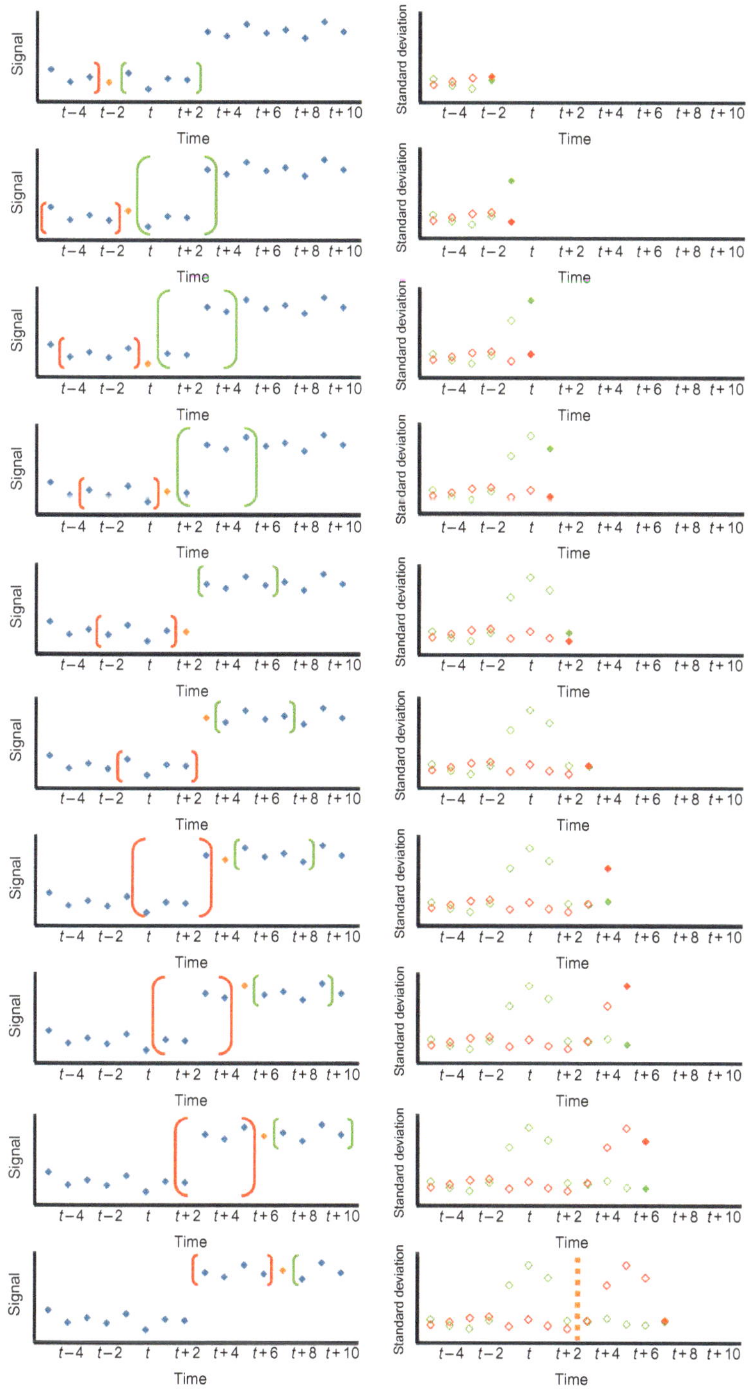

Fig. 6 See legend on next page.

Notably, both methods described here require user input to choose the appropriate thresholds. In the event that transitions are being missed (false negatives), the thresholds should be reduced. More often false transitions are identified. To test the significance of each transition, the FRET efficiency between each transition can be averaged, and the averages of adjacent FRET states can be subjected to a *t*-test. If two states are not statistically significantly different from one another (*p* level 0.05), then the transition between the two states can be identified as a false positive and discarded.

3.2.6 User Interaction and FRET-TACKLE

In the analysis described so far, the user is only required for the initial input parameters, such as smoothing windows and transition detection thresholds; the rest of the analysis can be completed by a computer in batch. Upon completion of the batch analysis, the computationally determined transitions can be verified, and any remaining false positives in each molecule's FRET time trace can be discarded by hand. This process, though tedious, can be crucial to recognizing patterns or detecting systematic errors in the computational approach.

Once the transitions have been determined, FRET-TACKLE (FRET Transition Analysis Coupled with Kinetic Lifetime Evaluation) analysis can extract the pertinent mechanistic information from the compiled data (DeRocco et al., 2014; Sass et al., 2010). In this method, distinct molecular conformations are identified by their characteristic FRET and kinetic lifetime properties. This approach allows molecular states with the same FRET but distinct kinetics to be distinguished.

3.3 Modeling DNA MMR Complex Kinetics From smFRET Measurements

Analysis of the distribution of FRET efficiencies, the characteristic lifetimes, and the preferred transitions reveals conformational, kinetic, and pathway preference information for the protein:DNA complexes being studied, which can provide mechanistic insights. Sass et al. used smFRET measurements of *Taq* MutS induced bending at a GT mismatch to deduce that as

Fig. 6 A novel transition detection method based on the Chung–Kennedy filter. "Forward" and "backward" predictor windows (*left column*, *green* and *red brackets*, respectively) are shown for successive time points (*blue diamonds*). Standard deviations for the depicted predictor windows (*right column*) reveal two identical peaks. Note that the midpoint between these peaks (between $t+2$ and $t+3$) identifies the time of the transition in the data.

many as six different conformations with different degrees of bending were sampled with complex kinetics (Sass et al., 2010). Some of these states had overlapping FRET levels, but were distinguishable by their lifetimes that varied by 20-fold and rates of intercoversion that differed by 2 orders of magnitude. From detailed modeling of the quantitative measurements of lifetimes of the states and their probabilities of making each possible transition, landmarks in the free energy landscape of the MutS interaction with a DNA mismatch could be deduced (DeRocco et al., 2014; Sass et al., 2010).

Using smFRET between MutS domain I donor and an acceptor 9 bases away from a DNA mismatch (Fig. 7A), Qiu et al. measured an ATP-

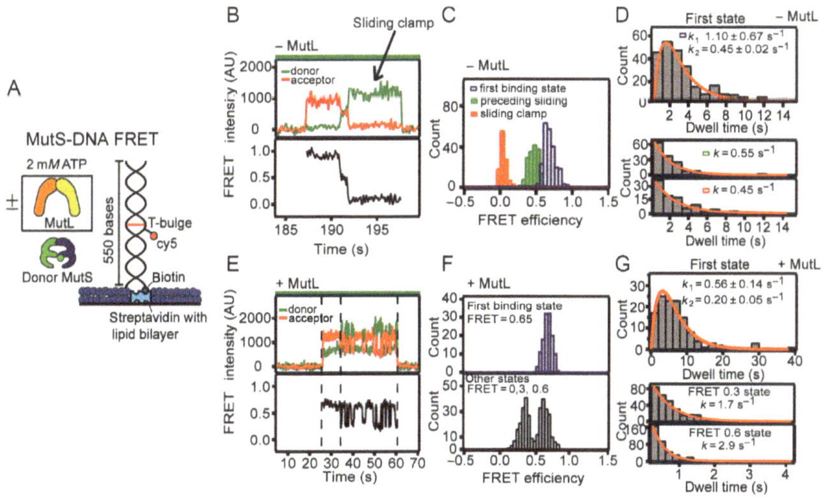

Fig. 7 smFRET measurements of the kinetics of MutS mismatch recognition and subsequent conformational changes. (A) Experimental scheme: 2 m*M* ATP, 10 n*M* Alexa555-MutS, Cy5-T-bulge DNA, and 200 n*M* unlabeled MutL (when present). Example time traces of donor and acceptor emission and calculated FRET in the absence (B) and presence (E) of MutL for events with transitions. FRET histograms for binding events with exactly one donor and one acceptor and with FRET transitions reveal three states in the absence of MutL (C) with dwell time distributions fit (*red line*) by a two-step model for the first state (D, *top panel*) and a one-step model for the middle state and last state (D, *middle and lower panels*). In the presence of MutL, FRET histograms for the first state (F, *top panel*) reveal a narrow peak, but the dwell time distributions (G, *top panel*) still require a fit (*red line*) with two steps. Histograms of the subsequent FRET states show two nonzero peaks (F, *lower panel*), and the dwell time distributions are fit well (*red line*) with a one-step model (G, *middle* and *lower panels*). *Numbers within panels* report rates obtained from the fits. *Adapted from Qiu, R., Sakato, M., Sacho, E. J., Wilkins, H., Zhang, X., Modrich, P., . . . Weninger, K. R. (2015). MutL traps MutS at a DNA mismatch.* Proceedings of the National Academy of Sciences of the United States of America, 112*(35), 10914–10919. doi:10.1073/pnas.1505655112.*

dependent sequence of conformational changes that occurred during conversion of mismatch-bound MutS into a state of MutS that leaves the mismatch and slides along the DNA (Qiu et al., 2012). MutS bound to a mismatch in a configuration that yielded FRET ~0.7 (Fig. 7B and C). A precise kinetic pathway of states was observed in the subset of molecules that eventually converted to sliding clamps in the presence of ATP.

Before sliding away from the mismatch, a transition to an intermediate state near FRET 0.5 occurred (Fig. 7B and C). Kinetic analysis of the lifetimes of the FRET 0.7 and FRET 0.5 lifetimes indicated that these states were significantly different. Measurements of intraprotein FRET between donor and label on the two DNA binding domains of the MutS dimer confirmed this two-step process (Qiu et al., 2012). The histogram of the lifetimes of many occurrences of the 0.5 FRET state fit well to a single exponential decay with lifetime of 0.6 s, suggesting a single-step conformational change accounted for the change in FRET level (Fig. 7D, lower). In contrast, the distribution of lifetimes of the 0.7 FRET state is clearly not a simple exponential decay (Fig. 7D, upper) (Qiu et al., 2015). The distribution rises at short times. It was well fit with a model of a two-step kinetic process that does not generate a conformational change leading to a distinct FRET value. Such a two-step process is characterized by lifetime distributions given by $k_1 * k_2 * (\exp(-k_1 * t) - \exp(-k_2 * t))/(k_2 - k_1)$, where k_1 and k_2 are the rates of the two distinct steps underling the process (Floyd, Harrison, & van Oijen, 2010; Yildiz et al., 2003). Fitting the experiments yields estimates of k_1 and k_2 as 1.1 ± 0.67 and $0.45 \pm 0.02\ \text{s}^{-1}$. Preliminary experiments indicate that one of these rates is affected by ATP concentration and the other is not, suggesting that one of the steps involves ATP binding from solution. Addition of MutL, the next protein in the repair signaling cascade that interacts with mismatch activated MutS, did not change the FRET level of the first state, but did alter the kinetic behavior (Fig. 7E–G). Subsequent FRET states and kinetics were also modified. In particular, MutL prevented MutS from sliding away from the mismatch. Rather, MutS was observed to always dissociate directly from the mismatch under those experimental conditions (Qiu et al., 2015).

4. COMPLEMENTARITY BETWEEN SINGLE MOLECULE AND ENSEMBLE KINETICS: AN MMR CASE STUDY

Ensemble kinetic measurements performed on the millisecond timescale of a single enzyme-catalyzed reaction cycle reveal the rates and order of multiple events, starting from substrate binding through transient conformational changes and catalysis and finally product release. For the MMR system, a combination of rapid-quench and stopped-flow experiments has outlined the kinetic mechanism employed by MutS as it uses ATPase activity to recognize and signal repair of mismatched bases in DNA. As with all ensemble methods, these rapid kinetic methods provide average measures of the population, which can obscure functionally significant distributions and dynamics of individual molecules, but they also yield detailed models of reaction pathways and enzyme mechanisms in solution. In this section, we use MutS to illustrate how single molecule and ensemble kinetic approaches complement each other to yield a richly textured view of the structure/dynamics–function relationships governing enzyme-catalyzed reactions.

4.1 DNA Binding and Bending by MutS

One key finding from the smFRET DNA binding measurements was that MutS forms a complex with mismatched DNA in bent conformation at a bimolecular rate constant of $1.3 \times 10^6\ M^{-1}\ s^{-1}$ (Sass et al., 2010). Stopped-flow experiments were also used to measure formation of the MutS:mismatched DNA complex, with different fluorescence-based assays monitoring MutS or DNA conformation during the reaction. MutS was mixed rapidly with DNA and the rates of signal change from fluorophore-labeled DNA or MutS were measured at increasing concentrations of the unlabeled reactant. Experiments with labeled DNA determined the following: (i) increase in fluorescence anisotropy of TAMRA-labeled DNA reported MutS:DNA complex formation, and linear dependence of the rate on MutS concentration yielded a binding rate constant of $6 \times 10^6\ M^{-1}\ s^{-1}$ (Jacobs-Palmer & Hingorani, 2007); (ii) increase in FRET between AF488 and AF594 fluorophores flanking the mismatch reported DNA bending in the MutS:DNA complex, and the initial linear rate dependence on MutS concentration yielded a similar binding/DNA bending rate constant of $4 \times 10^6\ M^{-1}\ s^{-1}$ (Fig. 8A and B, pink). Experiments with MutS labeled with IAANS on the DNA binding domain I also yielded a similar binding/MutS conformational change rate constant of $7 \times 10^6\ M^{-1}\ s^{-1}$ (Fig. 8A and B, blue) (Sharma et al., 2013).

The agreement between single molecule and ensemble rates (measured by different reporter assays) validated the design and setup of smFRET experiments, indicating that there were no perturbations, for example, due to surface immobilization or fluorophore choice/position. In terms of

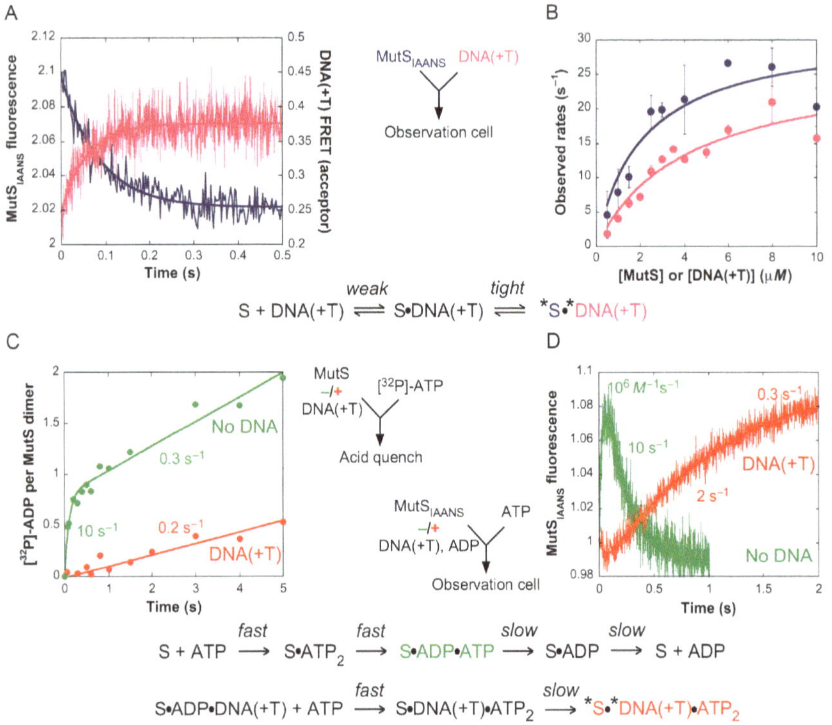

Fig. 8 Stopped-flow studies of MMR proteins. (A) In stopped-flow experiments, IAANS fluorescence quenching reports MutS domain I movement on DNA binding (0.1 μ*M* MutS$_{IAANS}$ mixed with 3 μ*M* DNA(+T); *blue trace*), and increasing FRET between AF488 and AF594 flanking the T-bulge reports DNA bending on MutS binding (0.03 μ*M* DNA(+T)$_{AF488\text{–}AF594}$ mixed with 3 μ*M* MutS; *pink trace*). (B) The rates of these conformational changes increase hyperbolically with titrant concentration to a maximum of ~25 s^{-1} and fit to a two-step binding model with initial rapid collision forming a weak complex followed by intramolecular conformational changes to form a tight complex ($K_D \sim 10$ n*M*). (C) Pre-steady-state rapid-quench experiments with MutS reveal a burst of ^{32}P-ADP formation at 10 s^{-1} (40°C) followed by slow steady state $k_{cat} = 0.3$ s^{-1} (*green trace*). The burst amplitude indicates rapid hydrolysis of one ATP per MutS dimer. Preincubation of MutS with T-bulge DNA results in the inhibition of the burst and a slow ATPase rate of 0.2 s^{-1} (*red trace*). (D) Stopped-flow traces of MutS$_{IAANS}$ mixed with ATP show that domains I rearrange concurrently with ATP binding (signal increase) and switch right back with ATP hydrolysis (signal decrease), whereas one ADP-bound MutS$_{IAANS}$–DNA(+T) complex shows a lag (likely reflecting ADP release) followed by slow change in MutS conformation (signal increase) to a stable ATP-bound MMR ready complex. *Adapted from Hingorani, M. M. (2016). Mismatch binding, ADP-ATP exchange and intramolecular signaling during mismatch repair.* DNA Repair (Amst), 38, *24–31. doi:10.1016/j.dnarep.2015.11.017.*

the mismatch recognition mechanism, the smFRET data indicated that MutS binding to DNA occurs simultaneously with DNA bending. However, unlike single molecule, ensemble experiments allow measurements of fast rates and at high reactant concentrations, which ultimately revealed a different mechanism. As shown in Fig. 8B, the observed rates of DNA binding/bending and MutS conformational change show a linear dependence until ~3 μ*M* of MutS or DNA titrant, but at higher concentrations the rates saturate at 20–25 s^{-1}. This hyperbolic relationship is indicative of a two-step mechanism in which initial collision between MutS and DNA is followed by isomerization of both molecules that determines formation of the final high-affinity mismatch recognition complex (K_D ~ 10 n*M*) (Sharma et al., 2013). On the flip side, while the ensemble data resolved a sequential induced-fit mechanism of mismatch recognition, understanding the nature of the conformational changes in MutS and DNA required single-molecule analysis. smFRET experiments with labeled MutS showed that the DNA binding domains close in toward the mismatch to stabilize bent DNA in the MutS: DNA complex (Qiu et al., 2012), and smFRET experiments with labeled DNA showed that while MutS samples DNA conformations with varying degrees of bending, the complex with highly bent DNA is preferred kinetically and energetically (Sass et al., 2010).

4.2 ATPase-Coupled MutS Actions in MMR

Ensemble equilibrium and transient kinetic measurements of ATP binding, hydrolysis, and product release have revealed the MutS ATPase mechanism and enabled understanding of how it drives MutS actions before and after mismatch recognition (reviewed in Hingorani, 2016). Pre-steady-state rapid-quench experiments mixing MutS with ^{32}P-ATP measured the stoichiometry and the rate of ATP hydrolysis (^{32}P-ADP formation) and determined that off the mismatch, MutS binds and hydrolyzes only one ATP per dimer rapidly and then undergoes a slow rate-limiting step that defines the ~40-fold slower steady-state rate of the reaction (k_{cat}) (Fig. 8C, green). Pre-steady-state stopped-flow experiments mixing MutS with ATP and MDCC-PBP (fluorophore-labeled phosphate binding protein reporter) measured the stoichiometry and the rate of phosphate release after ATP hydrolysis. The data were identical to those from ^{32}P-ATP hydrolysis measurements, showing that free or mismatch-bound MutS releases the phosphate product rapidly (Antony & Hingorani, 2004). In contrast, stopped-flow experiments with Mant-ADP (fluorophore-labeled ADP analog) showed ADP dissociating slowly from free MutS, indicating that ADP product release limits the k_{cat} following ATP hydrolysis and phosphate release (Geng et al., 2012). Thus, off the mismatch, MutS remains predominantly bound to (at least one) ADP. Similar experiments performed with MutS:mismatched DNA complex showed stimulation of ADP release associated with rapid ATP binding; however, ATP hydrolysis was strongly suppressed by ~30–50-fold (Fig. 8C, red). Thus, on the mismatch, MutS remains predominantly bound to (two) ATP (Antony & Hingorani, 2004). To summarize, prior to mismatch recognition the ATPase mechanism favors a MutS population in which at least one ADP is stably bound per dimer, whereas after mismatch recognition the mechanism favors a MutS:mismatched DNA complex population in which two ATP are stably bound per dimer.

Analogous transient kinetic measurements of IAANS-labeled MutS show that the protein adopts one conformation rapidly at the rate of ATP binding (increasing fluorescence) and another at the rate of ATP hydrolysis (decreasing fluorescence; Fig. 8D, green). These results indicate binary switching of MutS between ATP-bound and ADP-bound forms with the latter persisting in steady state prior to mismatch recognition. When MutS binds a mismatch, it releases ADP and binds ATP rapidly (as determined from the experiments described above), but the conformational change after ATP binding is much slower and involves multiple steps (Fig. 8D, red, increasing fluorescence). Since ATP hydrolysis is suppressed in the MutS: mismatched DNA complex, these results indicate rate-determining conversion of ATP-bound MutS into a postmismatch recognition form that can signal repair. The nature of the nucleotide-coupled changes in MutS conformation and dynamics was again revealed by single-molecule analysis. smFRET between DNA binding domains I of the MutS dimer reported transitions between low (domains open) and high (domains closed) FRET for nucleotide-free and ADP-bound protein in the absence of mismatched DNA (Qiu et al., 2012). Thus the ADP-bound form of MutS, which is predominant prior to mismatch recognition, as described above, exhibits conformational dynamics that enable it to enclose DNA within the binding cavity (Fig. 1) and probe the duplex for mismatched bases. smFRET experiments also revealed that domains I lock into a closed conformation upon mismatch recognition, effectively stalling MutS at the site. As the MutS:mismatched DNA complex releases ADP and binds ATP, domains I transition slowly into a more open, stable conformation that enables ATP-bound MutS to interact with MutL and initiate repair, or disengage from the mismatch and slide away within a few seconds if MutL is not available (Fig. 7)

(Qiu et al., 2012, 2015). Thus, ensemble measurements provided a framework to establish how MutS uses its ATPase reaction for mismatch recognition and initiation of repair. This information was not accessible by single-molecule methods due to a lack of reporter assays and inadequate temporal resolution. Single-molecule experiments provided complementary information on how ADP- and ATP-bound forms of MutS can recognize mismatches and license subsequent steps in the repair pathway, respectively. This information was not accessible by ensemble methods because of the inability to resolve conformational dynamics of individual MutS molecules on and off DNA.

5. CONCLUDING REMARKS

Here we have described smFRET methods to detect sequential conformational changes that occur when MMR proteins interact with a base–base mismatch or nucleotide insertion in DNA. By attaching donor and acceptor fluorophores on either DNA or protein, we developed assays sensitive to the degree of DNA bending, the relative positions of DNA binding domains in the MutS dimer, or the position of MutS on DNA relative to the mismatch. Automated edge detection in smFRET time traces enabled acquisition of sufficient quantitative data on conformational state transitions to statistically characterize the kinetic behavior of these macromolecules. We also outlined a combined single molecule and ensemble kinetics approach that provided more comprehensive insights into MutS structure/dynamics and catalytic mechanism during the reaction. Future extensions of smFRET experiments to include additional MMR proteins, three and four fluorophore labels yielding multiple FRET signal combinations (DeRocco et al., 2010; Lee et al., 2010), and applications in live cells (Huang, Nagy, Koide, Rock, & Koide, 2009; Li & Xie, 2011; Reyes-Lamothe, Sherratt, & Leake, 2010; Sakon & Weninger, 2010; Uphoff, Reyes-Lamothe, Garza de Leon, Sherratt, & Kapanidis, 2013) are underway to complete understanding of the mechanisms of action of all proteins in this critical DNA repair pathway.

ACKNOWLEDGMENTS

This work was supported by NIH Grants GM079480 and GM080294 (to D.A.E.), GM109832 (to D.A.E. and K.R.W.), and GM114743 (to M.M.H), and NSF Grant MCB-1022203 (to M.M.H.).

REFERENCES

Antony, E., & Hingorani, M. M. (2004). Asymmetric ATP binding and hydrolysis activity of the *Thermus aquaticus* MutS dimer is key to modulation of its interactions with mismatched DNA. *Biochemistry*, *43*(41), 13115–13128. http://dx.doi.org/10.1021/bi049010t.

Bignami, M., Casorelli, I., & Karran, P. (2003). Mismatch repair and response to DNA-damaging antitumour therapies. *European Journal of Cancer*, *39*(15), 2142–2149.

Blanco, M., & Walter, N. G. (2010). Analysis of complex single-molecule FRET time trajectories. *Methods in Enzymology*, *472*, 153–178. http://dx.doi.org/10.1016/S0076-6879(10)72011-5.

Brunger, A. T., Strop, P., Vrljic, M., Chu, S., & Weninger, K. R. (2011). Three-dimensional molecular modeling with single molecule FRET. *Journal of Structural Biology*, *173*(3), 497–505. http://dx.doi.org/10.1016/j.jsb.2010.09.004.

Choi, U. B., Strop, P., Vrljic, M., Chu, S., Brunger, A. T., & Weninger, K. R. (2010). Single-molecule FRET-derived model of the synaptotagmin 1-SNARE fusion complex. *Nature Structural & Molecular Biology*, *17*(3), 318–324. http://dx.doi.org/10.1038/nsmb.1763.

Choi, U. B., Weninger, K. R., & Bowen, M. E. (2012). Immobilization of proteins for single-molecule fluorescence resonance energy transfer measurements of conformation and dynamics. *Methods in Molecular Biology*, *896*, 3–20. http://dx.doi.org/10.1007/978-1-4614-3704-8_1.

Chung, S. H., & Kennedy, R. A. (1991). Forward-backward non-linear filtering technique for extracting small biological signals from noise. *Journal of Neuroscience Methods*, *40*(1), 71–86. http://dx.doi.org/10.1016/0165-0270(91)90118-J.

DeRocco, V. C., Anderson, T., Piehler, J., Erie, D. A., & Weninger, K. (2010). Four-color single-molecule fluorescence with noncovalent dye labeling to monitor dynamic multimolecular complexes. *Biotechniques*, *49*(5), 807.

DeRocco, V. C., Sass, L. E., Qiu, R., Weninger, K. R., & Erie, D. A. (2014). Dynamics of MutS-mismatched DNA complexes are predictive of their repair phenotypes. *Biochemistry*, *53*(12), 2043–2052. http://dx.doi.org/10.1021/bi401429b.

Erie, D. A., & Weninger, K. R. (2014). Single molecule studies of DNA mismatch repair. *DNA Repair (Amst)*, *20*, 71–81. http://dx.doi.org/10.1016/j.dnarep.2014.03.007.

Fedier, A., & Fink, D. (2004). Mutations in DNA mismatch repair genes: Implications for DNA damage signaling and drug sensitivity (review). *International Journal of Oncology*, *24*(4), 1039–1047.

Fishel, R. (2015). Mismatch repair. *Journal of Biological Chemistry*, *290*(44), 26395–26403. http://dx.doi.org/10.1074/jbc.R115.660142.

Floyd, D. L., Harrison, S. C., & van Oijen, A. M. (2010). Analysis of kinetic intermediates in single-particle dwell-time distributions. *Biophysical Journal*, *99*(2), 360–366. http://dx.doi.org/10.1016/j.bpj.2010.04.049.

Geng, H., Sakato, M., DeRocco, V., Yamane, K., Du, C., Erie, D. A., … Hsieh, P. (2012). Biochemical analysis of the human mismatch repair proteins hMutSalpha MSH2 (G674A)-MSH6 and MSH2-MSH6(T1219D). *Journal of Biological Chemistry*, *287*(13), 9777–9791. http://dx.doi.org/10.1074/jbc.M111.316919.

Genschel, J., & Modrich, P. (2003). Mechanism of 5′-directed excision in human mismatch repair. *Molecular Cell*, *12*(5), 1077–1086.

Genschel, J., & Modrich, P. (2009). Functions of MutLalpha, replication protein A (RPA), and HMGB1 in 5′-directed mismatch repair. *Journal of Biological Chemistry*, *284*(32), 21536–21544. http://dx.doi.org/10.1074/jbc.M109.021287.

Greenfeld, M., van de Meent, J. W., Pavlichin, D. S., Mabuchi, H., Wiggins, C. H., Gonzalez, R. L., Jr., & Herschlag, D. (2015). Single-molecule dataset (SMD):

A generalized storage format for raw and processed single-molecule data. *BMC Bioinformatics*, *16*, 3. http://dx.doi.org/10.1186/s12859-014-0429-4.
Groothuizen, F. S., Winkler, I., Cristovao, M., Fish, A., Winterwerp, H. H., Reumer, A., … Sixma, T. K. (2015). MutS/MutL crystal structure reveals that the MutS sliding clamp loads MutL onto DNA. *eLife*, *4*, e06744. http://dx.doi.org/10.7554/eLife.06744.
Ha, T., Enderle, T., Ogletree, D. F., Chemla, D. S., Selvin, P. R., & Weiss, S. (1996). Probing the interaction between two single molecules: Fluorescence resonance energy transfer between a single donor and a single acceptor. *Proceedings of the National Academy of Sciences of the United States of America*, *93*(13), 6264–6268.
Haran, G. (2004). Noise reduction in single-molecule fluorescence trajectories of folding proteins. *Chemical Physics*, *307*(2–3), 137–145. http://dx.doi.org/10.1016/j.chemphys.2004.05.017.
Heinen, C. D. (2016). Mismatch repair defects and Lynch syndrome: The role of the basic scientist in the battle against cancer. *DNA Repair (Amst)*, *38*, 127–134. http://dx.doi.org/10.1016/j.dnarep.2015.11.025.
Hingorani, M. M. (2016). Mismatch binding, ADP-ATP exchange and intramolecular signaling during mismatch repair. *DNA Repair (Amst)*, *38*, 24–31. http://dx.doi.org/10.1016/j.dnarep.2015.11.017.
Huang, J., Nagy, S. S., Koide, A., Rock, R. S., & Koide, S. (2009). A peptide Tag system for facile purification and single-molecule immobilization. *Biochemistry*, *48*(50), 11834–11836. http://dx.doi.org/10.1021/bi901756n.
Iyer, R. R., Pluciennik, A., Burdett, V., & Modrich, P. L. (2006). DNA mismatch repair: Functions and mechanisms. *Chemical Reviews*, *106*(2), 302–323. http://dx.doi.org/10.1021/cr0404794.
Jacobs-Palmer, E., & Hingorani, M. M. (2007). The effects of nucleotides on MutS-DNA binding kinetics clarify the role of MutS ATPase activity in mismatch repair. *Journal of Molecular Biology*, *366*(4), 1087–1098. http://dx.doi.org/10.1016/j.jmb.2006.11.092.
Kadyrov, F. A., Dzantiev, L., Constantin, N., & Modrich, P. (2006). Endonucleolytic function of MutLalpha in human mismatch repair. *Cell*, *126*(2), 297–308. http://dx.doi.org/10.1016/j.cell.2006.05.039.
Kadyrov, F. A., Holmes, S. F., Arana, M. E., Lukianova, O. A., O'Donnell, M., Kunkel, T. A., & Modrich, P. (2007). Saccharomyces cerevisiae MutLa is a mismatch repair endonuclease. *Journal of Biological Chemistry*, *282*, 37181–37190.
Karran, P., Offman, J., & Bignami, M. (2003). Human mismatch repair, drug-induced DNA damage, and secondary cancer. *Biochimie*, *85*(11), 1149–1160.
Kunkel, T. A., & Erie, D. A. (2005). DNA mismatch repair. *Annual Review of Biochemistry*, *74*, 681–710.
Kunkel, T. A., & Erie, D. A. (2015). Eukaryotic mismatch repair in relation to DNA replication. *Annual Review of Genetics*, *49*, 291–313. http://dx.doi.org/10.1146/annurev-genet-112414-054722.
Lakowicz, J. R. (2006). *Principles of fluorescence spectroscopy* (3rd ed.). . USA: Springer-Verlag.
Lamers, M. H., Perrakis, A., Enzlin, J. H., Winterwerp, H. H., de Wind, N., & Sixma, T. K. (2000). The crystal structure of DNA mismatch repair protein MutS binding to a G x T mismatch. *Nature*, *407*(6805), 711–717.
Lee, J., Lee, S., Ragunathan, K., Joo, C., Ha, T., & Hohng, S. (2010). Single-molecule four-color FRET. *Angewandte Chemie*, *49*(51), 9922–9925. http://dx.doi.org/10.1002/anie.201005402.
Li, G.-W., & Xie, X. S. (2011). Central dogma at the single-molecule level in living cells. *Nature*, *475*(7356), 308–315. http://dx.doi.org/10.1038/nature10315.
Liberti, S. E., Larrea, A. A., & Kunkel, T. A. (2013). Exonuclease 1 preferentially repairs mismatches generated by DNA polymerase alpha. *DNA Repair (Amst)*, *12*(2), 92–96. http://dx.doi.org/10.1016/j.dnarep.2012.11.001.

McCann, J. J., Choi, U. B., Zheng, L., Weninger, K., & Bowen, M. E. (2010). Optimizing methods to recover absolute FRET efficiency from immobilized single molecules. *Biophysical Journal*, *99*(3), 961–970. http://dx.doi.org/10.1016/j.bpj.2010.04.063.
McKinney, S. A., Joo, C., & Ha, T. (2006). Analysis of single-molecule FRET trajectories using hidden Markov modeling. *Biophysical Journal*, *91*(5), 1941–1951. http://dx.doi.org/10.1529/biophysj.106.082487.
Modrich, P., & Lahue, R. (1996). Mismatch repair in replication fidelity, genetic recombination, and cancer biology. *Annual Review of Biochemistry*, *65*, 101–133.
Morita, R., Nakane, S., Shimada, A., Inoue, M., Iino, H., Wakamatsu, T., … Kuramitsu, S. (2010). Molecular mechanisms of the whole DNA repair system: A comparison of bacterial and eukaryotic systems. *Journal of Nucleic Acids*, *2010*, 179594. http://dx.doi.org/10.4061/2010/179594.
Natrajan, G., Lamers, M. H., Enzlin, J. H., Winterwerp, H. H., Perrakis, A., & Sixma, T. K. (2003). Structures of *Escherichia coli* DNA mismatch repair enzyme MutS in complex with different mismatches: A common recognition mode for diverse substrates. *Nucleic Acids Research*, *31*(16), 4814–4821.
Nick McElhinny, S. A., Kissling, G. E., & Kunkel, T. A. (2010). Differential correction of lagging-strand replication errors made by DNA polymerases {alpha} and {delta}. *Proceedings of the National Academy of Sciences of the United States of America*, *107*(49), 21070–21075. http://dx.doi.org/10.1073/pnas.1013048107.
Obmolova, G., Ban, C., Hsieh, P., & Yang, W. (2000). Crystal structures of mismatch repair protein MutS and its complex with a substrate DNA. *Nature*, *407*(6805), 703–710. http://dx.doi.org/10.1038/35037509.
Pavlov, Y. I., Mian, I. M., & Kunkel, T. A. (2003). Evidence for preferential mismatch repair of lagging strand DNA replication errors in yeast. *Current Biology*, *13*(9), 744–748.
Peltomaki, P. (2016). Update on Lynch syndrome genomics. *Familial Cancer*, *15*, 385–393. http://dx.doi.org/10.1007/s10689-016-9882-8.
Pluciennik, A., Burdett, V., Lukianova, O., O'Donnell, M., & Modrich, P. (2009). Involvement of the beta clamp in methyl-directed mismatch repair in vitro. *The Journal of Biological Chemistry*, *284*(47), 32782–32791. http://dx.doi.org/10.1074/jbc.M109.054528.
Qiu, R., DeRocco, V. C., Harris, C., Sharma, A., Hingorani, M. M., Erie, D. A., & Weninger, K. R. (2012). Large conformational changes in MutS during DNA scanning, mismatch recognition and repair signalling. *The EMBO Journal*, *31*(11), 2528–2540. http://dx.doi.org/10.1038/emboj.2012.95.
Qiu, R., Sakato, M., Sacho, E. J., Wilkins, H., Zhang, X., Modrich, P., … Weninger, K. R. (2015). MutL traps MutS at a DNA mismatch. *Proceedings of the National Academy of Sciences of the United States of America*, *112*(35), 10914–10919. http://dx.doi.org/10.1073/pnas.1505655112.
Reyes-Lamothe, R., Sherratt, D. J., & Leake, M. C. (2010). Stoichiometry and architecture of active DNA replication machinery in *Escherichia coli*. *Science*, *328*(5977), 498–501. http://dx.doi.org/10.1126/science.1185757.
Sakon, J. J., & Weninger, K. R. (2010). Detecting the conformation of individual proteins in live cells. *Nature Methods*, 7(3), 203–205. http://dx.doi.org/10.1038/nmeth.1421.
Sass, L. E., Lanyi, C., Weninger, K., & Erie, D. A. (2010). Single-molecule FRET TACKLE reveals highly dynamic mismatched DNA-MutS complexes. *Biochemistry*, *49*(14), 3174–3190.
Schofield, M. J., Nayak, S., Scott, T. H., Du, C., & Hsieh, P. (2001). Interaction of *Escherichia coli* MutS and MutL at a DNA mismatch. *Journal of Biological Chemistry*, *276*(30), 28291–28299. http://dx.doi.org/10.1074/jbc.M103148200.
Sharma, A., Doucette, C., Biro, F. N., & Hingorani, M. M. (2013). Slow conformational changes in MutS and DNA direct ordered transitions between mismatch search,

recognition and signaling of DNA repair. *Journal of Molecular Biology, 425*(22), 4192–4205. http://dx.doi.org/10.1016/j.jmb.2013.08.011.

Shuang, B., Cooper, D., Taylor, J. N., Kisley, L., Chen, J., Wang, W., ... Landes, C. F. (2014). Fast step transition and state identification (STaSI) for discrete single-molecule data analysis. *Journal of Physical Chemistry Letters, 5*(18), 3157–3161. http://dx.doi.org/10.1021/jz501435p.

Stryer, L., & Haugland, R. P. (1967). Energy transfer: A spectroscopic ruler. *Proceedings of the National Academy of Sciences of the United States of America, 58*(2), 719–726.

Uphoff, S., Reyes-Lamothe, R., Garza de Leon, F., Sherratt, D. J., & Kapanidis, A. N. (2013). Single-molecule DNA repair in live bacteria. *Proceedings of the National Academy of Sciences of the United States of America, 110*(20), 8063–8068. http://dx.doi.org/10.1073/pnas.1301804110.

van de Meent, J. W., Bronson, J. E., Wiggins, C. H., & Gonzalez, R. L., Jr. (2014). Empirical Bayes methods enable advanced population-level analyses of single-molecule FRET experiments. *Biophysical Journal, 106*(6), 1327–1337. http://dx.doi.org/10.1016/j.bpj.2013.12.055.

Wang, H., Yang, Y., Schofield, M. J., Du, C., Fridman, Y., Lee, S. D., ... Erie, D. A. (2003). DNA bending and unbending by MutS govern mismatch recognition and specificity. *Proceedings of the National Academy of Sciences of the United States of America, 100*(25), 14822–14827. http://dx.doi.org/10.1073/pnas.2433654100.

Warren, J. J., Pohlhaus, T. J., Changela, A., Iyer, R. R., Modrich, P. L., & Beese, L. S. (2007). Structure of the human MutSalpha DNA lesion recognition complex. *Molecular Cell, 26*(4), 579–592.

Yildiz, A., Forkey, J. N., McKinney, S. A., Ha, T., Goldman, Y. E., & Selvin, P. R. (2003). Myosin V walks hand-over-hand: Single fluorophore imaging with 1.5-nm localization. *Science, 300*(5628), 2061–2065. http://dx.doi.org/10.1126/science.1084398.

CHAPTER ELEVEN

Single-Molecule Confocal FRET Microscopy to Dissect Conformational Changes in the Catalytic Cycle of DNA Topoisomerases

S. Hartmann, D. Weidlich, D. Klostermeier[1]
Institute for Physical Chemistry, University of Muenster, Muenster, Germany
[1]Corresponding author: e-mail address: dagmar.klostermeier@uni-muenster.de

Contents

Abstract

Molecular machines undergo large-scale conformational changes during their catalytic cycles that are linked to their biological functions. DNA topoisomerases are molecular machines that interconvert different DNA topoisomers and resolve torsional stress that is introduced during cellular processes that involve local DNA unwinding. DNA gyrase catalyzes the introduction of negative supercoils into DNA in an ATP-dependent

Methods in Enzymology, Volume 581
ISSN 0076-6879
http://dx.doi.org/10.1016/bs.mie.2016.08.013
© 2016 Elsevier Inc.
All rights reserved.

reaction. During its catalytic cycle, gyrase undergoes large-scale conformational changes that drive the supercoiling reaction. These conformational changes can be followed by single-molecule Förster resonance energy transfer (FRET). Here, we use DNA gyrase from *Bacillus subtilis* as an illustrative example to present strategies for the investigation of conformational dynamics of multisubunit complexes. We provide a brief introduction into single-molecule FRET and confocal microscopy, with a focus on practical considerations in sample preparation and data analysis. Different strategies in the preparation of donor–acceptor-labeled molecules suitable for single-molecule FRET experiments are outlined. The insight into the mechanism of DNA supercoiling by gyrase gained from single-molecule FRET experiment is summarized. The general strategies described here can also be applied to investigate conformational changes and their link to biological function of other multisubunit molecular machines.

ABBREVIATIONS

ADPNP 2′/3′-*O*-(*N*-methyl-anthraniloyl)-adenosine-5′-[(β,γ)-imido]triphosphate
ALEX alternating laser excitation
CTD C-terminal domain
FRET Förster resonance energy transfer
PIE pulsed interleaved excitation
TOPRIM topoisomerase-primase
WHD winged helix domain

1. INTRODUCTION

Molecular machines undergo large-scale conformational changes during their catalytic cycles that are linked to their biological functions. Many of these molecular machines are multisubunit complexes that act on DNA or RNA. DNA topoisomerases interconvert different DNA topoisomers and resolve torsional stress that is introduced during cellular processes that involve local DNA unwinding, such as DNA replication, transcription, recombination, and repair (reviewed in Vos, Tretter, Schmidt, & Berger, 2011). Topoisomerases are found in all kingdoms of life. To achieve changes in the topological state of the DNA, they have to cleave one or both strands of their DNA substrate, catalyzed by one or two catalytic tyrosine(s). Based on the number of strands cleaved during the reaction, topoisomerases are classified into type I (single-strand cleavage) and type II enzymes (double-strand cleavage). Type I topoisomerases change the linking number of the DNA in steps of one, i.e., they introduce or remove one supercoil per step, while type II topoisomerases change the linking number in steps of two. DNA gyrase is a special type II DNA topoisomerase that introduces negative supercoils into DNA in an ATP-dependent reaction (Gellert, Mizuuchi, O'Dea, & Nash, 1976). Supercoiling is thought to follow a strand-passage mechanism (Kampranis, Bates, & Maxwell, 1999), which predicts large-scale conformational changes of the enzyme during negative supercoiling (see Section 2).

Single-molecule Förster resonance energy transfer (smFRET) has proven to be a versatile tool to observe conformational changes in protein/DNA complexes (Cristovao et al., 2012; Hohlbein et al., 2013; Lamichhane, Berezhna, Gill, Van der Schans, & Millar, 2013; Nagy et al., 2015; Qiu et al., 2012; Rothwell et al., 2003; Santoso et al., 2010). FRET experiments measure the energy transfer from a donor fluorophore to an acceptor fluorophore, which shows a strong distance dependence in the nanometer range (2–10 nm). We have extensively used smFRET by confocal microscopy to dissect conformational changes in the catalytic cycle of molecular machines that modulate DNA and RNA structure (Andreou & Klostermeier, 2014; Aregger & Klostermeier, 2009; del Toro Duany, Ganguly, & Klostermeier, 2014; del Toro Duany, Jungblut, Schmidt, & Klostermeier, 2008; del Toro Duany & Klostermeier, 2011; del Toro Duany, Rudolph, & Klostermeier, 2011; Gubaev, Hilbert, & Klostermeier, 2009; Gubaev & Klostermeier, 2011, 2012; Harms, Andreou, Gubaev, & Klostermeier, 2014; Hilbert, Kebbel, Gubaev, & Klostermeier, 2011; Karow & Klostermeier, 2009, 2010; Lanz, Farhat, & Klostermeier, 2014; Lanz & Klostermeier, 2011, 2012; Linden, Hartmann, & Klostermeier, 2008; Theissen, Karow, Kohler, Gubaev, & Klostermeier, 2008). Here, we present strategies to investigate the conformational dynamics of multisubunit complexes, using DNA gyrase from *Bacillus subtilis* as an illustrative example. Section 1 provides a brief introduction into smFRET and confocal microscopy, with a focus on practical considerations in sample preparation and data analysis. Section 2 illustrates different strategies in the application of smFRET to investigate conformational changes in the catalytic cycle of DNA gyrase. We conclude by presenting the current understanding of the mechanism of DNA supercoiling by gyrase and provide a glimpse on future work.

1.1 Principle of FRET

Förster resonance energy transfer (FRET) is the transfer of energy from a fluorescent donor in the excited state to an acceptor in the ground state. The transfer of energy is radiationless: it occurs by the direct interaction of the donor and acceptor transition dipoles and does not involve photon emission by the donor and reabsorption by the acceptor. The phenomenon of FRET was first observed and quantitatively described by the German physicist Theodor Förster in the 1940s (Förster, 1948). The dipole–dipole interaction and the efficiency of energy transfer are strongly distance dependent, and FRET was therefore proposed to serve as a read-out for distances. Its suitability as a "spectroscopic ruler" within biomolecules was demonstrated in the 1970s (Stryer, 1978; Stryer & Haugland, 1967).

FRET occurs from the excited state of the donor and offers an alternative pathway for the donor to relax back into the ground state in addition to relaxation by nonradiative or radiative processes. FRET therefore kinetically competes with other relaxation mechanisms (Fig. 1A), and its efficiency E_{FRET} is determined by the ratio between the rate constant for FRET, k_{FRET}, and the sum of the rate constants for all relaxation processes:

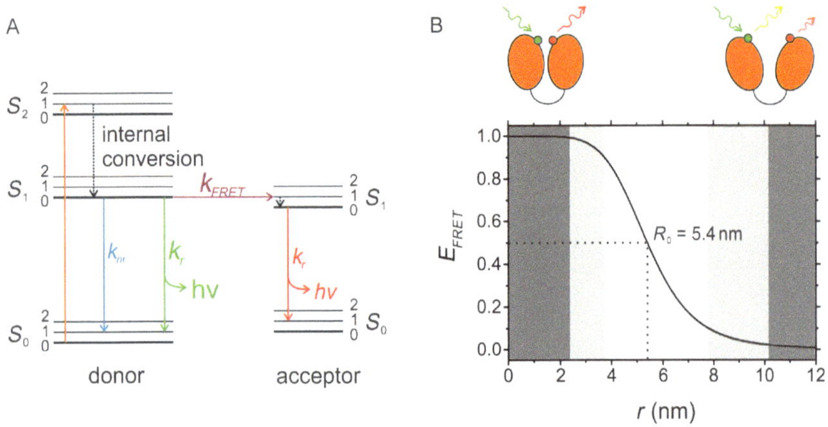

Fig. 1 FRET as a spectroscopic ruler. (A) Jablonski diagram illustrating kinetic competition between relaxation processes. Excitation of the donor fluorophore from the ground state S_0, either directly to S_1 or to higher excited states (*orange*), followed by a return to S_1 by internal conversion, populates the S_1 state. From the S_1 state, the donor can return to the ground state by nonradiative (*blue*) or radiative processes (*green*, donor fluorescence), or by energy transfer to the acceptor (*violet*). For the acceptor, only the radiative relaxation is shown (*red*, acceptor fluorescence). k_{FRET}: rate constant of Förster resonance energy transfer, k_{nr}: rate constant of relaxation by nonradiative processes, k_r: rate constant of relaxation by radiative processes. 0, 1, 2 indicate vibronic states. (B) Distance dependence of the FRET efficiency and sensitive ranges for the use of FRET as a nanometer ruler. Around the Förster distance R_0 for the used donor–acceptor pair, the energy transfer efficiency E_{FRET} changes strongly with the interdye distance. In the region shaded in *light gray*, changes in distance lead to only small changes in E_{FRET}, in the region shaded in *dark gray* E_{FRET} is essentially not distance dependent.

$$E_{FRET} = \frac{k_{FRET}}{k_{FRET} + k_r + k_{nr}} \tag{1}$$

In Eq. (1), k_r is the rate constant for radiative relaxation of donor molecules in the excited state, and k_{nr} is the rate constant for nonradiative relaxation processes. The rate constant k_{FRET} depends on the inverse sixth power of the distance r between the donor and the acceptor fluorophores:

$$k_{FRET} = \frac{1}{\tau_D}\left(\frac{R_0}{r}\right)^6 \tag{2}$$

τ_D is the lifetime of the excited state of the donor, and R_0 is the Förster radius or characteristic transfer distance at which the energy transfer efficiency is 50%. By combining Eqs. (1) and (2) with the definition of the donor lifetime τ_D

$$\tau_D = \frac{1}{k_r + k_{nr}} \tag{3}$$

the distance dependence of the FRET efficiency E_{FRET} is obtained as

$$E_{FRET} = \frac{R_0^6}{r^6 + R_0^6} \tag{4}$$

(Fig. 1B). The Förster radius R_0 depends on the spectral properties of the donor and acceptor pair, particularly on the quantum yield of the donor ϕ_D, the overlap integral J that measures the spectral overlap between donor emission and acceptor absorption spectrum, and the orientation factor κ^2 that is a measure for the relative orientation of the transition dipoles of donor and acceptor:

$${R_0}^6 = \frac{9 \ln 10 \phi_D \kappa^2}{128\pi^5 N_A n^4} \cdot J \tag{5}$$

N_A is Avogadro's number, n is the refractive index of the buffer. The overlap integral J can be calculated from the normalized emission of the donor F_D and the extinction coefficient of the acceptor ε_A in the region of spectral overlap (λ_{min} to λ_{max}) according to

$$J = \int_{\lambda_{min}}^{\lambda_{max}} F_D(\lambda)\varepsilon_A(\lambda)\lambda^4 d\lambda \tag{6}$$

As the spectral properties of fluorophores are strongly dependent on their local chemical environment, R_0 needs to be determined under the conditions of the particular measurement (see Section 1.2.1).

Förster radii of commonly used donor–acceptor pairs for FRET experiments are in the range of 5–6 nm, which renders them suitable to measure distances between approx. 3 and 9 nm (Fig. 1B). Conformational changes in multidomain and multisubunit proteins are associated with distance changes within this range, and are therefore amenable to FRET studies. The accessible distance range can be expanded in both directions. Lower distances can be measured by adding quenchers that reduce R_0 by reducing the quantum yield of the donor. Higher distances become accessible by using a relay of three dyes (Haustein, Jahnz, & Schwille, 2003; Watrob, Pan, & Barkley, 2003), or by combining one donor with multiple acceptor dyes, which leads to increased spectral overlap (far-FRET) (Krainer, Hartmann, & Schlierf, 2015).

The FRET efficiency is calculated from measured fluorescence intensities, corrected for fluorescence background. The fluorescence intensities of donor I_D and acceptor I_A upon excitation of the donor provide a measure for the number of photons that are transferred from the donor to the acceptor (I_A) relative to the amount of photons absorbed by the donor (I_D+I_A), and E_{FRET} can be calculated as

$$E_{FRET} = \frac{I_A}{I_A + I_D} \tag{7}$$

As an alternative to this ratiometric approach, FRET efficiencies can be calculated from donor quenching or from the sensitized emission of the acceptor (see Gubaev & Klostermeier, 2013, for a comparison of these approaches). In FRET experiments of large ensembles of molecules in solution, the measured intensities reflect average intensities over all molecules. As long as the sample is homogeneous and the donor–acceptor distance is identical for all molecules, this averaging does not pose any problem. However, for heterogeneous samples with multiple species that differ in donor–acceptor distances, the FRET efficiencies determined by Eq. (7) will be population-weighted averages over the entire ensemble. Contributions from minor species with low abundance will be hidden in the average value, and the average might not reflect the FRET efficiency of any of the main species present. As a consequence, the conversion of experimental FRET efficiencies into distances may lead to distances that are meaningless on the molecular level. These limitations are overcome in single-molecule

FRET experiments that allow for the calculation of FRET efficiencies for individual molecules, and directly reveal the distribution of FRET efficiencies over all species present in the sample without averaging (see Section 1.2.1).

1.2 Single-Molecule FRET Using Confocal Microscopy

FRET experiments at the single-molecule level can be performed by confocal microscopy either on freely diffusing molecules in solution or on surface-immobilized molecules in wide field, confocal scanning, or total internal reflection fluorescence microscopy. This review focuses on smFRET experiments by confocal microscopy. Total internal reflection microscopy is covered in several other chapters in this volume.

The basic principle of confocal microscopy was realized by Minsky in 1955 (Minsky, 1988) in a microscope with two small apertures, termed pinholes, one in the excitation path to limit the illumination of the sample and a second in the detection path to exclude off-focus signals. Modern confocal microscopes (Fig. 2A) use the same principle to selectively excite molecules in a small volume within the sample and to selectively detect their fluorescence. The excitation light, typically from high-energy lasers, is passed through an excitation filter and directed to the microscope via a dichroic mirror. The laser beam is focused by a high numerical aperture objective, in place of the first pinhole in Minsky's setup. The small volume (approx. 1 fL) that is illuminated within the sample is termed the confocal volume. To ensure that not more than one molecule is in the focus at any given time, picomolar concentrations of donor–acceptor-labeled molecules are used. Molecules that diffuse through the confocal volume are excited and produce bursts of fluorescence with a duration of a few milliseconds. This duration is limited by the diffusion time of the molecule through the confocal volume. The emitted fluorescence is collimated by the same objective (epifluorescence configuration) and passes the dichroic beam splitter. A pinhole selects for photons emitted from the focal plane. Photons from the donor and acceptor fluorophores are separated using a second dichroic mirror. For detection with single-molecule sensitivity, highly sensitive avalanche photodiode detectors are used. Each individual photon that arrives at the detectors is recorded with respect to its arrival time and the channel (donor, acceptor). From the arrival times of individual photons, donor and acceptor fluorescence intensities can be calculated as a function of time by integration over a defined time window, typically 0.5–1 ms.

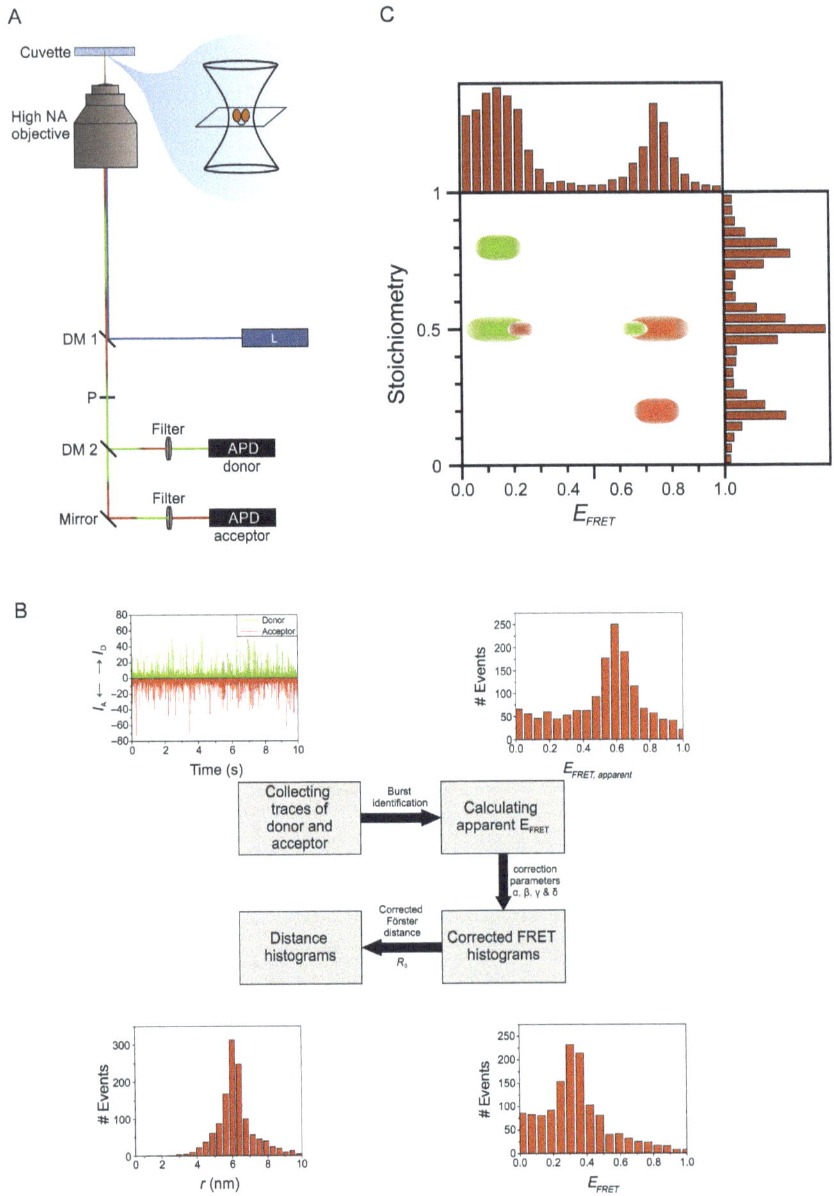

Fig. 2 Single-molecule FRET experiments by confocal microscopy and data analysis workflow. (A) Schematic depiction of a confocal microscope. The light from the excitation laser (*L*) is directed to the microscope via a dichroic mirror (DM1) and focused into a small volume (approx. 1 fL) within the sample by a high numerical aperture (NA) objective. In the confocal volume, the focused light excites all fluorophores present in the illuminated double cone. Fluorescent molecules diffusing through this illuminated

From these time traces, single-molecule events can be identified, and FRET efficiencies can be calculated (see Section 1.2.1). It is also possible to calculate auto- and cross-correlation curves to determine diffusion times. If pulsed lasers are used for excitation, the time increment Δt that has passed since the previous laser pulse is also recorded for each emitted photon. From the delay times between laser pulse and detected photons, fluorescence decays can be constructed to determine fluorescence lifetimes. If linearly polarized light is used for excitation, emitted photons can be selected for the polarization parallel and perpendicular to the polarization of the excitation light to calculate anisotropies and to gain insight into the mobility of the fluorescently labeled species. In the following, we will illustrate the workflow of data analysis to arrive at FRET efficiency histograms and distributions of interdye distances.

1.2.1 From Raw Data to Corrected Single-Molecule FRET Histograms and Interdye Distances

To calculate FRET efficiencies of single molecules from time traces of donor and acceptor fluorescence using Eq. (7), single-molecule events have to be identified in the first step of data analysis. Often, threshold criteria are

---

volume emit bursts of photons with a duration of a few milliseconds, limited by the diffusion time of the molecule. The emitted fluorescence is collimated by the same objective (epifluorescence configuration) and passes the dichroic mirror (DM1). After passing the detection pinhole (P), photons are separated into donor and acceptor spectral regions using a second dichroic mirror (DM2) and further selected by a bandpass filters. Avalanche photodiode (APD) detectors provide single-molecule sensitivity. (B) Workflow of data analysis. Time traces of donor and acceptor fluorescence are collected. Bursts originating from single molecules passing the confocal volume are selected using threshold criteria. From the integrated, background-corrected donor and acceptor fluorescence intensities from individual bursts, histograms of apparent FRET efficiencies, $E_{FRET,apparent}$, can be constructed. After determination of the correction parameters α–δ that correct for cross-talk between donor and acceptor channels, different excitation and detection efficiencies for donor and acceptor and direct excitation of the donor, corrected FRET histograms devoid of instrument effects are obtained. Corrected FRET efficiencies E_{FRET} can be converted into interdye distances *r* for the calculation of distance histograms. (C) Scheme of ALEX/PIE experiments and spectral sorting. By intermittently exciting the acceptor fluorophore, a stoichiometry parameter *S* is obtained that contains information on the dyes attached to the respective molecule. Two-dimensional plots of *S* vs. *E* allow for the distinction of donor-only (*green*) and acceptor-only labeled species (*red*) from species that carry both fluorophores, but have low or high FRET efficiencies (spectral sorting).

applied, and bursts with count rates or total numbers of photons above a threshold value are counted as single-molecule events. Alternatively, interphoton times can be used to identify bursts. Within bursts, times between two subsequently detected photons will be much shorter than outside bursts. Once bursts have been identified, the number of detected photons per time window is integrated over the duration of the burst to determine the donor and acceptor fluorescence intensities for each single-molecule event. These intensities are then corrected for background. The background signal can be quantified from count rate histograms. From the background-corrected donor and acceptor intensities for each single molecule, an apparent FRET efficiency can be calculated according to Eq. (7) (Fig. 2B). Changes in this apparent FRET efficiency are indicative of changes in interdye distances, and thus of conformational changes of the labeled molecule. However, these apparent FRET efficiencies are not suitable to derive absolute interdye distances. To calculate distances from FRET efficiencies, the measured intensities have to be corrected for instrument nonidealities, including cross-talk between donor and acceptor channels (α, β), different detection efficiencies in the spectral ranges of donor and acceptor fluorescence (γ), and direct excitation of the acceptor at the wavelength of donor excitation (δ). All of these effects can be determined in control experiments with molecules that carry either the donor fluorophore or the acceptor fluorophore. In the following equations (Eqs. 8–11), the subscript of the fluorescence intensities I indicates the fluorophore attached (d: donor, a: acceptor), the superscript denotes the detection channel (D: donor, A: acceptor). For the determination of the correction parameter α for cross-talk of donor photons into the acceptor channel, the molecule of interest carrying only a donor fluorophore is used. Fluorescence is excited at the excitation wavelength of the donor, and the fluorescence counts detected in the donor and acceptor channels are measured. Fluorescence signals are corrected for background (I_{buffer}). The correction parameter α is then calculated as

$$\alpha = \frac{I_d^A - I_{buffer}^A}{I_d^D - I_{buffer}^D} \tag{8}$$

By analogy, the correction factor β for cross-talk of acceptor photons into the donor channel is determined from the background-corrected signals in donor and acceptor channels using protein modified only with the acceptor fluorophore:

$$\beta = \frac{I_a^D - I_{buffer}^D}{I_a^A - I_{buffer}^A} \tag{9}$$

The direct excitation of the acceptor δ can be determined from the absorption of protein labeled only with donor (A_d, concentration c_d) or only with acceptor (A_a, concentration c_a) at the excitation wavelength of the donor, corrected for buffer contributions (A_{buffer}):

$$\delta = \frac{A_a - A_{buffer}}{A_d - A_{buffer}} \cdot \frac{c_d}{c_a} \tag{10}$$

Often, the absorption of the acceptor at the donor wavelength is very low. In this case, it may be more accurate to use fluorescence excitation spectra as a more sensitive read-out for absorption.

The correction factor γ for differences in detection efficiencies and quantum yields of donor and acceptor is determined from solutions of donor- and acceptor-only labeled species at equal concentrations. From the background-corrected intensities in donor and acceptor channels, γ is calculated as

$$\gamma = \frac{I_a^A - I_{buffer}^A}{\delta\left(I_d^D - I_{buffer}^D\right)} \tag{11}$$

For molecules that are labeled with donor and acceptor statistically by modifying the same reactive group with a mixture of reactive donor and acceptor derivatives, all possible combinations of differently labeled species are present in solution (D/−, D/D, D/A, A/D, A/A, A/−). The spectral properties of the dyes depend on the local environment, and both donor and acceptor may have different spectral properties depending on whether they are attached to one position or the other. Hence, correction parameters should be determined for both configurations (D/A and A/D) by performing the control experiments with D/− and −/A as well as −/D, and A/−. The spectral properties should ideally not depend on the attachment geometry, indicated by two similar sets of correction parameters. Site-specific dye attachment using two different functional groups has the advantage that one defined species (D/A) is generated, and correction parameters can be determined in this particular configuration (D/−, −/A) and from the construct used for donor–acceptor labeling.

With the correction parameters α–δ, corrected FRET efficiencies independent of the features of the particular microscope are obtained from the measured intensities:

$$E_{FRET} = \frac{(1+\beta\gamma\delta)\left(I_A - \frac{\alpha+\gamma\delta}{1+\beta\gamma\delta}I_D\right)}{(1+\beta\gamma\delta)\left(I_A - \frac{\alpha+\gamma\delta}{1+\beta\gamma\delta}I_D\right) + (\gamma+\gamma\delta)(I_D - \beta I_A)} \quad (12)$$

From the FRET efficiencies of all single molecules observed one after the other, FRET histograms can be constructed that directly reflect the distribution of FRET efficiencies and allow to identify different conformers and their relative populations (Fig. 2B).

For a conversion of the corrected FRET efficiencies into distances, the Förster radius R_0 needs to be known. Although R_0 values are tabulated for common donor–acceptor pairs, they also depend strongly on the spectral properties of donor and acceptor, and have to be determined under the conditions of each particular FRET experiment. To calculate R_0 (Eq. 5), the quantum yield of the donor ϕ_D and the overlap integral J need to be determined experimentally. The quantum yield of the donor ϕ_D, attached to the same position of the molecule of interest studied by FRET, can be measured relatively to a known quantum yield of a reference fluorophore or by absolute methods (Wurth, Grabolle, Pauli, Spieles, & Resch-Genger, 2013). The overlap integral $J(\lambda)$, which describes the spectral overlap between donor and acceptor (Eq. 6), can be calculated from the emission spectrum of the donor $F_D(\lambda)$, normalized to an area of one, and the wavelength dependence of the extinction coefficient of the acceptor $\varepsilon_A(\lambda)$. Instead of integrating the spectra, the overlap integral is typically approximated from spectra recorded with a step size of 1 nm by the sum in Eq. (13):

$$J = \sum_{\lambda_{min}}^{\lambda_{max}} F_D(\lambda)\varepsilon_A(\lambda)\lambda^4 \quad (13)$$

The orientation factor κ^2 can vary between zero and four, depending on the relative orientation of the dipoles of donor and acceptor. Provided that donor and acceptor dyes are flexibly attached and sample all possible relative orientations on the timescale of the fluorescence lifetime, κ^2 can be approximated to 2/3. Fluorescence anisotropy decays should be measured to confirm that the dyes indeed rotate freely (see Gubaev & Klostermeier, 2013, for more details). The refractive index n of the buffer is typically set to $n = 1.33$, the refractive index of water.

Once the Förster distance R_0 for the respective donor–acceptor-labeled molecule has been determined, FRET efficiencies can be converted into true interdye distances. A set of FRET-derived distances provides structural information and can be used to map global conformations or to generate structural models of multidomain proteins and large complexes (Karow & Klostermeier, 2010; Lanz & Klostermeier, 2011). Procedures to account for length and flexibility of linkers have been described elsewhere (Andrecka et al., 2009; Kalinin et al., 2012; Muschielok et al., 2008; Sindbert et al., 2011).

The raw data from FRET experiments obtained by confocal microcopy provide the necessary information to calculate fluorescence auto- or cross-correlation curves from the fluctuations of donor and acceptor fluorescence. The autocorrelation function $AC(\tau)$ is defined as

$$AC(\tau) = \frac{\langle F(t) \cdot F(t+\tau) \rangle}{\langle F(t) \rangle^2} \tag{14}$$

where $F(t)$ is the fluorescence intensity at a certain time and $F(t+\tau)$ is the fluorescence intensity at a time $t+\tau$, with τ in the range from microseconds to seconds. The concentration of fluorescently labeled molecules in the confocal volume is reflected in the amplitude of the autocorrelation curve, the inflection point of the obtained curve is the diffusion time of the molecule through the confocal volume. Changes in diffusion time can be used to confirm binding under single-molecule conditions.

The cross-correlation curve $CC(\tau)$ is obtained from a correlation of fluctuations F in donor and acceptor channels (F_D and F_A)

$$CC(\tau) = \frac{\langle F_D(t) \cdot F_A(t+\tau) \rangle}{\langle F_D(t) \rangle \langle F_A(t) \rangle} \tag{15}$$

The amplitude of the cross-correlation curve is a measure for molecules in which donor and acceptor fluorophores diffuse together and give rise to correlated signals in donor and acceptor intensity. The cross-correlation amplitude, compared to the amplitude of the autocorrelation curves, can therefore be used as a diagnostic tool to confirm that donor–acceptor-labeled species are indeed present in the sample.

If a pulsed laser is used for excitation and the time between a laser pulse and the next emitted photon is recorded during the experiment, fluorescence decays can be constructed and donor lifetimes can be calculated. Comparing FRET efficiencies calculated from lifetimes and from intensities allows to distinguish true changes in FRET efficiency from uncorrelated changes in donor or acceptor fluorescence not related to FRET and distance changes (Eggeling et al., 2001; Kühnemuth & Seidel, 2001).

1.2.2 Pulsed Interleaved Excitation as a Modification of Conventional Confocal Microscopy

Most labeling reactions lead to a significant fraction of molecules that carry only a donor fluorophore, and give rise to a peak at $E_{FRET}=0$ (or $E_{FRET}=-\delta$ after correction). Often, low FRET species are hidden under this zero-peak and may be difficult to identify. Pulsed interleaved excitation (PIE) and alternating laser excitation (ALEX) allow to probe for the presence of acceptor by direct excitation of the acceptor by a second laser. ALEX uses two continuous wave lasers and electrooptical modulators to create interleaved pulses on the microsecond to nanosecond time scale (Kapanidis et al., 2005; Laurence, Kong, Jager, & Weiss, 2005). In PIE, pulsed lasers are used for excitation alternating on the nanosecond time scale (Muller, Zaychikov, Brauchle, & Lamb, 2005). From the total signal in the donor and acceptor channels ($D+A$, superscript) after donor (subscript D_{ex}) or acceptor excitation (subscript A_{ex}), the stoichiometry S can be calculated as

$$S=\frac{F_{Dex}^{D+A}}{F_{Dex}^{D+A}+F_{Aex}^{D+A}} \tag{16}$$

Donor–acceptor-labeled molecules give a stoichiometry of $S=0.5$, for donor-only molecules S equals unity, and for acceptor-only molecules it is zero. In two-dimensional plots of the stoichiometry S vs. the FRET efficiency E_{FRET}, species with different stoichiometries but identical FRET efficiencies can be distinguished (Fig. 2C). The additional stoichiometry information allows to distinguish a lack of FRET due to missing acceptors and incomplete labeling or due to a distance beyond the range accessible by FRET. It also helps distinguish between donor-only molecules and molecules that carry donor and acceptor, but exhibit a low FRET efficiency. Furthermore, data analysis can be confined to molecules that carry both donor and acceptor.

1.2.3 Practical Considerations During Preparation of Multisubunit Complexes for Single-Molecule FRET Experiments

Single-molecule FRET studies of molecular machines that act on DNA can be performed using either labeled DNA or labeled protein. Donor and acceptor fluorophores can be incorporated into chemically synthesized

DNA, and larger DNA molecules can be assembled by enzymatic ligation. Proteins, especially multidomain and multisubunit proteins, pose more challenges during labeling.

The first step toward a donor–acceptor-labeled protein is the identification of suitable attachment sites for fluorophores. Apart from geometric considerations to ensure that the fluorophores report on the process to be studied, reactive groups for fluorophore attachment should be surface exposed to achieve high yields of labeled products. To prevent interference with protein function, these sites should be outside conserved motifs. Structural information can be used to identify suitable sites; in the absence of structural information, homology models may provide orientation (see Sections 2.2–2.6). Once suitable sites for fluorophore attachment have been identified, an adequate strategy for their attachment has to be chosen (Fig. 3). Although other labeling chemistries have been used, the introduction of cysteines at the site of interest and the selective modification of their thiol groups by fluorophores functionalized with maleimide groups is the most common approach. If the protein of interest contains intrinsic cysteines that are solvent accessible and react with maleimides, a cysteine-free variant of the protein needs to be generated first, following standard site-directed mutagenesis procedures. The most common replacement for cysteine is the isosteric amino acid serine, but alanine or the hydrophobic amino acids leucine, isoleucine, or valine may also be suitable. Once wild-type-like activity has been confirmed for the cysteine-free variants, these constructs serve as a starting point for the introduction of cysteines for fluorescent labeling at the desired sites by site-directed mutagenesis, and donor and acceptor maleimides are coupled statistically to these side chains. Such a reaction gives all possible combinations of D/D, D/−, D/A, A/D, A/A, A/− and unlabeled species (Fig. 3A). In this case, correction parameters and Förster distances need to be determined for both possible configurations (D/A and A/D; see Section 1.2.1) to ensure similar photophysical properties. The incorporation of unnatural amino acids by suppression of amber stop codons (reviewed in Wang, Xie, & Schultz, 2006) has expanded the spectrum of reactive groups available for fluorescent labeling, including ketones (Brustad, Lemke, Schultz, & Deniz, 2008), alkynes (Tyagi & Lemke, 2013), azides (Agard, Baskin, Prescher, Lo, & Bertozzi, 2006), or allyls (Lee et al., 2014). Two unnatural amino acids introduced into the protein of interest can be selectively coupled to donor and acceptor dyes in the presence of intrinsic cysteines, circumventing the time-consuming removal of larger numbers of cysteines, and providing an alternative labeling strategy

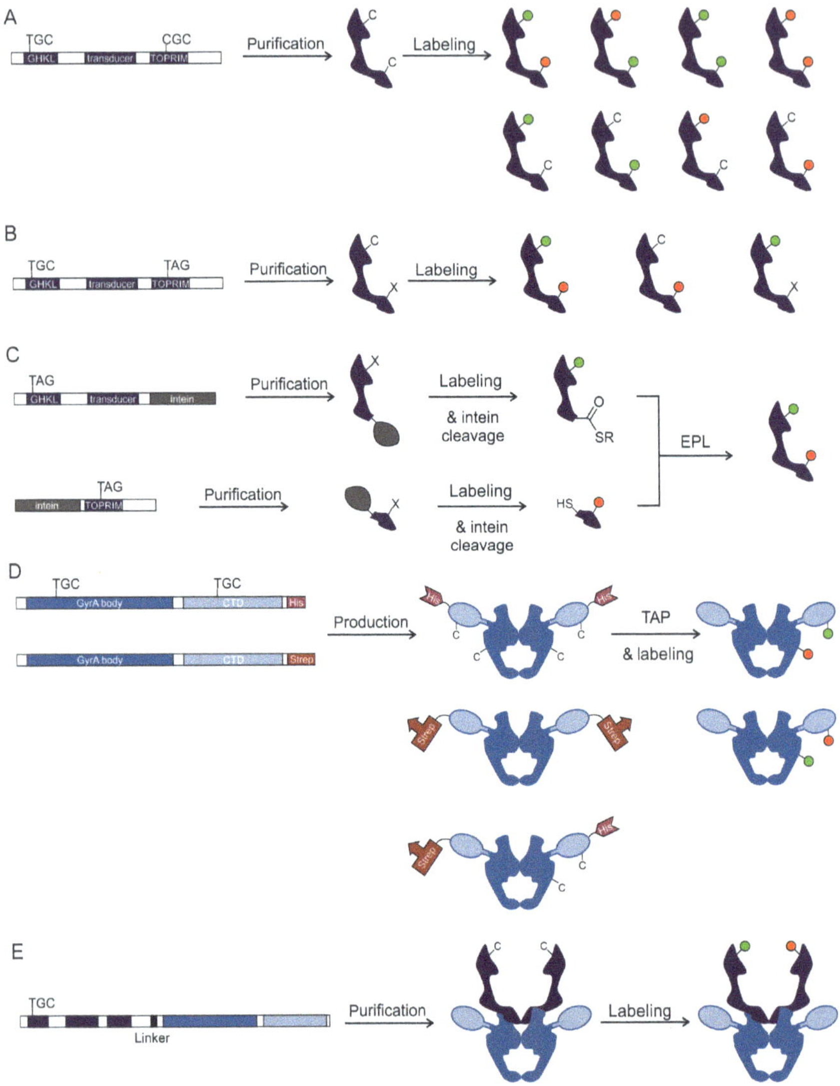

Fig. 3 Donor-acceptor labeling schemes. (A) Introduction of cysteine codons (TGC, CGC) by site-directed mutagenesis, and labeling of two cysteines (C) with maleimide derivatives of donor (*green circle*) and acceptor dyes (*red circle*) yields a mixture of labeled species (here: GyrB). (B) Labeling with different chemistries, for example, by coupling the donor to a cysteine (C) and the acceptor to an unnatural amino acid (X) incorporated by amber stop codon (TAG) suppression, gives a single donor–acceptor-labeled species (here: GyrB), if the labeling reaction is complete. For incomplete labeling reactions, one donor-only- and one acceptor-only-labeled species is generated. (C) Defined attachment of donor and acceptor fluorophores can also be achieved by separate production of two domains of the protein of interest (here: GyrB)

if intrinsic cysteines are required for function (Lee et al., 2016). More importantly, incorporation of an unnatural amino acid in combination with a cysteine or incorporation of two different unnatural amino acids with different functional groups for fluorophore attachment offers the possibility to combine two different chemistries to achieve site-specific labeling and generation of one defined D/A species (Tyagi & Lemke, 2013) (Fig. 3B). For labeling strategies based on the introduction of unique functional groups by incorporation of unnatural amino acids, amber stop codons for suppression are introduced at the desired position by site-directed mutagenesis. Although the functional groups of unnatural amino acids can be modified using click-chemistry reactions that occur rapidly and produce high yields, the reaction conditions (low pH, reactions at high concentrations of denaturant and subsequent refolding, Cu-catalyzed reactions) often have adverse effects on protein integrity.

A different approach to achieve specific labeling with the same chemistry for donor and acceptor coupling is the use of intein fusion proteins that are labeled with donor and acceptor fluorophore separately (Fig. 3C). The full-length protein is then generated by intein-mediated expressed protein ligation (reviewed in Blaschke, Silberstein, & Muir, 2000; Karow, Theissen, & Klostermeier, 2007). Fusions of each part with a split intein that forms a functional intein upon mixing that catalyzes its own excision with concomitant ligation of the two parts of the protein of interest can also be used (Bachmann, Matern, Schutz, & Mootz, 2015; Sun, Yang, & Liu, 2004; Volkmann & Liu, 2009; Wu, Hu, & Liu, 1998). In oligomeric protein complexes, it might be of interest to selectively monitor a conformational change within one of the subunits. In this case, this subunit can be generated and labeled separately, and the complex can be reconstituted from the individual

as intein fusions and separate labeling, followed by cleavage of the intein fusion proteins and expressed protein ligation (EPL) to reconstitute the full-length protein. (D) Generation of dimers (here: GyrA) labeled only in one subunit by coproduction of a cysteine-free subunit and a subunit with two cysteines or other groups for fluorophore attachment, both fused to different affinity tags (His, His-tag; Strep, Strep-tag), and tandem-affinity purification (TAP) of the asymmetric dimer, followed by fluorescent labeling. If two different functional groups are combined, a defined labeled species is obtained. (E) Generation of a fusion protein (here, fusion of GyrB and GyrA to GyrBA) to prevent subunit dissociation at single-molecule concentrations. The fusion protein contains two cysteines (C) that can be statistically labeled with donor- and acceptor-fluorophore maleimides (also generating the species D/D, D/−, A/−, A/A, −/−).

subunits (Fig. 3D). For dimeric proteins, asymmetric labeling of only one subunit can be achieved by coproduction of a wild-type copy and a variant containing functional groups for fluorophore attachment, both carrying different affinity tags for purification (Gubaev, Weidlich, & Klostermeier, 2016; Lanz & Klostermeier, 2011). The desired heterodimer is then isolated by tandem-affinity purification and labeled (see Section 2.4). More complex labeling schemes can in principle be realized in multistep procedures that combine these approaches.

Single-molecule studies are performed at picomolar concentrations of the labeled species to ensure that only one molecule resides within the focal volume at any given time. Multisubunit proteins must be stable under these conditions to be amenable to single-molecule FRET. To prevent dissociation of the heterotetrameric gyrase during the experiment, fusion proteins that contain two subunits on one polypeptide chain have been instrumental (Gubaev & Klostermeier, 2011) (Fig. 3E; see Section 2.5).

Independent of the type of reaction used for labeling, reaction conditions should be optimized to achieve high labeling efficiencies. For statistical labeling with mixtures of donor and acceptor, the formation of high fractions of molecules carrying only the donor fluorophores needs to be avoided to minimize the donor-only peak in FRET histograms. Labeling efficiencies can be determined photometrically. As the spectral properties of fluorophores may change upon coupling to the protein of interest, it is advisable to cleave the protein with proteinase K before the absorption measurement. The labeled protein has to be purified by size-exclusion, ion exchange, or affinity chromatography to remove any fluorescent contaminations, particularly by free dye. Irrespective of the labeling strategy, it has to be confirmed that the fluorophores do not interfere with protein activity.

To minimize fluorescent impurities leading to high fluorescent background in single-molecule experiments, all buffers are treated with active charcoal. Loss of protein during the measurement because of surface adsorption can be limited by passivation of cuvettes with bovine serum albumin or an inactive, unlabeled variant of the protein of interest.

1.2.4 Monitoring Substrate-Induced Conformational Changes in smFRET Experiments

Conformational changes in molecular machines occur upon binding of ligands. DNA gyrase interacts with DNA and nucleotide, and both ligands affect the conformation of the protein (reviewed in Gubaev & Klostermeier, 2014). Analyzing the effect ligand binding on protein conformation under single-molecule conditions requires high ligand concentrations to shift the equilibrium toward complex formation. At a typical concentration of the donor–acceptor-labeled protein of 100 p*M*, a DNA concentration of 500 n*M* is required to achieve >90% saturation with K_d values in the 50 n*M* range. Nucleotide affinities are often even lower, with K_d values in the high micromolar range, and hence millimolar concentrations of nucleotide are required. When substrates are added at high concentrations, it is particularly important that they be free of fluorescent impurities.

In the presence of the cognate substrate that is turned over, the protein of interest samples all conformational states associated with substrate turnover during the single-molecule experiment. FRET histograms therefore represent the steady-state distribution of these conformers. To populate individual reaction intermediates, and to unambiguously assign FRET efficiencies to certain functional states, substrate analogs that mimic natural substrates but are not turned over can be employed. To investigate conformational changes in the nucleotide cycle of DNA gyrase, we have used the ATP analogs ADPNP (adenosine 5′-(β,γ-imido)triphosphate), $ADP{\cdot}BeF_x$ and $ADP{\cdot}MgF_x$ to populate different intermediates of the nucleotide cycle and to detect their FRET efficiency distributions in histograms (Gubaev & Klostermeier, 2011, 2012) (see Section 2.5). To study the series of effects of DNA substrate on gyrase conformation step by step, we used DNA substrates of different lengths (see Sections 2.4 and 2.5). As an alternative to using substrate analogs and model substrates, turnover can be prevented by using protein variants that are deficient in catalyzing individual steps, exemplified by ATPase-deficient or DNA cleavage-deficient gyrase variants, which leads to the accumulation of intermediates (see Sections 2.4 and 2.5).

2. DISSECTING CONFORMATIONAL CHANGES IN THE CATALYTIC CYCLE OF GYRASE BY SINGLE-MOLECULE FRET

Gyrase belongs to the family of type IIA topoisomerases that cleave both strands of the DNA and catalyzes the introduction of negative supercoils into DNA at the expense of ATP (Gellert et al., 1976). The two subunits GyrA and GyrB assemble into the active heterotetrameric enzyme $GyrA_2GyrB_2$ (Higgins, Peebles, Sugino, & Cozzarelli, 1978) (Fig. 4). The GyrB subunit is a monomer in solution and consists of an N-terminal ATPase domain of the GHKL family (Dutta & Inouye, 2000) followed by a transducer domain and the C-terminally located topoisomerase-primase

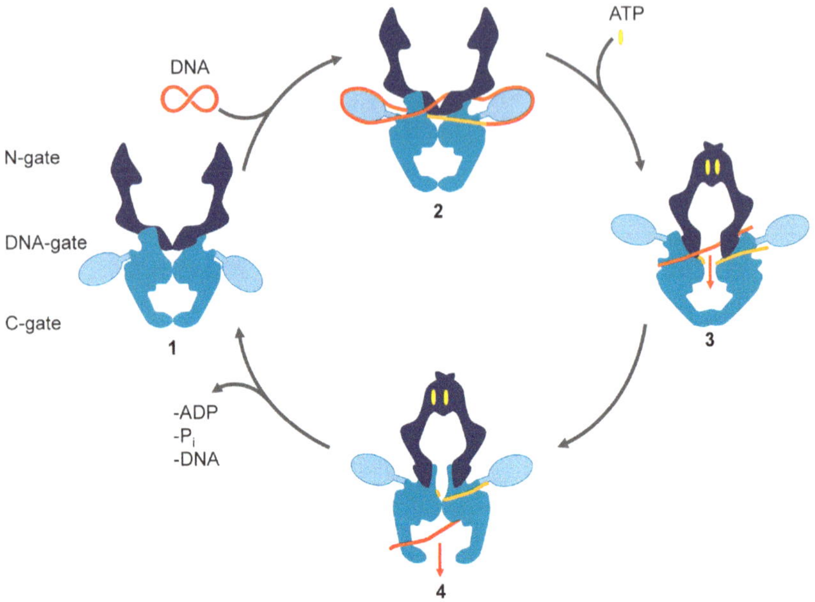

Fig. 4 Strand passage model for DNA supercoiling by gyrase "before FRET." DNA gyrase consists of two GyrA (*light blue*) and two GyrB (*dark blue*) subunits that form the active A_2B_2 heterotetramer (1) with three protein/protein interfaces, the N-gate, DNA-gate, and C-gate that delimit an upper and a lower cavity. The C-terminal domains of GyrA (*light blue ovals*) are involved in DNA wrapping around the enzyme. DNA supercoiling by strand passage occurs by binding of a double-stranded G-segment (*orange line*) at the DNA-gate (2), capture of a second double-stranded DNA, the T-segment (*red line*) in the upper cavity by nucleotide-induced closure of the N-gate (3), G-segment cleavage, opening of the DNA-gate, and passage of the T-segment through the gap in the G-segment (4), closure of the DNA-gate, religation of the G-segment, and exit of the T-segment through an open C-gate. ATP hydrolysis reopens the N-gate for subsequent catalytic cycles. Each catalytic cycle converts a positive DNA crossing into a negative crossing and changes the linking number in steps of two.

(TOPRIM) domain (Aravind, Leipe, & Koonin, 1998; Wigley, Davies, Dodson, Maxwell, & Dodson, 1991). GyrA is a stable dimer and comprises an N-terminal domain with a winged helix domain (WHD) harboring the catalytic tyrosine for DNA cleavage, a tower domain and a coiled-coil domain, as well as the C-terminal domain (CTD) (Corbett, Shultzaberger, & Berger, 2004; Horowitz & Wang, 1987; Morais Cabral et al., 1997). The overall structure of gyrase shows a twofold symmetry. GyrB and GyrA form three protein–protein interfaces, termed N-gate, DNA-gate, and C-gate, which delimit two cavities (Fig. 4). The N-terminal domains of GyrB form the N-gate by dimerization upon nucleotide binding (Williams, Howells, & Maxwell, 2001), the DNA-gate is built by the TOPRIM domain of GyrB and the WHD from GyrA (Morais Cabral et al., 1997; Schoeffler, May, & Berger, 2010), and a small, globular domain at the end of a coiled-coil structure of GyrA forms the third gate, the C-gate (Berger, Gamblin, Harrison, & Wang, 1996; Morais Cabral et al., 1997).

Gyrase is believed to catalyze DNA supercoiling by a strand-passage mechanism (Kampranis et al., 1999). In the postulated supercoiling cycle (Fig. 4), a DNA (G-segment) is bound at the DNA-gate of gyrase and wrapped around the CTDs. A second DNA, the T-segment, is positioned on top of the G-segment in the first cavity between N-gate and DNA-gate. The G-segment is cleaved by the catalytic tyrosines, and the T-segment is transported through the gap in the G-segment into the second cavity between DNA- and C-gate. G-segment ligation and release of the T-segment complete the catalytic cycle. The transfer of the T-segment through G-segment and DNA-gate is thought to be mediated by opening and closing of gates in a temporally coordinated manner.

2.1 A Toolbox to Identify Conformational Changes in the Catalytic Cycle of Gyrase

We have established a toolbox to analyze conformational changes in the catalytic cycle of DNA gyrase in smFRET experiments by confocal microscopy, and to isolate individual steps. DNAs of different length can be used as model substrate to differentiate between G-segment binding at the DNA-gate and by the CTDs, and T-segment capture. By promoting the population of cleavage complexes using phosphorothioate-modified DNA (Deweese, Burgin, & Osheroff, 2008; Gubaev et al., 2009) or by suppressing their population using cleavage-deficient gyrase (Y → F substitution) (Gubaev et al., 2009), events that are coupled to DNA cleavage can be analyzed. With ATPase-deficient variants as well as nucleotide analogs (Gubaev & Klostermeier, 2011), defined nucleotide states can be populated. Labeling of the DNA with donor–acceptor fluorophores allows to address conformational changes of the DNA (Gubaev et al., 2009), while labeling of gyrase at the N-, DNA-, or C-gate can be used to monitor conformational changes of the enzyme (Gubaev et al., 2009; Gubaev & Klostermeier, 2011, 2012; Lanz et al., 2014; Lanz & Klostermeier, 2011, 2012) (Fig. 5). Preparation of heterodimers by coproducing different subunits and purification by tandem-affinity purification enables the asymmetric labeling of gyrase within one subunit (Gubaev et al., 2016). Fusion

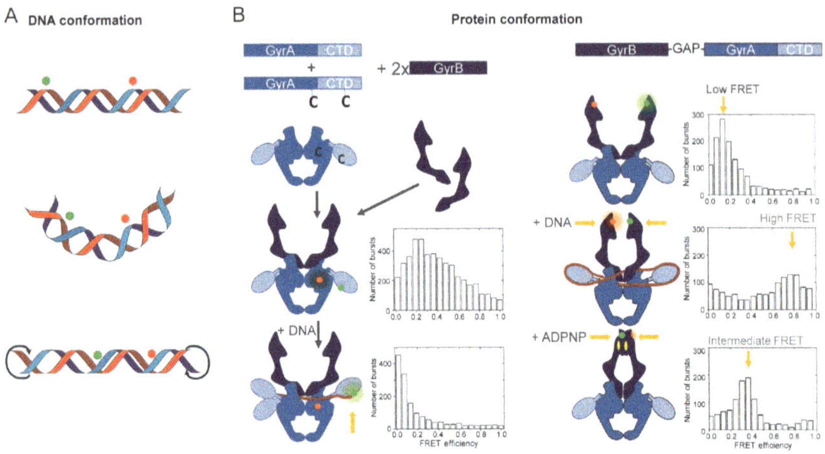

Fig. 5 Strategies to monitor conformational changes of DNA and gyrase during the supercoiling cycle. (A) A donor–acceptor-labeled DNA (*top*) to monitor conformational changes of the G-segment. Changes in FRET between donor (*green circle*) and acceptor dyes (*red circle*) report on bending (*middle*) or twisting (*bottom*) of the DNA. (B) By introducing donor and acceptor fluorophores into gyrase, protein conformational changes can be followed. (*Left*) Generation of an asymmetric gyrase with cysteines for fluorescent labeling in only one of the two GyrA subunits to monitor movement of the CTD upon DNA binding. (*Right*) Generation of dimeric gyrase from a GyrBA fusion protein to ensure formation of complete gyrase complexes and to prevent subunit dissociation in smFRET experiments on conformational changes of the N-gate. FRET histograms for gyrase only (low FRET, open N-gate), for gyrase in complex with DNA (high FRET, narrowed N-gate), and for gyrase in complex with the nonhydrolyzable ATP analog ADPNP (intermediate FRET, closed N-gate) are shown. *Orange arrows* highlight the conformational changes.

of subunits excludes dissociation of the tetrameric enzyme during single-molecule experiments (Gubaev & Klostermeier, 2011, 2012). In the following, we will illustrate how these tools can be used in combination to investigate specific events in the catalytic cycle of gyrase by single-molecule FRET using confocal microscopy.

2.2 The Conformation of DNA Bound to Gyrase: Donor–Acceptor-Labeled DNA

To investigate the mechanism of supercoiling by gyrase via smFRET either the DNA or the protein can be labeled. We have used a 60 bp DNA with a preferred cleavage site for gyrase in the center that is flanked by a donor (Alexa 488) and an acceptor fluorophore (Alexa 546 or Alexa 555) as a model for a G-segment (Fig. 5A) (Gubaev et al., 2009). In smFRET experiments, this DNA shows FRET efficiencies (corrected for instrument nonidealities) of 0.53 ± 0.01 (Alexa 488/Alexa 546) and 0.42 ± 0.01 (Alexa 488/Alexa 555), respectively, corresponding to interdye distances of 6.7 and 7.0 nm that are in agreement with B-form DNA (Gubaev et al., 2009). When this DNA was bound to gyrase, a broader FRET histogram with two FRET distributions was observed, with ~15% in a low FRET state ($E_{FRET} = 0.14 \pm 0.01$) and ~85% in a higher FRET state ($E_{FRET} = 0.30 \pm 0.01$). This observation points to two different conformations of DNA bound to gyrase (Gubaev et al., 2009). The appearance of the lower FRET state could be linked to cleavage in two ways: (1) by observation of a shift in the distribution to the lower FRET state in the presence of phosphorothioate-modified DNA that is less efficiently religated and leads to a higher fraction of cleavage complexes, and (2) by the disappearance of the lower FRET state when a cleavage-deficient Y123F variant of gyrase was used.

2.3 DNA Conformation vs. Gyrase Conformation: Donor and Acceptor Dyes at the DNA-Gate Report on DNA-Gate Conformation

The lower and higher FRET species of DNA bound to gyrase might also report on the conformation of the DNA-gate. To directly monitor the conformation of the DNA-gate, donor and acceptor fluorophores were introduced at the DNA-gate by introducing one cysteine in GyrA, which provides two cysteines in the GyrA dimer for fluorophore attachment on each side of the gate by statistical labeling (Fig. 5B) (Gubaev et al., 2009). The ratio of donor and acceptor dyes in the labeling reaction was optimized to minimize donor-only species and maximize the fraction of gyrase molecules that carry donor and acceptor fluorophores. Gyrase was incubated with plasmid and ATP, and single-molecule FRET experiments during the DNA supercoiling reaction were performed. The FRET histogram, representing all possible gyrase conformers sampled throughout the catalytic cycle, showed a single peak at FRET efficiencies close to one, consistent with a closed DNA-gate (Gubaev et al., 2009). In comparison with the FRET distributions of bound DNA, it becomes evident that the two FRET states of the DNA represent two different conformations of DNA bound to gyrase, whereas the DNA-gate is predominantly closed during supercoiling. Short excursions into a conformation with an open DNA-gate that are not significantly populated might be missed in FRET histograms. However, also within single bursts, no excursions to a low FRET state were observed (at a time resolution on the order of 0.5–1 ms) (Gubaev et al., 2009).

2.4 Orientation and Movement of the GyrA CTDs: Labeling Only One GyrA Subunit

The GyrA subunit harbors a CTD that is essential for negative DNA supercoiling by gyrase. The CTDs are a distinctive feature of gyrase, and are built of six blades that form a β-pinwheel. Blade one contains the conserved seven amino acid sequence motif Q(R/K)RGG(K/R)G, the GyrA-box that is connected to blade number six (Corbett et al., 2004; Ward & Newton, 1997). Deleting the CTDs from gyrase abolishes negative DNA supercoiling, but gyrase lacking the CTDs is now able to relax DNA at the expense of ATP hydrolysis (Gubaev et al., 2009; Kampranis & Maxwell, 1996), a reaction catalyzed by eukaryotic topo II. DNA binding and bending by the isolated CTDs has implicated the CTDs in wrapping of DNA around gyrase (Corbett et al., 2004; Reece & Maxwell, 1991). In the absence of a crystal structure of full-length GyrA, the localization of the CTDs relative to the body of the GyrA dimer during the supercoiling cycle remained unclear, and SAXS data were controversially interpreted (Baker, Weigand, Maar-Mathias, & Mondragon, 2010; Costenaro, Grossmann, Ebel, & Maxwell, 2005). Placing the CTDs relative to the remainder of GyrA is in principle possible by introducing a donor on the CTD and an acceptor fluorophore on the GyrA body at different positions, measurement of the pairwise FRET efficiencies, and conversion of FRET efficiencies into interdye distances that are used as restraints in the generation of structural models. GyrA is a stable dimer at single-molecule conditions, and introduction of one cysteine each in the GyrA body and the CTD leads to GyrA dimers and gyrase with four cysteines, two in each subunit. Statistical labeling would then lead to a heterogeneous population of species, and a set of interdye distances that are difficult, if not impossible to assign. We therefore generated GyrA heterodimers consisting of one cysteine-free subunit and one subunit with a cysteine on the CTD and a second cysteine on the GyrA body. Four solvent-exposed and nonconserved residues, K570 and K594 in blade two, D695 in blade four, and E726 in blade five, were chosen for the introduction of cysteines, based on homology models using the CTD crystal structure of the *Escherichia coli* gyrase CTD as a template (Ruthenburg, Graybosch, Huetsch, & Verdine, 2005), and on sequence comparisons among CTDs. By the same rationale, T140, E211, and E250 in the GyrA body were chosen for fluorescent labeling. The two GyrA subunits were coproduced, the heterodimer was purified by affinity chromatography (Lanz & Klostermeier, 2011, 2012) and statistically labeled with a mixture of donor (Alexa Fluor 488) and acceptor (Alexa Fluor 546) dyes. To minimize the fraction of donor-only labeled proteins, a ratio of 1:3:4 of protein: donor:acceptor was used. All variants were tested for supercoiling activity (Lanz & Klostermeier, 2011).

First, the position of the CTDs with respect to the GyrA body was investigated for the isolated GyrA dimer by measuring the pairwise FRET efficiencies and inter-converting them into interdye distances. Well-defined histograms with unimodal FRET distributions with FRET efficiencies from intermediate (0.3–0.5) to high (0.6–0.8) were obtained. The derived distances were used to place the structure of the isolated CTDs (Ruthenburg et al., 2005) relative to the structure of the GyrA dimer lacking the CTDs (Morais Cabral et al., 1997) in a configuration that was in agreement with the experimentally derived distances. The distances were consistent with a symmetric position of the CTDs, facing downward close to the GyrA body (Lanz & Klostermeier, 2011), in agreement with a prior SAXS study on *E. coli* GyrA (Costenaro et al., 2005). In the meantime, highly refined procedures for structural modeling using FRET-derived distance information have been reported (Kalinin et al., 2012; Muschielok et al., 2008).

Once the CTD had been located in the GyrA dimer, the same set of experiments was performed with gyrase by adding GyrB subunits to the labeled GyrA dimer, and with the gyrase–DNA complex, formed by incubation of the labeled GyrA dimer with GyrB and plasmid DNA. In the context of the gyrase heterotetramer, decreased FRET efficiencies pointed to rearrangements during gyrase assembly that result in increased distances between CTDs and the GyrA body (Lanz & Klostermeier, 2011). Binding of DNA to gyrase in single-molecule experiments was evidenced by an increased diffusion time in autocorrelation curves. FRET efficiencies in the presence of DNA were significantly reduced, in many cases to almost zero. From the FRET-derived distances, a model for the CTD position in the gyrase/DNA complex was obtained in which the CTDs have moved upwards and further away from the GyrA body (Lanz & Klostermeier, 2011). A later cryoelectron microscopy structure of *Thermus thermophilus* gyrase bound to DNA confirmed this interpretation (Papillon et al., 2013).

The repositioning of the CTDs was further dissected using linear DNAs of different lengths (37–110 bp). The shortest DNA serves as a G-segment only, whereas longer DNAs bind at the DNA-gate and contain flanking regions that emanate from the DNA-gate and can establish increasing contacts with the CTDs, up to complete wrapping. Using gyrase with one GyrA T140C K594C subunit, the effect of DNA binding on the position of the CTDs was tested using the decrease in FRET as a signal. A clear shift of the FRET efficiency to the value characteristic of the upward position was observed from a minimal length of the DNA of 48 bp (Lanz & Klostermeier, 2011). According to affinity data, this is the shortest DNA that shows higher affinity binding to gyrase than to gyrase lacking the CTDs, which means it is the shortest DNA able to establish contacts with the CTDs (Lanz & Klostermeier, 2011). Thus, smFRET experiments in combination with affinity data revealed that the CTDs undergo a DNA-induced repositioning and identified first contacts of DNA emanating from the DNA-gate with the CTDs as the trigger for this movement. Using heterodimeric GyrA labeled on one subunit and lacking the GyrA-box in both subunits (GyrA_ΔAbox/GyrA_ΔAbox T140C K594C) to reconstitute gyrase, we showed that the DNA-induced upward movement of the CTDs still occurs, although this enzyme is not capable of DNA supercoiling (Lanz & Klostermeier, 2012).

2.5 N-Gate Conformational Changes: A GyrBA Fusion Protein to Prevent Subunit Dissociation

Once a G-segment is bound to gyrase, capture of a T-segment by ATP-induced N-gate closure was believed to be the next step in the supercoiling cycle. The previously described smFRET experiments using labeled GyrA subunits relied on the *in situ* assembly of active gyrase heterotetramer from purified subunits. In all of these experiments, the GyrA dimer was stable at picomolar concentrations, and the equilibrium was driven toward gyrase formation by adding high concentrations of GyrB (8 μ*M*). Further increase of the GyrB concentration did not lead to changes in FRET efficiencies (Gubaev et al., 2009). For investigations of the conformation of the N-gate, the GyrB subunits have to be labeled. Sites for labeling were selected based on the same rationale as explained above for GyrA and the CTDs (see Section 2.4). Single-molecule conditions now dictated that the GyrB concentration must be in the picomolar range. Adding an excess of GyrA would drive the equilibrium toward GyrB binding, but with a high likelihood to form incomplete $GyrA_2GyrB$ complexes. Indeed, a mixture of donor and acceptor-labeled GyrB subunits with GyrA did not show FRET, indicating that the heterotetramer is not significantly populated. To ensure the formation of gyrase at single-molecule conditions and to prevent subunit dissociation during the experiment, we exploited the stability of the GyrA dimer and the twofold symmetry of the enzyme, and fused GyrB to GyrA, linked by a three amino acid linker glycine–alanine–proline (Gubaev & Klostermeier, 2011; Trigueros & Roca, 2002). The resulting GyrBA fusion protein assembled into stable dimers that showed wild-type-like DNA supercoiling activity and were stable under single-molecule conditions (Gubaev & Klostermeier, 2011; Lanz et al., 2014; Lanz & Klostermeier, 2011).

2.5.1 Nucleotide-Induced Conformational Changes at the N-Gate

To monitor conformational changes at the N-gate, cysteines were introduced into the GyrB part at positions S7, E17, and E131, leading to two cysteines across the N-gate in the GyrBA dimer that were then statistically labeled with donor and acceptor fluorophores (Fig. 5B). For all three constructs, unimodal, but broad histograms with low to intermediate FRET efficiencies were obtained, in agreement with an open N-gate with appreciable conformational flexibility (Gubaev & Klostermeier, 2011).

The addition of the nonhydrolyzable ATP analog ADPNP led to an increase in FRET efficiencies for all three constructs, in agreement with nucleotide-induced N-gate closure (Gubaev & Klostermeier, 2011). A narrowing of the FRET distributions points toward a decrease in flexibility of the N-gate in the closed state. We next used ADP·BeF$_x$ and ADP·MgF$_x$ to mimic the nucleotide states before and after ATP hydrolysis, and ADP to populate the nucleotide state after phosphate release. The high FRET state indicative of a closed N-gate was only observed in the presence of the ground state ATP analog ADP·BeF$_x$, whereas the low FRET state corresponding to the open N-gate was observed in the presence of ADP·MgF$_x$ and ADP (Gubaev & Klostermeier, 2011). Together, these results show that the N-gate closes upon ATP binding and reopens upon hydrolysis of both ATP molecules.

2.5.2 DNA-Induced Conformational Changes at the N-Gate

After validating that the N-gate of dimeric gyrase formed by two GyrBA subunits acts as an ATP-operated clamp using three different labeling geometries, we used these constructs to investigate the effect of DNA on the conformational state of the N-gate. For gyrase labeled at positions 17 and 131, plasmid DNA binding led to a drastic increase in FRET efficiency, as observed for nucleotide-dependent N-gate closure. Interestingly, for gyrase labeled at position 7 the FRET efficiency for the DNA-bound state was clearly distinguishable from the FRET efficiency of nucleotide-bound gyrase with a closed N-gate. Distance restraints from all three experiments are in agreement with a narrowed state of the N-gate in which the GyrB arms have come closer, but not yet formed a stable dimer (Gubaev & Klostermeier, 2011). These experiments identified a hitherto unknown intermediate in the catalytic cycle of DNA gyrase. To place N-gate narrowing with the sequence of events in the catalytic cycle, we then tested the effect of linear DNAs of different lengths, ranging from 60 to 139 bp, on N-gate closure (Gubaev & Klostermeier, 2011). A minimum of 110 bp was required to trigger N-gate narrowing upon binding, in contrast to the 48 bp DNA sufficient for the DNA-induced movement of the CTDs (Lanz & Klostermeier, 2011) (see Section 2.4), which revealed that CTD movement and N-gate narrowing are two separate events (Gubaev & Klostermeier, 2011). Interestingly, the N-gate of gyrase lacking the CTDs or only the GyrA-box did not show narrowing upon DNA binding, implicating the CTDs in the formation of this intermediate (Lanz & Klostermeier, 2012). Altogether, these observations are consistent with DNA having to wrap entirely around the CTDs to induce the transition to a narrowed N-gate.

To probe for the population of gyrase with a narrowed N-gate during DNA supercoiling, single-molecule FRET experiments were performed in the presence of ATP, using gyrase labeled at position 7 which allowed to distinguish open, narrowed, and closed N-gates by different FRET efficiencies (Gubaev & Klostermeier, 2011). ATP is turned over by gyrase, and FRET histograms reflect a steady-state population of all relevant intermediates, depending on the nature of rate-limiting steps. Under these steady-state conditions, the predominant population is that of gyrase with a narrowed or open N-gate. The fraction with a closed N-gate is less populated, indicating that N-gate closing is a transient event.

2.6 DNA-Gate and C-Gate: The Difficult Interpretation of a Lack of FRET Changes

Using similar rationales as described for labeling of gyrase on the CTDs, the GyrA body, and on GyrB, we also generated cysteine variants to probe for opening of the DNA- and C-gates. smFRET measurements always showed high FRET efficiencies, consistent with closed gates. However, crystal structures of the *B. subtilis* GyrA dimer (Rudolph & Klostermeier, 2013) indicate that these gates can open in principle. A lack of change in FRET efficiencies for a suspected conformational change can have different reasons. The labeling geometry may simply not be sensitive to gate opening because the distances between the labels are too similar in closed and open states. The crystal structures indicate a shearing movement of the gates that does not lead to large distance changes between equivalent residues on each subunit. Alternatively, the opening of the DNA- and C-gate may occur transiently, and the brief excursions to the resulting low FRET state are missed because of the limited time resolution (0.5–1 ms) of the experiment. We can therefore not conclude from these experiments that the DNA- and C-gates do not open during the catalytic cycle.

2.7 Current Picture of the DNA Supercoiling Cycle

We have probed conformational changes in the catalytic cycle of gyrase by single-molecule FRET experiments with donor/acceptor-labeled DNA substrates and donor/acceptor-labeled gyrase. The gyrase supercoiling cycle (Fig. 6; reviewed in Gubaev & Klostermeier, 2014) starts with binding of a G-segment at the DNA-gate, G-segment distortion, and cleavage (Gubaev et al., 2009). The CTDs help distort the bound G-segment (Gubaev et al., 2009). DNA regions flanking the G-segment contact the CTDs, and induce their upward movement (Lanz & Klostermeier, 2011, 2012). When DNA wraps completely around the CTDs, a narrowing of the N-gate is induced (Gubaev & Klostermeier, 2011). DNA distortion and N-gate narrowing are linked to DNA cleavage (Gubaev et al., 2009; Gubaev & Klostermeier, 2011), whereas the CTD movement can occur independently of and possibly prior to cleavage (Lanz & Klostermeier, 2011). ATP binding to the GyrB subunits causes N-gate closure (Gubaev & Klostermeier, 2011, 2012). The twisting motion of the GyrB subunits observed in the cryo-EM structure (Papillon et al., 2013) can also be probed by smFRET (A. Gubaev & D. Klostermeier, unpublished data). N-gate narrowing, twisting, and N-gate closure are three mechanistically distinct events. Once the N-gate has closed, the T-segment is fixed above the G-segment in the upper cavity of gyrase. We have so far not been able to detect DNA- or C-gate opening during the supercoiling reaction by smFRET. Hydrolysis of both ATP leads to reopening of the N-gate (Gubaev & Klostermeier, 2011) and resets gyrase for subsequent catalytic cycles. The smFRET studies summarized here have revealed a cascade of DNA- and nucleotide-induced conformational changes at the beginning of the supercoiling reaction, and have substantially refined and expanded the mechanistic model of DNA supercoiling that is currently found in text books (Fig. 4).

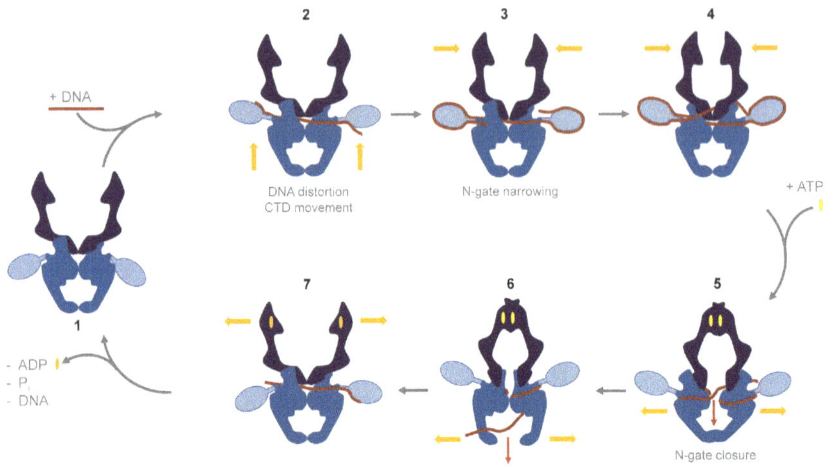

Fig. 6 Refined model for conformational changes of gyrase during negative DNA supercoiling. The G-segment binds to the DNA-gate of gyrase (1) and is distorted in a process that is coupled to cleavage and requires the CTDs (2). First contacts of the DNA emanating from the DNA-gate with the CTDs induce an upward movement of the CTDs (2). DNA that is wrapped around the CTDs completely triggers a narrowing of the N-gate (3). A flanking region of the DNA is placed into the upper cavity of as the T-segment (4). ATP binding to the GyrB subunits then leads to N-gate closure (5) and capture of the T-segment in the upper cavity of gyrase that can pass through the gap in the cleaved G-segment upon DNA-gate opening. Opening of the C-gate (6) then allows the T-segment to leave the enzyme. The G-segment is religated (7). After ATP hydrolysis and release of phosphate and ADP, gyrase is ready to engage in further supercoiling cycles.

3. CONCLUSIONS AND OUTLOOK

smFRET has contributed substantially to our understanding the mechanism of ATP-dependent DNA supercoiling by gyrase. Despite the refined picture of conformational changes at the beginning of the supercoiling, several questions are still open. For example, DNA- and C-gate opening have not been detected yet. Although we have gained some insight into the order of individual events by using DNAs of different lengths, or by identifying the dependence of conformational changes on the presence of the CTDs or on DNA cleavage, the temporal coordination of these conformational changes during the supercoiling reaction is still not clear. However, the toolbox illustrated in this review enables us to study the temporal coordination by generating gyrase that contains more than two fluorescent dyes in adequate positions to monitor two conformational changes at the same time. smFRET experiments on immobilized DNA gyrase could provide kinetic information on conformational changes complementary to the information obtained by confocal microscopy. The role of twofold symmetry for the gyrase mechanism could be addressed by studies of asymmetric enzymes. For example, the question of whether both CTDs move up simultaneously during DNA supercoiling or if they may follow a hand-over-hand mechanism has not been answered. The rationales and biophysical and biochemical tools we have outlined here using gyrase as an example can also be applied to other type II and type I topoisomerases to understand the role of conformational dynamics for different reactions catalyzed by different enzymes. A comparative study of the conformational changes and their coordination in other type IIA topoisomerases would be particularly interesting to understand how a common core structure with twofold symmetry can catalyze different reactions. The general strategies will also be useful to investigate conformational changes in other multisubunit molecular machines and their link to biological function.

ACKNOWLEDGMENTS

We thank all previous and current members of the laboratory for their contributions to the work described here.

Funding: Work in our laboratory was supported by the Swiss National Science Foundation and the Deutsche Forschungsgemeinschaft.

REFERENCES

Agard, N. J., Baskin, J. M., Prescher, J. A., Lo, A., & Bertozzi, C. R. (2006). A comparative study of bioorthogonal reactions with azides. *Chemistry & Biology*, *1*, 644–648.

Andrecka, J., Treutlein, B., Arcusa, M. A., Muschielok, A., Lewis, R., Cheung, A. C., et al. (2009). Nano positioning system reveals the course of upstream and nontemplate DNA within the RNA polymerase II elongation complex. *Nucleic Acids Research*, *37*, 5803–5809.

Andreou, A. Z., & Klostermeier, D. (2014). eIF4B and eIF4G jointly stimulate eIF4A ATPase and unwinding activities by modulation of the eIF4A conformational cycle. *Journal of Molecular Biology*, *426*, 51–61.

Aravind, L., Leipe, D. D., & Koonin, E. V. (1998). Toprim—A conserved catalytic domain in type IA and II topoisomerases, DnaG-type primases, OLD family nucleases and RecR proteins. *Nucleic Acids Research*, *26*, 4205–4213.

Aregger, R., & Klostermeier, D. (2009). The DEAD-box helicase YxiN maintains a closed conformation during ATP hydrolysis. *Biochemistry*, *48*, 10679–10681.

Bachmann, A. L., Matern, J. C., Schutz, V., & Mootz, H. D. (2015). Chemical-tag labeling of proteins using fully recombinant split inteins. *Methods in Molecular Biology*, *1266*, 145–159.

Baker, N. M., Weigand, S., Maar-Mathias, S., & Mondragon, A. (2010). Solution structures of DNA-bound gyrase. *Nucleic Acids Research*, *39*, 755–766.

Berger, J. M., Gamblin, S. J., Harrison, S. C., & Wang, J. C. (1996). Structure and mechanism of DNA topoisomerase II. *Nature*, *379*, 225–232.

Blaschke, U. K., Silberstein, J., & Muir, T. W. (2000). Protein engineering by expressed protein ligation. *Methods in Enzymology*, *328*, 478–496.

Brustad, E. M., Lemke, E. A., Schultz, P. G., & Deniz, A. A. (2008). A general and efficient method for the site-specific dual-labeling of proteins for single molecule fluorescence resonance energy transfer. *Journal of the American Chemical Society*, *130*, 17664–17665.

Corbett, K. D., Shultzaberger, R. K., & Berger, J. M. (2004). The C-terminal domain of DNA gyrase A adopts a DNA-bending beta-pinwheel fold. *Proceedings of the National Academy of Sciences of the United States of America*, *101*, 7293–7298.

Costenaro, L., Grossmann, J. G., Ebel, C., & Maxwell, A. (2005). Small-angle X-ray scattering reveals the solution structure of the full-length DNA gyrase A subunit. *Structure*, *13*, 287–296.

Cristovao, M., Sisamakis, E., Hingorani, M. M., Marx, A. D., Jung, C. P., Rothwell, P. J., et al. (2012). Single-molecule multiparameter fluorescence spectroscopy reveals directional MutS binding to mismatched bases in DNA. *Nucleic Acids Research*, *40*, 5448–5464.

del Toro Duany, Y. D., Ganguly, A., & Klostermeier, D. (2014). Differential contributions of the latch in Thermotoga maritima reverse gyrase to binding of single-stranded DNA before and after ATP hydrolysis. *Biological Chemistry*, *395*, 83–93.

del Toro Duany, Y., Jungblut, S. P., Schmidt, A. S., & Klostermeier, D. (2008). The reverse gyrase helicase-like domain is a nucleotide-dependent switch that is attenuated by the topoisomerase domain. *Nucleic Acids Research*, *36*, 5882–5895.

del Toro Duany, Y., & Klostermeier, D. (2011). Nucleotide-driven conformational changes in the reverse gyrase helicase-like domain couple the nucleotide cycle to DNA processing. *Physical Chemistry Chemical Physics*, *13*, 10009–10019.

del Toro Duany, Y., Rudolph, M. G., & Klostermeier, D. (2011). The conformational flexibility of the helicase-like domain from Thermotoga maritima reverse gyrase is restricted by the topoisomerase domain. *Biochemistry*, *50*, 5816–5823.

Deweese, J. E., Burgin, A. B., & Osheroff, N. (2008). Using 3′-bridging phosphorothiolates to isolate the forward DNA cleavage reaction of human topoisomerase IIalpha. *Biochemistry*, *47*, 4129–4140.

Dutta, R., & Inouye, M. (2000). GHKL, an emergent ATPase/kinase superfamily. *Trends in Biochemical Sciences*, *25*, 24–28.

Eggeling, C., Berger, S., Brand, L., Fries, J. R., Schaffer, J., Volkmer, A., et al. (2001). Data registration and selective single molecule analysis using multi-parameter fluorescence detection. *Journal of Biotechnology*, *86*, 163–180.

Förster, T. (1948). Zwischenmolekulare Energiewanderung und Fluoreszenz. *Annalen der Physik*, *437*, 55–75.

Gellert, M., Mizuuchi, K., O'Dea, M. H., & Nash, H. A. (1976). DNA gyrase: An enzyme that introduces superhelical turns into DNA. *Proceedings of the National Academy of Sciences of the United States of America*, *73*, 3872–3876.

Gubaev, A., Hilbert, M., & Klostermeier, D. (2009). The DNA gate of *Bacillus subtilis* gyrase is predominantly in the closed conformation during the DNA supercoiling reaction. *Proceedings of the National Academy of Sciences of the United States of America*, *106*, 13278–13283.

Gubaev, A., & Klostermeier, D. (2011). DNA-induced narrowing of the gyrase N-gate coordinates T-segment capture and strand passage. *Proceedings of the National Academy of Sciences of the United States of America*, *108*, 14085–14090.

Gubaev, A., & Klostermeier, D. (2012). Potassium ions are required for nucleotide-induced closure of gyrase N-gate. *The Journal of Biological Chemistry*, *287*, 10916–10921.

Gubaev, A., & Klostermeier, D. (2013). Fluorescence resonance energy transfer as a tool to investigate RNA structure and folding. In D. Klostermeier & C. Hammann (Eds.), *RNA structure and folding* (pp. 181–213). Berlin: de Gruyter.

Gubaev, A., & Klostermeier, D. (2014). The mechanism of negative DNA supercoiling: A cascade of DNA-induced conformational changes prepares gyrase for strand passage. *DNA Repair (Amst)*, *16*, 23–34.

Gubaev, A., Weidlich, D., & Klostermeier, D. (2016). DNA gyrase with a single catalytic tyrosine can catalyze DNA superociling by a nicking-closing mechanism. *Nucleic Acids Research*. pii: gkw740.

Harms, U., Andreou, A. Z., Gubaev, A., & Klostermeier, D. (2014). eIF4B, eIF4G and RNA regulate eIF4A activity in translation initiation by modulating the eIF4A conformational cycle. *Nucleic Acids Research*, *42*, 7911–7922.

Haustein, E., Jahnz, M., & Schwille, P. (2003). Triple FRET: A tool for studying long-range molecular interactions. *Chemphyschem*, *4*, 745–748.

Higgins, N. P., Peebles, C. L., Sugino, A., & Cozzarelli, N. R. (1978). Purification of subunits of Escherichia coli DNA gyrase and reconstitution of enzymatic activity. *Proceedings of the National Academy of Sciences of the United States of America*, *75*, 1773–1777.

Hilbert, M., Kebbel, F., Gubaev, A., & Klostermeier, D. (2011). eIF4G stimulates the activity of the DEAD box protein eIF4A by a conformational guidance mechanism. *Nucleic Acids Research*, *39*, 2260–2270.

Hohlbein, J., Aigrain, L., Craggs, T. D., Bermek, O., Potapova, O., Shoolizadeh, P., et al. (2013). Conformational landscapes of DNA polymerase I and mutator derivatives establish fidelity checkpoints for nucleotide insertion. *Nature Communications*, *4*, 2131.

Horowitz, D. S., & Wang, J. C. (1987). Mapping the active site tyrosine of *Escherichia coli* DNA gyrase. *The Journal of Biological Chemistry*, *262*, 5339–5344.

Kalinin, S., Peulen, T., Sindbert, S., Rothwell, P. J., Berger, S., Restle, T., et al. (2012). A toolkit and benchmark study for FRET-restrained high-precision structural modeling. *Nature Methods*, *9*, 1218–1225.

Kampranis, S. C., Bates, A. D., & Maxwell, A. (1999). A model for the mechanism of strand passage by DNA gyrase. *Proceedings of the National Academy of Sciences of the United States of America*, *96*, 8414–8419.

Kampranis, S. C., & Maxwell, A. (1996). Conversion of DNA gyrase into a conventional type II topoisomerase. *Proceedings of the National Academy of Sciences of the United States of America*, *93*, 14416–14421.

Kapanidis, A. N., Laurence, T. A., Lee, N. K., Margeat, E., Kong, X., & Weiss, S. (2005). Alternating-laser excitation of single molecules. *Accounts of Chemical Research*, *38*, 523–533.

Karow, A. R., & Klostermeier, D. (2009). A conformational change in the helicase core is necessary but not sufficient for RNA unwinding by the DEAD box helicase YxiN. *Nucleic Acids Research*, *37*, 4464–4471.

Karow, A. R., & Klostermeier, D. (2010). A structural model for the DEAD box helicase YxiN in solution: Localization of the RNA-binding domain. *Journal of Molecular Biology*, *402*, 629–637.

Karow, A. R., Theissen, B., & Klostermeier, D. (2007). Authentic interdomain communication in an RNA helicase reconstituted by expressed protein ligation of two helicase domains. *FEBS Journal*, *274*, 463–473.

Krainer, G., Hartmann, A., & Schlierf, M. (2015). farFRET: Extending the range in single-molecule FRET experiments beyond 10 nm. *Nano Letters*, *15*, 5826–5829.

Kühnemuth, R., & Seidel, C. A. M. (2001). Principles of single molecule multiparameter fluorescence spectroscopy. *Single Molecules*, *2*, 251–254.

Lamichhane, R., Berezhna, S. Y., Gill, J. P., Van der Schans, E., & Millar, D. P. (2013). Dynamics of site switching in DNA polymerase. *Journal of the American Chemical Society*, *135*, 4735–4742.

Lanz, M. A., Farhat, M., & Klostermeier, D. (2014). The acidic C-terminal tail of the GyrA subunit moderates the DNA supercoiling activity of *Bacillus subtilis* gyrase. *The Journal of Biological Chemistry*, *289*, 12275–12285.

Lanz, M. A., & Klostermeier, D. (2011). Guiding strand passage: DNA-induced movement of the gyrase C-terminal domains defines an early step in the supercoiling cycle. *Nucleic Acids Research*, *39*, 9681–9694.
Lanz, M. A., & Klostermeier, D. (2012). The GyrA-box determines the geometry of DNA bound to gyrase and couples DNA binding to the nucleotide cycle. *Nucleic Acids Research*, *40*, 10893–10903.
Laurence, T. A., Kong, X., Jager, M., & Weiss, S. (2005). Probing structural heterogeneities and fluctuations of nucleic acids and denatured proteins. *Proceedings of the National Academy of Sciences of the United States of America*, *102*, 17348–17353.
Lee, T. C., Kang, M., Kim, C. H., Schultz, P. G., Chapman, E., & Deniz, A. A. (2016). Dual unnatural amino acid incorporation and click-chemistry labeling to enable single-molecule FRET studies of p97 folding. *Chembiochem*, *17*, 981–984.
Lee, Y. J., Kurra, Y., Yang, Y., Torres-Kolbus, J., Deiters, A., & Liu, W. R. (2014). Genetically encoded unstrained olefins for live cell labeling with tetrazine dyes. *Chemical Communications (Cambridge, England)*, *50*, 13085–13088.
Linden, M. H., Hartmann, R. K., & Klostermeier, D. (2008). The putative RNase P motif in the DEAD box helicase Hera is dispensable for efficient interaction with RNA and helicase activity. *Nucleic Acids Research*, *36*, 5800–5811.
Minsky, M. (1988). Memoir on inventing the confocal microscope. *Scanning*, *10*, 128–138.
Morais Cabral, J. H., Jackson, A. P., Smith, C. V., Shikotra, N., Maxwell, A., & Liddington, R. C. (1997). Crystal structure of the breakage-reunion domain of DNA gyrase. *Nature*, *388*, 903–906.
Muller, B. K., Zaychikov, E., Brauchle, C., & Lamb, D. C. (2005). Pulsed interleaved excitation. *Biophysical Journal*, *89*, 3508–3522.
Muschielok, A., Andrecka, J., Jawhari, A., Bruckner, F., Cramer, P., & Michaelis, J. (2008). A nano-positioning system for macromolecular structural analysis. *Nature Methods*, *5*, 965–971.
Nagy, J., Grohmann, D., Cheung, A. C., Schulz, S., Smollett, K., Werner, F., et al. (2015). Complete architecture of the archaeal RNA polymerase open complex from single-molecule FRET and NPS. *Nature Communications*, *6*, 6161.
Papillon, J., Menetret, J. F., Batisse, C., Helye, R., Schultz, P., Potier, N., et al. (2013). Structural insight into negative DNA supercoiling by DNA gyrase, a bacterial type 2A DNA topoisomerase. *Nucleic Acids Research*, *41*, 7815–7827.
Qiu, R., DeRocco, V. C., Harris, C., Sharma, A., Hingorani, M. M., Erie, D. A., et al. (2012). Large conformational changes in MutS during DNA scanning, mismatch recognition and repair signalling. *The EMBO Journal*, *31*, 2528–2540.
Reece, R. J., & Maxwell, A. (1991). The C-terminal domain of the Escherichia coli DNA gyrase A subunit is a DNA-binding protein. *Nucleic Acids Research*, *19*, 1399–1405.
Rothwell, P. J., Berger, S., Kensch, O., Felekyan, S., Antonik, M., Wohrl, B. M., et al. (2003). Multiparameter single-molecule fluorescence spectroscopy reveals heterogeneity of HIV-1 reverse transcriptase: Primer/template complexes. *Proceedings of the National Academy of Sciences of the United States of America*, *100*, 1655–1660.
Rudolph, M. G., & Klostermeier, D. (2013). Mapping the spectrum of conformational states of the DNA- and C-gates in Bacillus subtilis gyrase. *Journal of Molecular Biology*, *425*, 2632–2640.
Ruthenburg, A. J., Graybosch, D. M., Huetsch, J. C., & Verdine, G. L. (2005). A superhelical spiral in the Escherichia coli DNA gyrase A C-terminal domain imparts unidirectional supercoiling bias. *The Journal of Biological Chemistry*, *280*, 26177–26184.
Santoso, Y., Joyce, C. M., Potapova, O., Le Reste, L., Hohlbein, J., Torella, J. P., et al. (2010). Conformational transitions in DNA polymerase I revealed by single-molecule FRET. *Proceedings of the National Academy of Sciences of the United States of America*, *107*, 715–720.

Schoeffler, A. J., May, A. P., & Berger, J. M. (2010). A domain insertion in *Escherichia coli* GyrB adopts a novel fold that plays a critical role in gyrase function. *Nucleic Acids Research, 38*, 7830–7844.
Sindbert, S., Kalinin, S., Nguyen, H., Kienzler, A., Clima, L., Bannwarth, W., et al. (2011). Accurate distance determination of nucleic acids via Forster resonance energy transfer: Implications of dye linker length and rigidity. *Journal of the American Chemical Society, 133*, 2463–2480.
Stryer, L. (1978). Fluorescence energy transfer as a spectroscopic ruler. *Annual Review of Biochemistry, 47*, 819–846.
Stryer, L., & Haugland, R. P. (1967). Energy transfer: A spectroscopic ruler. *Proceedings of the National Academy of Sciences of the United States of America, 58*, 719–726.
Sun, W., Yang, J., & Liu, X. Q. (2004). Synthetic two-piece and three-piece split inteins for protein trans-splicing. *The Journal of Biological Chemistry, 279*, 35281–35286.
Theissen, B., Karow, A. R., Kohler, J., Gubaev, A., & Klostermeier, D. (2008). Cooperative binding of ATP and RNA induces a closed conformation in a DEAD box RNA helicase. *Proceedings of the National Academy of Sciences of the United States of America, 105*, 548–553.
Trigueros, S., & Roca, J. (2002). A GyrB-GyrA fusion protein expressed in yeast cells is able to remove DNA supercoils but cannot substitute eukaryotic topoisomerase II. *Genes to Cells, 7*, 249–257.
Tyagi, S., & Lemke, E. A. (2013). Genetically encoded click chemistry for single-molecule FRET of proteins. *Methods in Cell Biology, 113*, 169–187.
Volkmann, G., & Liu, X. Q. (2009). Protein C-terminal labeling and biotinylation using synthetic peptide and split-intein. *PloS One, 4*, e8381.
Vos, S. M., Tretter, E. M., Schmidt, B. H., & Berger, J. M. (2011). All tangled up: How cells direct, manage and exploit topoisomerase function. *Nature Reviews. Molecular Cell Biology, 12*, 827–841.
Wang, L., Xie, J., & Schultz, P. G. (2006). Expanding the genetic code. *Annual Review of Biophysics and Biomolecular Structure, 35*, 225–249.
Ward, D., & Newton, A. (1997). Requirement of topoisomerase IV parC and parE genes for cell cycle progression and developmental regulation in *Caulobacter crescentus*. *Molecular Microbiology, 26*, 897–910.
Watrob, H. M., Pan, C. P., & Barkley, M. D. (2003). Two-step FRET as a structural tool. *Journal of the American Chemical Society, 125*, 7336–7343.
Wigley, D. B., Davies, G. J., Dodson, E. J., Maxwell, A., & Dodson, G. (1991). Crystal structure of an N-terminal fragment of the DNA gyrase B protein. *Nature, 351*, 624–629.
Williams, N. L., Howells, A. J., & Maxwell, A. (2001). Locking the ATP-operated clamp of DNA gyrase: Probing the mechanism of strand passage. *Journal of Molecular Biology, 306*, 969–984.
Wu, H., Hu, Z., & Liu, X. Q. (1998). Protein trans-splicing by a split intein encoded in a split DnaE gene of Synechocystis sp. PCC6803. *Proceedings of the National Academy of Sciences of the United States of America, 95*, 9226–9231.
Wurth, C., Grabolle, M., Pauli, J., Spieles, M., & Resch-Genger, U. (2013). Relative and absolute determination of fluorescence quantum yields of transparent samples. *Nature Protocols, 8*, 1535–1550.

CHAPTER TWELVE

Probing the Conformational Landscape of DNA Polymerases Using Diffusion-Based Single-Molecule FRET

J. Hohlbein*,†,1, A.N. Kapanidis‡,1

*Laboratory of Biophysics, Wageningen University and Research, Wageningen, The Netherlands
†Microspectroscopy Centre, Wageningen University and Research, Wageningen, The Netherlands
‡Clarendon Laboratory, University of Oxford, Oxford, United Kingdom
[1]Corresponding authors: e-mail address: johannes.hohlbein@wur.nl; kapanidis@physics.ox.ac.uk

Contents

Abstract

Monitoring conformational changes in DNA polymerases using single-molecule Förster resonance energy transfer (smFRET) has provided new tools for studying fidelity-related mechanisms that promote the rejection of incorrect nucleotides before DNA synthesis. In addition to the previously known open and closed conformations of DNA polymerases, our smFRET assays utilizing doubly labeled variants of *Escherichia coli* DNA polymerase I were pivotal in identifying and characterizing a partially closed conformation as a primary checkpoint for nucleotide selection. Here, we provide a comprehensive overview of the methods we used for the conformational analysis of wild-type DNA polymerase and some of its low-fidelity derivatives; these methods include strategies

Methods in Enzymology, Volume 581
ISSN 0076-6879
http://dx.doi.org/10.1016/bs.mie.2016.08.023
© 2016 Elsevier Inc.
All rights reserved.

for protein labeling and our procedures for solution-based single-molecule fluorescence data acquisition and data analysis. We also discuss alternative single-molecule fluorescence strategies for analyzing the conformations of DNA polymerases in vitro and in vivo.

ABBREVIATIONS

ALEX alternating laser excitation
BVA burst variance analysis
DNA deoxyribonucleic acid
dNTP deoxyribonucleoside triphosphate
E. coli *Escherichia coli*
FRET Förster resonance energy transfer
KF Klenow fragment
PDA probability distribution analysis
Pol polymerase
RNA ribonucleic acid
smFRET single-molecule Förster resonance energy transfer
TIRF total internal reflection fluorescence
wt wild type

1. INTRODUCTION

DNA polymerases play a pivotal role in DNA replication and DNA repair by catalyzing the stepwise addition of deoxyribonucleotides to a growing DNA chain. The complexity of the cellular environment and resultant requirements on adaptability and specificity led to a profound differentiation of DNA polymerases (Baker & Kornberg, 1992; Hübscher, Spadari, Villani, & Maga, 2010a, 2010b), exemplified by a staggering number of 18 different DNA polymerases functioning in human cells (García-Gómez, Reyes, Martínez-Jiménez, et al., 2013). Here, we discuss studies of the prokaryotic DNA polymerase I (hereafter Pol I), which we have been using as a model system to investigate the influence of the polymerase' conformational landscape on the fidelity of DNA synthesis by means of single-molecule fluorescence techniques (Fig. 1).

The *Escherichia coli* Pol I was the first DNA polymerase discovered (Lehman, Bessman, Simms, & Kornberg, 1958) and features three enzymatic activities: the 5′-to-3′ polymerase activity, which is responsible for DNA synthesis; the 3′-to-5′ exonuclease activity, which allows the removal of

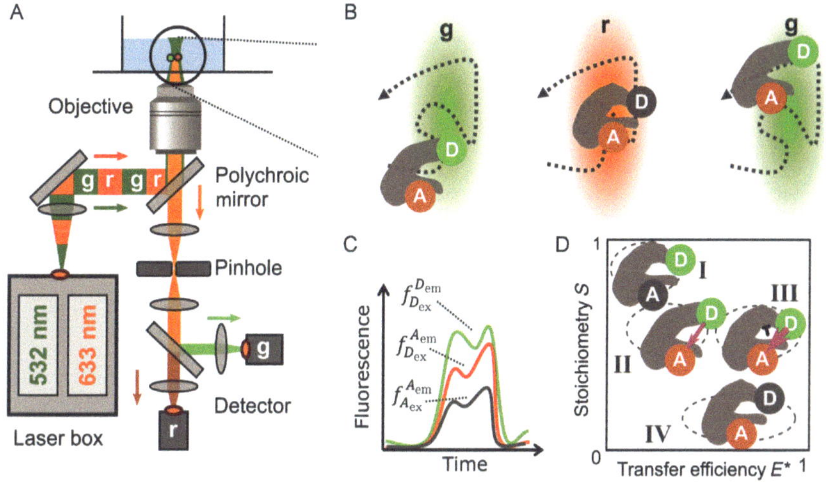

Fig. 1 Implementation of single-molecule FRET in confocal microscopy. (A) Schematic overview of a confocal setup for smFRET detection featuring alternating laser excitation. The green and red laser light is collimated, reflected by a polychroic mirror, and focused with an objective of high numerical aperture to a femtoliter-sized excitation spot in the sample volume. Fluorescence, originating from fluorophores covalently attached to DNA polymerases diffusing through the confocal spot, is collected by the same objective and spatially filtered with a pinhole before the fluorescence light is spectrally split into a green (donor) and a red (acceptor) detection channel. (B) During the transit of the doubly labeled DNA polymerase through the focus, short (~μs) periods of green (g) and red (r) excitation alternate faster than the corresponding diffusion time (1–3 ms for an underfilled objective). Upon direct excitation of the donor, some energy can be transferred to the acceptor via FRET. The existence of the acceptor is probed during red excitation. (C) In the recorded time traces, every identified burst is characterized by three photon numbers: First, the number of photons in the donor channel after donor excitation ($f_{D_{ex}}^{D_{em}}$); second, the number of photons in the acceptor channel after donor excitation ($f_{D_{ex}}^{A_{em}}$); and third, the number of photons in the acceptor channel after direct acceptor excitation ($f_{A_{ex}}^{A_{em}}$). (D) The three photon numbers are used to calculate the values for the transfer efficiency E^* and for the stoichiometry S for each individual burst allowing to plot a two-dimensional E^*-S histogram. Since DNA polymerases bearing only a single photoactive fluorophore have stoichiometry values around $S=1$ (for donor-only species) or $S=0$ (for acceptor-only species), species with two photoactive fluorophores can be clearly identified and selected for further data analysis.

wrongly incorporated bases; and the 5′-to-3′ exonuclease activity, which allows the polymerase to remove flaps of downstream DNA created as a result of strand displacement by Pol I. Much of our structural and mechanistic knowledge on Pol I comes from studies of a cleavage product of full-length Pol I, known as Klenow fragment (KF), which lacks the

5′-to-3′ exonuclease activity (Joyce & Grindley, 1983; Joyce, Potapova, DeLucia, et al., 2008). As expected for an enzyme involved in gap filling and strand displacement of RNA primers in DNA replication, Pol I features a remarkably high fidelity: insertion errors are extremely rare and occur only at a rate of 1 in $\sim 10^6$ additions (Bebenek, Joyce, Fitzgerald, & Kunkel, 1990; Eckert & Kunkel, 1991). The overall structure of Pol I (KF) resembles the form of a human right hand and consists of four subdomains: the 3′-to-5′ exonuclease, the thumb, the palm, and the fingers subdomain, with the latter playing an important role in nucleotide selection and incorporation (Fig. 2A).

Superimposing the high-resolution cocrystal structures of a KF homologue binary structure (DNA polymerase + DNA) with the ternary structure (DNA polymerase + DNA + complementary nucleotide) revealed a large conformational change of the fingers subdomain upon nucleotide addition (Fig. 2A) (Johnson et al., 2003). This inferred conformational change has been the starting point of our work, which has been largely driven by the question how exactly conformational changes contribute to polymerase fidelity. In the following, we will describe the individual steps necessary to address the research question that did not only require the optimization of existing biochemical protocols for labeling DNA polymerases with fluorophores but also the development of new tools for single-molecule spectroscopy. We will also link our work to recent advances in the field of single-molecule fluorescence detection, providing the interested reader with a brief overview over alternative approaches to study molecular machines and DNA polymerases at the single-molecule level.

2. SINGLE-MOLECULE FÖRSTER RESONANCE ENERGY TRANSFER

The quest for understanding biology on the molecular level has led researchers to develop sophisticated instrumentation and analytical methods. Nuclear magnetic resonance spectroscopy (Foster, McElroy, & Amero, 2007; Wüthrich, 2001) and X-ray crystallography (Ilari & Savino, 2008), for example, have been pivotal in providing structural information with atomic resolution, and recent advances such as time-resolved X-ray crystallography (Freudenthal, Beard, Shock, & Wilson, 2013; Schotte, Lim, Jackson, et al., 2003) and cryo-EM (Bai, McMullan, & Scheres, 2015) added new ways of studying protein function and assembly with high spatial resolution. Increasingly, these structural techniques are being complemented

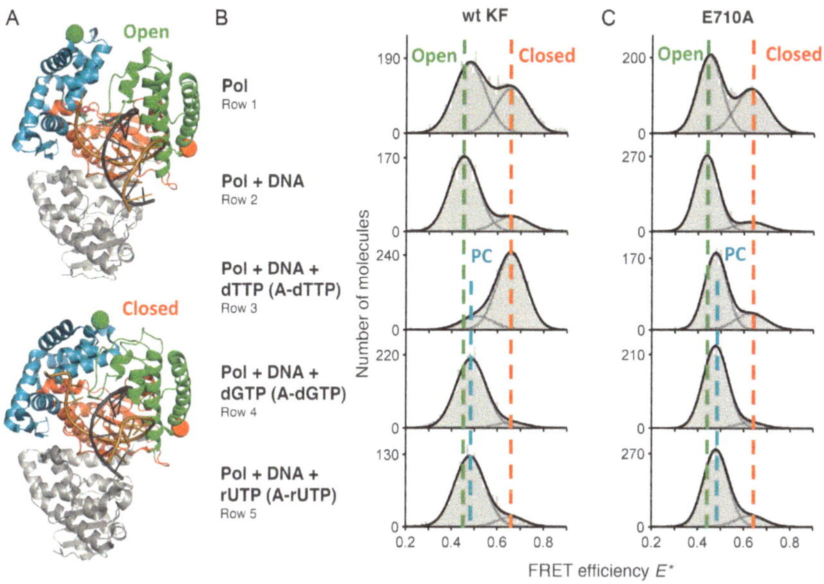

Fig. 2 Conformational equilibrium landscapes of DNA polymerase 1 (Klenow fragment). (A) The open Pol-DNA binary complex (PDB accession code 1L3U) and the closed Pol-DNA-dNTP (correct) ternary complex (PDB accession code 1LV5) are illustrated using structural data from *Bacillus stearothermophilus* (Bst) DNA polymerase (Johnson, Taylor, & Beese, 2003). The fingers subdomain is colored *blue*, the thumb *green*, the palm subdomain *red*, and the 3′–5′ exonuclease *gray*. The DNA template strand is colored *brown* and the DNA primer strand *black*. The amino acids Y766 (following the Pol I KF numbering scheme) and E710 are colored *magenta* and *cyan*, respectively. The labeling positions are indicated as a *red sphere* for the acceptor (residue 550) and a *green sphere* for the donor (residue 744). (B) DNA Pol I (KF) wild type: Unliganded enzymes populate both open and closed conformation of the fingers (row 1). In the binary complex, the addition of 100 n*M* DNA (with A as the templating base) shifts the conformational equilibrium toward the open state (row 2). The ternary complex with DNA and correct nucleotides (1 m*M* dTTP) shows most molecules in the closed conformation (row 3), whereas the ternary complex formed with incorrect nucleotides (1 m*M* rUTP or 1 m*M* dGTP) adopts a partially closed conformation (rows 4 and 5). (C) DNA Pol I (KF) E710A: In contrast to the wild type, the polymerase is not able to recognize dTTP as the correct nucleotide. Instead, the conformational equilibrium is similar to those for the incorrect nucleotides. *Adapted from Hohlbein, J., Aigrain, L., Craggs, T. D., et al. (2013). Conformational landscapes of DNA polymerase I and mutator derivatives establish fidelity checkpoints for nucleotide insertion.* Nature Communications, 4, *2131. http://dx.doi.org/10.1038/ncomms3131.*

by single-molecule Förster resonance energy transfer (smFRET), a powerful technique that allows analyzing dynamic interactions within and between biomolecular complexes under physiologically relevant conditions. Specifically, the ability of smFRET to resolve static and dynamic heterogeneity in biological pathways (Farooq, Fijen, & Hohlbein, 2013; Kapanidis & Strick, 2009) makes smFRET an excellent tool for studying time-dependent sample distributions, conformational dynamics, intermediate states, and prechemistry reaction kinetics in DNA polymerases.

Förster resonance energy transfer (FRET) describes the distance-dependent, nonradiative energy transfer from a donor fluorophore to an acceptor chromophore via a dipole–dipole interaction (Förster, 1946, 1948). For FRET to occur, three basic conditions need to be fulfilled: (i) the spectra for donor emission and acceptor absorption must overlap, (ii) donor and acceptor must be in close proximity (<10 nm), and (iii) the relative orientation of the donor and the acceptor transition dipole moments must allow transfer of energy (Lakowicz, 2006). The FRET transfer efficiency E can be experimentally determined from the donor fluorescence lifetime in the absence of the acceptor (τ_D) and in the presence of the acceptor (τ_T). The transfer efficiency E is inversely proportional to sixth power of the distance R between the two transition dipoles according to

$$E = 1 - \frac{\tau_T}{\tau_D} = \frac{R_0^6}{R_0^6 + R^6}. \tag{1}$$

R_0 is known as the Förster radius and represents the distance corresponding to an energy transfer of 50% between donor and acceptor. As R_0 is related to the properties of the fluorophores and the relative orientation of their dipole moments, great care should be taken with using published values; we also strongly recommend measuring the fluorescence anisotropy of fluorophores attached to proteins of interest. High values of the fluorescence anisotropy for certain fluorophores attached at particular labeling positions can cause severe complications in data analysis as we will discuss further below.

To detect fluorescence emitted from single fluorophores, we employed diffusion-based confocal microscopy (Fig. 1A) which requires dilute solutions of fluorescently labeled species typically at 10–100 pM concentration (Deniz, Dahan, Grunwell, et al., 1999). The molecules diffuse through a femtoliter-sized excitation and detection volume (confocal volume) formed by a microscope objective with high numerical aperture (typically

N.A. > 1.2). Laser excitation of the donor can lead to energy transfer to a nearby acceptor, causing the latter to fluoresce. The emitted fluorescence is then spectrally separated from the excitation laser(s) and spatially filtered by a pinhole eliminating intensity contributions from outside the focus. For FRET detection, the fluorescence is split into two detection channels, representing the spectral range of the donor and acceptor emission, respectively. In our configuration, the apparent FRET efficiency E^* for every burst, representing the passage of one molecule through the focus (Fig. 1B and C), is calculated using the number of photons detected in the donor channel after donor excitation $\left(f_{D_{ex}}^{D_{em}}\right)$ and the number of photons detected in the acceptor channel after donor excitation $\left(f_{D_{ex}}^{A_{em}}\right)$ as

$$E^* = \frac{f_{D_{ex}}^{A_{em}}}{f_{D_{ex}}^{D_{em}} + f_{D_{ex}}^{A_{em}}}. \tag{2}$$

In our conformational analysis, we have not corrected our E^* values for background noise (which should not exceed 2000 photons per second per detection channel), spectral crosstalk of the donor into the acceptor emission channel, and the instrument-dependent detection efficiencies of the dyes, since we were mainly interested in detecting relative changes in distances between fluorophores. If necessary, however, such corrections can be easily performed using our alternating laser excitation (ALEX) data and related procedures (Hohlbein, Craggs, & Cordes, 2014; Kapanidis, Lee, Laurence, et al., 2004).

In ALEX, short periods of donor excitation alternate with short periods of direct acceptor excitation to probe the presence of a fluorescently active acceptor fluorophore in the confocal volume (Fig. 1B and C). ALEX provides an additional photon count for each burst, $f_{A_{ex}}^{A_{em}}$, representing the number of photons in the acceptor channel after direct excitation of the acceptor, allowing us to calculate the (raw) stoichiometry for each burst:

$$S^{raw} = \frac{f_{D_{ex}}^{A_{em}} + f_{D_{ex}}^{D_{em}}}{f_{D_{ex}}^{A_{em}} + f_{D_{ex}}^{D_{em}} + f_{A_{ex}}^{A_{em}}}. \tag{3}$$

The stoichiometry of donor–acceptor species was tuned (by adjusting the S-related photon counts via setting the ratio of laser excitation intensities) such that we can use the corresponding two-dimensional $E^* - S^{raw}$ histogram (Fig. 1D) to separate donor–acceptor species from species bearing only one active fluorophore (namely, donor-only and acceptor-only species).

For the experimental results in the following sections, although the use of ALEX is not strictly required (mainly due to the high-FRET efficiencies corresponding to the conformational states involved), it does help to identify and solve potential problems in data acquisition and analysis. For example, the shape of species forming the $E^* - S^{raw}$ histogram allows detecting premature photobleaching (Nir, Michalet, Hamadani, et al., 2006), a feature that can lead to apparent peak shifts of FRET efficiencies in single-laser excitation schemes if not identified correctly.

We performed diffusion-based smFRET at room temperature (22°C) using a home-built confocal microscope equipped with 20 kHz ALEX between a 532-nm (operating at a laser power of 240 μW) and a 638-nm laser (operating at 60 μW), coupled to a 60×, 1.35 NA, UPLSAPO 60XO (Olympus) oil-immersion objective. The diameter of the collimated laser light before entering the objective was set to ~3 mm, thereby significantly underfilling the microscope objective. This arrangement led to an expanded excitation focus which, in combination with the given laser powers, our fluorophores of choice (Cy3B as the donor and ATTO647N as the acceptor) and the burst identification discussed below, gave us well-defined, narrow histograms of FRET efficiencies. Both fluorophores feature comparable quantum yields of ~0.7 (Gust, Zander, Gietl, et al., 2014), high photostability, and an R_0 of ~6.5 nm. For the identification of bursts, we used algorithms described elsewhere (Eggeling, Berger, Brand, et al., 2001; Nir et al., 2006). In short, we used the arrival times from the $f_{A_{ex}}^{A_{em}}$ photon stream and searched for intervals with $L = 12$ successive photons that have at least $M = 7$ photons in a time window of length $T = 500$ μs centered around their own arrival time to identify bursts from acceptor-bearing species. The bursts were further filtered by applying an additional threshold of $f_{A_{ex}}^{A_{em}} > 40$ photons, and we selected the bursts of species bearing both donor and acceptor by applying another threshold of $0.60 < S^{raw} < 0.85$. Due to the large amount of recorded data, we relied on batch processing of the experimental data. Our tools for data analysis are available on request, although a more polished software suite for data analysis of confocal data has recently been published (Ingargiola, Lerner, Chung, Weiss, & Michalet, 2016).

Each set of measurements contained daily reference measurements of a DNA standard ($E^* \sim 0.5$) (Holden, Uphoff, Hohlbein, et al., 2010) to monitor any subtle changes in the setup alignment. Doubly labeled Pol I (KF) derivatives described below were present at <100 pM in single-molecule observation buffer consisting of 40 mM 4-(2-hydroxyethyl)-1-piperazineethanesulfonic acid–NaOH, pH 7.3, 10 mM $MgCl_2$, 1 mM dithiothreitol, 100 mg/mL bovine serum albumin, 5% (vol/vol) glycerol, and 1 mM mercaptoethylamine.

3. PURIFICATION AND SITE-SPECIFIC LABELING OF POL I (KF)

Single-molecule FRET is now widely applied to study molecular interactions and dynamics (Deniz, Mukhopadhyay, & Lemke, 2008; Farooq et al., 2013; Hohlbein, Gryte, Heilemann, & Kapanidis, 2010; Kim & Ha, 2013; Moerner, 2007; Preus & Wilhelmsson, 2012; Roy, Hohng, & Ha, 2008; Walter, Huang, Manzo, & Sobhy, 2008; Weiss, 1999). The general applicability of the technique, however, depends on preparing fluorescently labeled species of highest quality and purity. In our experience, the development of robust protocols for site-specific labeling represents one of the biggest challenges, especially as we opted to label the enzyme with both a donor and an acceptor fluorophore. Starting from the crystal structures, it is up to the researcher to identify labeling positions that maximize the expected distance change during the conformational transition while preserving the enzymatic activity.

A viable strategy to select appropriate labeling sites is to identify nonconserved residues by performing a BLAST analysis of Pol I derivatives of various prokaryotic species. We identified residue K550 (on the thumb subdomain, Fig. 2A) and residue L744 (on the fingers subdomain, Fig. 2A) as sites for labeling and mutated the residues into cysteines (K550C and L744C) which can be labeled on their thiol side chains with maleimide derivatives of fluorophores. From the above mentioned crystal structures of Pol I, we determined the distances between the C_β positions of residues 550 and 744 to be 5.2 nm in the open and 4.2 nm in the closed conformation.

We further checked the Pol I sequence and structure for existing surface-exposed cysteine residues in the wild-type protein, and decided to mutate C907 into a serine (C907S). Two additional modifications were necessary: first, we used the D424A modification (to abolish the 5′-to-3′ exonuclease activity of Pol I KF). Second, we added a N-terminal His_6 tag for protein purification by affinity chromatography on Ni-NTA agarose (Joyce et al., 2008). Single-turnover kinetic data of the purified polymerases for complementary deoxyribonucleoside triphosphate (dNTP) incorporation confirmed enzymatic activities comparable to unmodified enzymes.

The initial labeling strategy we employed (Santoso, Joyce, Potapova, et al., 2010) was based on stochastic labeling of two Cys residues, a "nonorthogonal" chemical labeling approach, since both donor and acceptor fluorophores were expected to have similar reactivities toward the surface-exposed cysteines. In control experiments with singly labeled KF derivatives, however, we identified that labeling the fingers subdomain (L744C) with ATTO647N led to a high fluorescence anisotropy of the fluorophore, indicating undesired interactions between the hydrophobic fluorophore and surrounding surface of Pol I KF. To minimize potential problems with certain labeling configurations, we developed a novel labeling strategy which allowed us to improve the labeling bias and generate a highly pure doubly labeled KF isomer (~99% 550-ATTO647N, 744-Cy3B (Hohlbein et al., 2013)). The labeling reaction took place in presence of a dideoxy-terminated duplex DNA primer-template and complementary dNTP to bias the conformational equilibrium, as we will discuss later, toward the closed conformation. After the addition of ATTO647N maleimide at 1.2-fold molar excess and an 1-h incubation at 22°C, a 3.4-fold molar excess of Cy3B maleimide was added and incubated for a further 16 h at 4°C (for further details on the protein purification, see Hohlbein et al. (2013)). We determined the molarities of protein and dye labels from absorbance spectra and measured the relative amounts of Cy3B and ATTO647N at positions 550 and 744 using partial digestion with chymotrypsin, as described (Santoso, Joyce, et al., 2010). After fractionation on SDS-PAGE, the Cy3B and ATTO647N fluorescence used to estimate the fraction of the donor–acceptor molecules had the desired 550-ATTO647N, 744-Cy3B labeling pattern.

In summary, even though newly developed orthogonal labeling strategies involving unnatural amino acids (Chakraborty, Wang, Ebright, & Ebright, 2010; Chatterjee, Sun, Furman, et al., 2013; Lang & Chin, 2014) offer additional advantages (e.g., naturally occurring cysteines do not have to get removed), the careful utilization of slight labeling biases can be sufficient to obtain doubly labeled samples of high quality and well-defined labeling stoichiometry. In any case, we recommend careful characterization of singly and doubly labeled species using fluorescence spectroscopy, time-resolved fluorescence anisotropy, and biochemical functional assays.

4. MONITORING THE CONFORMATIONAL LANDSCAPE OF POL I (KF)

Our first smFRET report on doubly labeled Pol I (KF) supported the general applicability of the assay for monitoring conformational landscapes (Santoso, Joyce, et al., 2010). Due to the substantial improvements in our labeling scheme, we will mainly discuss our results from a later reference (Hohlbein et al., 2013). Our data revealed that the fingers-closing mechanism of Pol I does not only occur in a ternary complex of polymerase, (nonextendable) DNA, and a correct nucleotide but also in binary complexes composed of polymerase and DNA and even in the unliganded enzyme (Fig. 2B). As discussed above, we probed the conformational landscape by labeling the mobile part of the fingers subdomain with a donor, and a fairly static position on the thumb subdomain with an acceptor fluorophore. Consequently, a high FRET efficiency $\left(E^{*}_{\text{closed}} \sim 0.65\right)$ indicated a "fingers-closed" conformation, while a low FRET efficiency marks an open conformation $\left(E^{*}_{\text{open}} \sim 0.45\right)$. Thus, Pol I is able to switch between different conformational states, though it depends on the complex which state is preferred. Binary complexes were mainly found in the open conformation, whereas ternary complexes with correct nucleotides were mostly found in the closed conformation.

Our results suggested that fingers closing is no indication for successful incorporation of a nucleotide and that other mechanisms are in place to check nucleotides for their compatibility to the base of the templating DNA strand before incorporation. This model was supported by the fact that the fingers do not close completely in ternary complexes formed with mispaired dNTPs or when ribonucleotides were added to the buffer solution (Hohlbein et al., 2013; Santoso, Joyce, et al., 2010). Under these conditions, we observed molecules showing a mean FRET efficiency that was only slightly higher than the one indicating the open conformation ($E^{*}_{\text{inter}} \sim 0.49$ instead of $E^{*}_{\text{open}} \sim 0.45$, Fig. 2B). Seemingly, Pol I is able to detect a mispair before full fingers closing would occur. This finding suggests that one or more intermediate states exist, in which the incoming nucleotide is previewed and subsequently rejected or incorporated in the DNA. In 2011, a published crystal of structure Pol I revealed the presence of an intermediate (ajar) conformation of the fingers subdomain when bound to an incorrect nucleotide (Wu & Beese, 2011). Interestingly, this intermediate conformation corresponds to fingers closing along the open-to-closed transition, leading to a distance change of ~2 Å between the sites carrying the FRET probes, in perfect agreement with our observations of $\Delta E^{*} = 0.04$ (equaling 20% of a total conformational change corresponding to the fingers movement by one nanometer). Additional smFRET-based studies have verified the presence of an intermediate conformation in ternary complexes with incorrect nucleotides (Berezhna, Gill, Lamichhane, & Millar, 2012; Rothwell, Allen, Sisamakis, et al., 2013).

We further studied mutator derivatives of Pol I (KF) in which key residues of the active site were altered (Hohlbein et al., 2013). Ternary complexes of these low-fidelity mutants with complementary dNTPs adopted the same conformation as the ternary complexes formed with noncomplementary nucleotides (Fig. 2C). As the specific mutator phenotype correlated well with reduced affinity for complementary dNTPs, our work established the partially closed conformation as a primary checkpoint for nucleotide selection.

5. ANALYZING THE CONFORMATIONAL LANDSCAPES FOR NUCLEOTIDE SELECTION OF POL I (KF)

In the previous section, we showed data of the conformational landscapes after saturating the equilibria with a high concentration of DNA (100 n*M*, roughly two orders of magnitude above the dissociation constant of Pol I binding to recessed DNA (Markiewicz, Vrtis, Rueda, & Romano, 2012)) for the binary complex, as well as with DNA (100 n*M*) and dNTPs (1 m*M*) for the ternary complexes studied. To calculate equilibrium and dissociation constants for nucleotide binding on one hand, and verify the existence of the intermediate-FRET state (which differs marginally from the peak position of the open low-FRET state), we performed rigorous sets of titration experiments.

5.1 Nucleotide Titrations and Triplicate Measurements

In any of the titration experiments, we started with a measurement of the binary complex for ~30 min to obtain several thousand FRET bursts in total. We then prepared a fresh sample, in which a small concentration of dNTPs was added and measured the new sample for a similar time. We used nucleotide concentrations of (0, 0.1, 0.3, 1, 3, 10, 30) μ*M* for titrations of A-dTTP to Pol I (KF) + DNA and (0, 0.01, 0.03, 0.1, 0.3, 1, 2) m*M* for dNTP titrations with noncomplementary nucleotides or when using Pol I (KF) derivatives (Fig. 3A). The spacing of the concentrations was chosen such that we cover regions in which large changes of the observables discussed below are expected while keeping the measurement time on an acceptable level. As we are dealing with concentrations spanning many

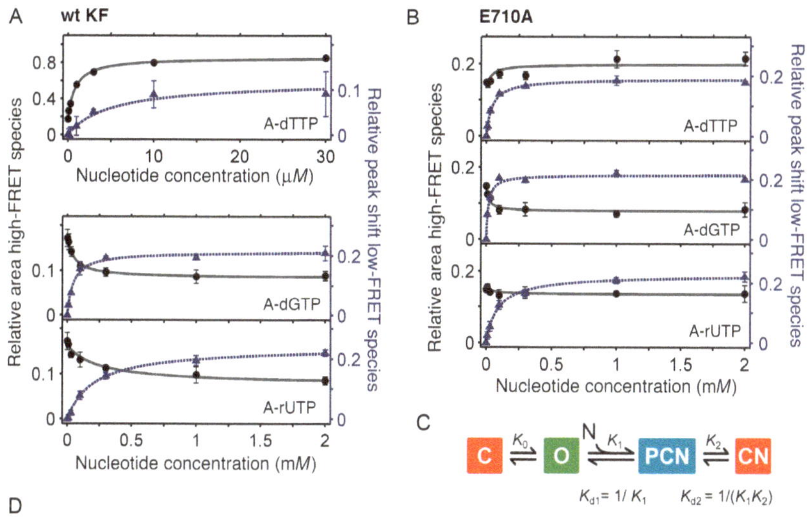

D

Complex	Constants	wt KF	E710A
A-dTTP	K_{d1} (μ*M*)	5.1 ± 0.6	43 ± 10
	K_2	6.1 ± 0.5	0.25 ± 0.01
	K_{d2} (μ*M*)	0.83 ± 0.12	170 ± 40
A-dGTP	K_{d1} (μ*M*)	46 ± 5	16 ± 2
	K_2	0.091 ± 0.005	0.084 ± 0.006
A-rUTP	K_{d1} (μ*M*)	210 ± 24	92 ± 8
	K_2	0.083 ± 0.007	0.16 ± 0.00

Fig. 3 Determining equilibrium constants for wt KF and E710A using FRET observables associated with the formation of ternary complexes. (A and B) We plotted the fraction of molecules in the high-FRET conformation (*black circles*) and the normalized shift of the mean E^* of the low-FRET species (*blue triangles*) as a function of nucleotide concentration. Data points with error bars are represented as mean ± s.e.m., derived from three independent measurements. The data were globally fitted using a four-state equilibrium model (C), which describes the nucleotide concentration-dependent FRET observables. K_0 represents the equilibrium between open and closed forms of the DNA-Pol binary complex (O: open, C: closed). Upon addition of nucleotides (N), the nucleotide-bound partially closed (PCN) and closed (CN) states are formed sequentially. K_1 and K_2 describe nucleotide-dependent equilibria and allow the calculation of dissociation constants K_{d1} and K_{d2}. (D) Table summarizing the numeric results obtained from fitting the data in (A) and (B) using the model described in (C). *Adapted from Hohlbein, J., Aigrain, L., Craggs, T. D., et al. (2013). Conformational landscapes of DNA polymerase I and mutator derivatives establish fidelity checkpoints for nucleotide insertion.* Nature Communications, 4, *2131. http://dx.doi.org/10.1038/ncomms3131.*

orders of magnitude, the purity of the dNTPs has to be checked as even smallest contaminations could severely influence the results of the titration experiments. Each titration experiment was completed in a single day. All titrations were repeated independently and on different days for three times to improve the statistical quality of the data and to allow us calculating error bars representing mean $\pm$ s.e.m.

We fitted each stoichiometry-filtered E^* FRET histograms to double-Gaussian functions that were constrained by a maximum width of $\sigma_{max}(E^*) \leq 0.07$ and reported on peak positions and relative occupancies (Fig. 2B). We note that fitting with Gaussian functions represents an approximation, as it ignores the potential influence of dynamic averaging on the shape of the distribution of transfer efficiencies. However, we found the approximation to be remarkably robust especially as it is straightforward to extract meaningful FRET observables. For the titration experiments, we plotted the relative peak shift $(E^*_{obs} - E^*_{open})/(E^*_{closed} - E^*_{open})$ of the low-FRET Gaussian and the fractional occupancy of the high-FRET Gaussian as a function of nucleotide concentration. We fixed the position of the high-FRET Gaussians to the value obtained from a free fit of the binary complex or a saturated ternary complex with complementary dNTPs; however, we did not fix the position of the low-FRET Gaussian.

5.2 Derivation of the Multistate Equilibrium Model

To determine the parameters governing the equilibria between the species observed in the FRET histograms (Figs. 2 and 3), we employed a four-state model (Fig. 3C) and derived equations for the two concentration-dependent FRET observables: the peak position of the low-FRET Gaussian and the fraction of the population in the high-FRET Gaussian. The binary complex (Pol + DNA) is in dynamic equilibrium between open (O, low-FRET) and closed (C, high-FRET) conformations, and is described by the equilibrium constant $K_0 = [C]/[O]$. Binding of a nucleotide (N) to the open conformation leads to the formation of the intermediate-FRET ternary complex (IN) that is described by the association constant $K_1 = [IN]/[O][N]$. The intermediate-FRET ternary complex can undergo an additional conformational transition to the closed ternary complex (CN, high FRET), governed by the equilibrium constant $K_2 = [CN]/[IN]$.

The corresponding dissociation constants are defined as $K_{d1} = 1/K_1$ for the formation of the intermediate-FRET state and $K_{d2} = 1/(K_1 K_2)$ for the formation of the high-FRET (fully closed) state. The total polymerase concentration is $[\text{Total}] = [\text{O}] + [\text{C}] + [\text{IN}] + [\text{CN}] = [\text{O}] + K_0[\text{O}] + K_1[\text{O}][\text{N}] + K_1K_2[\text{O}][\text{N}]$, and the fractions of each species as a function of nucleotide concentration are $[\text{O}]/[\text{Total}] = 1/(1 + K_0 + K_1[\text{N}] + K_1K_2[\text{N}])$, $[\text{C}]/[\text{Total}] = K_0/(1 + K_0 + K_1[\text{N}] + K_1K_2[\text{N}])$, $[\text{IN}]/[\text{Total}] = K_1[\text{N}]/(1 + K_0 + K_1[\text{N}] + K_1K_2[\text{N}])$, and $[\text{CN}]/[\text{Total}] = K_1K_2[\text{N}]/(1 + K_0 + K_1[\text{N}] + K_1K_2[\text{N}])$. Therefore, the two FRET observables are given by (a) the normalized peak shift of the low-FRET Gaussian $\left(E^*_{\text{obs}} - E^*_{\text{open}}\right)/\left(E^*_{\text{closed}} - E^*_{\text{open}}\right) = \gamma_{\max}[\text{N}]/(K_{\text{d1}} + [\text{N}])$, where $\gamma_{\max}$ corresponds to the normalized peak shift at saturating nucleotide concentration, and (b) the fraction of high-FRET species $([\text{C}] + [\text{CN}])/[\text{Total}] = (K_0 + K_1K_2[\text{N}])/(1 + K_0 + K_1[\text{N}] + K_1K_2[\text{N}])$.

We used a global fitting approach to fit both FRET observables simultaneously as a function of nucleotide concentration to determine the equilibrium constants K_0, K_1, and K_2 (which then allowed us to calculate dissociation constants) and the parameter $\gamma_{\max}$. The results for Pol I KF and E710A show that binding of the complementary nucleotide A-dTTP represented by K_{d1} is tight for wild-type Pol I (KF) and weak for Pol I (E710A) (Fig. 3D). In fact, the weak binding of A-dTTP for E710A is on a similar level as the binding of noncomplementary nucleotides for both Pol I (KF) and E710A.

In conclusion, we note that the analysis of the titration experiments requires experimental data of very high quality as the expected changes in the mean peak positions as well as changes in the occupancies of the FRET species differ in most cases only marginally between different nucleotide concentrations. In our case, data acquisition in triplicates provided means of confirming the robustness of the assays and is therefore highly recommended.

5.3 Characterizing Conformational Dynamics in Diffusion-Based smFRET Experiments

For the titration experiments described above, we analyzed the FRET distributions solely using the peak height and the peak position of the particular FRET species. However, additional information on the dynamics of interconverting FRET species can be obtained from analyzing the width and the overall shape of the entire FRET distribution. The required statistical framework has been developed by various groups (Antonik, Felekyan, Gaiduk, & Seidel, 2006; Felekyan, Sanabria, Kalinin, et al., 2013; Gopich & Szabo, 2010; Kalinin, Felekyan, Antonik, & Seidel, 2007; Kalinin, Felekyan,

Valeri, & Seidel, 2008; Nir et al., 2006; Santoso, Torella, & Kapanidis, 2010) and is known as Probability Distribution Analysis (PDA), as the methods aim at recapitulating FRET distributions based on the experimentally obtained distributions of photon counts.

In our simplified implementation, we used a dynamic and a static model of only two species being present. Albeit our model of nucleotide selection populates up to four different species (Fig. 3C), we focused our analysis on situations in which only two species are dominant as it is the case for the unliganded polymerase, the binary complex, and the saturated ternary complexes.

The static model of two species is fully described by two peak positions E_i^*, two excess widths σ_i (accounting for additional broadening of $\langle E_i^* \rangle$ due to acceptor photophysics (Kalinin et al., 2007; Nir et al., 2006)), and two relative occupancies N_i. In comparison, the dynamic model replaces the relative occupancies with two rate constants: k_1 for the open-to-closed transition and k_{-1} for the closed-to-open transition. We optimized parameters for both models using a Least Mean Square algorithm implemented in MATLAB (software available on request).

For the unliganded Pol I (KF), both models fit the experimental FRET distributions very well (Fig. 4A), raising the question how to decide between the models and how to prove the hypothesized dynamic conformational changes. To solve this issue, we developed Burst Variance Analysis (BVA), which directly evaluates FRET distributions for the presence of dynamic heterogeneity (Torella, Holden, Santoso, et al., 2011). In BVA, the experimentally measured standard deviation of the FRET efficiency (s_{E^*}) for individual transits through the confocal spot is compared to the standard deviation predicted by the static and dynamic models. In particular, the FRET efficiency is calculated for every five photons detected after excitation of the donor. Each burst will contain many of these five-photon windows, and the standard deviation calculated over all windows of one burst will indicate whether the system is likely to be dynamic beyond the value expected for a pure shot-noise-based photon distribution in one of the two channels for donor or acceptor detection. Similar approaches have been presented elsewhere (Felekyan et al., 2013; Tomov, Tsukanov, Masoud, et al., 2012). Our data revealed that BVA based on the dynamic model perfectly describes the experimental data showing that the unliganded polymerase populates (at least) two conformations showing dynamics on the millisecond time scale (Fig. 4B). We strongly recommend to perform BVA also on static FRET samples (such

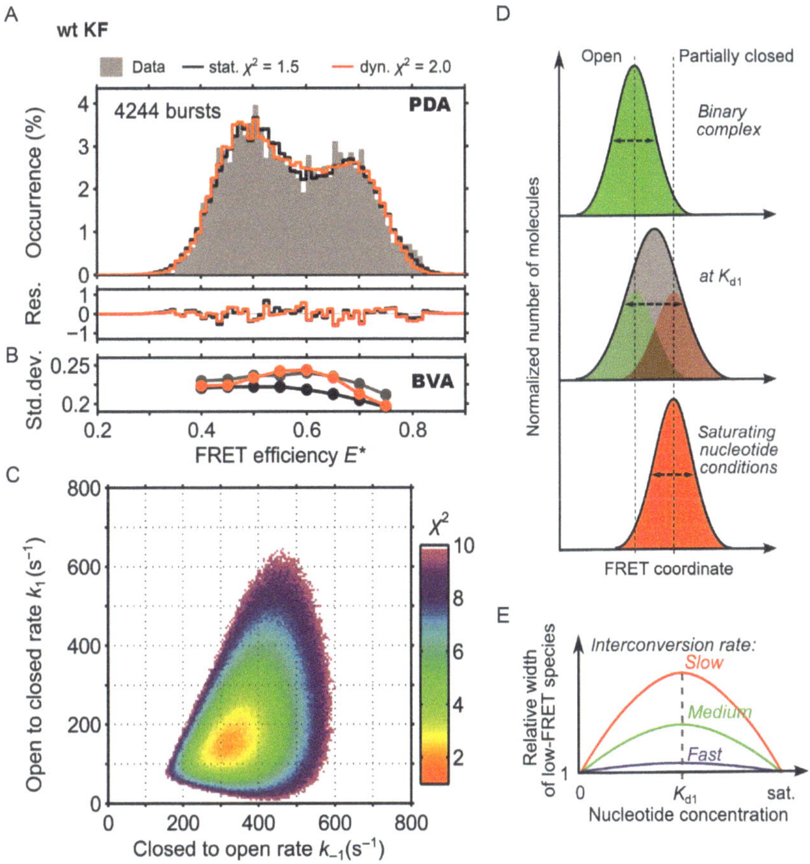

Fig. 4 Computational techniques for analyzing conformational landscapes from smFRET data. (A–C) Analysis of Pol I wt KF using Probability Distribution Analysis (PDA), Burst Variance Analysis (BVA), and a rate-dependent PDA χ^2-minimization surface. (A) We fitted the E^* histogram of unliganded wt KF using PDA, assuming either a static model of two nonconverting species (*black line*) or a dynamic model in which the polymerase is allowed to undergo conformational changes between a closed or an open conformation (*red line*). The fit with the dynamic model gave $k_1 = 150\ s^{-1}$ for the open-to-close transition rate and $k_{-1} = 320\ s^{-1}$ for reversed transition. (B) Both models describe the data reasonably well ($\chi^2 \leq 2$), but the BVA analysis suggests that the dynamic model (*red line*) is to be favored as it matches the experimental data (*gray line*) much better. (C) The PDA χ^2-minimization surface uses the mean FRET positions and excess widths from the dynamic fit, plotting the $\chi^2 < 10$ values as a function of the rates. Best fits are obtained for $k_1 \sim 160\ s^{-1}$ and $k_{-1} \sim 320\ s^{-1}$. (D and E) Width analysis for characterizing fast conformational dynamics of closely neighbored FRET species. (D) At nucleotide concentrations around K_{d1}, both O and PCN states (see Fig. 3C) are equally populated, and the FRET distribution for the low-FRET species is a convolution of both states (*gray* distribution in the *middle panel*). At concentrations far below or

(Continued)

as doubly labeled DNA) to ensure that photophysical fluctuations, as well as mismatches in the green and red detection volumes, are not misinterpreted as conformational dynamics of a protein. We further note that the dynamics which can be probed with PDA should be in the order of milliseconds, i.e., comparable to the average diffusion time of a single fluorescent molecule through the confocal volume. If the dynamics are much faster, they are averaged out temporarily, whereas if they are too slow, every passage will be either in one or the other conformation. As a rule of thumb, the more the measured distribution deviates from being able to be fitted with a limited number of Gaussians, the easier it is to obtain reliable values from PDA. We therefore decided to perform PDA using the FRET efficiencies E_i^* obtained from the initial dynamic PDA fit (Fig. 4A), and plotted the χ^2 values as a function of k_1 and k_{-1} (Fig. 4C). For Pol I (KF), the minimization surface forms a distinct well at $k_1 = 160\ \mathrm{s}^{-1}$ and $k_{-1} = 320\ \mathrm{s}^{-1}$ ($\chi^2 < 2$) (Hohlbein et al., 2013).

5.4 Width Analysis for Estimating Rate Constants

As PDA requires the resolvability of two FRET species to obtain robust solutions, we developed a new technique termed width analysis to estimate upper and lower boundaries for the rates of interconversion of otherwise unresolvable species (Fig. 4D). At nucleotide concentrations equal to the dissociation constant K_{d1} (at which [O] equals [IN]), the interconversion rates between the open conformation O and the intermediate (partially closed) conformation IN, $k_{O \to IN}$ and $k_{IN \to O}$, are equal. Therefore, these rates also control the width of the low-FRET species distribution: if $k_{O \to IN}$ and $k_{IN \to O}$ are high and lead to multiple interconversions during the transit of the complex through the detection volume (~3 ms (Santoso, Joyce, et al., 2010)), the FRET averaging between O and IN will yield a small FRET width for the low-FRET species comparable to the width of the binary complex or partially closed ternary complexes at saturated nucleotide

Fig. 4—Cont'd above K_{d1}, the width of the low-FRET Gaussian resembles that of either the open conformation (*upper panel*) or the partially closed conformation (*lower panel*). (E) Even though we cannot resolve the individual FRET states at K_{d1}, any observed broadening of the FRET species indicates conformational dynamics at timescales slower than the corresponding diffusion time through the confocal focus (~3 ms). *Adapted from Hohlbein, J., Aigrain, L., Craggs, T. D., et al. (2013). Conformational landscapes of DNA polymerase I and mutator derivatives establish fidelity checkpoints for nucleotide insertion.* Nature Communications, 4, *2131. http://dx.doi.org/10.1038/ncomms3131.*

concentrations. In contrast, slower interconversions will lead to an increase of the FRET width peaking around K_{d1}. We normalized the FRET width of low-FRET species at different nucleotide concentrations vs the width of the corresponding binary complex, and we observed a broadening of the low-FRET species around the K_{d1} for most, if not all, nucleotides (Hohlbein et al., 2013). To estimate the conformational rates, we performed the FRET width analysis of simulated data as a function of interconversion rates. The simulations represented the four-state model (Fig. 3C), and we used four fixed rates (corresponding to fixed rate constants $k_{O \to C}$, $k_{C \to O}$, $k_{IN \to CN}$, and $k_{CN \to IN}$ as determined with PDA) and varied $k_{O \to IN}$ and $k_{IN \to O}$. The FRET width for each simulation was normalized vs the width of the corresponding binary complex. To further account for FRET broadening beyond the shot-noise limit, we added noise to each burst, a technique similar to the dithering technique described previously (Nir et al., 2006); this additional broadening does not change the position of the individual species. The width of the low-FRET species at a nucleotide concentration equal to K_{d1} remains constant for high IN-to-O rates ($\geq 300\ s^{-1}$), whereas for rates of $<300\ s^{-1}$, there is a significant width increase around K_{d1}. We thus proposed a rate of $300\ s^{-1}$ as an upper limit for the IN-to-O transition (Hohlbein et al., 2013).

5.5 Calculating the Free-Energy Landscape of Ternary Complexes

In order to construct the free-energy landscape for the prechemistry conformations of wild-type Pol I (KF) in the presence of nucleotides, we were able to calculate the relative Gibbs free energies ΔG for the partially closed conformation and the fully closed conformation (Hohlbein et al., 2013). For details on the procedure, the reader is referred to the original publication. Here we would like to emphasize that the free-energy landscape relies especially on the correct determination of the equilibrium constants between the conformational states and thereby the FRET species. If we compare, for example, older data of Pol I (KF) recorded using the first labeling procedure, in which a fraction of polymerases (around 20%) had unfavorable labeling geometry (Santoso, Joyce, et al., 2010), with newer data from the improved labeling scheme (Hohlbein et al., 2013), it becomes clear that any sophisticated data analysis requires excellent sample preparation. In fact, we believe that careful characterization of the sample using established and often ensemble-based techniques is a prerequisite.

6. FUTURE EXTENSIONS OF SINGLE-MOLECULE FRET ANALYSIS OF DNA POLYMERASE CONFORMATIONS

A severe drawback of diffusion-based smFRET measurements using labeled DNA polymerases is the limited observation time of the species in the confocal volume, which limits the context and timescales at which structural states and conformational changes can be probed. Several techniques have been described that effectively increase the observation time. First, increasing the effective viscosity of the solution, for example, by using long PEG polymers or polyacrylamide gels (Santoso & Kapanidis, 2009) will prolong the diffusion time, albeit it is important to ensure that the equilibrium distributions are not affected by the increased viscosity. Second, fluorescence recurrence analysis can be used to identify successive bursts that originate from one and the same diffusing molecule (Hoffmann, Nettels, Clark, et al., 2011). As a result, the conformational history of a species can be monitored over tens of milliseconds. A third approach used the wide-field detection of geometrically confined species as they diffuse in microfluidic channels (Tyagi, VanDelinder, Banterle, et al., 2014).

These approaches, however, rarely provide observation times longer than 10–100 ms, and thus cannot be used to monitor conformational changes occurring in the longer millisecond to seconds or even minute timescale. Furthermore, the analysis of repeated conformational changes within a single protein, either as part of a thermally driven fluctuation or due to multiple reaction turnovers, is hampered. To access longer timescales, tethering of individual polymerase molecule or its complexes on a solid support (typically a glass coverslip) is required. Although confocal scanning can be used to individually localize and detect immobilized complexes (Gouridis, Schuurman-Wolters, Ploetz, et al., 2015), we recommend the use of wide-field detection schemes, such as total internal reflection fluorescence (TIRF) microscopy, as they allow the highly parallelized detection of fluorescence from donor and acceptor labeled species over extended time periods (minutes).

Although the temporal resolution of TIRF-based smFRET is generally lower (~10 ms with the current FRET pairs and available instrumentation), strategies for considerably higher time resolution have been recently demonstrated (Farooq & Hohlbein, 2015; Juette, Terry, Wasserman, et al., 2016), paving the way for studying conformational changes over a large range of timescales using the same instrumentation. However, establishing an assay with a fully functional surface-attached protein requires careful surface passivation, selection of immobilization handles, and functional assays that ensure that immobilization does not affect the protein conformational equilibria (which can be examined through comparisons with distributions obtained using solution-based smFRET) and its reaction kinetics. For an example of looking directly at conformational states of Pol I (KF), we refer the reader to a TIRF-based study of several binary and ternary KF complexes with DNA and nucleotides (Evans, Hohlbein, Craggs, Aigrain, & Kapanidis, 2015), which also features an extensive supplementary information describing the experiments in greater detail.

Instead of using smFRET solely to monitor relative changes in distances between the donor and the acceptor fluorophores, we have also started to employ accurate FRET to define and explore novel DNA conformations in binary DNA-Pol complexes (Craggs et al. unpublished). The technical key points have been discussed elsewhere (Brunger, Strop, Vrljic, et al., 2011; Craggs & Kapanidis, 2012; Hohlbein et al., 2014; Kalinin, Peulen, Sindbert, et al., 2012) and include the consecutive measurement of multiple distances between labeling positions on both DNA polymerase and DNA, the precise implementation of FRET corrections, and modeling of dye conformations in combination with rigid-body docking and molecular dynamics simulations.

Finally, technical advances in the delivery of fluorescent biomolecules into cells as well as in single-molecule imaging and tracking in vivo are now permitting analysis of protein conformations inside living cells using smFRET (Sustarsic & Kapanidis, 2015): Microinjection has been used to deliver FRET constructs in eukaryotic (König et al., 2015; Sakon & Weninger, 2010); further, we have successfully demonstrated electroporation as a means of delivering doubly labeled constructs in live *E. coli* bacteria (Crawford, Torella, Aigrain, et al., 2013; Plochowietz, Crawford, & Kapanidis, 2014; Sustarsic, Plochowietz, Aigrain, et al., 2014). Using a doubly labeled Pol I (KF) electroporated into *E. coli*, we observed a FRET distribution consistent with time-averaged FRET fluctuations between the two main fingers conformations (Crawford et al., 2013). Further improvements in temporal resolution, image analysis, and site-specific labeling should allow us to visualize conformational changes at the single-molecule level as they perform their reactions in real time in their natural context, opening new exciting avenues for mechanistic analysis of biological processes in living cells.

ACKNOWLEDGMENTS

J.H. acknowledges support from a Marie Curie Career Integration Grant (#630992). A.N.K. was supported by the European Research Council (starter grant 261227), the UK BBSRC (grants BB/H01795X/1 and BB/J00054X/1), and a Wellcome Trust Investigatorship (110164/Z/15/Z). We also acknowledge our collaborators Cathy Joyce and Nigel Grindley who helped us establish single-molecule fluorescence assays of DNA polymerase, and Yusdi Santoso, Joseph Torella, Tim Craggs, Geraint Evans, Louise Aigrain, Ludovic Le Reste, Seamus Holden, and Pooya Soolizadeh, who worked on many aspects of our single-molecule studies of DNA polymerase in the Kapanidis laboratory over the past few years.

Conflict of Interest: The authors declare that they have no conflict of interest.

REFERENCES

Antonik, M., Felekyan, S., Gaiduk, A., & Seidel, C. A. M. (2006). Separating structural heterogeneities from stochastic variations in fluorescence resonance energy transfer distributions via photon distribution analysis. *The Journal of Physical Chemistry B*, *110*, 6970–6978. http://dx.doi.org/10.1021/jp057257.

Bai, X., McMullan, G., & Scheres, S. H. W. (2015). How cryo-EM is revolutionizing structural biology. *Trends in Biochemical Sciences*, *40*, 49–57. http://dx.doi.org/10.1016/j.tibs.2014.10.005.

Baker, T. A., & Kornberg, A. (1992). *DNA replication* (2nd ed.). . New York: W.H. Freeman.

Bebenek, K., Joyce, C. M., Fitzgerald, M. P., & Kunkel, T. A. (1990). The fidelity of DNA synthesis catalyzed by derivatives of *Escherichia coli* DNA polymerase I. *The Journal of Biological Chemistry*, *265*, 13878–13887.

Berezhna, S. Y., Gill, J. P., Lamichhane, R., & Millar, D. P. (2012). Single-molecule Förster resonance energy transfer reveals an innate fidelity checkpoint in DNA polymerase I. *Journal of the American Chemical Society*, *134*, 11261–11268. http://dx.doi.org/10.1021/ja3038273.

Brunger, A. T., Strop, P., Vrljic, M., et al. (2011). Three-dimensional molecular modeling with single molecule FRET. *Journal of Structural Biology*, *173*, 497–505. http://dx.doi.org/10.1016/j.jsb.2010.09.004.

Chakraborty, A., Wang, D., Ebright, Y. W., & Ebright, R. H. (2010). Azide-specific labeling of biomolecules by Staudinger-Bertozzi ligation phosphine derivatives of fluorescent probes suitable for single-molecule fluorescence spectroscopy. *Methods in Enzymology*, *472*, 19–30. http://dx.doi.org/10.1016/S0076-6879(10)72018-8.

Chatterjee, A., Sun, S. B., Furman, J. L., et al. (2013). A versatile platform for single- and multiple-unnatural amino acid mutagenesis in *Escherichia coli*. *Biochemistry (Moscow)*, *52*, 1828–1837. http://dx.doi.org/10.1021/bi4000244.

Craggs, T. D., & Kapanidis, A. N. (2012). Six steps closer to FRET-driven structural biology. *Nature Methods*, *9*, 1157–1158. http://dx.doi.org/10.1038/nmeth.2257.

Crawford, R., Torella, J. P., Aigrain, L., et al. (2013). Long-lived intracellular single-molecule fluorescence using electroporated molecules. *Biophysical Journal*, *105*, 2439–2450. http://dx.doi.org/10.1016/j.bpj.2013.09.057.

Deniz, A. A., Dahan, M., Grunwell, J. R., et al. (1999). Single-pair fluorescence resonance energy transfer on freely diffusing molecules: Observation of Förster distance dependence and subpopulations. *Proceedings of the National Academy of Sciences of the United States of America*, *96*, 3670–3675. http://dx.doi.org/10.1073/pnas.96.7.3670.

Deniz, A. A., Mukhopadhyay, S., & Lemke, E. A. (2008). Single-molecule biophysics: At the interface of biology, physics and chemistry. *Journal of the Royal Society, Interface*, *5*, 15–45. http://dx.doi.org/10.1098/rsif.2007.1021.

Eckert, K. A., & Kunkel, T. A. (1991). DNA polymerase fidelity and the polymerase chain reaction. *Genome Research*, *1*, 17–24. http://dx.doi.org/10.1101/gr.1.1.17.

Eggeling, C., Berger, S., Brand, L., et al. (2001). Data registration and selective single-molecule analysis using multi-parameter fluorescence detection. *Journal of Biotechnology*, *86*, 163–180.

Evans, G. W., Hohlbein, J., Craggs, T., Aigrain, L., & Kapanidis, A. N. (2015). Real-time single-molecule studies of the motions of DNA polymerase fingers illuminate DNA synthesis mechanisms, Nucleic Acids Research. *43*, 5998–6008. http://dx.doi.org/10.1093/nar/gkv547.

Farooq, S., Fijen, C., & Hohlbein, J. (2013). Studying DNA–protein interactions with single-molecule Förster resonance energy transfer. *Protoplasma*, *251*, 317–322. http://dx.doi.org/10.1007/s00709-013-0596-6.

Farooq, S., & Hohlbein, J. (2015). Camera-based single-molecule FRET detection with improved time resolution. *Physical Chemistry Chemical Physics*, *17*, 27862–27872. http://dx.doi.org/10.1039/C5CP04137F.

Felekyan, S., Sanabria, H., Kalinin, S., et al. (2013). Chapter two—Analyzing Förster resonance energy transfer with fluctuation algorithms. In S. Y. Tetin (Ed.), *Methods in enzymology: Vol. 519* (pp. 39–85): Academic Press.

Förster, T. (1946). Energiewanderung und Fluoreszenz. *Naturwissenschaften*, *33*, 166–175.

Förster, T. (1948). Zwischenmolekulare Energiewanderung und Fluoreszenz. *Annals of Physics*, *2*, 55–75.

Foster, M. P., McElroy, C. A., & Amero, C. D. (2007). Solution NMR of large molecules and assemblies. *Biochemistry (Moscow)*, *46*, 331–340. http://dx.doi.org/10.1021/bi0621314.

Freudenthal, B. D., Beard, W. A., Shock, D. D., & Wilson, S. H. (2013). Observing a DNA polymerase choose right from wrong. *Cell*, *154*, 157–168. http://dx.doi.org/10.1016/j.cell.2013.05.048.

García-Gómez, S., Reyes, A., Martínez-Jiménez, M. I., et al. (2013). PrimPol, an archaic primase/polymerase operating in human cells. *Molecular Cell*, *52*, 541–553. http://dx.doi.org/10.1016/j.molcel.2013.09.025.

Gopich, I. V., & Szabo, A. (2010). FRET efficiency distributions of multistate single molecules. *The Journal of Physical Chemistry B*, *114*, 15221–15226. http://dx.doi.org/10.1021/jp105359z.

Gouridis, G., Schuurman-Wolters, G. K., Ploetz, E., et al. (2015). Conformational dynamics in substrate-binding domains influences transport in the ABC importer GlnPQ. *Nature Structural & Molecular Biology*, *22*, 57–64. http://dx.doi.org/10.1038/nsmb.2929.

Gust, A., Zander, A., Gietl, A., et al. (2014). A starting point for fluorescence-based single-molecule measurements in biomolecular research. *Molecules*, *19*, 15824–15865. http://dx.doi.org/10.3390/molecules191015824.

Hoffmann, A., Nettels, D., Clark, J., et al. (2011). Quantifying heterogeneity and conformational dynamics from single molecule FRET of diffusing molecules: Recurrence analysis of single particles (RASP). *Physical Chemistry Chemical Physics*, *13*, 1857. http://dx.doi.org/10.1039/c0cp01911a.

Hohlbein, J., Aigrain, L., Craggs, T. D., et al. (2013). Conformational landscapes of DNA polymerase I and mutator derivatives establish fidelity checkpoints for nucleotide insertion. *Nature Communications*, *4*, 2131. http://dx.doi.org/10.1038/ncomms3131.

Hohlbein, J., Craggs, T. D., & Cordes, T. (2014). Alternating-laser excitation: Single-molecule FRET and beyond. *Chemical Society Reviews*, *43*, 1156–1171. http://dx.doi.org/10.1039/c3cs60233h.

Hohlbein, J., Gryte, K., Heilemann, M., & Kapanidis, A. N. (2010). Surfing on a new wave of single-molecule fluorescence methods. *Physical Biology*, 7, 31001. http://dx.doi.org/10.1088/1478-3975/7/3/031001.

Holden, S. J., Uphoff, S., Hohlbein, J., et al. (2010). Defining the limits of single-molecule FRET resolution in TIRF microscopy. *Biophysical Journal, 99*, 3102–3111. http://dx.doi.org/10.1016/j.bpj.2010.09.005.

Hübscher, U., Spadari, S., Villani, G., & Maga, G. (2010a). History and discovery of DNA polymerases. In *DNA polymerases* (pp. 1–58): World Scientific. http://dx.doi.org/10.1142/9789814299176_0001.

Hübscher, U., Spadari, S., Villani, G., & Maga, G. (2010b). DNA polymerases in the three kingdoms of life: Bacteria, archaea and eukaryotes. In *DNA polymerases* (pp. 59–83): World Scientific. http://dx.doi.org/10.1142/9789814299176_0001.

Ilari, A., & Savino, C. (2008). Protein structure determination by X-ray crystallography. In J. M. Keith (Ed.), *Bioinformatics* (pp. 63–87). Totowa, NJ: Humana Press.

Ingargiola, A., Lerner, E., Chung, S., Weiss, S., & Michalet, X. (2016). FRETBursts: An open source toolkit for analysis of freely-diffusing single-molecule FRET. *PLoS ONE, 11*, e0160716. http://dx.doi.org/10.1371/journal.pone.0160716.

Johnson, S. J., Taylor, J. S., & Beese, L. S. (2003). Processive DNA synthesis observed in a polymerase crystal suggests a mechanism for the prevention of frameshift mutations. *Proceedings of the National Academy of Sciences of the United States of America, 100*, 3895–3900. http://dx.doi.org/10.1073/pnas.0630532100.

Joyce, C. M., & Grindley, N. D. (1983). Construction of a plasmid that overproduces the large proteolytic fragment (Klenow fragment) of DNA polymerase I of *Escherichia coli*. *Proceedings of the National Academy of Sciences of the United States of America, 80*, 1830–1834.

Joyce, C. M., Potapova, O., DeLucia, A. M., et al. (2008). Fingers-closing and other rapid conformational changes in DNA polymerase I (Klenow fragment) and their role in nucleotide selectivity. *Biochemistry (Moscow), 47*, 6103–6116. http://dx.doi.org/10.1021/bi7021848.

Juette, M. F., Terry, D. S., Wasserman, M. R., et al. (2016). Single-molecule imaging of non-equilibrium molecular ensembles on the millisecond timescale. *Nature Methods, 13*, 341–344. http://dx.doi.org/10.1038/nmeth.3769.

Kalinin, S., Felekyan, S., Antonik, M., & Seidel, C. A. M. (2007). Probability distribution analysis of single-molecule fluorescence anisotropy and resonance energy transfer. *The Journal of Physical Chemistry. B, 111*, 10253–10262. http://dx.doi.org/10.1021/jp072293p.

Kalinin, S., Felekyan, S., Valeri, A., & Seidel, C. A. M. (2008). Characterizing multiple molecular states in single-molecule multiparameter fluorescence detection by probability distribution analysis. *The Journal of Physical Chemistry. B, 112*, 8361–8374. http://dx.doi.org/10.1021/jp711942q.

Kalinin, S., Peulen, T., Sindbert, S., et al. (2012). A toolkit and benchmark study for FRET-restrained high-precision structural modeling. *Nature Methods, 9*, 1218–1225. http://dx.doi.org/10.1038/nmeth.2222.

Kapanidis, A. N., Lee, N. K., Laurence, T. A., et al. (2004). Fluorescence-aided molecule sorting: Analysis of structure and interactions by alternating-laser excitation of single molecules. *Proceedings of the National Academy of Sciences of the United States of America, 101*, 8936–8941. http://dx.doi.org/10.1073/pnas.0401690101.

Kapanidis, A., & Strick, T. (2009). Biology, one molecule at a time. *Trends in Biochemical Sciences, 34*, 234–243. http://dx.doi.org/10.1016/j.tibs.2009.01.008.

Kim, H., & Ha, T. (2013). Single-molecule nanometry for biological physics. *Reports on Progress in Physics, 76*, 16601. http://dx.doi.org/10.1088/0034-4885/76/1/016601.

König, I., Zarrine-Afsar, A., Aznauryan, M., et al. (2015). Single-molecule spectroscopy of protein conformational dynamics in live eukaryotic cells. *Nature Methods, 12*, 773–779. http://dx.doi.org/10.1038/nmeth.3475.

Lakowicz, J. R. (2006). *Principles of fluorescence spectroscopy* (3rd ed.). Berlin: Springer.

Lang, K., & Chin, J. W. (2014). Bioorthogonal reactions for labeling proteins. *ACS Chemical Biology*, *9*, 16–20. http://dx.doi.org/10.1021/cb4009292.

Lehman, I. R., Bessman, M. J., Simms, E. S., & Kornberg, A. (1958). Enzymatic synthesis of deoxyribonucleic acid. I. Preparation of substrates and partial purification of an enzyme from *Escherichia coli*. *The Journal of Biological Chemistry*, *233*, 163–170.

Markiewicz, R. P., Vrtis, K. B., Rueda, D., & Romano, L. J. (2012). Single-molecule microscopy reveals new insights into nucleotide selection by DNA polymerase I. *Nucleic Acids Research*, *40*, 7975–7984. http://dx.doi.org/10.1093/nar/gks523.

Moerner, W. E. (2007). New directions in single-molecule imaging and analysis. *Proceedings of the National Academy of Sciences of the United States of America*, *104*, 12596–12602. http://dx.doi.org/10.1073/pnas.0610081104.

Nir, E., Michalet, X., Hamadani, K. M., et al. (2006). Shot-noise limited single-molecule FRET histograms: Comparison between theory and experiments. *The Journal of Physical Chemistry. B*, *110*, 22103–22124. http://dx.doi.org/10.1021/jp063483n.

Plochowietz, A., Crawford, R., & Kapanidis, A. N. (2014). Characterization of organic fluorophores for in vivo FRET studies based on electroporated molecules. *Physical Chemistry Chemical Physics*, *16*, 12688–12694. http://dx.doi.org/10.1039/C4CP00995A.

Preus, S., & Wilhelmsson, L. M. (2012). Advances in quantitative FRET-based methods for studying nucleic acids. *ChemBioChem*, *13*, 1990–2001. http://dx.doi.org/10.1002/cbic.201200400.

Rothwell, P. J., Allen, W. J., Sisamakis, E., et al. (2013). dNTP-dependent conformational transitions in the fingers subdomain of Klentaq1 DNA polymerase: Insights into the role of the "nucleotide-binding" state. *The Journal of Biological Chemistry*, *288*, 13575–13591. http://dx.doi.org/10.1074/jbc.M112.432690.

Roy, R., Hohng, S., & Ha, T. (2008). A practical guide to single-molecule FRET. *Nature Methods*, *5*, 507–516. http://dx.doi.org/10.1038/nmeth.1208.

Sakon, J. J., & Weninger, K. R. (2010). Detecting the conformation of individual proteins in live cells. *Nature Methods*, *7*, 203–205. http://dx.doi.org/10.1038/nmeth.1421.

Santoso, Y., Joyce, C. M., Potapova, O., et al. (2010). Conformational transitions in DNA polymerase I revealed by single-molecule FRET. *Proceedings of the National Academy of Sciences of the United States of America*, *107*, 715–720. http://dx.doi.org/10.1073/pnas.0910909107.

Santoso, Y., & Kapanidis, A. N. (2009). Probing biomolecular structures and dynamics of single molecules using in-gel alternating-laser excitation. *Analytical Chemistry*, *81*, 9561–9570. http://dx.doi.org/10.1021/ac901423e.

Santoso, Y., Torella, J. P., & Kapanidis, A. N. (2010). Characterizing single-molecule FRET dynamics with probability distribution analysis. *ChemPhysChem*, *11*, 2209–2219. http://dx.doi.org/10.1002/cphc.201000129.

Schotte, F., Lim, M., Jackson, T. A., et al. (2003). Watching a protein as it functions with 150-ps time-resolved X-ray crystallography. *Science*, *300*, 1944–1947. http://dx.doi.org/10.1126/science.1078797.

Sustarsic, M., & Kapanidis, A. N. (2015). Taking the ruler to the jungle: Single-molecule FRET for understanding biomolecular structure and dynamics in live cells. *Current Opinion in Structural Biology*, *34*, 52–59. http://dx.doi.org/10.1016/j.sbi.2015.07.001.

Sustarsic, M., Plochowietz, A., Aigrain, L., et al. (2014). Optimized delivery of fluorescently labeled proteins in live bacteria using electroporation. *Histochemistry and Cell Biology*, *142*, 113–124. http://dx.doi.org/10.1007/s00418-014-1213-2.

Tomov, T. E., Tsukanov, R., Masoud, R., et al. (2012). Disentangling subpopulations in single-molecule FRET and ALEX experiments with photon distribution analysis. *Biophysical Journal*, *102*, 1163–1173. http://dx.doi.org/10.1016/j.bpj.2011.11.4025.

Torella, J. P., Holden, S. J., Santoso, Y., et al. (2011). Identifying molecular dynamics in single-molecule FRET experiments with burst variance analysis. *Biophysical Journal, 100*, 1568–1577. http://dx.doi.org/10.1016/j.bpj.2011.01.066.

Tyagi, S., VanDelinder, V., Banterle, N., et al. (2014). Continuous throughput and long-term observation of single-molecule FRET without immobilization. *Nature Methods, 11*, 297–300. http://dx.doi.org/10.1038/nmeth.2809.

Walter, N., Huang, C., Manzo, A., & Sobhy, M. (2008). Do-it-yourself guide: How to use the modern single-molecule toolkit. *Nature Methods, 5*, 475–489. http://dx.doi.org/10.1038/nmeth.1215.

Weiss, S. (1999). Fluorescence spectroscopy of single biomolecules. *Science, 283*, 1676–1683. http://dx.doi.org/10.1126/science.283.5408.1676.

Wu, E. Y., & Beese, L. S. (2011). The structure of a high fidelity DNA polymerase bound to a mismatched nucleotide reveals an "ajar" intermediate conformation in the nucleotide selection mechanism. *The Journal of Biological Chemistry, 286*, 19758–19767. http://dx.doi.org/10.1074/jbc.M110.191130.

Wüthrich, K. (2001). The way to NMR structures of proteins. *Nature Structural & Molecular Biology, 8*, 923–925. http://dx.doi.org/10.1038/nsb1101-923.

CHAPTER THIRTEEN

Methods for Investigating DNA Accessibility with Single Nucleosomes

M.D. Gibson[1], M. Brehove[1], Y. Luo[2], J. North[3], M.G. Poirier[4]

The Ohio State University, Columbus, OH, United States

[4]Corresponding author: e-mail address: poirier.18@osu.edu

Contents

Abstract

Nucleosomes are the fundamental organizing unit of all eukaryotic genomes. Understanding how proteins gain access to DNA-binding sites located within nucleosomes is important for understanding DNA processing including transcription, replication, and repair. Single-molecule total internal reflection fluorescence (smTIRF) microscopy measurements can provide key insight into how proteins gain and maintain access

[1] Authors contributed equally.

[2] Current address: Wyss Institute for Biologically Inspired Engineering Harvard University, Boston MA, 02115, United States.

[3] Current address: Department of Microbiology, The Ohio State University, Columbus Ohio, 43210, United States.

ISSN 0076-6879

http://dx.doi.org/10.1016/bs.mie.2016.08.014

© 2016 Elsevier Inc. All rights reserved.

to DNA sites within nucleosomes. Here, we describe methods for smTIRF experiments including the preparation of fluorophore-labeled nucleosomes, the smTIRF system, data acquisition, analysis, and controls. These methods are presented for investigating transcription factor binding within nucleosomes. However, they are applicable for investigating the binding of any site-specific DNA-binding protein within nucleosomes.

1. INTRODUCTION

The fundamental organizing unit of all eukaryotic genomes is the nucleosome, which contains ~146 bp of DNA wrapped around a histone protein octamer core (Luger, Mader, Richmond, Sargent, & Richmond, 1997). Nucleosomes are repeatedly assembled along genomic DNA to form long chromatin fibers (Kornberg, 1974, 1977). Both nucleosome and chromatin structures control DNA accessibility to complexes that regulate DNA processes, including transcription, repair, and replication (Almouzni & Cedar, 2016; Li, Carey, & Workman, 2007; Sinha & Peterson, 2009). Nucleosome and chromatin structures are dynamic (Blossey & Schiessel, 2011; Chereji & Morozov, 2015; Venkatesh & Workman, 2015; Widom, 2001), providing transient exposure of buried DNA target sites to DNA-binding proteins. In addition, numerous factors regulate nucleosome and chromatin structural dynamics including histone posttranslational modifications (PTMs) (Bowman & Poirier, 2015), proteins that specifically bind these modifications (Lalonde, Cheng, & Cote, 2014; Musselman, Khorasanizadeh, & Kutateladze, 2014), chromatin-remodeling complexes (Narlikar, Sundaramoorthy, & Owen-Hughes, 2013), and chromatin architectural proteins (Li & Reinberg, 2011). Yet, how nucleosome and chromatin dynamics are controlled to regulate DNA accessibility is not well understood.

There are a number of approaches used to investigate DNA accessibility within nucleosomes. One important approach is the use of DNA endonucleases. Numerous studies have used DNase I, which cleaves DNA with significant sequence bias (Baek, Sung, & Hager, 2012; Lu & Richardson, 2004; Shibata & Crawford, 2009). Therefore, DNaseI digestions result in a digestion pattern, which changes when the DNA is wrapped within nucleosomes. This provides insight into where nucleosomes restrict DNA accessibility. Alternatively, restriction endonucleases can be used to quantitatively measure accessibility to their specific DNA target sequence located within a nucleosome (Anderson & Widom, 2000; Polach & Widom, 1995, 1999). More recent approaches use fluorescence resonance energy transfer (FRET) to quantify nucleosome and chromatin dynamics, including DNA unwrapping, nucleosome repositioning, nucleosome disassembly, and DNA–protein binding within the nucleosome (Andrews, Chen, Zevin, Stargell, & Luger, 2010; Li, Levitus, Bustamante, & Widom, 2005; Li & Widom, 2004; North et al., 2012; Park, Dyer, Tremethick, & Luger, 2004; Tims & Widom, 2007). These ensemble measurements have provided important insight into how nucleosomes and chromatin regulate DNA accessibility.

An advantage of using fluorescence and FRET to detect nucleosome structural changes is that fluorescence can be detected at the single molecule level (Ha et al., 1996, 1999; Nie & Zare, 1997; Schutz, Trabesinger, & Schmidt, 1998). This has enabled the application of single-molecule total internal reflection fluorescence (smTIRF) microscopy to investigate nucleosome dynamics including chromatin remodeling (Blosser, Yang, Stone, Narlikar, & Zhuang, 2009; Deindl et al., 2013), nucleosome unwrapping (Neumann et al., 2009; Wei, Falk, Black, & Lee, 2015), transcription factor (TF) binding within nucleosomes (Luo, North, Rose, & Poirier, 2014), nucleosome assembly (Lee, Lee, Yue, & Lee, 2015), nucleosome gapping (Ngo & Ha, 2015), and heterochromatin protein 1 (HP1) interactions with nucleosomes (Kilic, Bachmann, Bryan, & Fierz, 2015). These single-molecule studies provide important mechanistic insight into the regulation of DNA accessibility by determining the distribution of nucleosome structural states and detecting transient states that are difficult to detect by ensemble measurements. Furthermore, smTRIF measurements simultaneously determine each equilibrium transition rate between nucleosome states with minimal sample in contrast to ensemble rate measurement techniques like stopped-flow (Tims & Widom, 2007). This type of data can then be modeled to provide further insight into the structural dynamics that regulates DNA accessibility (Chen & Bundschuh, 2014).

Here, we report the smTIRF methods for investigating the interactions of DNA-binding proteins with DNA sites located within and without single nucleosomes. We focus on TFs, but these methods are applicable to any protein that binds DNA sites within nucleosomes. We detect TF binding within nucleosomes by detecting the reduction in FRET efficiency resulting from the TF binding to its site within the nucleosome and trapping the nucleosome in a partially unwrapped state (Fig. 1A–C). We separately use single-molecule protein-induced fluorescence enhancement (smPIFE) (Hwang, Kim, & Myong, 2011) to detect TF binding to nucleosome-free DNA (Fig. 1D). Together, these measurements enable quantification of nucleosome dynamics and how they are regulated to control DNA accessibility.

Application of smTIRF to quantify protein–DNA binding within a nucleosome involves four key components: (i) nucleosome sample preparation, (ii) smTIRF instrumentation, (iii) single-molecule nucleosome data

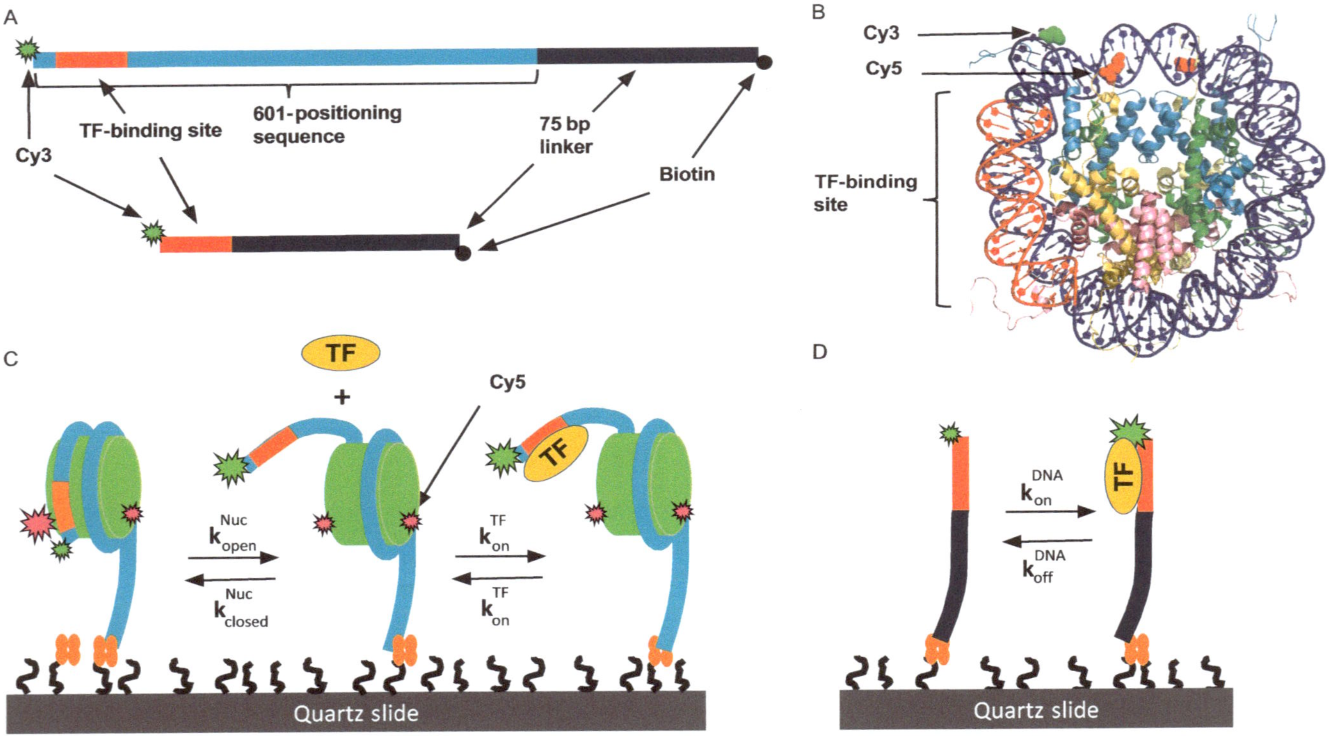

Fig. 1 (A) Diagram of DNA constructs used for single-molecule experiments measuring transcription factor association and dissociation rates while binding to sites within nucleosomes (*top*) and to naked DNA (*bottom*). A biotin/streptavidin linkage at one end affixes the molecule to the surface of a quartz slide and a donor Cy3 molecule on the other is excited through a TIR evanescent field. (B) Crystal structure of the nucleosome core particle (PDB ID: 1KX5). The TF-binding site is highlighted at bases 8–27 in *red* with the positions of the Cy3/Cy5 FRET pair as *green* and *red spheres*, respectively. (C) Three state model of TF binding within a nucleosome. A normally wrapped nucleosome in a high FRET state undergoes thermal fluctuations to transition from state 1 to 2 at which point the TF-binding site is exposed and a TF can bind and lock the nucleosome in an unwrapped, low FRET state. (D) Two state model of a TF binding to free DNA. The protein binding changes the local environment of the attached Cy3 molecule and increases its quantum efficiency, detectable as an increase in fluorescence intensity from the molecule.

acquisition and processing, (iv) comparison of single-molecule and ensemble measurements. The detailed nucleosome sample preparation presented here is modified and optimized from established protocols (Luger, Rechsteiner, & Richmond, 1999a, 1999b) for efficient and cost-effective incorporation of fluorescent labels for use in ensemble and smTIRF measurements. The smTIRF microscopy and flow cell preparation are largely based on previously published protocols (Roy, Hohng, & Ha, 2008) with modifications for the study of TF binding to nucleosomes and nucleosome-free DNA. Fully detailed methods for smTIRF and ensemble data analysis of TF binding and nucleosome dynamics are provided in Luo, North, and Poirier (2014). Therefore, we outline the single-molecule system, sample preparation, data acquisition, and analysis with emphasis on important details specific to smTIRF measurements of TF binding to nucleosomes and nucleosome-free DNA.

2. PREPARATION OF FLUOROPHORE-LABELED NUCLEOSOMES FOR SINGLE-MOLECULE FLUORESCENCE MEASUREMENTS

The preparation and labeling strategy of nucleosomes for smTIRF measurements is designed to help ensure that a majority of the nucleosome population is homogenous and not affected by surface immobilization. The strategy used for our TF-binding experiments is shown in Fig. 1. Our approach involves three key aspects: (i) inclusion of a DNA extension, which projects the nucleosomes away from the smTIRF flow cell surface; (ii) inclusion of one fluorophore label on either H2A or H2B, which allows for partially disassembled nucleosomes to be detected since the H2A–H2B heterodimer is the first part of the nucleosome to disassociate; (iii) purification of the nucleosomes, which minimizes the amount of free DNA, depositioned nucleosomes, and DNA–histone aggregates in the sample. We describe several methods for preparing fluorophore-labeled histone octamer, which allows for incorporation of multiple fluorophores or other modifications into the histone octamer. We then describe the preparation of fluorophore-labeled DNA and nucleosomes.

The refolding and labeling methods presented here are broader than what is required to perform the TF-binding experiments shown in Fig. 1. For such experiments whole octamer can simply be labeled, purified, and incorporated into nucleosomes. However, alternative labeling strategies may require multiple labels on the histone octamer, which requires either labeling unfolded histones, or separately labeling refolded H2A/H2B heterodimer and H3/H4 tetramer.

2.1 Histone Octamer, Tetramer, and Heterodimer Refolding

Histones are expressed and purified as previously described (Luger et al., 1999a, 1999b) and stored at −80°C for later refolding. There are two methods of octamer refolding: (i) refolding full octamer directly and (ii) refolding dimer and tetramer separately before combining them into full octamer. The appropriate method depends on the number of labels to be incorporated into the histone octamer.

Required for these protocols:

- Unfolding buffer
  - 7 *M* Guanidine–HCl (Biomedicals 820539)
  - 10 m*M* Tris–HCl, pH 7.5 (Sigma T1503)
  - 10 m*M* Dithiothreitol (DTT, Sigma 43816)
- Refolding buffer
  - 2 *M* NaCl (Sigma S9888)
  - 10 m*M* Tris–HCl, pH 7.5 (Sigma T1504)
  - 1 m*M* ethylenediaminetetraacetic acid (EDTA, Sigma E1644), pH 8.0
  - 2 m*M* 2-mercaptoethanol (BME, Sigma M3145). Add just before use, leave out for size exclusion elution buffer

2.1.1 Octamer Refolding

Directly refolding the histone octamer with all four histones is the recommended approach if the octamer is to be labeled with only one fluorophore species, which minimizes the number of purification steps. For this we recommend using the well-studied *Xenopus laevis* histones (Luger et al., 1999a, 1999b) which contain a single cysteine at H3(C110). Mutation of H3(C110) to an alanine minimally impacts nucleosome structure and dynamics (Ferreira, Flaus, & Owen-Hughes, 2007; Shogren-Knaak et al., 2006) but does slightly impact nucleosome unwrapping (Shimko, North, Bruns, Poirier, & Ottesen, 2011), which should be taken into account if using H3(C110A) vs H3(C110)—containing nucleosomes. For detecting nucleosome unwrapping by TF binding, we recommend using H2A (K119C) as the labeling site.

1. Refolding begins by resuspending lyophilized histone proteins. Resuspend aliquots of H3(C110A), H2A(K119C), H2B, and H4 in unfolding buffer to no more than 5 mg/mL to avoid aggregation. Incubate on ice for 1 h, thoroughly mix, and centrifuge at 20,000 × *g* to pellet aggregates before determining concentration of each histone on a UV–VIS spectrophotometer blanked against the same unfolding buffer used to

resuspend the histones. If the concentration is much greater than 5 mg/mL, dilute the solution with unfolding buffer to 5 mg/mL. Incubate on ice for 15 min, thoroughly mix, centrifuge again, and remeasure the concentration.

2. Based on the concentration of each histone, combine equal moles (not mass) of H3 and H4 with 10% molar excess H2A and H2B giving a 1/1/1.1/1.1 molar ratio of H3/H4/H2A/H2B. The excess H2A and H2B help ensure complete incorporation of H3–H4 tetramer into full octamer, given separation of tetramer from octamer is difficult in subsequent purification steps. Unused, unfolded histones may be flash frozen in liquid nitrogen and stored at −80°C. Keep record of stock concentration as UV absorption measurements using other unfolding buffer preparations as the blank will not be accurate due to variations in guanidine–HCl concentration.
3. Unfolded histones will be refolded through double dialysis. We have found that double dialysis results in a sufficiently slow rate in change of buffer concentration to maximize efficient octamer refolding. Table 1 shows the proper volumes and recommended tubes for making a microdialysis chamber as described previously (Thastrom, Lowary, & Widom, 2004). These microdialysis chambers are placed within a 25 cm × 50 mm, piece of 6000–8000MWCO dialysis tubing filled with about 50 mL of unfolding buffer.
4. Dialyze the 50 mL dialysis bag containing the microdialysis chambers against three changes of 2 L refolding buffer for at least 6 h per change with the last change going overnight.
5. Reclaim the now refolded histones by first placing the microdialysis chamber membrane-up on a sheet of parafilm. Next remove any excess

Table 1 Recommendations for Inner Dialysis Chambers for Use with a Range of Final Sample Volumes

Sample Volume	Microdialysis Chamber
50 μL	BrandTech 0.5-mL PCR tube (781310)
250 μL	Eppendorf 1.5-mL PCR tube (022364120)
>500 μL	10 mm 6000–8000MWCO dialysis tubing (Spectrum Labs 1332645)

Double dialysis slows the rate of buffer exchange and is appropriate for histone octamer refolding and nucleosome reconstitution. In which case, the inner membrane will be placed inside a larger outer membrane.

liquid on top of the dialysis membrane and then use a pipette to pierce the membrane and remove the sample within.

6. Purification of refolded octamer is necessary to remove aggregates and free histones. Purification is performed by size exclusion chromatography on a GE Superdex 200 10/300 GL column preequilibrated with 2 column volumes of refolding buffer. The column is capable of baseline resolution between octamer and dimer, but not between octamer and tetramer. Elute at 0.5 mL/min for 1 column volume in refolding buffer without BME and collect 0.5 mL fractions. Full octamer will elute at about 15 mL followed by dimer at roughly 18 mL.

   Note: Refoldings smaller than 250 μg total mass may be completely lost during purification, whereas resolution will begin to decrease for octamer refoldings larger than 4000 μg. In the latter case multiple injections can be performed or use the larger GE HiLoad Superdex 200 16/600 PG column.

7. Analyze the fractions by SDS-PAGE on a 12% gel and stain with coomassie Brilliant Blue. Pool fractions containing equal concentrations of H3, H2A, H2B, and H4 into an Amicon 30k MWCO centrifugal filter (Millipore UFC-203024) and concentrate to a final concentration of roughly 10–30 μM. Measure octamer concentration by UV–VIS spectrophotometer at 276 nm blanked against refolding buffer without BME using an absorbance coefficient of $\varepsilon = 48{,}825\ \text{cm}^{-1}\text{M}^{-1}$ (Luger et al., 1999a, 1999b). Concentrated octamer can be stored on ice at 4°C for a few months. For longer storage, add glycerol to a final concentration of 40% and store at −20°C.

2.1.2 Octamer Preparation with Separately Refolded Heterodimer and Tetramer

H2A/H2B dimer and H3/H4 tetramer can be separately refolded and then combined to form octamer before final purification. This allows the tetramer and dimer to be labeled separately. In addition, it allows for the same fluorophore-labeled tetramer or heterodimer to be matched with different histone proteins such as variants, mutants, or different PTMs and ensure the labeling efficiency is identical between samples.

1. Unfold and quantify each histone as in Section 2.1.1.
2. Combine only the tetramer-forming histones, H3 and H4, to a tube in a 1:1 molar ratio. If a specific histone is more precious, add up to 10% excess of the other to ensure better incorporation of the more precious

histone into the tetramer. Refold H2A and H2B into heterodimer using the same method in a different tube.

3. Dialyze as described in Section 2.1.1.
4. Labeling reactions may now be carried out separately on the tetramer and dimer. Purify away excess label by gel filtration on GE Superdex 200 10/300 column as described in Section 2.1.1. Pool desired fraction and concentrate to about 50 μ*M* using Amicon 30k MWCO centrifugal filters. The detailed labeling and purification protocol will be described in Section 2.2.1.
5. Combine dimer and tetramer in a 2.2:1 molar ratio. If either is labeled with fluorophores, be sure to take the absorbance of the fluorophore at 276 nm into account when calculating the protein's molar concentration. If the change in histone extinction coefficient is unknown, it may be necessary to add even more dimer to ensure an excess of dimer upon purification. However, too much dimer can reduce resolution during purification.
6. Place the combined histones on ice and incubate for at least 1 h.
7. Purify the combined tetramer and dimer by size exclusion chromatography on a GE Superdex 200 10/300 GL as in Section 2.1.1.
8. Analyze the fractions by SDS-PAGE on a 12% gel. Pool positive fractions, concentrate, and store as described earlier.

2.2 Fluorophore Labeling of the Histone

There are two general labeling strategies for detecting structural changes within the nucleosome with FRET. One approach is to label the DNA around a nucleosome with both FRET donor and acceptor to probe nucleosome structural changes (Lee & Lee, 2012; Ngo & Ha, 2015), while the alternative is to attach one label to the DNA and the other to a histone protein (Fig. 1). We find it beneficial to place the acceptor fluorophore on the H2A–H2B heterodimer (e.g., H2A(K119C)) because this allows for direct determination of which molecules contain two heterodimers and hence a full octamer, since heterodimers dissociate from nucleosomes before the H3–H4 tetramer (Bohm et al., 2011; Oohara & Wada, 1987). Moreover, The $(\text{Ratio})_A$ method used for determining FRET efficiencies is not impacted by the labeling efficiency of the acceptor fluorophore (Clegg, 1992), which can vary between 70% and 95%.

There are three approaches for labeling the histone octamer: (i) directly label refolded octamer; (ii) label refolded tetramer and/or heterodimer, which can then be combined to form the histone octamer; (iii) label unfolded single histones, which can then be refolded into the histone octamer. Here, we only discuss maleimide labeling of unique cysteine residues. However, many other chemistries are possible with histones, which have enabled the incorporation of a range of modifications including electron paramagnetic spin labels (Bowman, Ward, El-Mkami, Owen-Hughes, & Norman, 2010; Gao et al., 2013) and PTMs (Fierz & Muir, 2012; Howard, Yu, Gardner, Shimko, & Ottesen, 2015). These chemistries can often be combined with fluorophore labeling by using one of the three approaches for labeling histone octamer (Brehove et al., 2015; Musselman et al., 2013).

We have found that labeling folded histone octamer, tetramer, or heterodimer yields higher labeling efficiency over unfolded, individual histones. We therefore recommend labeling folded structures when possible. Fully formed octamer requires the fewest purifications and so is the preferred method for singly labeled octamer. If two labels are required with one on either H2A or H2B and the other on H3 or H4, separately labeling dimer and tetramer will likely yield the best results. Finally, if more than one label is required within the heterodimer or tetramer, histones must be labeled individually in unfolding conditions.

2.2.1 Refolded Octamer, Tetramer, and Heterodimer Labeling

Histones can be labeled site-specifically on cysteines after being refolded into octamer, dimer, or tetramer.

Buffers required for this protocol:

- 200 m*M* tris(2-carboxyethyl)phosphine (TCEP, Sigma C4706) pH to 7.1 with NaOH, store at −80°C
- PIPES buffer
  - 5 m*M* Sodium PIPES, pH 6.1 (Sigma P6757)
  - 2 *M* NaCl
- 2 *M* HEPES (Sigma H3375) pH to 7.1 with NaOH
- Anhydrous DMF (Sigma 227056)
- 1 *M* dithiothreitol (DTT)

1. Refold 0.1–1 mg of each histone into histone octamer, H3/H4 tetramer, or H2A/H2B dimer as described in Section 2.1.2. Do not purify refolded histone complexes until after labeling. Refolded histone stocks should be stored above 1.5 mg/mL.
2. Reduce histones by adding 200 m*M* TCEP pH 7.1 to 10 m*M* final concentration and incubate on ice for 30 min.

3. Dialyze histones in an appropriate microdialysis chamber as indicated by Table 1 against three changes of at least 100-fold excess volume of PIPES buffer for at least 3 h per change, with the last change going overnight. This promotes the free thiol state of the cysteine and removes any reducing agents which can lower final labeling efficiency.
4. Reclaim histones and measure the absorbance of the octamer/tetramer/dimer at 276 nm. For octamer $\varepsilon = 48{,}825\ \text{cm}^{-1}\text{M}^{-1}$ (Luger et al., 1999a, 1999b); for dimer $\varepsilon = 13{,}035\ \text{cm}^{-1}\text{M}^{-1}$; for tetramer $\varepsilon = 22{,}371\ \text{cm}^{-1}\text{M}^{-1}$ (Godfrey, Baxevanis, & Moudrianakis, 1990).
5. Dilute histone octamer to 1.5–3 mg/mL (tetramer or dimer to 0.75–1.5 mg/mL) with 5 m*M* PIPES pH 6.1, 2 *M* NaCl.
6. Place sample in an air-tight vial fitted with a needle inlet and outlet. Purge oxygen from sample using a slow and steady stream of argon for 30 min at 4°C which has been prepurified over an oxygen trap. Ensure that the sample does not bubble due to the argon stream, leading to sample degradation.
7. At the same time, purge with argon an aliquot of 2 *M* HEPES pH 7.1 as earlier, however, in this case directly bubble argon through the buffer for more efficient oxygen removal.
8. Equilibrate maleimide-functionalized fluorophore (GE Healthcare PA25031 for cy5-labeled histones) to room temperature.
9. Dissolve 0.25 mg of maleimide-functionalized fluorophore in 14 μL of anhydrous DMF to 18 mg/mL final concentration.
10. When ready to label, remove samples and HEPES for argon stream and seal vials. Equilibrate to room temperature for 5–10 min. At this point, it is necessary to act quickly. Oxygen is leaking back into the sample and HEPES buffer.
11. To the histone sample add 2 *M* HEPES pH 7.1 to 100 m*M* final concentration to rapidly raise the pH. Immediately begin adding maleimide-functionalized fluorophore in 2 μL additions mixing well between additions to limit micelle formation of the hydrophobic label until a 7.5-fold molar excess of label is reached. Unused label can be dried and stored at 4°C in a desiccator.
12. Incubate at room temperature for 1 h on a lab rotisserie, and then transfer sample and rotisserie to 4°C overnight.
13. Quench the reaction with 10 m*M* final concentration DTT and store at 4°C on ice until purification by size exclusion chromatography on a Superdex 200 GL column as described in Section 2.1.1.

2.2.2 Labeling Unfolded Histones

Individual histones can be site-specifically labeled at a single cysteine mutation under unfolding conditions.

Required for this protocol:

- 7 *M* Guanidine–HCL (Biomedicals 820539)
- 200 m*M* TCEP, pH 7.1 (see Section 2.2.1)
- PIPES-guanidine buffer
  - 5 m*M* Sodium PIPES, pH 6.1 (Sigma P6757)
  - 2 *M* Guanidine–HCl
- 2 *M* HEPES (Sigma H3375) pH to 7.1 with NaOH
- Anhydrous dimethylformamide (DMF, Sigma 227056)
- TU1000 buffer
  - 7 *M* Urea
  - 20 m*M* Tris–HCl, pH 9.0 (Sigma T1503)
  - 1 m*M* EDTA, pH 8.0
  - 1 *M* NaCl
  - Deionize with mixed bed resin and filter; add BME to 1 m*M* final concentration
- 5 m*M* BME in dH_2O

1. Resuspend 1–5 mg of lyophilized histones in 7 *M* guanidine to 4–5 mg/mL final concentration.
2. Reduce histones as in Section 2.2.1 with TCEP (final concentration 10 m*M*) and incubate on ice for 30 min.
3. Dialyze histones against three changes of at least 100-fold volume excess PIPES-guanidine buffer for 2–3 h per change with the last change going overnight.
4. Measure the concentration of the histones and dilute to 150 μ*M* with PIPES-guanidine buffer.
5. Purge both the histones and HEPES under argon as in Section 2.2.1.
6. Equilibrate maleimide-functionalized fluorophore (GE Healthcare PA25031 for cy5-labeled histones) to room temperature.
7. Dissolve 0.25 mg of maleimide-functionalized fluorophore in 14 μL of anhydrous DMF to 18 mg/mL final concentration.
8. Add fluorophore as in Section 2.2.1 in 2 μL additions until a 7.5-fold molar excess of label is reached
9. Allow the reaction to proceed on a lab rotisserie for 1 h at room temperature before transferring to 4°C overnight as in Section 2.2.1.
10. Quench the reaction with 10 m*M* final concentration DTT and purify by size exclusion chromatography on a Sephadex G-25 column

equilibrated with TU1000 buffer at room temperature. Elute sample at 1 mL/min collecting 0.5 mL fractions.

11. Analyze the fractions by SDS-PAGE on a 12% gel. Bands corresponding to free dye and labeled histone are visible by the naked eye or by fluorescence imaging.
12. Pool selected fractions and dialyze against 500 mL of dH_2O with BME at 4°C with at least three changes.
13. Reclaim labeled histones, aliquot, and vacuum concentrate until dry. Store at −80°C.

2.3 Preparation of Labeled DNA

To detect nucleosomes trapped by a DNA-binding protein in partially unwrapped states by FRET, the DNA needs to be labeled so that the fluorophore is positioned within the Förster radius of the histone label in the fully wrapped state. The DNA label then needs to shift further away from the histone label in the partially wrapped state. Labeled DNA constructs used for TF-binding experiments are shown in Fig. 1A. Both the fluorophore label and biotin label are attached to oligonucleotides that are then incorporated into duplex DNA via the polymerase chain reaction (PCR). A bacterial plasmid that contains a nucleosome-positioning sequence (NPS) such as the Widom 601 sequence (Lowary & Widom, 1998) should be used as the template for the PCR. An NPS helps ensure that the DNA label is correctly positioned relative to the histone octamer fluorophore such as H2A(K119C). It is important to include additional DNA between the NPS and the biotin label to extend the nucleosome away from the flow cell surface for smTIRF microscopy. We and others have found that 75 bp are sufficient (Blosser et al., 2009; Luo, North, & Poirier, 2014; Luo, North, Rose, et al., 2014). We typically purchase the biotin-labeled DNA primer, whereas for the fluorophore-labeled primer, we label a purchased amine-containing primer. We recommend attaching the donor molecule, Cy3, to the DNA primer. Depending on the required location of the fluorophore on the DNA, it is attached via an amine-functionalized six carbon linker on either the 5′-end or to a modified thymine in the DNA primer. The primer labeling procedure is identical regardless of labeling position, but internal thymine labels tend to have lower labeling efficiencies. However, because the labeled DNA primer is purified away from the unlabeled primer by reversed-phase HPLC, the final labeling efficiency is nearly 100%.

This protocol is similarly used to prepare DNA for PIFE experiments that are used to detect protein binding to a target site in nucleosome-free DNA. The key aspects to the design of the DNA for PIFE experiments are that a Cy3 fluorophore is typically used and the fluorophore needs to be positioned within 1–3 bp from the binding site. Additionally, a Cy5 fluorophore is placed on the end of the DNA outside of the Förster radius as an internal control for ensemble and single-molecule PIFE measurements (see Fig. 1D).

Required for this protocol:

- 98% Ethanol (Sigma 270741)
- 75% Ethanol, −20°C
- 5 *M* NaCl
- 0.1 *M* sodium tetraborate buffer (Sigma B9876) pH to 8.5 with HCl, store at −20°C
- Anhydrous DMF
- 0.5 × TE (5 m*M* Tris–HCl, 0.5 m*M* EDTA, pH 8.0)
- Buffer A
  - 0.1 *M* Triethylamine (Sigma T0086)
  - 1 m*M* EDTA
  - pH to 7.0 with concentrated acetic acid
- Acetonitrile, HPLC grade

1. Ethanol precipitate the DNA primer three times prior to labeling to remove any free amine which may be present after synthesis. Remove excess ethanol from the DNA pellet after the third precipitation by vacuum concentration before resuspension with 0.1 *M* sodium tetraborate buffer to a DNA concentration of 400 μ*M*.
2. Dissolve *N*-hydroxy-succinimide (NHS)-functionalized fluorophore in 15 μL anhydrous DMF per 300 n*M* of the NHS fluorophore and mix thoroughly.
3. Gradually add NHS-functionalized fluorophore in 5 μL additions to primers in 0.1 *M* sodium tetraborate buffer with thorough mixing. A 5 molar excess of labels it typically sufficient to achieve maximum labeling efficiency.
4. Incubate on lab rotisserie overnight at room temperature.
5. Quench the reaction by three ethanol precipitations with resuspension in 0.5 × TE. After the three precipitations no free dye should be visible in the supernatant.
6. Purify labeled from unlabeled primers by reversed-phase HPLC using a C18 column (Vydac C18 218TP54). Elute at 1 mL/min flow rate with

a gradient from 90% Buffer A and 10% Acetonitrile to 70% Buffer A and 30% Acetonitrile over 20 min. Track the absorbance of the fluorophore at its peak wavelength and the DNA absorbance at 260 nm to distinguish labeled and unlabeled primer elution.

7. Collect and pool fractions that absorb at both the fluorophore peak and 260 nm DNA peak. Remove acetonitrile by vacuum concentration.
8. Ethanol precipitate the purified, labeled DNA primer and resuspend in 0.5 × TE.
9. The labeled oligo can now be used to generate the desired DNA construct through standard PCR techniques.
10. Extract DNA product from PCR via phenol–chloroform extraction and purify to homogeneity via ion exchange chromatography on a GenPak-Fax (Waters) or MonoQ (GE Lifescience) column following manufacturers' protocols.

2.4 Nucleosome Reconstitutions with Fluorophore-Labeled Histone Octamer and DNA

Nucleosomes are reconstituted with DNA and histone octamer by salt dialysis. There are multiple methods for salt dialysis including the stepwise method (Polach & Widom, 1995), the pump method (Luger et al., 1999a, 1999b), and the double dialysis method (Thastrom et al., 2004). We recommend using the double dialysis method as we described for histone octamer refolding. This produces a gradual change in salt concentrations similar to the pump method, but with the ease of making just a few buffer changes. Select an appropriate microdialysis chamber from Table 1, keeping in mind that losses during the reconstitution and purification can be up to 50–75%.

Required for this protocol:

- 0.5 × TE (5 m*M* Tris–HCl, 0.5 m*M* EDTA, pH 8.0)
- High Salt Buffer
  - 5 m*M* Tris–HCl pH 8.0
  - 0.5 m*M* EDTA
  - 1 m*M* Benzamidine (BZA, Sigma 12072)
  - 2 *M* NaCl
- Low Salt Buffer
  - 5 m*M* Tris–HCl pH 8.0

  - 0.5 m*M* EDTA
  - 1 m*M* BZA
- 0.3 × TBE
  - 27 m*M* Tris
  - 27 m*M* boric acid
  - 0.6 m*M* EDTA pH 8.0

1. Prepare 3 × 4 L of low salt buffer and 100 mL high salt buffer for later dialysis; store at 4°C.
2. Combine purified histone octamer and DNA in a 1:1.25 octamer:DNA molar ratio. A minimum of 1 pmol/μL of octamer should be used for efficient reconstitutions. Keep DNA concentrations below 5 μ*M* to avoid forming aggregates.

   Note: When calculating the concentration of labeled octamer, it is beneficial to not subtract the contribution of the fluorophore's absorbance at 276 nm. Because the local environment and rotational freedom of a fluorophore impacts its absorbance spectrum, the precise contribution to the 276 nm absorbance peak is not known. Additional histone octamer beyond the given ratio results in poorly formed nucleosomes or aggregation of histones onto DNA and loss of sample. Therefore, it is safer to overestimate histone concentration than to underestimate it.
3. Place combined DNA and histones into the selected microdialysis chamber using 6000–8000MWCO dialysis tubing as dialysis membrane, and place that microdialysis chamber into a larger 50 mm 6000–8000MWCO dialysis bag for double dialysis. Fill 50 mm bag with 50 mL of high salt buffer and dialyze against three changes of 4 L low salt buffer for at least 6 h per change with the last one going overnight.
4. Reclaim the reconstitution mixture. At this stage the sample is a combination of well-positioned nucleosomes, aggregates, depositioned nucleosomes, free DNA, and possibly tetrasomes, and hexasomes. Purification helps ensure a well-defined homogenous sample for further single molecule work.
5. Purification is performed by sucrose gradient. Gradients are made using a BioComp Gradient Master from 5% to 30% sucrose in 0.5 × TE using Beckman-Coulter thin-wall polypropylene ultracentrifuge tubes, 13.2 mL total volume. Gradients are then balanced and nucleosomes are loaded onto the top of the prepared gradients. The gradients are then

centrifuged for 22 h at 41 k rpm, 4°C in a Beckman-Coulter Ti-41 swinging bucket rotor.

6. To fractionate the gradients after spinning, insert a 1-mm glass capillary tube attached to 1 mm I.D. tubing into the bottom of the centrifuge tube. Connect the tubing to a peristaltic pump and pull at 1 mL/min fractionating into 0.4 mL fractions. In this way, the earliest fractions correspond to molecules with the highest sedimentation velocities. Aggregates will imbed themselves in the centrifuge tube; nucleosomes will sediment to fractions ~7–14 and free DNA to ~17–24.
7. Run unpurified nucleosome sample and each fraction on a 5% native polyacrylamide gel with 0.3 × TBE to determine the location of well-formed nucleosomes. Pool fractions containing nucleosomes while selecting against depositioned nucleosomes, hexasomes, tetrasomes, and free DNA (see Fig. 2).
8. Buffer exchange and concentrate nucleosomes into 0.5 × TE using an Amicon 30k MWCO centrifugal filter. This step should be performed within 24 h of purification.
9. Run free DNA, unpurified nucleosome sample, and newly purified nucleosomes on a 5% native polyacrylamide gel in 0.3 × TBE to verify the purification process. The unpurified nucleosomes should contain excess DNA, which ensures that the nucleosome band represents well-formed nucleosomes rather than aggregates; the purified nucleosomes should consist of only a single band representing a homogenous nucleosome sample.

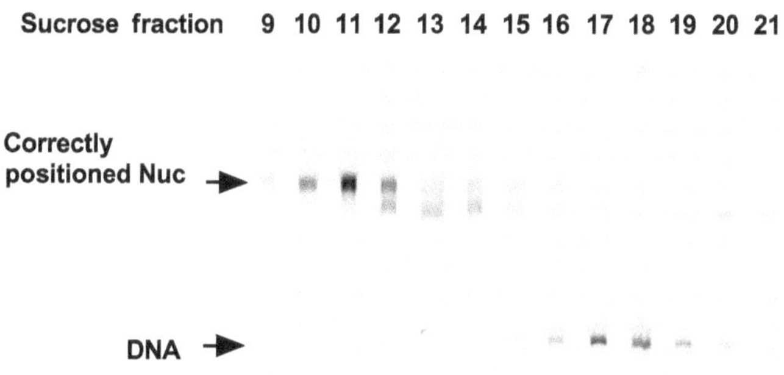

Fig. 2 5% Native polyacrylamide gel imaging Cy3 of the fractions taken from a nucleosome sucrose gradient purification (lower fractions show faster sucrose mobility). The DNA construct was 222 bp as shown in Fig. 1. Fractions 9–11 were pooled and concentrated for future experiments. Unpooled bands represent contaminants with lower FRET efficiencies (data not shown) such as hexasomes, depositioned nucleosomes, or free DNA.

3. ENSEMBLE FLUORESCENCE MEASUREMENTS

We recommend performing ensemble fluorescence measurements to accompany any smTIRF measurements on protein binding to nucleosomes and nucleosome-free DNA. Ensemble measurements accurately determine the average binding constant for a protein to its target site within a nucleosome and nucleosome-free DNA, which can be directly compared to the smTIRF measurements (see Fig. 3). This comparison helps show that surface immobilization and dilute conditions are not significantly impacting smTIRF measurements and nucleosome behavior. Ensemble fluorescence measurements are performed in a fluorometer such as a Fluoromax 4 (Horiba) in 50 μL cuvettes. Experimental buffer conditions are similar to those used in the single-molecule imaging buffer. We have found the oxygen-scavenging system and triplet state quenchers have no effect for our experiments and so we do not include them in our ensemble studies.

For FRET experiments of protein binding to recognition sites within nucleosomes, FRET efficiency is calculated through the $(\text{Ratio})_A$ method (Clegg, 1992) at varying binding concentrations (0–100 μ*M*) and 2–10 n*M* nucleosome concentrations. The method requires emission scans separately exciting the donor and acceptor. Nucleosome FRET efficiency is then given by

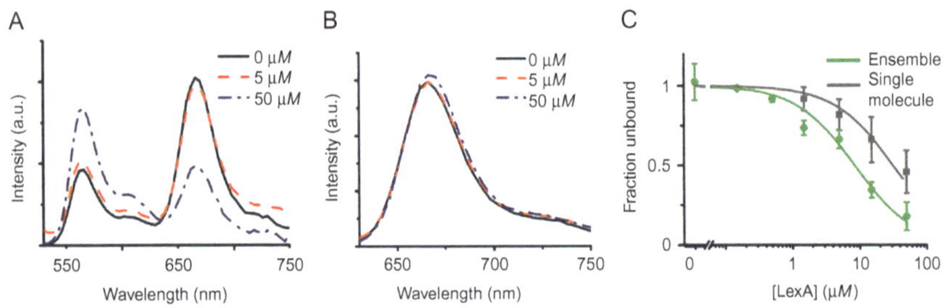

Fig. 3 (A) Fluorescence spectra of Cy3/Cy5 emission with excitation (510 nm) of nucleosomes labeled as in Fig. 1 taken in bulk using a fluoromax 4 (Horiba Scientific). The efficiency of the FRET interaction is dependent on LexA concentration and decreases with increasing LexA concentration from 0 μ*M* (*black*) to 5 μ*M* (*red*) to 50 μ*M* (*blue*) LexA. (B) Fluorescence spectra of the same nucleosomes while directly exciting Cy5 at 610 nm. The concentration of nucleosomes does not change with LexA concentration and so the spectra are constant across LexA concentrations. (C) Comparison between single molecule (*gray*) and ensemble (*green*) LexA mediated stabilization of unwrapped nucleosomes.

$$E = \frac{2\left(\varepsilon^{A}_{Aex}F^{A}_{Dex}/F^{A}_{Aex} - \varepsilon^{A}_{Dex}\right)}{\varepsilon^{D}_{Dex}d^{+}}.$$

Fluorophore emission (F) superscripts refer to the donor (D) and acceptor (A) fluorophores and subscripts to the excitation frequencies for donor (Dex) and acceptor (Aex). Hence, F^{A}_{Aex} is the fluorescence emission of the acceptor when directly excited, and F^{A}_{Dex} is the fluorescence emission of the acceptor when the donor is excited. ε^{A}_{Aex}, ε^{A}_{Dex}, and ε^{D}_{Dex} are the molar extinction coefficients of acceptor and donor at the acceptor and donor excitation wavelengths. d^{+} is the donor labeling efficiency, which for DNA prepared by these methods is nearly 1. A prefactor of 2 reflects the presence of two acceptor labeling sites per nucleosome, but only one undergoes high FRET efficiency with the donor.

Protein binding to its recognitions site on nucleosome-free DNA can also be measured in bulk by PIFE by quantifying the enhancement of Cy3 fluorescence emission for increasing protein concentrations around the dissociation constant (K_d) and with Cy3-labeled DNA concentration at a concentration below the K_d in 1 mL cuvettes. Small variations in DNA concentration can add noise to a PIFE signal, so it is beneficial to include an additional fluorophore (e.g., Cy5) on the opposite DNA end that does not undergo PIFE and whose fluorescence intensity can be used to control for varying concentration.

For comparison to smTIRF data, normalize the ensemble FRET efficiency and PIFE data from 0 to 1 (Fig. 3C, plot 1—PIFE data), which represents the average fraction of nucleosomes and nucleosome-free DNA, respectively, that are unbound at a given protein concentration. The fraction of nucleosomes in a bound or unbound state is directly comparable to the smTIRF measurement of the total fraction of time all nucleosomes spend in the bound or unbound state because both measurements are taken in equilibrium. The same is true for nucleosome-free DNA measures. See Section 6 for further discussion of the smTIRF and smPIFE measurements.

4. SINGLE-MOLECULE TOTAL INTERNAL REFLECTION FLUORESCENCE MICROSCOPY

We use a quartz prism-based smTIRF microscope for the single-molecule fluorescence measurements as described in Roy et al. (2008), which we outline later. However, objective-based TIRF with micromirrors also provides high-quality single-molecule fluorescence measurements with nucleosome samples (Kilic et al., 2015).

The experimental equipment used is shown in Fig. 4, but conceptually it can be divided into three groups: the excitation path, the microscope body, and the emission path. The excitation path spatially aligns lasers toward the microscope body in order to excite the fluorophores. An arbitrary number of lasers may be used to excite various fluorophores, but for a two-fluorophore FRET system (and for PIFE measurements), we use two 50 mW CW diode pumped lasers (CrystaLaser) at 532 and 638 nm. The power output is variable and adjusted to balance signal-to-noise ratio (SNR) and fluorescence lifetime. Typical powers are about 25 mW for 532 nm and 13 mW for 638 nm. Laser output is cleaned using Semrock Brightline filters 531/22 (532 nm laser) and 640/14 (638 nm laser) (Fig. 4; F1 and F2). Lasers are aligned with a Semrock Brightline 580 nm dichroic long pass filter (Fig. 4; D1) and focused using a 250 mm focal length plano-convex lens (ThorLabs) (Fig. 4; L1) into a Pellin-Broca prism (Melles Griot) atop a $1'' \times 3'' \times 1$ mm quartz slide (G. Finkenbeiner) to produce a TIRF field. The prism is held in place by a custom made mount.

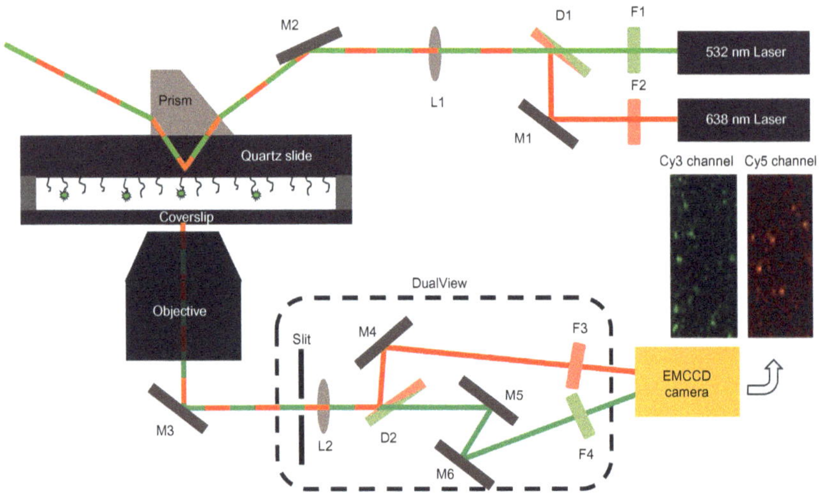

Fig. 4 Schematic diagram of the prism-based single molecule TIRF microscope setup. The laser paths are filtered through bandpass filters F1, F2. Beam paths converge on dichroic D1 and are focused through lens L1 into the prism at a supercritical angle. This illuminates the samples attached to the quartz slide. Fluorescence from the samples is collected through the objective and split into emission channels by the DualView optical assembly and projected onto different halves the EMCCD camera.

The microscope body is an IX71 inverted microscope (Olympus). The emission signal is acquired through a 60× water immersion objective (UPlanSApo 60×/1.20w, Olympus). The long working distance allows for imaging through the thick flow cell to the surface of the quartz slide on which an evanescent TIRF field excites immobilized, fluorophore-labeled samples.

Fluorescence emission light is directed unfiltered through the microscope body's camera port. Once light exits the microscope body, it moves to a DualView two-channel simultaneous-imaging system (Optical Insights) which houses a dichroic mirror (Chroma Technology, T635lpxr) (Fig. 4; D2) that spectrally splits the fluorescence light into donor and acceptor channels. The donor and acceptor channels are filtered through bandpass filters, Chroma Technology D585/30 and D680/35, respectively (Fig. 4; F3 and F4), and steered toward two halves of a PhotonMax electron multiplying (EM) CCD camera (Princeton Instruments). The DualView is adjusted to ensure subpixel resolution alignment of the donor emission and acceptor emission images on each half of the CCD. Images are collected with WinView (Roper Scientific) using 2 × 2 pixel binning and variable exposure times, depending on the specifics of the rates being studied.

5. FLOW CELL PREPARATION AND SINGLE-MOLECULE DATA ACQUISITION

Experiments are performed in microchannel flow cells with eight channels per microscope slide. Each flow cell surface is coated with polyethylene glycol (PEG) to pacify the slide surface against nonspecific interactions. A fraction of the PEG is labeled with a biotin molecule so that a biotin-streptavidin-biotin linkage tethers biotin-labeled nucleosomes and DNA to the surface. The methods used for preparation of these flow cells and experimental methodologies are based on previously described methods (Roy et al., 2008) and outlined later.

5.1 Flow Cell Preparation

Flow cells are formed from a three-layer stack as shown in Fig. 5. Eight channels are cut in Parafilm (Bemis PM966) which is then sandwiched between a passivated 50 × 24 × 0.16 mm glass coverslip (Fisher 12-544-E) and a 1″ × 3″ × 1 mm quartz slide (G. Finkenbeiner) with drilled inlet and outlet holes. These holes may be drilled with a CO_2 laser engraver

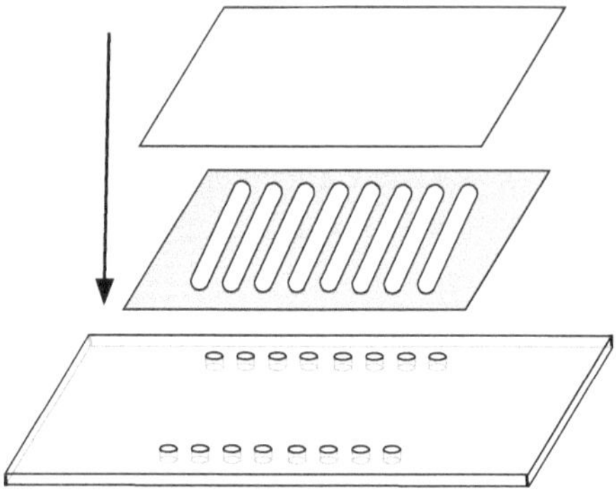

Fig. 5 Diagram showing the assembly of the flow cells. The quartz slide has 16 (0.029 inch diameter) holes, two for each of the channels. Parafilm forms the horizontal boundaries of the channels and holds the slide and coverslip together to form the completed flow cell.

or Starlight 115005 diamond drill bit to a diameter of 0.029 inches. Carefully clean and dry all glassware before use in functionalizing flow cell components.

5.1.1 Clean Slides and Coverslips

Pacification and functionalization of slides and coverslips begins with cleaning any organics from the surfaces. Slides and coverslips are mounted vertically and separate in a custom made Teflon holder throughout the cleaning process.

Required for this protocol:

- Chromasolv toluene (Sigma 34866)
- Chromasolv ethanol (Sigma 270741)
- Concentrated H_2SO_4 (Sigma 320501)
- 50% (w/v) H_2O_2 (Sigma 516813)
- NaOH pellets (Sigma 306576)

1. Sonicate the quartz slides and coverslips in toluene and then ethanol for 20 min each.
2. Thoroughly rinse in dH_2O three times to remove organics.
3. Put on acid-safe personal protection equipment (gown, gloves, and face shield) in preparation for piranha cleaning. In a clean beaker on a heat resistant surface in an acids fume hood, add 2 parts H_2SO_4 followed by 1 part fresh 50% H_2O_2 to cover the slides, typically 450 mL total. Vigorous bubbling and heating indicate a successful reaction. When the reaction has completed after about 30 min, remove the slides and rinse three times with dH_2O as before.

4. In a new beaker sonicate in 1 *M* Filtered NaOH for 20 min.
5. Rinse three times with dH_2O as earlier.
6. Dry slides overnight at 75°C.

5.1.2 Silanize Slides and Coverslips

Amine-functionalized silane is covalently bonded to hydroxyl groups on the quartz and glass present after cleaning to serve as functional groups for subsequent PEG derivatization. The slides and coverslips should be placed in the Teflon holders used in Section 5.1.1 for this protocol.

Required for this protocol:

- 2% v/v (3-aminopropyl)triethoxysilane (MP Biomedicals 215476680)
  - (lasts 3 months if stored in refrigerated desiccator)
- Acetone (Sigma 650501)

1. In a dry beaker cover the slides with 500 mL of water-free acetone. Fresh acetone from an unopened, factory sealed bottle is essential to eliminate water contamination which quenches the silane reaction.
2. Swirl the slides for 10 min vigorously without exposing them to air.
3. While swirling, add 8 mL silane drop-wise over the course of 1 min. Allow the reaction to proceed for 2 min.
4. Quench the reaction by transferring the slides to a beaker containing 50% dH_2O 50% acetone.
5. Rinse with dH_2O three times as in Section 5.1.1.
6. Incubate at 75°C for 6 h.

5.1.3 Coat Slides and Coverslips with PEG

Functionalized PEG is covalently bound to the amino-silane derivatized surface and is responsible for both surface pacification and biotin functionalization for sample immobilization. This is accomplished through a mixture of 5000 Da average molecular weight monofunctional PEG-Succinimidyl Valerate (mPEG-SVA) and biotin-PEG-SVA (bioPEG-SVA).

Required for this protocol:

- 5000 Da monofunctional-PEG-SVA (Laysan Bio, mPEG-SVA-5000)
  - Stored in 50 mg aliquots at −80°C for 1 year
- 5000 Da biotin-PEG-SVA (Laysan Bio, Biotin-PEG-SVA-5000)
  - Stored in 5 mg aliquots at −80°C for 1 year
- 0.1 *M* Potassium tetraborate (Sigma P5754)
  - pH to 8.4 with HCl and store at −20°C

1. Dissolve 2 mPEG-SVA aliquots into 450 μL of 0.1 *M* Potassium Tetraborate Buffer each to make 10% (w/w) solutions. Suspend 1 bioPEG-SVA aliquot in 200 μL 0.1 M Potassium Tetraborate Buffer to make a 2.5% (w/w) solution. Vortex each to fully suspend PEG powder. The solution should be clear or slightly cloudy.
2. Add 20 μL of bioPEG-SVA to each 10% mPEG-SVA solution yielding 1:100 bioPEG:mPEG mass ratio. Vortex to mix and centrifuge at 20,000 × *g* for 3 min to pellet undissolved solids. Transfer supernatant to a clean 1.5 mL tube.
3. Remove all amino-silane derivatized quartz slides from the slide holder and place on a clean glass plate.
4. Pipette 50 μL of bioPEG/mPEG solution onto the center of each slide. Slowly place one coverslip centered on top of the quartz slide. Ensure no air bubbles are formed.
5. Track slide/coverslip pairs and the sides receiving PEG treatment for later reassembly.
6. Cover against dust and let sit at room temperature for 1 h.
7. Carefully separate the coverslip from the quartz slide and wash in dH_2O to remove extra PEG solution by washing in a dH_2O bath.
8. Air dry the slides and coverslips vertically in the Teflon slide holder from Section 5.1.1.
9. Clean and dry the glass plate.
10. Repeat step 4–7 above ensuring the treatment is performed on the same side as before with the same pair.
11. Incubate at 37°C until dry.

5.1.4 Assemble Flow Cells with Parafilm

Cleaned, PEG-functionalized slides and coverslips are finally assembled into fully formed flow cells by attaching them with a strip of Parafilm (American National Can) with flow channels cut into it and adhered with heat.

1. Prepare Parafilm by cutting channels in line with previously drilled inlet/outlet holes on the quartz slides.
2. Place one PEGylated quartz slide on a clean, dry glass plate with the PEG side facing up. Place a single sheet of prepared Parafilm with paper backing removed on the quartz slide and align with the inlets/outlets holes on the slide. Cover with a PEGylated coverslip, PEG side down. Flow cell assembly is shown in Fig. 5. Press the coverslip slightly to temporarily adhere it to the slide.
3. Heat a glass plate to 90°C on a hot plate.

4. Place one flow cell assembly at a time on the hot glass, quartz slide down. Heat just long enough for the Parafilm to turn partially clear. Gently press the coverslip down to permanently adhere it to the Parafilm. Do not heat long enough to completely melt the Parafilm to avoid channel distortion and slide cracking.
5. Store flow cells at room temperature in low humidity. They are good for at least 1 month. Store under dry nitrogen in a desiccator for longer term storage.
6. Flow cells are dismantled by soaking in acetone for 15 min and gently prying the coverslip and slide apart. Quartz slides can be reused repeatedly.

5.2 Single-Molecule TIRF Data Acquisition

Single-molecules experiments are typically performed at room temperature. Samples are kept separate on ice until just before use and imaging buffers are prepared just before for use in a single flow channel. The protocols later were used to acquire data for TFs binding within nucleosomes. Only minimal modification is necessary to repurpose the protocols for use with PIFE measurements on nucleosome-free DNA as noted. Also, when injecting liquid into flow cells always absorb excess liquid from the outlet and avoid injecting and passing air bubbles through the flow cell.

5.2.1 Washing Flow Cells and Blocking to Prevent Nonspecific Interactions

Prior to binding of the nucleosomes or DNA to the surface, flow cells are first hydrated with Wash buffer and further pacified with Blocking buffer containing BSA. The Wash buffer varies based on the specific experiment. It is designed to match the conditions in which the experiment takes place including pH, ionic strength, viscosity, etc. Normally, a surfactant such as Tween-20 is also included in the Wash buffer to help prevent nonspecific sticking. The Wash buffer described here is used for experiments that measure TF binding inside of nucleosomes and to nucleosome-free DNA.

Required for this protocol:

- Wash buffer
  - 10 m*M* Tris–HCl, pH 8.0
  - 130 m*M* NaCl
  - 10% (v/v) glycerol (Sigma G5516)
  - 1.5 × (0.0075%, v/v) Tween-20 (Sigma P1379)
- Blocking buffer
  - 1 mg/mL Ultrapure BSA (Ambion AM2616) in wash buffer

1. Hydrate the flow cell with 100–200 μL Wash buffer for 2 min.
2. Inject 20 μL of Blocking buffer into the flow cell.
3. Incubate the flow cell for 2 min.

5.2.2 Attaching Nucleosomes to the Flow Cell Surface

After the flow cell is further pacified with Blocking buffer, it must be prepared to specifically attach nucleosomes. This is done by first depositing streptavidin on the biotin-PEG molecules present on the flow cell surfaces. Streptavidin then provides an attachment point for biotinylated DNA. There is a higher density of biotin-PEG attachment points than required for a reasonable density of attached molecules for single-molecule imaging. Therefore, sample density is controlled directly through sample concentration. This allows for increasing or decreasing density as appropriate to the experiment in question rather than requiring preparation of slides with varying biotin-PEG density. A sample concentration of 10 p*M* typically yields a good molecule density (roughly 1000 molecules per field of view), but this concentration can be varied as needed.

Required for this protocol:

- Wash buffer (see Section 5.2.1)
- 20 μg/mL Streptavidin (Sigma S4762) in wash buffer
- Nucleosome or DNA sample sample at 20 n*M* in 0.5 × TE

1. Wash blocking buffer from flow cell with 100–200 μL of Wash buffer.
2. Inject 20 μL of Streptavidin in wash buffer and incubate for 2 min.
3. Meanwhile dilute biotinylated, fluorescent nucleosome, or DNA sample to 1–10 p*M* in 20 μL Wash buffer. The optimal sample concentration to yield a final desired molecule distribution on the flow cell surface will need to be empirically determined for each sample. Note nucleosome concentrations of 1–10 p*M* in solution are not stable and need to be diluted just before surface attachment.
4. Wash Streptavidin from flow cell with 100–200 μL of Wash buffer.
5. Immediately inject 20 μL of biotinylated, fluorescent nucleosomes as in Fig. 1A.
6. Incubate for 5 min. During incubation prepare 50 μL imaging buffer for injection into the flow cell.

5.2.3 Preparation of Imaging Buffer

A glucose oxidase/catalase-based oxygen-scavenging system (Gloxy) is used in the imaging buffer to increase fluorescence lifetime before photobleaching. In addition to this system, triplet state quenchers are used to improve fluorophore stability and inhibit fluorophore blinking. These quenchers are Cyclooctatetraene and 3-Nitrobenzyl alcohol (COT/NBA), Trolox (both used for Cy3/Cy5 FRET experiments), and BME (used for Cy3 PIFE experiments). We recommend using BME instead of Trolox and COT/NBA in the PIFE experiments because BME is an extremely effective triplet state quencher for Cy3 and easier to prepare. However, BME promotes a long lived Cy5 dark state and so COT/NBA and Trolox should be used for FRET measurements.

Required for this protocol:

- Immersion oil (Cargille Type FF, index of refraction matched to the quartz slide)
- Wash Buffer (see Section 5.2.1)
- 1 × Trolox buffer (For Nucleosome FRET)
  - Dissolve 10 mg Trolox (Sigma 238813) in 10 mL of 50 m*M* Tris–HCl pH 8.0 for 1 h on a shaker table at 37°C. Check pH and adjust to 8.0 with 1 *M* NaOH if necessary. Always prepare fresh.
- 1 × Tris Buffer (For DNA PIFE)
  - 50 m*M* Tris–HCl pH 8.0
- 1 × Preimaging buffer
  - 0.75 × Trolox buffer (for FRET) or 0.75 × Tris Buffer (for PIFE)
  - 10% glycerol
  - 130 m*M* NaCl
  - 0.075% Tween-20
  - 1% (v/v) BME (for PIFE only)
- 40% glucose (Sigma G8270)
- 25 × Gloxy
  - 37.5 mg/mL Glucose Oxidase (Sigma G2133)
  - 1830 μg/mL Catalase (Sigma C3155)
  - 10 m*M* Tris–HCl pH 8.0
  - 50 m*M* NaCl
  - Centrifuge for 5 min at 15,000 × *g* and filter by 0.2 μm filter; store at −20°C. Make dilutions of Gloxy to 8 × in 10 m*M* Tris–HCl pH 8.0, 50 m*M* NaCl and keep on wet ice on the day of experiment.

- 1000 × COT/NBA
  - 11.5% Cyclooctatetraene (COT, Sigma 138924)
  - 13.5% Nitrobenzyl alcohol (NBA, Sigma 146056)
  - 75% Ethanol
  - Store at −20°C; make dilutions to 25 × in dH_2O and keep on wet ice on the day of experiment.
- TF or other protein of interest diluted to 10 × final concentration in appropriate buffer.
- Final Imaging Buffer
  - 0.75 × Preimaging buffer
  - 1.6% glucose
  - 1 × COT/NBA (not used for PIFE)
  - 0.3 × Gloxy
  - 1 × TF or protein of interest

1. Wash unbound sample from the flow cell with 100–200 μL of wash buffer.
2. Inject 50 μL imaging buffer into the channel.
3. Clean the top surface of the quartz slide using a Kimwipe or lens paper.
4. Place a small drop of immersion oil at the center of the flow channel. The diameter of the drop on a clean surface should be a little smaller than the width of the channel. Too much or too little immersion oil will adversely affect the TIRF field through formation of diffraction patterns.

5.2.4 Data Acquisition

After a flow cell is prepared it must be mounted to the microscope, a TIRF field must be established and optimized, and signal must be confirmed before data acquisition begins. Flow cells last on the order of 30 min to 1 h so time is critical. For typical nucleosome dynamics, enough time is available to generate a sample set of several thousand molecules before the imaging buffer loses efficacy.

1. Mount the flow cell with the microscope coverslip down (immersion oil up) to the microscope body. Lower the prism mount and roughly align the immersion oil with the objective so that the prism, once lowered, is wetted in a pool of immersion oil, optically coupling it to the quartz slide. Place the prism in its holder and gently lower it onto the slide, ensuring no air bubbles are trapped between the two. Slide the prism back and forth to disperse immersion oil.

2. Using bright field illumination, raise the objective to focus through the flow cell onto the bottom surface of the quartz slide.
3. Change from bright field to TIRF illumination and change the microscope output from the eyepiece to the DualView two-channel simultaneous-imaging system and CCD camera.
4. Turn on each laser successively to ensure the generation of a good TIRF field.
5. Place the camera in a real-time video readout mode and refocus the image.
6. Adjust the location of the excitation beam to give uniform illumination over the entire field of view.
7. When the acceptor is directly excited, only the acceptor channel should show emission. When the donor channel is excited, emission in the acceptor channel indicates FRET. In the case of nucleosomes prepared as in Fig. 1B, more than half of the points visible during acceptor excitation will exhibit high FRET efficiency.
8. Specifics of data acquisition are highly dependent on the experiment and experimental setup, but typical settings are a high EM gain, roughly 50–200 ms exposure time, a 2000 frame acquisition window and 2×2 pixel binning of a 512×512 CCD pixel image. EM gain is set to maximize SNR, exposure time is minimized to catch fast fluctuations while maximizing SNR, and the total acquisition time should be long enough to allow for significant photobleaching of the field of view.
9. For FRET measurements:
   a. Begin data acquisition by first directly exciting the acceptor fluorophore without donor excitation for 25–50 frames. These frames will be used to construct a high SNR average image to identify points of interest during data analysis.
   b. Switch laser sources from acceptor exciting to donor exciting and allow the remainder of the acquisition window to elapse.
10. For PIFE measurements:
    a. Simply excite the fluorophore and acquire data for the full acquisition window.
11. Save the data and proceed to a new, unphotobleached location within the flow cell for another video acquisition.
12. Repeat steps 7–13 until enough data is collected or the imaging buffer degrades (roughly 1 h). After data are collected, this protocol may be repeated on the other channels of the flow cell to allow for up to eight independent experiments.

6. SELECTION AND ANALYSIS OF DATA TO EXTRACT LIFETIME INFORMATION

Data, once collected, should be stored in some form of lossless, bitwise format for further analysis and future reference. We use Tiff stacks as they are universal, compressible, and extensible. Stored data are then analyzed by lab written software (available on request), the flow of which is discussed later.

1. Molecules are first identified by the presence of an acceptor fluorophore. During acquisition, the acceptor is directly excited for a number of frames. The acceptor half of these frames are averaged together to produce a low background image which is then processed by well-established flood filling local maxima identification techniques. Points of local minima are also detected and stored for later use.
2. Identified molecules must then be scrutinized for the presence of both fluorophores during the FRET phase of the experiment. To do this, traces are formed as the average of the molecules maxima and the 8 pixels that surrounding it. A commensurate trace is formed from the same location on the donor half of the image to form two trace channels per molecule. These traces measure the total donor and acceptor emission of each molecule in time. Traces are then split into 100 frame chunks (skipping the acceptor excitation frames) and the median is calculated for each. The median and standard deviation of the background are also calculated from the local minima data. If a given trace chunk is a user specified number of standard deviations (typically 3) above the noise floor, the molecule is marked as likely fluorescing in the given fluorophore's channel. This is repeated for each trace chunk for each channel for each molecule. Typical fractions of identified molecules which fluoresce in the donor channel are 80% while roughly 40% of identified molecules appear to have any trace chunk which fluoresces in the acceptor channel during donor excitation.
3. Molecules which have been selected as fluorescing in each channel during the previous phase are then further analyzed to determine a likelihood that the trace shows binding and unbinding events. This is done by calculating the correlation coefficient between the two fluorophore channels. FRET events show anticorrelated donor and acceptor signals and a user specified level (typically −0.25) is used as a cutoff for molecules which are likely to evince unwrapping fluctuations. For a good nucleosome sample, 80–90% of molecules which FRET also fluctuate

or about 35% of all identified molecules. *Note*: This step is not required for PIFE experiments.

4. Traces likely exhibiting fluctuations can be semiautomatically selected for further processing, but it is necessary to cut off the acceptor excitation portion of the experiment, any photobleaching events and to truncate traces appropriately to reject spurious event information such as fluorescent debris momentarily increasing one signal without a commensurate change in the other. Summed together, these potential confounding signals require human input to individually select for sections of traces before moving on to the computationally intensive determination of idealized FRET states (see Fig. 6A).

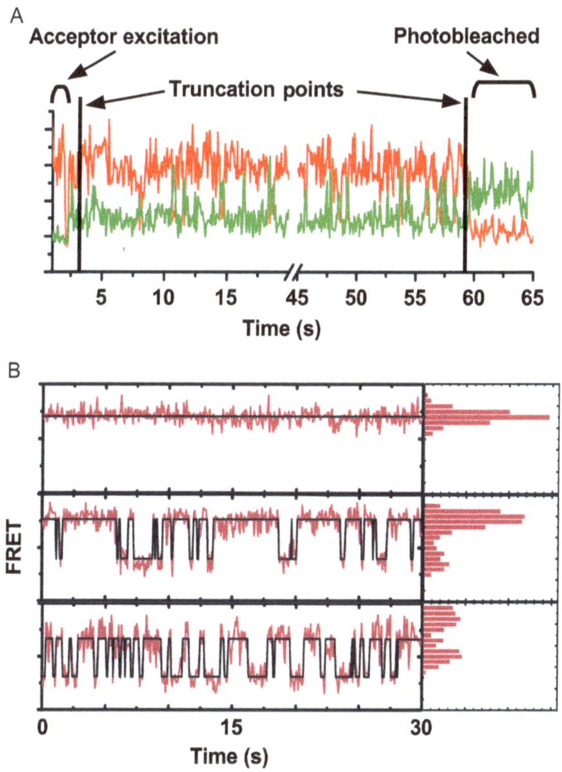

Fig. 6 (A) Untruncated data showing donor (*green*) and acceptor (*red*) traces. The acceptor excitation phase and donor photobleaching event must be truncated before analysis. (B) FRET data (*purple*) and idealized traces output from vbFRET (*black*) for 0 μ*M* (*top*), 5 μ*M* (*middle*), and 50 μ*M* (*bottom*) LexA binding to its target site in the entry–exit region of the nucleosome.

5. vbFRET (Bronson, Fei, Hofman, Gonzalez, & Wiggins, 2009), a MATLAB package provided by Dr. Ruben Gonzalez, is used to generate idealized FRET (or PIFE) traces from selected, truncated traces. In preparation for import into vbFRET, the background of each fluorescent signal is subtracted and a pseudo FRET trace is calculated for each molecule using $E = I_A/(I_A + I_D)$, or the intensity of the acceptor divided by the total intensity of the molecule at a given time. In order to determine binding and unbinding rates, only dwell times are necessary outputs from vbFRET, in which case the actual FRET value is arbitrary. Example FRET traces and the idealized fits output from vbFRET are shown in Fig. 6B. PIFE measurements may be scaled from 0 to 1 to be visible in the vbFRET graphical interface, but this is not necessary. Molecules which fit to a two-state model (one for a fully wrapped nucleosome and one for an unwrapped nucleosome) are selected and dwell time histograms are produced as shown in Fig. 7A and B. In order to

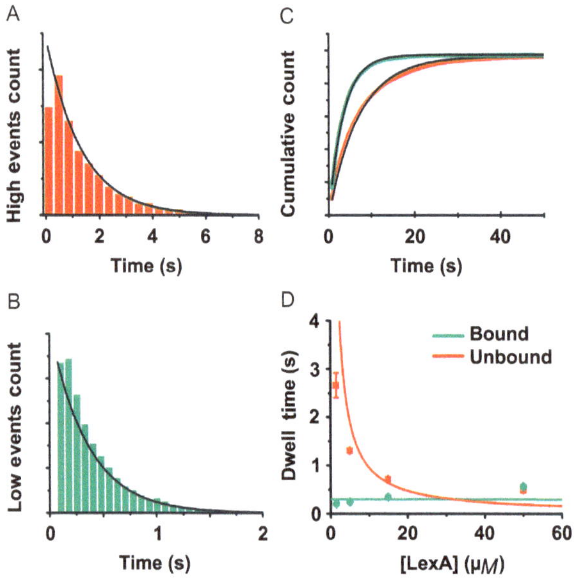

Fig. 7 Lifetime analysis of idealized trace data. (A) Example binned high state dwell time histogram (*red*) with single-exponential decay fit (*black*). (B) Same as (A) but with low state dwell times (*cyan*). (C) Cumulative sum lifetime in the high (*red*) and low (*cyan*) states along with fits to a cumulative sum single-exponential distribution (*black*) to find characteristic dwell times. (D) Characteristic dwell times are plotted as a function of TF concentration. Low state dwell times (*cyan*) show no concentration dependence whereas high state dwell times (*red*) show an inverse concentration dependence.

use all the information in our dataset and avoid using an arbitrary histogram bin size, cumulative sum distributions are produced and fit to find characteristic dwell times as shown in Fig. 7C. The characteristic dwell times correspond to the binding and unbinding rates for each concentration. It is important to show that the high state dwell times are inversely proportional to protein concentration to confirm the fluctuations are in fact a result of protein binding and that low state dwell times do not depend on protein concentration (Fig. 7D). Additionally, it is possible to calculate from these idealized traces an excellent analog to bulk fluorescence measurement of the fraction of nucleosomes unbound. The fraction of time spent in the unbound state (for FRET this is the high FRET state and for PIFE this is the low intensity state) averaged over all molecules directly measures the fraction of nucleosomes unbound at a given concentration of TF. The two should match as in Fig. 3C; If not, there may be an issue selecting molecules or at some other point in the single-molecule pipeline.

7. CONCLUSIONS

Here, we presented smTIRF methods for investigating DNA–protein binding to target sites within the nucleosome and duplex DNA. The smTIRF microscope and the flow cell preparation follow established protocols with minimal modifications, while the preparation of fluorophore-labeled nucleosomes varies more significantly from established protocols. Therefore, we focused on the details for preparing nucleosomes for smTIRF studies to help laboratories that are interested in investigating DNA–protein interactions with nucleosomes and chromatin. While we have used these methods to investigate TF binding to target sites within single nucleosomes, nucleosome arrays and duplex DNA (Luo, North, & Poirier, 2014; Luo, North, Rose, et al., 2014), these methods are applicable to investigating any DNA-binding protein to sites within nucleosomes and DNA. Furthermore, nucleosomes contain a wide variation in histone proteins that include histone variants and PTMs. The methods for nucleosome preparation have been presented so it is straightforward to include histone variants and histones containing specific PTMs in these smTIRF studies.

ACKNOWLEDGMENTS

The authors thank Jennifer Ottesen and members from the Poirier Lab for helpful discussions. We are grateful for the advice from Rubin Gonzalez on setting up the smTIRF system, flow cell preparation, and data analysis. This work was fund by the National Institutes of Health GM083055 (M.G.P.) and an Ohio State University Comprehensive Cancer Center—The James Pelotonia fellowship (J.A.N.).

REFERENCES

Almouzni, G., & Cedar, H. (2016). Maintenance of epigenetic information. *Cold Spring Harbor Perspectives in Biology*, *8*(5), pii:8:a019372.

Anderson, J. D., & Widom, J. (2000). Sequence and position-dependence of the equilibrium accessibility of nucleosomal DNA target sites. *Journal of Molecular Biology*, *296*(4), 979–987.

Andrews, A. J., Chen, X., Zevin, A., Stargell, L. A., & Luger, K. (2010). The histone chaperone Nap1 promotes nucleosome assembly by eliminating nonnucleosomal histone DNA interactions. *Molecular Cell*, *37*(6), 834–842.

Baek, S., Sung, M. H., & Hager, G. L. (2012). Quantitative analysis of genome-wide chromatin remodeling. *Methods in Molecular Biology*, *833*, 433–441.

Blosser, T. R., Yang, J. G., Stone, M. D., Narlikar, G. J., & Zhuang, X. (2009). Dynamics of nucleosome remodelling by individual ACF complexes. *Nature*, *462*(7276), 1022–1027.

Blossey, R., & Schiessel, H. (2011). The dynamics of the nucleosome: Thermal effects, external forces and ATP. *The FEBS Journal*, *278*(19), 3619–3632.

Bohm, V., Hieb, A. R., Andrews, A. J., Gansen, A., Rocker, A., Toth, K., et al. (2011). Nucleosome accessibility governed by the dimer/tetramer interface. *Nucleic Acids Research*, *39*(8), 3093–3102.

Bowman, G. D., & Poirier, M. G. (2015). Post-translational modifications of histones that influence nucleosome dynamics. *Chemical Reviews*, *115*(6), 2274–2295.

Bowman, A., Ward, R., El-Mkami, H., Owen-Hughes, T., & Norman, D. G. (2010). Probing the (H3-H4)2 histone tetramer structure using pulsed EPR spectroscopy combined with site-directed spin labelling. *Nucleic Acids Research*, *38*(2), 695–707.

Brehove, M., Wang, T., North, J., Luo, Y., Dreher, S. J., Shimko, J. C., et al. (2015). Histone core phosphorylation regulates DNA accessibility. *The Journal of Biological Chemistry*, *290*(37), 22612–22621.

Bronson, J. E., Fei, J., Hofman, J. M., Gonzalez, R. L., Jr., & Wiggins, C. H. (2009). Learning rates and states from biophysical time series: A Bayesian approach to model selection and single-molecule FRET data. *Biophysical Journal*, *97*(12), 3196–3205.

Chen, C., & Bundschuh, R. (2014). Quantitative models for accelerated protein dissociation from nucleosomal DNA. *Nucleic Acids Research*, *42*(15), 9753–9760.

Chereji, R. V., & Morozov, A. V. (2015). Functional roles of nucleosome stability and dynamics. *Briefings in Functional Genomics*, *14*(1), 50–60.

Clegg, R. M. (1992). Fluorescence resonance energy transfer and nucleic acids. *Methods in Enzymology*, *211*, 353–388.

Deindl, S., Hwang, W. L., Hota, S. K., Blosser, T. R., Prasad, P., Bartholomew, B., et al. (2013). ISWI remodelers slide nucleosomes with coordinated multi-base-pair entry steps and single-base-pair exit steps. *Cell*, *152*(3), 442–452.

Ferreira, H., Flaus, A., & Owen-Hughes, T. (2007). Histone modifications influence the action of Snf2 family remodelling enzymes by different mechanisms. *Journal of Molecular Biology*, *374*(3), 563–579.

Fierz, B., & Muir, T. W. (2012). Chromatin as an expansive canvas for chemical biology. *Nature Chemical Biology*, *8*(5), 417–427.

Gao, M., Nadaud, P. S., Bernier, M. W., North, J. A., Hammel, P. C., Poirier, M. G., et al. (2013). Histone H3 and H4 N-terminal tails in nucleosome arrays at cellular concentrations probed by magic angle spinning NMR spectroscopy. *Journal of the American Chemical Society, 135*(41), 15278–15281.

Godfrey, J. E., Baxevanis, A. D., & Moudrianakis, E. N. (1990). Spectropolarimetric analysis of the core histone octamer and its subunits. *Biochemistry, 29*(4), 965–972.

Ha, T., Enderle, T., Ogletree, D. F., Chemla, D. S., Selvin, P. R., & Weiss, S. (1996). Probing the interaction between two single molecules: Fluorescence resonance energy transfer between a single donor and a single acceptor. *Proceedings of the National Academy of Sciences of the United States of America, 93*(13), 6264–6268.

Ha, T., Ting, A. Y., Liang, J., Caldwell, W. B., Deniz, A. A., Chemla, D. S., et al. (1999). Single-molecule fluorescence spectroscopy of enzyme conformational dynamics and cleavage mechanism. *Proceedings of the National Academy of Sciences of the United States of America, 96*(3), 893–898.

Howard, C. J., Yu, R. R., Gardner, M. L., Shimko, J. C., & Ottesen, J. J. (2015). Chemical and biological tools for the preparation of modified histone proteins. *Topics in Current Chemistry, 363*, 193–226.

Hwang, H., Kim, H., & Myong, S. (2011). Protein induced fluorescence enhancement as a single molecule assay with short distance sensitivity. *Proceedings of the National Academy of Sciences of the United States of America, 108*(18), 7414–7418.

Kilic, S., Bachmann, A. L., Bryan, L. C., & Fierz, B. (2015). Multivalency governs HP1alpha association dynamics with the silent chromatin state. *Nature Communications, 6*, 7313.

Kornberg, R. D. (1974). Chromatin structure: A repeating unit of histones and DNA. *Science, 184*(4139), 868–871.

Kornberg, R. D. (1977). Structure of chromatin. *Annual Review of Biochemistry, 46*, 931–954.

Lalonde, M. E., Cheng, X., & Cote, J. (2014). Histone target selection within chromatin: An exemplary case of teamwork. *Genes & Development, 28*(10), 1029–1041.

Lee, J. Y., & Lee, T. H. (2012). Effects of DNA methylation on the structure of nucleosomes. *Journal of the American Chemical Society, 134*(1), 173–175.

Lee, J. Y., Lee, J., Yue, H., & Lee, T. H. (2015). Dynamics of nucleosome assembly and effects of DNA methylation. *The Journal of Biological Chemistry, 290*(7), 4291–4303.

Li, B., Carey, M., & Workman, J. L. (2007). The role of chromatin during transcription. *Cell, 128*(4), 707–719.

Li, G., Levitus, M., Bustamante, C., & Widom, J. (2005). Rapid spontaneous accessibility of nucleosomal DNA. *Nature Structural & Molecular Biology, 12*(1), 46–53.

Li, G., & Reinberg, D. (2011). Chromatin higher-order structures and gene regulation. *Current Opinion in Genetics & Development, 21*(2), 175–186.

Li, G., & Widom, J. (2004). Nucleosomes facilitate their own invasion. *Nature Structural & Molecular Biology, 11*(8), 763–769.

Lowary, P. T., & Widom, J. (1998). New DNA sequence rules for high affinity binding to histone octamer and sequence-directed nucleosome positioning. *Journal of Molecular Biology, 276*(1), 19–42.

Lu, Q., & Richardson, B. (2004). DNaseI hypersensitivity analysis of chromatin structure. *Methods in Molecular Biology, 287*, 77–86.

Luger, K., Mader, A. W., Richmond, R. K., Sargent, D. F., & Richmond, T. J. (1997). Crystal structure of the nucleosome core particle at 2.8 Å resolution. *Nature, 389*(6648), 251–260.

Luger, K., Rechsteiner, T. J., & Richmond, T. J. (1999a). Expression and purification of recombinant histones and nucleosome reconstitution. *Methods in Molecular Biology, 119*, 1–16.

Luger, K., Rechsteiner, T. J., & Richmond, T. J. (1999b). Preparation of nucleosome core particle from recombinant histones. *Methods in Enzymology, 304*, 3–19.

Luo, Y., North, J. A., & Poirier, M. G. (2014a). Single molecule fluorescence methodologies for investigating transcription factor binding kinetics to nucleosomes and DNA. *Methods, 70*(2–3), 108–118.
Luo, Y., North, J. A., Rose, S. D., & Poirier, M. G. (2014b). Nucleosomes accelerate transcription factor dissociation. *Nucleic Acids Research, 42*(5), 3017–3027.
Musselman, C. A., Gibson, M. D., Hartwick, E. W., North, J. A., Gatchalian, J., Poirier, M. G., et al. (2013). Binding of PHF1 Tudor to H3K36me3 enhances nucleosome accessibility. *Nature Communications, 4*, 2969.
Musselman, C. A., Khorasanizadeh, S., & Kutateladze, T. G. (2014). Towards understanding methyllysine readout. *Biochimica et Biophysica Acta, 1839*(8), 686–693.
Narlikar, G. J., Sundaramoorthy, R., & Owen-Hughes, T. (2013). Mechanisms and functions of ATP-dependent chromatin-remodeling enzymes. *Cell, 154*(3), 490–503.
Neumann, H., Hancock, S. M., Buning, R., Routh, A., Chapman, L., Somers, J., et al. (2009). A method for genetically installing site-specific acetylation in recombinant histones defines the effects of H3 K56 acetylation. *Molecular Cell, 36*(1), 153–163.
Ngo, T. T., & Ha, T. (2015). Nucleosomes undergo slow spontaneous gaping. *Nucleic Acids Research, 43*(8), 3964–3971.
Nie, S., & Zare, R. N. (1997). Optical detection of single molecules. *Annual Review of Biophysics and Biomolecular Structure, 26*, 567–596.
North, J. A., Shimko, J. C., Javaid, S., Mooney, A. M., Shoffner, M. A., Rose, S. D., et al. (2012). Regulation of the nucleosome unwrapping rate controls DNA accessibility. *Nucleic Acids Research, 40*(20), 10215–10227.
Oohara, I., & Wada, A. (1987). Spectroscopic studies on histone-DNA interactions. II. Three transitions in nucleosomes resolved by salt-titration. *Journal of Molecular Biology, 196*(2), 399–411.
Park, Y. J., Dyer, P. N., Tremethick, D. J., & Luger, K. (2004). A new fluorescence resonance energy transfer approach demonstrates that the histone variant H2AZ stabilizes the histone octamer within the nucleosome. *The Journal of Biological Chemistry, 279*(23), 24274–24282.
Polach, K. J., & Widom, J. (1995). Mechanism of protein access to specific DNA sequences in chromatin: A dynamic equilibrium model for gene regulation. *Journal of Molecular Biology, 254*(2), 130–149.
Polach, K. J., & Widom, J. (1999). Restriction enzymes as probes of nucleosome stability and dynamics. *Methods in Enzymology, 304*, 278–298.
Roy, R., Hohng, S., & Ha, T. (2008). A practical guide to single-molecule FRET. *Nature Methods, 5*(6), 507–516.
Schutz, G. J., Trabesinger, W., & Schmidt, T. (1998). Direct observation of ligand colocalization on individual receptor molecules. *Biophysical Journal, 74*(5), 2223–2226.
Shibata, Y., & Crawford, G. E. (2009). Mapping regulatory elements by DNaseI hypersensitivity chip (DNase-Chip). *Methods in Molecular Biology, 556*, 177–190.
Shimko, J. C., North, J. A., Bruns, A. N., Poirier, M. G., & Ottesen, J. J. (2011). Preparation of fully synthetic histone H3 reveals that acetyl-lysine 56 facilitates protein binding within nucleosomes. *Journal of Molecular Biology, 408*(2), 187–204.
Shogren-Knaak, M., Ishii, H., Sun, J. M., Pazin, M. J., Davie, J. R., & Peterson, C. L. (2006). Histone H4-K16 acetylation controls chromatin structure and protein interactions. *Science, 311*(5762), 844–847.
Sinha, M., & Peterson, C. L. (2009). Chromatin dynamics during repair of chromosomal DNA double-strand breaks. *Epigenomics, 1*(2), 371–385.
Thastrom, A., Lowary, P. T., & Widom, J. (2004). Measurement of histone-DNA interaction free energy in nucleosomes. *Methods, 33*(1), 33–44.
Tims, H. S., & Widom, J. (2007). Stopped-flow fluorescence resonance energy transfer for analysis of nucleosome dynamics. *Methods, 41*(3), 296–303.

Venkatesh, S., & Workman, J. L. (2015). Histone exchange, chromatin structure and the regulation of transcription. *Nature Reviews. Molecular Cell Biology*, *16*(3), 178–189.

Wei, S., Falk, S. J., Black, B. E., & Lee, T. H. (2015). A novel hybrid single molecule approach reveals spontaneous DNA motion in the nucleosome. *Nucleic Acids Research*, *43*(17), e111.

Widom, J. (2001). Role of DNA sequence in nucleosome stability and dynamics. *Quarterly Reviews of Biophysics*, *34*(3), 269–324.

CHAPTER FOURTEEN

Single-Molecule Fluorescence Studies of Fast Protein Folding

Z. Wang*, L.A. Campos†, V. Muñoz*,†,1
*Bioengineering Program, School of Engineering, University of California at Merced, Merced, CA, United States
†Centro Nacional de Biotecnología, Consejo Superior de Investigaciones Científicas, Madrid, Spain
[1]Corresponding author: e-mail address: vmunoz3@ucmerced.edu

Contents

Abstract

Watching the conformational wanderings of protein molecules during their search for their native structure is a holy grail for protein folding experimentalists. Such capability is essential to provide reality checks to the complex mechanistic heterogeneity that theory and molecular simulations predict. Single-molecule fluorescence resonance energy transfer (SM-FRET) is an attractive technique to meet that end, but its time resolution was insufficient for the microsecond motions of folding proteins. The temporal resolution of SM-FRET pivots on how quickly photons emitted by an individual molecule can be collected in sufficient numbers as to minimize statistical shot noise. Recent multipronged advances in that front have led us to gain first access to fast folding dynamics. Center stage is the implementation of photochemical strategies to quickly recover

Methods in Enzymology, Volume 581
ISSN 0076-6879
http://dx.doi.org/10.1016/bs.mie.2016.09.011
© 2016 Elsevier Inc.
All rights reserved.

fluorophores from long-lasting dark states. Improvements in optics and novel procedures for data analysis in probabilistic terms are additional contributions. In parallel, noninvasive methods for fluorescent labeling and immobilization of proteins have also been implemented. Here we discuss all these exciting developments, providing experimental guidelines and procedural details for the implementation of fast SM-FRET experiments on proteins.

1. INTRODUCTION

Single-molecule fluorescence (SMF) offers the opportunity of resolving the stochastic fluctuations in conformation and binding of individual biomolecules (Weiss, 1999). The application of the Förster resonance energy transfer (FRET) principle, in which fluorescence signals are correlated to nanoscale distances between segments of a macromolecule to measure conformation or between two molecules to measure binding (Stryer & Haugland, 1967), is particularly attractive. The FRET experiment involves labeling the biomolecules with two fluorophores: one that acts as a donor and another as an acceptor. In the single-molecule version (SM-FRET) a microscope is used to illuminate a miniscule sample volume and collect photons emitted from the sole molecule present at a given time (Ha et al., 1996). Separate detection of donor and acceptor photons allows determining the FRET efficiency (E), a parameter that is converted into the distance between the two dyes, thus probing discrete conformational changes and binding events as a function of time in steady-state conditions. SM-FRET has become increasingly popular in molecular biophysics, with applications that target the dynamics and function of nucleic acids (Seidel & Dekker, 2007; Zhuang, 2005), protein folding (Ferreon & Deniz, 2011; Haran, 2012; Michalet, Weiss, & Jager, 2006; Schuler & Eaton, 2008; Schuler & Hofmann, 2013; Schuler, Soranno, Hofmann, & Nettels, 2016), and function (Hofmann, 2014; Smiley & Hammes, 2006).

Here we focus on the applications to investigate protein folding, a problem for which attaining single-molecule resolution is a key. Theory indicates that proteins fold by diffusively searching for their biologically functional 3D structure on a hyperdimensional energy landscape in which an energy gradient toward the native state (the folding funnel) acts as a driving force (Bryngelson, Onuchic, Socci, & Wolynes, 1995). Despite the overall funnel, the detailed landscape topography should be corrugated, resulting in myriads of distinct microscopic folding routes and complex mechanisms (Dinner,

Sali, Smith, Dobson, & Karplus, 2000; Onuchic, Nymeyer, Garcia, Chahine, & Socci, 2000). Such intricate microscopic complexity is also observed in advanced atomistic computer simulations of protein folding (Bowman, Voelz, & Pande, 2011; Brooks, 1998; Shaw et al., 2010), but is in stark contrast with conventional folding experiments, which produce strikingly simple behavior that resembles a two-state process (Jackson, 1998). Recent experimental developments have begun to unveil key deviations from two-state folding, both for proteins that fold ultrafast and thus adhere to the downhill folding scenario (Garcia-Mira, Sadqi, Fischer, Sanchez-Ruiz, & Muñoz, 2002; Li, Oliva, Naganathan, & Muñoz, 2009; Yang & Gruebele, 2003), and for proteins with structures organized in subdomains that populate discrete folding intermediates (Englander & Mayne, 2014). These new observations emphasize the need to experimentally probe the fine topography of folding landscapes and resolve the conformational motions of proteins during their search for the native structure (Bartlett & Radford, 2009).

The application of SM-FRET to protein folding is challenging. Labeling proteins with bulky fluorescent dyes is not trivial, especially doing so without perturbing folding. Dyes linked to proteins experience elevated photosensitivity, decreasing the count rates that are obtained from a single molecule before the dyes become inactive. Such low count rates normally limit the resolution of the technique to milliseconds (Roy, Hohng, & Ha, 2008). Despite such technical difficulties, major strides have been made in protein SM-FRET, including experiments that measure the shifting populations of folded and unfolded states during protein denaturation (Brustad, Lemke, Schultz, & Deniz, 2008; Chung et al., 2011; Deniz et al., 2000; Hoffmann et al., 2011; Huang, Ying, & Fersht, 2009; Jager, Nir, & Weiss, 2006; Liu et al., 2012; Oikawa, Kamagata, Arai, & Takahashi, 2015; Schuler, Lipman, & Eaton, 2002), the dimensions of intrinsically disordered proteins (Banerjee, Mitrea, Kriwacki, & Deniz, 2016; Moosa, Ferreon, & Deniz, 2015), the dynamics of unfolded and folded states (Aznauryan, Nettels, Holla, Hofmann, & Schuler, 2013; Campos et al., 2013; Kuzmenkina, Heyes, & Nienhaus, 2005, 2006; Laurence, Kong, Jager, & Weiss, 2005; Merchant, Best, Louis, Gopich, & Eaton, 2007; Nettels et al., 2009; Rosenkranz, Schlesinger, Gabba, & Fitter, 2011; Sherman & Haran, 2006; Voelz et al., 2012), the population of folding intermediates (Lindhoud, Pirchi, Westphal, Haran, & van Mierlo, 2015; Rieger, Kobitski, Sielaff, & Nienhaus, 2011; Stockmar, Kobitski, & Nienhaus, 2016; Wang & Lu, 2015), and even mean folding transition path times (Chung & Eaton, 2013; Chung, Louis, & Eaton, 2009; Chung, McHale, Louis, & Eaton, 2012). Other exciting applications include detecting conformational changes in folded proteins (Lamboy, Kim, Dembinski, Ha, & Komives, 2013; Marcinowski et al., 2011), chaperone-induced unfolding transitions (Kahra et al., 2011; Kellner et al., 2014; Lee et al., 2016), and conformational changes upon binding (Cooper et al., 2015).

The major challenge is, however, how to reach the time resolution necessary to resolve folding dynamics. But, what are the relevant temporal regimes? The timescales of folding vary widely, with elementary conformational motions such as hydrophobic collapse, α-helix formation, and loop closure happening in nanoseconds, and complete folding taking sometimes hours or even days (Muñoz & Cerminara, 2016). However, the key timescale appears to be in the microseconds range. For instance, various empirical estimates of the folding speed limit (or the preexponential factor (τ_0) in the folding rate expression: $\tau = \tau_0 \exp\left(-\Delta G^{U\dagger}/RT\right)$) render values near 1 μs, including estimates based on conformational dynamics of chemically denatured proteins (Jones et al., 1993), size-scaling arguments (Naganathan & Muñoz, 2005), the observation of biphasic kinetics in fast folding proteins (Yang & Gruebele, 2003), and the relaxation time of downhill folding (Li et al., 2009). By the same token, although protein folding rates span many orders of magnitude (Naganathan & Muñoz, 2005), there is a growing group of studied domains that fold up in just microseconds (De Sancho & Muñoz, 2011). The fastest folders are optimal for SM-FRET experiments because they are likely to follow the downhill scenario in which folding dynamics closely track the fine topography of the underlying energy landscape (Eaton, 1999). Fast folding also facilitates comparison with results from atomistic computer simulations (Muñoz & Cerminara, 2016). For barrier-crossing folding the key timescale is the one that determines the transition paths, which are the conformational excursions that take the protein over the folding barrier (Chung & Eaton, 2013). Transition paths occur very rarely, but when they occur they only take microseconds (Chung & Eaton, 2013; Chung et al., 2009, 2012). Conformational motions on folded proteins (Bahar, Lezon, Yang, & Eyal, 2010) and folding coupled to binding processes also take place in microseconds (Henzler-Wildman & Kern, 2007; Muñoz, Campos, & Sadqi, 2016; Prigozhin & Gruebele, 2013; Shammas, Crabtree, Dahal, Wicky, & Clarke, 2016).

It is thus clear that to effectively measure the distribution of microscopic folding routes and mechanisms the time resolution of SM-FRET experiments needs to be extended well beyond 1 ms. Over the last few years we have witnessed multiple developments that bring the resolution of SM-FRET close to the timescales of protein folding dynamics. Such advances involve (1) methods to increase the photon count rates of fluorescent dyes, (2) approaches to increase the collection efficiency of SM-FRET microscopes, and (3) improved procedures for the analysis of photon trajectories. We discuss such advances here, providing practical recipes for the application to fast folding. Our ultimate goal is for this article to serve as a guide for the implementation of fast SM-FRET in any laboratory with access to an SMF microscope and expertise in protein chemistry.

2. SMF TECHNIQUES

Fluorescence-based detection of individual molecules entails being able to detect/count single photons and to selectively illuminate a miniscule volume to target only one molecule and avoid light contamination from other sources. In practical terms, there are two basic optical setups to achieve SMF detection: confocal microscopy and total internal reflection microscopy.

2.1 Confocal Fluorescence Microscopy

A confocal microscope for SM-FRET uses a high numerical aperture (N.A.) objective to focus the excitation beam and also to collect the photons emitted by the single molecule (Fig. 1). The basis of the confocal technique is the pinhole, which is placed in the emission path to remove any light coming from out of the illumination focus. Several photoavalanche detectors register the arrival of individual photons with picosecond resolution selecting color and/or polarization through filters and dichroic mirrors. The photons are time-tagged and stored on a multichannel counter/timer device for subsequent analysis.

The lateral resolution of any optical microscope is set by the diffraction limit of the objective (i.e., $d = 1.22 \times \lambda/\text{N.A.}$). The objective's N.A. determines the maximum angle at which the impinging light can be observed or focused according to

$$\text{N.A.} = n \times \sin(\theta) \tag{1}$$

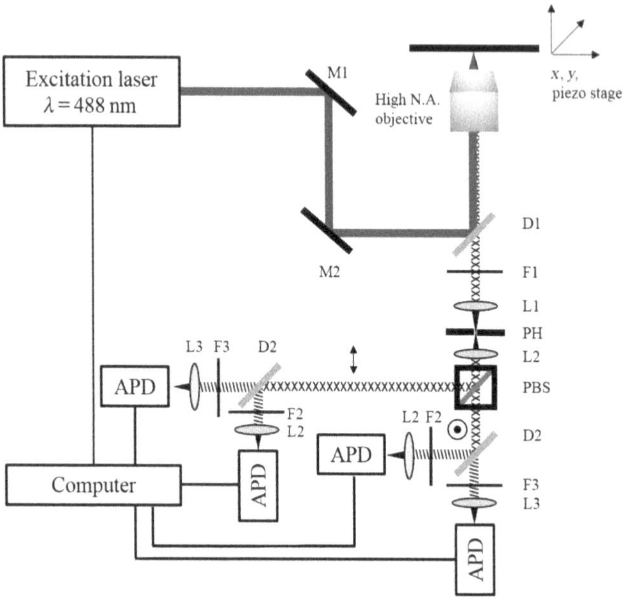

Fig. 1 Diagram of a confocal microscopy setup for single-molecule FRET. *D1* and *D2*, dichroic mirror; *F1*, emission filter; *F2* and *F3*, band pass filter; *L1*, tube lens; *L2–3*, achromatic lens; *M1*, *M2*, 100% reflection mirror; *PH*, pinhole; *PBS*, polarizer beam splitter; *APD*, avalanche photo diode.

where n is the refractive index, and θ is the maximal half-angle of the light cone that can enter or exit the objective. For single-molecule experiments one must choose an objective with a high N.A. (i.e., between 1.35 for water immersion and 1.65 for high refractive index synthetic oil immersion), which results in lateral resolutions slightly higher than the wavelength of incident light (e.g., ~300 nm for green light). In the confocal setup the axial resolution (along the z-axis) is drastically enhanced by the pinhole. For instance, for a pinhole with diameter of 1 Airy unit, the axial resolution z is

$$z = \frac{1.4 \times n \times \lambda}{(\text{N.A.})^2} \tag{2}$$

Therefore, using a high N.A. objective the resulting confocal illumination volume follows a three-dimensional Gaussian intensity distribution (Hess & Webb, 2002) that occupies only 1 fL. The small z eliminates any light emitted by molecules above or below the focal plane, thus permitting the selective observation of individual molecules, either tethered to the glass surface or passing through the illumination volume by free diffusion. In the free diffusion SM-FRET experiment one just needs to perform measurements at a protein concentration that is sufficiently dilute to guarantee a nearly negligible probability (<1%) of having more than one molecule at any given time (e.g., ~30 p*M* or lower). Single molecules are then identified from the bursts of emitted photons that are intermittently registered. Immobilization-based SM-FRET measurements require the protein molecules to be chemically tethered to the glass slide at a low density. The microscope stage is used to scan along the *x*–*y* plane and identify the locations of individual molecules for their visualization.

2.2 Total Internal Reflection Fluorescence Microscopy

An alternative approach takes advantage of the total internal reflection fluorescence (TIRF) phenomenon. TIRF is attained by making a collimated beam approach the illumination surface at such a closed angle that results in total reflection and the generation of an evanescent field that penetrates only ~100 nm into the solution (Axelrod, 1990, 2003). Only molecules that are attached to the surface become excited, resulting in wide-field excitation with extremely low background fluorescence. A TIRF microscope can be implemented using an objective (OTIR) or a prism (PTIR) as the optical means to achieve total reflection (Fig. 2). On OTIR, photons emitted by

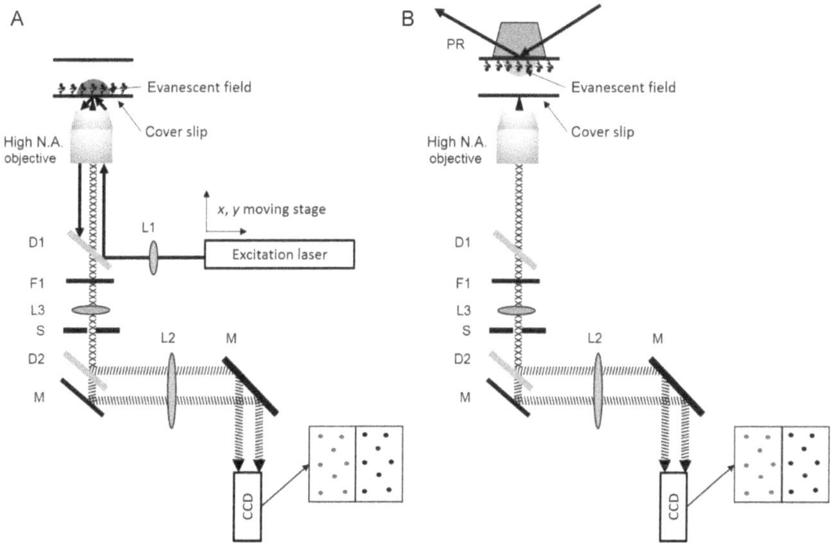

Fig. 2 Diagrams of total internal reflection fluorescence microscopy setups for single-molecule FRET using an objective, OTIR (A), or a prism, PTIR (B). *D1* and *D2*, dichroic mirror; *F1*, emission filter; *L1–3*, achromatic lens; *M*, 100% reflection mirror; *S*, slits.

the molecules present in the ultrathin illuminated slice are back-collected by the same objective, separated onto donor and acceptor photons by dichroic filters, and then delivered to the detector. On PTIR, a prism is mounted on top of the sample of an inverted microscope. The excitation beam is delivered to the prism with a high angle to produce an evanescent field toward the sample. Emitted photons are collected by a long-working distance water-immersion objective and delivered to the detector.

The advantages of TIRF are its very high axial resolution (even better than for confocal microscopy) and ability to resolve multiple individual molecules at the same time. TIRF measurements require the molecules to be surface immobilized. For fast SM-FRET TIRF microscopes are also limited in temporal resolution to the readout time of the EMCCD cameras that are used as detectors. EMCCD cameras offer very high resolution in the x–y plane (~0.5 megapixels), detect single photons, and have low dark counts, and thus are well suited for wide-field single-molecule imaging. However, their readout times are <400 frames per second, or up to 3500 frames per second (i.e., ~300 μs resolution) by selecting a small detector area (e.g., 32 × 32 pixels), thus sacrificing most of the x–y resolution. In general, TIRF microscopy is an ideal technique for applications that are not too fast but benefit from its wide-field illumination, such as particle tracking or simultaneous monitoring of multiple molecules.

2.3 Pulsed Interleaved Excitation

One of the major artifacts in SM-FRET experiments is the formation of dark states of the dyes, particularly the acceptor, which leads to erroneous estimates of E. Pulsed interleaved excitation (PIE) was invented to circumvent this problem using two synchronized pulsed lasers for exciting donor and acceptor in alternating fashion (Muller, Zaychikov, Brauchle, & Lamb, 2005). The first laser pulse excites the donor and records the emission of both dyes to measure E. The second pulse comes after a time interval longer than the dyes' lifetimes and excites directly the acceptor to determine whether the molecule carries a photoactive acceptor. Implementing PIE on a confocal SM-FRET microscope requires two pulsed lasers and a synchronization device. Like in any standard confocal SM-FRET instrument, the fluorescence emitted by both dyes is collected by the objective and separated into channels. The only difference is the use of a double dichroic to reject the excitation light from both lasers.

3. ENGINEERING PROTEINS FOR SM-FRET APPLICATIONS

Performing SM-FRET experiments on proteins requires careful engineering of the subject molecule to introduce a FRET pair that is sensitive to the conformational change of interest and often times also to add an immobilization tag. Such protein engineering is far from trivial. The design of the placeholders for the fluorophores must be based on careful analysis of the protein native 3D structure and a judicious choice of dyes. Minimizing possible protein–dye interactions is also important, especially since proteins have relatively low stability (~10–30 kJ/mol; Jackson, 1998) and even weak interactions can drastically change the folding behavior and also the dyes' properties. In contrast to nucleic acids and peptides, proteins need to be produced by recombinant means. Therefore, protein modifications must be done in-house. Here we provide some general guidelines for choosing suitable fluorophores, performing the specific labeling of proteins, and immobilizing them.

3.1 Labeling of Proteins With SM-FRET Fluorescent Dyes

A suboptimal selection of donor and acceptor fluorophores, or an inappropriate location in the protein, could jeopardize the interpretation of the single-molecule data in terms of folding–unfolding transitions. It is thus useful to define some criteria for the selection and placing of the FRET pair and describe simple procedures for specific labeling and purification.

3.1.1 Selecting the Pair of Fluorophores for Fast-Folding Applications

- The key parameter for a FRET pair is the characteristic Förster distance (R_0), or the distance that results in 50% energy transfer. R_0 is defined as

$$R_0 = \left(\frac{9 \cdot (\ln 10) \cdot k^2 \cdot Q_D \cdot J}{128 \cdot \pi^5 \cdot N_A \cdot n^4} \right)^{1/6} \tag{3}$$

  where k^2 is the dipole orientation factor (2/3 when there is complete dynamic averaging of the relative orientation of the two dyes within the fluorescence lifetime), Q_D is the donor quantum yield, N_A is Avogadro's number, n is the refractive index, and J is the spectral overlap integral between donor's emission and acceptor's excitation. Eq. (3)

permits to determine R_0 empirically assuming $k^2 = 2/3$, which is a reasonable assumption if at least one of the dyes is linked to an unstructured, flexible segment of the protein.

– R_0 should be intermediate between the distances separating the dyes in the end states (e.g., native and unfolded) to maximize the change in E upon unfolding. E depends on the inter-dye distance r as defined by

$$E = \frac{1}{1 + \left(\frac{r}{R_0}\right)^6} \quad (4)$$

The six power dependence with r limits the sensitivity range to distances between ~1.5 nm below and above R_0. The inter-dye distances can be estimated from the native 3D structure (pdb file) and from a polymer model for the chemically denatured unfolded state (e.g., for a freely jointed chain of N residues with $b \sim 0.9$ nm).

– The dyes should have maximum brightness: quantum yields as close to 1 as possible and donor with high extinction coefficient (>50,000 M^{-1} cm^{-1}). It is also convenient to select dyes with similar quantum yields to facilitate the interpretation of fluorescence signals in terms of E.
– Emission spectra of the two dyes should not overlap to facilitate selection of donor and acceptor photons. As a rule of thumb, one should choose dyes with less than 5% leak through onto the opposite detection channel, which depends on the emission spectra for both dyes and the dichroic mirror cutoff.

Over the last years many pairs of fluorophores have been utilized for SM-FRET of protein folding (Table 1). However, the most commonly utilized combination by far is Alexa488 as donor and Alexa594 as acceptor. This pair has an R_0 of ~6 nm for the free dyes, which is well suited for detecting folding–unfolding on single-domain proteins. Moreover, both dyes are relatively polar (anions) and thus have a somewhat reduced tendency to interact with partially folded and unfolded conformations of proteins.

3.1.2 Deciding Where to Place the Dyes on the Protein

As first basic rule the dyes should never be inserted in positions that may interfere with the protein hydrophobic core (Fig. 3). Surface positions with potential steric clashes or interactions with neighboring residues (especially charged residues) should also be avoided since the dyes may perturb the

Table 1 Properties of the Fluorescent Dyes Used for SM-FRET in Protein Folding

Name	Dye Family	Absorption Maximum (nm)	Extinction Coefficient (cm^{-1} M^{-1})	Emission Maximum (nm)	Quantum Yield	Lifetime (ns)
Alexa488[a]	Rhodamine	495	73,000	519	0.92	4.1
Alexa546[a]	Rhodamine	556	112,000	573	0.79	4.1
Alexa555[a]	Rhodamine	555	155,000	565	0.10	0.3
Alexa568[a]	Rhodamine	578	88,000	603	0.69	3.6
Alexa594[a]	Rhodamine	590	92,000	617	0.66	3.9
Alexa647[a]	Cyanine	650	270,000	668	0.33	1.0
Atto532[b]	Rhodamine	532	115,000	552	0.90	3.8
Atto633[b]	Carbopyronine	630	130,000	651	0.64	3.3
Atto647[b]	Carbopyronine	647	120,000	667	0.20	2.4
Atto647N[b]	Carbopyronine	646	150,000	664	0.65	3.5
Cy3[c]	Cyanine	548	150,000	562	0.04	<0.3
Cy5[c]	Cyanine	646	250,000	664	0.27	1.0
TMR[a]	Rhodamine	552	80,000	575	—	—

[a]Data from ThermoFisher Scientific.
[b]Data from Atto-tec.
[c]Data from Amersham Biosciences.

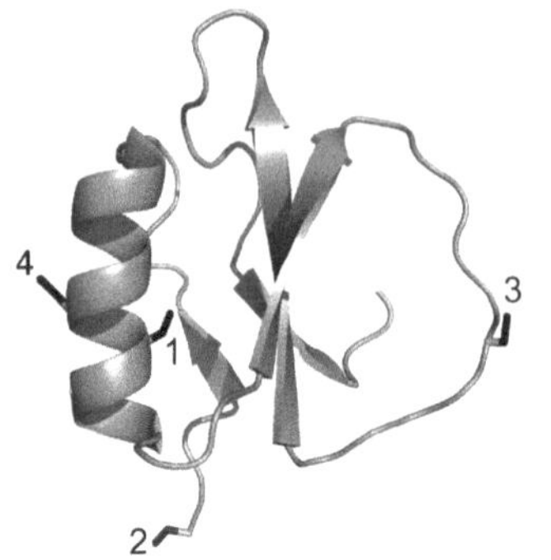

Fig. 3 Target positions for labeling proteins with fluorophores for SM-FRET. Examples for labeling options shown on the structure of model protein with numeric labels indicating the order of discussion in the text: *1*, core positions; *2*, protein ends; *3*, disordered loops; *4*, exposed positions in secondary structure elements.

folded state or become restricted in motion. The safest option is to attach the dyes to the C- and N-termini and separate them from the 3D structure using short flexible spacers (segments of 4–6 small, polar amino acids such as glycine, serine, alanine, and asparagine). An alternative is to place the dye on a long unstructured loop provided that there is enough room and flexibility for the dye to point outward. If the experiments require insertion of a fluorophore within a secondary structure element, a fully solvent-exposed position should be identified, ensuring there are no charged residues in i, $i+3$ and i, $i+4$ (or i, $i-3$ and i, $i-4$) positions for α helices, or i, $i+2$ (i, $i-2$) positions for β strands. The latter approach is, however, still structurally invasive and can perturb the protein. In general, the dye should not replace a residue that is interacting with others in the native structure (making hydrogen bonds, electrostatic or hydrophobic interactions). After labeling is completed, it is essential to determine experimentally whether the attached dyes perturb the protein. Chemical denaturation equilibrium experiments will highlight any change in protein stability and kinetic measurements any changes in folding and/or unfolding rates.

3.1.3 Protocols for Labeling Proteins With Fluorescent Dyes

The preferred labeling strategy for protein folding SM-FRET relies on small organic fluorophores (Table 1) that are chemically attached to the protein through a chemical reaction with the side chain of a given amino acid type (Fig. 4) (see Boutureira & Bernardes, 2015; Spicer & Davis, 2014) rather than fusions to green fluorescent protein variants. The most common approach is the reaction of a cysteine thiol with maleimide (left in Fig. 4).

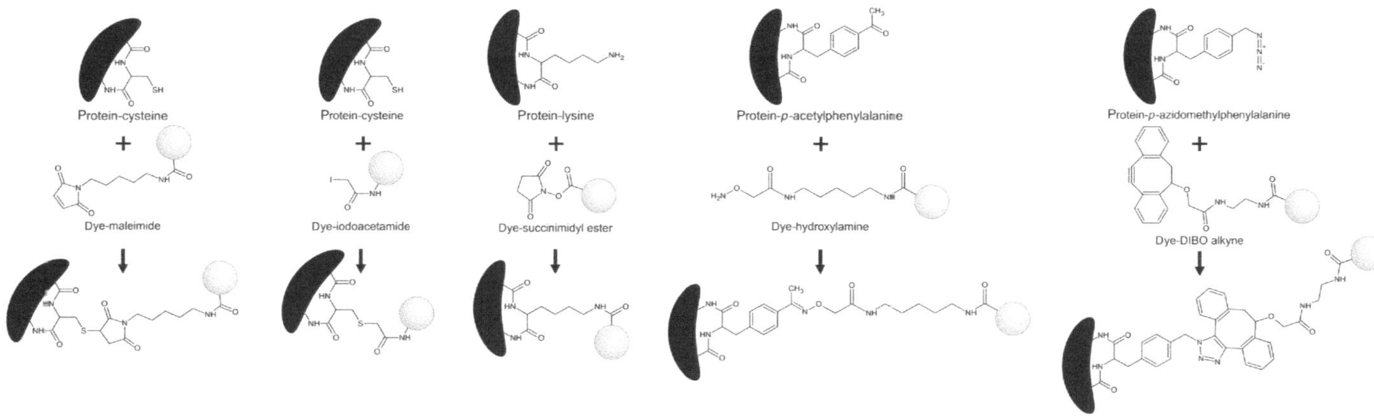

Fig. 4 Strategies for specific labeling of proteins with SM-FRET fluorophores (from *left* to *right*): cysteine reactions with maleimides or iodoacetamides; lysine reaction with succinimidyl esters; *p*-acetylphenylalanine reaction with hydroxylamines; *p*-azidomethylphenylalanine reaction with DIBO alkynes.

Cysteine is a relatively rare amino acid, especially on nonsecreted proteins, which facilitates specific labeling. An existing cysteine can be employed if it is at an adequate position for labeling (see Section 3.1.2). More frequently, cysteines are introduced at selected locations by site-directed mutagenesis. It is advisable to eliminate any other existing cysteine that is not involved in native disulphide bonds, which can be done substituting them with serines. The maleimide reaction is very specific and is performed in mild conditions. Sometimes, one of the two cysteines is more reactive, allowing stereospecific labeling of donor and acceptor by labeling in two steps (see below). Most usually, however, the reaction yields a mix of dyes labeled in both cysteines as well as protein labeled with only donor, only acceptor, or with none. Reaction by-products can be eliminated by protein purification, but the mix of labeling positions is often unavoidable. Nevertheless, if both dyes experience rotational freedom on the protein the determination of E should be still accurate as in such case it only depends on the relative distance.

A reaction of thiols with iodoacetamide is also possible (Fig. 4) but less common because iodoacetamides also react with unprotonated histidines, tyrosines, methionines, and the N-terminal amine group. Iodoacetamides are also very sensitive to light, making it harder to work with them. Even more rarely used is the reaction of primary amines with succinimidyl esters (Fig. 4), which react with the epsilon amino group of lysines and the N-terminal amine group, because the high number of labeling targets often impedes specific labeling. Here we include one-step and two-state protocols for labeling proteins using the most common maleimide–thiol reaction.

Double labeling in one step

1. Mix the protein with two free cysteines with donor and acceptor maleimide dyes at a ratio of 1:1:1 in buffer at neutral pH that includes 2 mM TCEP for maintaining the thiols fully reduced. For cysteines that exhibit low reactivity use denaturing conditions (6 M guanidinium hydrochloride) to maximize the solvent exposure of the thiols
2. Incubate at room temperature for 2 h or at 4°C overnight
3. Stop the reaction adding 2-mercaptoethanol to 20 mM and incubating for 2 h
4. Purify the labeled protein from reaction by-products by a combination of size-exclusion and anionic exchange chromatography or by reverse-phase chromatography (if the protein withstands the freeze-dry procedure)
5. Characterize the chromatographic fractions by absorbance, fluorescence, and mass spectroscopy to identify fractions with equimolar amounts of

donor and acceptor. Pool selected fractions and buffer exchange them to water using ultrafiltration

6. Prepare small aliquots of the sample and store at −80°C

Double labeling in two steps

1. Mix the protein with two free cysteines with the first maleimide dye at a ratio of 1:1 in buffer at neutral pH adding 2 m*M* TCEP
2. Incubate at room temperature for 2 h or at 4°C overnight
3. Stop the reaction adding 2-mercaptoethanol to 20 m*M* for 2 h
4. Purify the singly labeled protein from reaction by-products as described above
5. Pool fractions containing single labeled protein and buffer exchange them to the buffer for the second reaction (reaction buffer plus 6 *M* guanidine hydrochloride)
6. Perform steps 2–6 of the previous procedure

A more recent approach exploits the Schultz method for introducing nonnatural amino acids on proteins (Wang, Brock, Herberich, & Schultz, 2001). The catalog of available nonnatural amino acids for use with this method has been greatly expanded, including some that are intrinsically fluorescent and others ready for dye functionalization (Boutureira & Bernardes, 2015; Spicer & Davis, 2014). The method uses one stop codon (commonly UAG), the suppressor tRNA CUA, and an engineered UAA tRNA synthetase that loads the nonnatural amino acid onto the suppressor tRNA, which will in turn insert the amino acid during translation at the UAG codon. For SM-FRET applications the phenylalanine derivatives *p*-acetylphenylalanine and *p*-azidomethylphenylalanine, which react with hydroxylamine and DIBO-alkyne-modified dyes, respectively, are particularly attractive (Brustad et al., 2008; Cooper et al., 2015; Lee et al., 2016) (Fig. 5).

Insertion of *p*-acetylphenylalanine into a protein

1. Cotransform the plasmid containing the protein of interest carrying the amber codon and the plasmid pEvol (Young, Ahmad, Yin, & Schultz, 2010)
2. Select cells transformed with both plasmids using chloramphenicol (selection for pEvol) plus the antibiotic for the plasmid carrying the protein of interest
3. Grow an overnight preinoculum in LB medium with both antibiotics at 37°C
4. Next morning, pour the preinoculum in 1 L of LB with both antibiotics and 0.4 g/L of *p*-acetylphenylalanine and grow it at 37°C until reaching an OD of 0.6

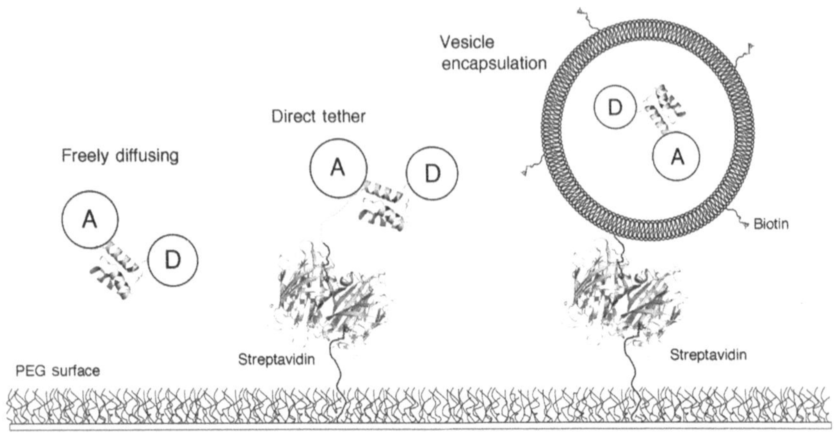

Fig. 5 Strategies for SM-FRET imaging of protein folding: free diffusion (*left*), immobilization via a direct tether (*center*), and vesicle encapsulation (*right*).

5. Induce expression of protein, synthetic tRNA synthetase, and suppressor tRNA CUA adding 4 g/L of arabinose and the required amount of inducer for the plasmid containing the protein of interest
6. Harvest the cells after 6 h of induction
7. Purify the protein containing *p*-acetylphenylalanine following its standard purification procedure

Labeling of *p*-acetylphenylalanine with a hydroxylamine derivatized dye

1. Mix the commercially available hydroxylamine derivative of the selected dye and the protein containing *p*-acetylphenylalanine at a 5:1 molar ratio in mildly acidic buffer (e.g., acetate at pH 4)
2. Incubate the reaction mixture during 72 h at 37°C
3. After 72 h eliminate unreacted dye by buffer exchange to water or any buffer at neutral pH by ultrafiltration
4. Confirm reaction efficiency with absorbance and mass spectroscopy
5. If a second labeling reaction is required (e.g., a maleimide–cysteine labeling for the second dye), prepare the solution for the next step and proceed as described above. Otherwise, aliquot and store at −80°C

Finally, the combination of various labeling methods, such as a selective protocol for multiple maleimide–thiol reactions (Jager, Michalet, & Weiss, 2005), chemical synthesis of protein segments and further ligation through click-chemistry (Deniz et al., 2000), and/or incorporation of nonnatural amino acids, opens the opportunity to perform more complex labeling schemes for multicolor SM-FRET experiments, something that is of great interest for protein folding. Such multicolor techniques have been recently applied to investigate protein folding with relative success (Gambin & Deniz, 2010), although this is still a developing area.

3.2 Surface Immobilization

Often times it is convenient to immobilize the labeled protein onto the surface to track it for as long as needed. The difficulty is to avoid nonspecific interactions between the very sticky labeled protein and the glass surface. The most straightforward procedure uses recombinant DNA technology to attach to one end of the protein an immobilization tag that binds to a chemical group present on a suitably derivatized glass surface. In general, the glass surface should be coated with a polyethyleneglycol (PEG) layer that acts as a spacer to avoid protein–surface interactions.

There are multiple solutions depending on the immobilization tag and how it is attached. A simple strategy uses a track of histidines (histidine tag) that binds to a PEG-coated surface also derivatized with *N*-[(3-trimethoxysilyl)propyl]ethylenediamine triacetic acid loaded with Ni^{2+} (Chung et al., 2009; Chung, Louis, & Eaton, 2010; Ha et al., 1999). However, the most widely employed technique exploits the ultra-strong interaction (K_d of 10^{-15} *M*) between biotin and the avidin protein family (e.g., streptavidin or neutravidin, with four biotin-binding sites) (Chung & Eaton, 2013; Chung et al., 2009, 2012; Kuzmenkina et al., 2005) (Fig. 5). The protein is produced in biotinylated form by fusing its gene to a biotinylation tag placed at the N-terminus. Another biotinylation procedure inserts the nonnatural amino acid *p*-azido-L-phenylalanine, as described above, which is then reacted with biotin-alkyne (Seo et al., 2011). The biotinylated protein is then labeled with both fluorophores and tethered to a PEG/biotin-PEG-coated glass surface by adding streptavidin/neutravidin, which uses its multiple binding sites to link protein and surface. One advantage of the biotin–streptavidin system is that it forms a long spacer. More recent approaches extend that feature by combining the biotin–streptavidin linkage, His tag binding, and antibodies to form large quaternary complexes (Pal, Lesoine, Lieb, Novotny, & Knauf, 2005).

3.2.1 Biotin-Mediated Immobilization (Derived From Chung et al., 2009)

Protein biotinylation

1. Clone the gene for the protein of interest including the biotin-binding sequence GLNDIFEAQ**K**IEWHE on the N-terminus (biotin will be linked to the lysine highlighted in bold). The plasmid should be

cotransformed with a plasmid containing the gene for BirA expression, a biotin ligase (from Avidity LLC)
2. Select colonies by plating on LB medium containing the selection antibiotics for the two plasmids
3. Grow an overnight preinoculum (1.5 mL) on LB with both antibiotics at 37°C
4. Next morning, pour the preinoculum in 1 L of LB with both antibiotics and incubate at 37°C until reaching an OD of 0.8. Add 50 μ*M* of D-biotin half an hour before that point (at an OD of ~0.5)
5. Induce protein and biotin ligase expression with the required inducer(s)
6. Harvest the cells 6 h after induction
7. Purify the protein of interest and check insertion of biotin by mass spectrometry

Surface immobilization

1. Use glass coverslips previously derivatized with a combination of PEG and biotin-PEG or commercial PEG-coated glass surfaces with biotin in its surface (Microsurfaces Inc.). Wash coverslips thoroughly with milliQ water and let them dry
2. Incubate the surface with streptavidin solution (typically 25 μg/mL) for 5 min
3. Wash several times with working buffer
4. Add the protein solution (e.g., at ~100 p*M*) prepared in the same buffer
5. Wash several times with the working buffer

An alternative to a direct tether is to encapsulate single protein molecules into phospholipid vesicles, which are then tethered to the glass surface (Fig. 5). This method, developed by Haran and coworkers (Boukobza, Sonnenfeld, & Haran, 2001), has been used for protein folding studies in several occasions (Liu, Mazouchi, & Gradinaru, 2010; Rhoades, Cohen, Schuler, & Haran, 2004; Rhoades, Gussakovsky, & Haran, 2003). Vesicle encapsulation is allegedly a much milder procedure, and permits to probe the protein while is freely diffusing but contained in a volume much smaller than the confocal volume (the average vesicle volume is about 2000 times smaller than the confocal volume), thus guaranteeing uniform irradiation.

3.2.2 A Protocol for Vesicle-Mediated Protein Immobilization

Preparation of loaded vesicles

1. Vesicles are made with white egg phosphatidylcholine (PC) and biotinyl cap phosphoethanolamine (Biotinyl PE). Dissolve PC at 20 mg/mL and biotinyl PE at 5 mg/mL in *tert*-butanol or chloroform. Mix thoroughly

250 μL of PC solution and 3 μL of PE solution in an Eppendorf tube, freeze the solution in liquid nitrogen, and lyophilize to remove the solvents

2. Prepare the working buffer, add active charcoal, and shake for several hours. Finally, pass the solution through a 0.2-μm filter
3. Resuspend the lyophilized phospholipids in 450 μL of charcoal-cleaned working buffer by vortexing and ultrasonicating for 15 min
4. Take 200 μL of the lipid solution and add 10 μL of labeled protein solution prepared at 1 μ*M*, leave for 5 min
5. Pass the lipid–protein solution back and forth 40 times through a 0.1-μm extruder (commercially available or alternatively build one using a 0.1-μm filter and two 1-mL syringes) to produce 100-nm diameter vesicles
6. Eliminate nonencapsulated protein from the loaded vesicles using microspin columns (GE Healthcare)

Vesicle immobilization

1. Use glass coverslips derivatized with biotin on its surface, wash them thoroughly with milliQ water, and let them dry
2. Add streptavidin solution (e.g., 25 μg/mL) to the coverslip and incubate for 5 min
3. Wash several times with working buffer
4. Add the loaded protein vesicle solution and incubate for 5 min
5. Wash several times with buffer

4. OPTIMIZING TIME-RESOLUTION VS CONFORMATIONAL DYNAMICS

The temporal resolution of an SM-FRET measurement should be at least 5–10 times better than the intrinsic relaxation time of the process under study to avoid dynamic averaging (blurring) (Gopich & Szabo, 2005, 2007; Torella, Holden, Santoso, Hohlbein, & Kapanidis, 2011). Confocal SM-FRET microscopes detect photons in picoseconds, but the temporal resolution is in fact determined by the time it takes to collect enough photons to measure E with sufficiently accuracy as defined by the shot noise expression

$$\sigma_{SN}^2 = \frac{E(1-E)}{N_{ph}} \tag{5}$$

According to Eq. (5), using 40 photons (N_{ph}) to determine an $E=0.5$ results in an uncertainty of ±0.08, which still is relatively high. The time resolution of SM-FRET is then set by the minimum time it takes to collect about 40 photons (or eventually ~25 photons, or σ ±0.10, if the process of interest only involves a few states (2 or 3) with widely different E).

Practically, the combination of shot noise, instrument collection efficiency, and the fluorescence emission rates of organic dyes sets the maximum temporal resolution. Moreover, ideal emission conditions are rarely met since organic dyes experience photo-induced deactivation when are irradiated at the high intensities that are required to maximize excitation. Accordingly, conventional SM-FRET experiments result on photon count rates of ~100 kHz, restricting the practical resolution to milliseconds. When one is interested in resolving faster processes, it becomes imperative to maximize the photon count rates and/or slow down the dynamics of the process itself by changing the conditions of the medium.

4.1 Strategies to Increase Photon Count Rates

4.1.1 Instrument Collection Efficiency

Maximizing the photon count rate on an SM-FRET experiment starts with the optimization of instrumental parameters such as optical alignment, configuration of filters and dichroic mirrors, pinhole size, and focus on the detector's sensitive area. One can further increase count rates by making some judicious choices about several components of the instrument. In this regard, it is useful to consider how various factors play out in the overall photon output. We do it here for illustration using A488 as the donor fluorophore (Table 1) and assuming a confocal configuration.

The first step involves calculating the probability that A488 will absorb a photon. Such probability depends on the absorption cross-section (σ) which is related to the molar extinction coefficient by equation (Lakowicz, 1983)

$$\varepsilon = \frac{N_0}{1000 \cdot \ln 10} \cdot \sigma = 2.616 \cdot 10^{20} \cdot \sigma \tag{6}$$

Assuming the dye occupies a spherical volume with radius r, the absorption cross-section becomes

$$\sigma = p \cdot \pi \cdot r^2 \tag{7}$$

where p is the probability that light impinging on the dye will be absorbed. For A488 ($\varepsilon = 73{,}000\ M^{-1}\ \mathrm{cm}^{-1}$ and stokes radius ~5.8×10^{-8} cm; Heyman & Burt, 2008) the probability of excitation is 0.0264.

The second step requires determining the probability of a photon hitting the dye, which depends on the excitation irradiance. Using Planck's equation we find that each 488-nm photon carries an energy of 4.07×10^{-19} J, and thus a continuous (CW) laser set to an output power of 20×10^{-6} W delivers to the sample 4.9×10^4 photons per nanosecond. If a pulsed laser is used for excitation (nowadays more common), the energy is delivered as a high-frequency train (e.g., every 12.5 ns or at 80 MHz) of picosecond pulses. For an 80 MHz repetition rate and 20×10^{-6} W of output power the pulsed laser packs 6.1×10^5 photons per pulse. Only photons that cross the small fraction of the area occupied by the dye (f_{area}) will be susceptible to absorption (Fig. 6, left). Such area is defined by

$$f_{area} = \frac{\pi \cdot r^2 \cdot N_0 \cdot C}{1000} \tag{8}$$

Because in a free diffusion SM-FRET experiment protein concentration is roughly 25 pM, the fractional area is $\sim 1.6 \times 10^{-4}$, implying that for the pulsed laser described above nearly 100 photons hit the dye during each pulse. Those conditions ensure excitation saturation during each pulse (100 photons $\times$ 0.0264 excitation probability converts to $\sim$2.5 excitations per pulse). The quantum yield of A488 is 0.92, and thus the actual photon emission rate from a single molecule will be $80 \times 0.92 = 73.6$ MHz.

However, the instrument will collect only a small fraction of the emitted photons (the count rate). One key factor for setting the collection efficiency is the objective N.A., which determines the size of the collection light cone (CLC). The excited molecule emits photons spreading spherically from the focal point $\left(A_{sphere} = 4 \cdot \pi \cdot r^2\right)$, but the objective can only collect light from the cap that intersects with the CLC (Fig. 6, right), which is calculated as

$$A_{cap} = 2 \cdot \pi \cdot r^2 \cdot (1 - \cos\theta) \tag{9}$$

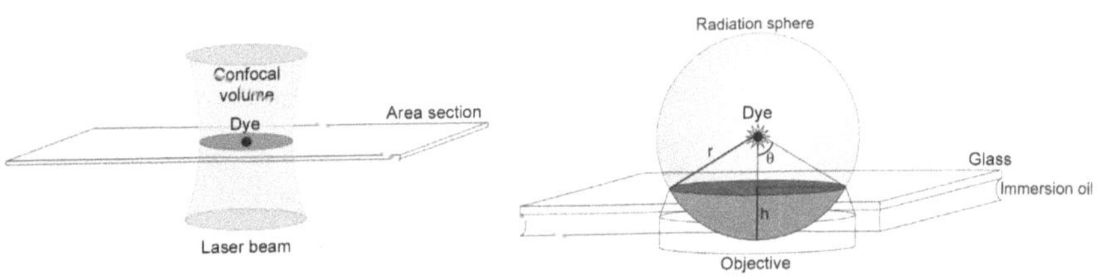

Fig. 6 Optical parameters that determine the photon count rate on an SM-FRET experiment. (*Left*) Representation of the area fraction corresponding to the molecule inside the confocal volume. (*Right*) Representation of the intersect between the spherical irradiation of fluorescence photons from the focal point and the collection light cone of the objective.

where θ is the maximal half-angle of the objective. For an $100\times$ oil-immersion objective with N.A. 1.40 and working distance of 0.13 mm the maximal half-angle is 68° (calculated from Eq. 1) and the ratio between intersect cap and total emitted spherical surface is 0.30 (or a 30% collection efficiency). An SM-FRET instrument has many other losses, including reflections within the multilens configuration of modern high N.A. objectives (~80 light throughput for single-coated lenses), transmittance of filters and dichroics, and the quantum efficiency of the photoavalanche detectors (~70%) (Michalet et al., 2014). When all these losses are computed, the collection efficiency of an SM-FRET confocal microscope turns out to be of only ~2% (Gell, Brockwell, & Smith, 2006; Sisamakis, Valeri, Kalinin, Rothwell, & Seidel, 2010).

Putting all these factors together we obtain a maximal photon count rate of ~1.5 MHz, or 1.5 photons per microsecond, which indicates that the maximally attainable count rate sets the limit in temporal resolution to ~20 μs, only sufficient to resolve processes with relaxation times longer than ~100 μs. It is also important to note that such estimated count rate does not take into account dye blinking (see next section).

The question is whether there are any instrument-based strategies to increase the count rate beyond that limit. The first idea that comes to mind is the use of a CW source. Pulsed lasers are very convenient to determine fluorescence lifetimes in situ and to use for PIE (see Section 2.3). However, their repetition rate is slower than the fluorescence lifetime of the dyes, and thus it is the pulse train what limits the maximum excitation rate. Pulsed lasers also compress an enormous amount of photons into extremely short times being much more likely to result in multiphoton excitation, which is the preamble for dye blinking and bleaching. An alternative CW source set at 150×10^{-6} W also reaches saturation conditions by delivering 3.7×10^{5} photons per nanosecond (52 photons actually hitting the molecule or 1.37 excitations), and since the fluorescence lifetime of the dyes is 4–5 ns, the donor can be excited at ~200 MHz (2.7-fold increase over the fastest pulsed laser). Besides, the higher CW power still corresponds to a much lower photon density at excitation, thereby resulting in less photodamage (i.e., even higher effective rates). Summarizing, true CW excitation is highly preferable for fast SM-FRET.

A second option is to increase the objective's N.A. The 1.4 N.A. oil-immersion objective discussed above has an absolute numerical aperture of $1.4/1.51=0.9272$ relative to a theoretical maximum of 1 (90° half-angle). This is a pretty good performance, but there are new oil-immersion objectives that reach N.A. values of 1.45 or even 1.49 (Mattheyses, Simon, & Rappoport, 2010; Vale, 2008). For an absolute N.A. of 1.49/1.51 the absolute numerical aperture is almost maximal (0.987), resulting in a half-angle >80° and an improvement of 33% in collection efficiency. Here it is important to emphasize that the key parameter is N.A./*n* rather than just N.A. Another useful strategy is to incorporate one of the new generation objectives with multilayer coated lenses, which can greatly increase the brightness of the objective to values >95% (Abramowitz, Spring, Keller, & Davidson, 2002). Therefore, a confocal SM-FRET microscope optimally configured for fast applications (CW excitation and highest N.A./*n*) may render a maximal photon count rate of ~5 MHz, leading to a theoretical limit in time resolution of ~5 μs.

4.1.2 Minimizing Fluorophore Bleaching and Blinking

At the high irradiance required for saturating excitation it becomes essential to minimize dye photophysics (Gust et al., 2014). Fluorescent dyes emit a photon from the single state (S_1) in just a few nanoseconds, but once on S_1, they can also transition to a nonfluorescent state (dark state) through conversion to the triplet state (T_1) by intersystem crossing, formation of radical states, or other nonradiative process (Fig. 7). T_1 decays very slowly (i.e., microseconds to hours), impeding subsequent excitation for long periods of time. Although these transitions are rare, their probability builds up when the dye is subjected to multiple excitations. Dyes on T_1 also react with oxygen radicals or reactive oxygen compounds (ROS) like hydroxyl radicals ($^{\bullet}OH$), hydrogen peroxide (H_2O_2), and superoxide radicals ($^{\bullet}O_2^-$), thereby leading to formation of a radical ion state and the irreversible loss of fluorescence (photobleaching) (Fig. 7). The singlet state of O_2, which is naturally generated as subproduct from the interaction between two triplets, is also highly reactive (Grewer & Brauer, 1994).

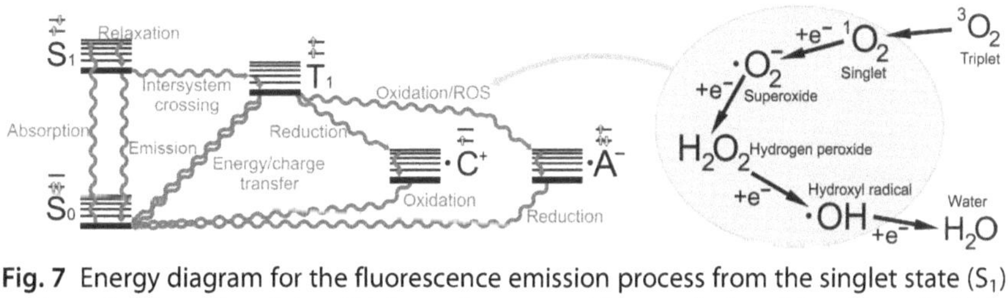

Fig. 7 Energy diagram for the fluorescence emission process from the singlet state (S_1) and mechanisms for triplet (T_1) formation, quenching, and photobleaching via ROS-mediated oxidation (*gray circle*).

It is therefore obvious that maximizing count rates involves finding ways to eliminate triplet buildup. The mechanisms for T_1 quenching are energy transfer, charge transfer, oxidation, or reduction. The first two mechanisms return the dye to S_0, making it readily available for new excitation. The other two require an additional coupled redox process for recovery. O_2 is one of the most efficient triplet quenchers available, operating via internal conversion (Eq. 10) or energy transfer (Eq. 11) (Grewer & Brauer, 1994)

$$^3M_1 + ^3O_2 \leftrightarrow M_0 + ^3O_2 \tag{10}$$

$$^3M_1 + ^3O_2 \leftrightarrow M_0 + ^1O_2 \tag{11}$$

Because O_2 is also naturally found in solution at relatively high concentration, it is in principle an ideal candidate for minimizing triplet formation with the caveat that it can also generate highly reactive singlet O_2 (Eq. 11). Alternative energy transfer triplet quenchers are 1–3 dienes (Turro, Ramamurthy, & Scaiano, 2010); naphthalene, azulene, or cyclooctatetraene (Widengren, Chmyrov, Eggeling, Lofdahl, & Seidel, 2007; Zheng et al., 2012); and Ni^{2+} (Glembockyte, Lincoln, & Cosa, 2015). Aliphatic thiols (i.e., the thiolate form) quench triplets via geminate recombination (Holzmeister, Gietl, & Tinnefeld, 2014). Several reducing agents that are commonly used in SM-FRET as photoprotectors, such as glutathione, 2-mercaptoethanol, and cysteamine (also known as mercaptoethylamine), can be included in this category as well (Campos et al., 2011; Nettels et al., 2009; Song, Varma, Verhoeven, & Tanke, 1996).

Oxidizing and reducing agents quench triplets by transforming the dye into an ionic radical. The ionic radical does not spontaneously convert to S_0 and cross-reacts with other ionic radicals bleaching the fluorophore. Therefore, one must combine them as a redox pair to recover the dye's ground state (Vogelsang et al., 2008). Examples used in SM-FRET are the oxidizers methyl viologen, nitrobenzyl alcohol (NBA), and trolox quinone, and the reducers *N*-propyl gallate, *p*-phenylenediamine, 1,4-diazabicyclo[2,2,2] octane (DABCO), trolox, 2-mercaptoethanol, cysteamine, and ascorbic acid (Campos et al., 2011; Cooper, Uhm, Tauzin, Poddar, & Landes, 2013; Dittrich & Schwille, 2001; Rasnik, McKinney, & Ha, 2006; Vogelsang et al., 2008; Widengren et al., 2007; Zheng et al., 2012).

Reducing photobleaching is also important as a way to increase observation times. A commonly used strategy for nucleic acid SM-FRET involves removing O_2 from the solution to impede formation of ROS (Aitken, Marshall, & Puglisi, 2008; Bernas, Zarebski, Dobrucki, & Cook, 2004).

The problem is that without O_2 the dyes remain in the triplet state for very long times, decreasing the photon count rate by 40- to 50-fold relative to experiments in the presence of dissolved O_2 (Campos et al., 2011). For fast SM-FRET applications is thus a much better option to keep oxygen in solution combining it with compounds that directly eliminate ROS, such as general reducers (Aitken et al., 2008; Rasnik et al., 2006) or specific free radical scavengers (Campos et al., 2011). The combination of dissolved O_2, the triplet quencher trolox, and the radical scavenger cysteamine is extremely efficient for rhodamine derivatives such as A488 and A594, yielding count rates near 1 MHz (Campos et al., 2011; Chung et al., 2012). For a rhodamine-based donor and a cyanine-based acceptor the combination of methyl viologen as triplet quencher and ascorbic acid as reducer has been found to be highly effective (Chung et al., 2012).

The performance of photoprotection cocktails is ultimately constrained by the intrinsic aqueous solubility of the components because their concentration determines the quenching collision rate and the number of quenching cycles. It is rather common that faster quenchers also have poorer solubility. For instance, trolox has a near diffusion-controlled quenching rate but its maximum solubility is only 2 m*M*, whereas the high solubility of ascorbic acid (>200 m*M*) is penalized by an inferior quenching rate. A practical solution might be to mix multiple components looking for additive effects.

An attractive alternative is to covalently link quencher and dye to accelerate the quenching rate by physical proximity, resulting in a "self-healing" fluorophore (Altman et al., 2012; Zheng et al., 2012, 2014). The main issue in this case is to devise chemical procedures for synthesizing quencher-derivatized dyes for labeling proteins. This problem has been recently surmounted, developing a method based on the introduction of nonnatural amino acids (van der Velde et al., 2016). The self-healing fluorophores, however, still underperform the best solution cocktails (van der Velde et al., 2016), probably because each molecule of the self-healing fluorophore can undergo just one quenching cycle. This suggests that a hybrid approach combining self-healing fluorophores and a cocktail in solution for regeneration may be a winning practical strategy.

4.2 Strategies to Slow Down Protein Folding Dynamics

The maximal count rates that have been obtained experimentally so far are ~1 MHz (Campos et al., 2011), which are insufficient to reach the 1 μs folding speed limit. However, protein folding dynamics depend on temperature (Naganathan, Doshi & Muñoz, 2007) and on the viscosity of surrounding solvent (Ansari, Jones, Henry, Hofrichter, & Eaton, 1992; Cellmer, Henry, Hofrichter, & Eaton, 2008; Jas, Eaton, & Hofrichter, 2001). One can thus exploit this property to slow down folding and make it accessible to the current resolution of SM-FRET. Temperature and solvent viscosity rate effects are in fact sizable. Fast folding proteins increase their rate with temperature by ~1 kJ/mol per protein residue (Naganathan et al., 2007), or a 20-fold change for a 60-residue domain going from physiological to just above freezing temperatures. With this strategy the folding time of fast folders BBL and gpW could be increased from the few microseconds of their midpoint temperature (Fung, Li, Godoy-Ruiz, Sanchez-Ruiz, & Muñoz, 2008; Li et al., 2009) down to ~200 μs, permitting to resolve their folding dynamics by SM-FRET (Campos et al., 2011) (see Section 5) or relaxation dispersion NMR (Sanchez-Medina et al., 2014), respectively. For proteins with little internal friction, conformational dynamics scale inversely proportional with solvent viscosity (Cellmer et al., 2008). Practically, this means that folding slowdowns of 10-fold are feasible by adding high concentrations of viscogens such as glycerol, a trick that was instrumental to resolve the fast folding dynamics of a WW domain using SM-FRET (Chung et al., 2012).

5. ANALYZING PHOTON TRAJECTORIES FROM SM-FRET EXPERIMENTS

Optimal SM-FRET data acquisition and analysis are also critical factors when striving for maximal temporal resolution. Early instruments used data acquisition cards that registered the counts at fixed sampling rates. The output data consisted on a registry of donor and acceptor counts per time bin (e.g., every 1 or 0.1 ms). Modern instruments include time-correlated single photon counting electronics that registers the arrival of each photon with picosecond resolution. The output data is a photon arrival trajectory (PAT) consisting of a sequence of photons with tags for arrival time and detected channel (Fig. 8). PAT data allows for more sophisticated analytical procedures based on probabilistic methods, which reduce the reliance on shot noise statistics. In this section we discuss both types of data acquisition and analysis, emphasizing their weaknesses and strengths in terms of time resolution.

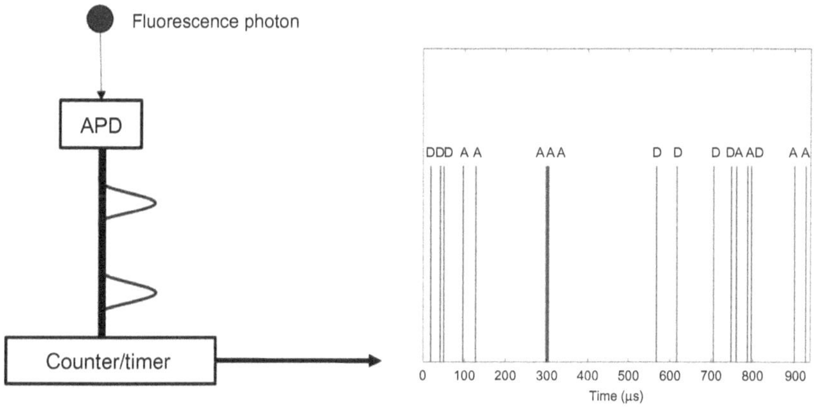

Fig. 8 Recording of photon arrival trajectories using time-correlated single photon counting electronics.

5.1 Analysis of Time-Binned Data

The simplest way to analyze SM-FRET data involves grouping donor and acceptor counts into fixed time intervals (bins). Binned data is then converted into E using equation:

$$E = \frac{I_A}{I_D + I_A} \tag{12}$$

where I_A and I_D are the number of acceptor and donor photons per time bin, respectively. At this point it is possible to introduce corrections to account for differences in detection efficiency for donor and acceptor photons, the probability of leak through to the opposite channel, and the quantum yields of the dyes (Campos et al., 2013; Chung et al., 2011; Gopich & Szabo, 2007; Ha et al., 1999). Estimates of E with Eq. (12) are independent of changes in donor quantum yield (see Campos et al., 2013 for a simple demonstration). It is important to note, however, that whereas such corrections are meant to provide more accurate determinations of E, they also artificially broaden the overall distribution (Gopich & Szabo, 2007). An alternative procedure to minimize statistical artifacts consists on randomly dropping photons from either the donor or the acceptor counts with probabilities according to the required correction (Gopich & Szabo, 2007; Nir et al., 2006). Practically, one can ignore such corrections altogether and treat the data simply as apparent E.

Eq. (12) applied to data from an immobilized molecule results in an FRET trajectory that is analyzed in terms of molecular states and their transitions. To identify molecular states one can calculate the FRET efficiency histogram (FEH) from the trajectory, and correlate the observed peaks with species (Fig. 9). The peak positions indicate the mean E for each species, which serve as references to identify molecular transitions in the trajectory. The distribution of dwell times for each species provides the microscopic rate coefficients (Sisamakis et al., 2010). Often times, however, fluctuations in E are large (i.e., shot noise and/or population of multiple states), making the identification of molecular states and transitions not a clear-cut case. An option is to interpret the FRET trajectory with methods devised to identify states and transitions such as hidden Markov models and maximum likelihood methods (Chung et al., 2009; Gopich & Szabo, 2007; Jung & Dickson, 2009; Lee, 2009; Liu, Park, Dahmen, Chemla, & Ha, 2010).

For a free diffusion experiment binned data shows bursts of high counts when molecules diffuse through the illumination volume (Fig. 10). Bins with the highest counts are selected by applying a photon threshold, which guarantees the presence of a molecule and a relatively uniform shot noise (per Eq. 5). Selected bins are then converted onto an FEH from which the number of molecular species, their average E, and relative populations are obtained (Sisamakis et al., 2010). Despite their apparent simplicity, FEHs are extremely difficult to interpret quantitatively. The apparent E distribution includes shot noise (Gopich & Szabo, 2005, 2011). Moreover, in real experiments the dyes inevitably bleach or blink, further distorting the distribution. The measured count rates also fluctuate stochastically due to

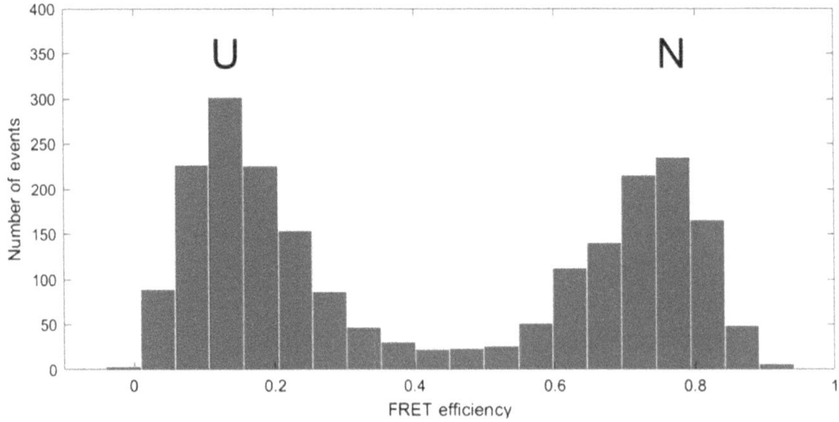

Fig. 9 FRET efficiency histogram obtained from time-binned SM-FRET data.

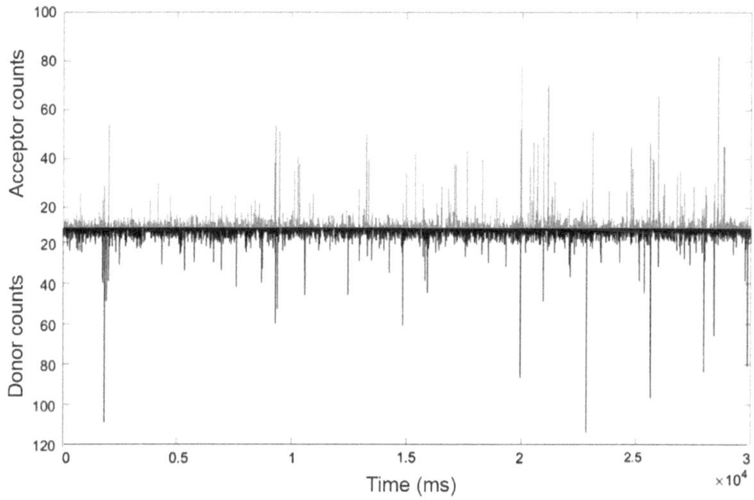

Fig. 10 Count rates for the donor (*black*) and acceptor (*gray*) channels from a free diffusion SM-FRET experiment. The passing of a single molecule is manifested as a photon burst (>40 photons).

the Brownian dynamics of the protein molecule while it traverses the confocal volume (Gopich & Szabo, 2007). Accordingly, many researchers have worked on developing procedures for improving FEH analysis, including empirical (Antonik, Felekyan, Gaiduk, & Seidel, 2006; Kalinin, Sisamakis, Magennis, Felekyan, & Seidel, 2010; Nir et al., 2006; Sisamakis et al., 2010) and theoretical methods (Chung et al., 2011; Gopich & Szabo, 2003, 2005).

Of particular significance for fast SM-FRET applications are the effects caused by the intrinsic dynamics of the molecule. For molecular processes that are either much slower or faster than the binning time (i.e., <1/10 or >10), the FEH represents a static distribution of species or the shot-noise-broadened, time-averaged mean E, respectively. However, when the molecular timescale is comparable to the binning time, the molecule has time to interconvert between states, resulting in dynamically blurred FEHs (Gopich & Szabo, 2005, 2011). In this case the underlying dynamic information can be extracted from a series of FEHs obtained with different binning times using a suitable kinetic model. For a two-state process, the resulting FEH can be quantitatively approximated as the sum of three Gaussian curves, two corresponding to the physical states and one with intermediate E that represents the interconverting molecules (Gopich & Szabo, 2007).

5.2 Probabilistic Analysis of PATs

The patterns of photon colors and arrival times contain a wealth of kinetic information that is lost after binning. In principle, from the correlations between a given photon and the ones arriving before and after one could ascertain, at least in probabilistic terms, whether the molecule has undergone a transition. Because inter-photon times can be shorter than 1 μs (see Section 3.1), decoding such information could increase the time resolution of SM-FRET. Gopich and Szabo developed a method to decode photon trajectories in terms of transitions between the states defined by a given kinetic model (Gopich & Szabo, 2009). The basic idea is to determine the probability that a PAT arises from any given set of kinetic parameters of the model (i.e., a choices of rate coefficients). One can then iteratively find the rate coefficients that maximize such probability or likelihood (maximum likelihood analysis, MLA) (Gopich & Szabo, 2009). This approach requires defining a priori the kinetic model for which to determine the rate coefficients, or alternatively, it can be used to identify the better model by comparing maximum likelihoods for various models, or by employing a flexible model (see below).

To illustrate the GS-MLA we employ a two-state folding kinetic model defined by the rate matrix:

$$\mathbf{K} = \begin{pmatrix} -k_u & k_f \\ k_u & -k_f \end{pmatrix} \tag{13}$$

in which k_f and k_u are the rate constants for protein folding and unfolding, respectively. The likelihood that a particular PAT emerges from the set of rate coefficients, species, and E values defined by the kinetic model is computed as:

$$L_t = 1^T \prod_{i=2}^{N} [\mathbf{F}(c_i) \exp(\mathbf{K}\tau_i)] \mathbf{F}(c_1) p_{eq} \tag{14}$$

where the calculation is run over the N photons contained in the trajectory, $\mathbf{K}$ is the rate matrix of Eq. (13), τ_i is the inter-photon time between the arrival of the $i-1$th and ith photons, and p_{eq} is a column vector defining the equilibrium probabilities for all the species. $\mathbf{F}(c_i)$ is a matrix that depends on the color of the ith photon (donor or acceptor) and defines the probabilities of that photon being emitted by all of the model species:

$$\begin{aligned} \mathbf{F}(\text{acceptor}) &= E \\ \mathbf{F}(\text{donor}) &= \mathbf{I} - E \end{aligned} \tag{15}$$

where **I** is the identity matrix and E is a diagonal matrix with the FRET efficiency values for all the states defined in the model. The likelihood varies with model parameters and thus one must optimize the parameter set to maximize L_t. Introducing data from multiple PATs is straightforward because the total likelihood is the product of L_t for all photon trajectories, which can be maximized in logarithmic form to reduce computer precision overflow issues.

GS-MLA has been used to extract folding and unfolding rate coefficients from SM-FRET data of the fast folding proteins α_3D and villin headpiece subdomain (Chung & Eaton, 2013; Chung et al., 2011). The method has also been used to investigate the folding transition path times of the slow folder GB1 and a mildly fast folding WW domain using viscogens to slow down protein dynamics (see Section 3.2) and comparing the maximum likelihood for two-state and three-state kinetic models (Chung & Eaton, 2013; Chung et al., 2012). The three-state model represented transition paths that were actually resolved (i.e., measurable residence times at the barrier top), whereas the two-state model represented seemingly instantaneous paths (Chung & Eaton, 2013; Chung et al., 2012). With this simple analysis the authors could resolve the mean folding transition path time for one of the proteins, yielding a value of ~10 μs, and estimate an upper bound for the other.

The major limitation for GS-MLA is the need to know a priori the kinetic model (rate matrix). Meeting this requirement is not always feasible or even desirable, particularly in experiments that investigate fast folding dynamics. There are, however, statistical mechanical models that represent the kinetics of protein folding as diffusion on a one-dimensional free energy surface (De Sancho, Doshi, & Muñoz, 2009; De Sancho & Muñoz, 2011; Muñoz, 2014; Naganathan et al., 2007). The 1D free energy surface represents the projection of the folding energy landscape onto a single order parameter for folding, thus ensuring consistency with energy landscape theory (Best & Hummer, 2010; Bryngelson et al., 1995; Inanami, Terada, & Sasai, 2014; Muñoz & Eaton, 1999; Suzuki, Noel, & Onuchic, 2011; Weinkam, Zimmermann, Romesberg, & Wolynes, 2010; Zimmermann, Nicolaus, Neuert, & Blank, 2010) and comparison with computer simulations (Gelman & Gruebele, 2014; Henry, Best, & Eaton, 2013). This type of model accommodates any folding scenario (two-state, three-state, marginal barrier, downhill) by changing its adjustable parameters, and thus it can be used generically (De Sancho & Muñoz, 2011). Ramanathan and Muñoz have recently combined the 1D free energy surface model with GS-MLA (Ramanathan & Muñoz, 2015). These authors investigated the performance of GS-MLA in recovering the original folding surface and diffusion coefficient from synthetic PAT data generated by stochastic simulations under varying conditions (i.e., number of photons, photon count rates, and background noise) (Ramanathan & Muñoz, 2015). The performance was remarkable and relatively insensitive to the photon count rate (Ramanathan & Muñoz, 2015). A major result was the ability of the procedure to produce robust estimates of the folding free energy surface and diffusion coefficient even when the count rates were only 10 times higher than the folding relaxation time. This characteristic implies an approximately fivefold increase in time resolution relative to binned data. The approach can be implemented in-house by following the detailed recipe provided in the article (Ramanathan & Muñoz, 2015) implemented in a simple programming routine on any professional data analysis software such as Matlab or Igor. We should also emphasize that even though the original analysis focused on fast folding applications, the diffusive rate matrix approach is generic and thus it can be used with any 1D free energy surface representation, including models of protein binding and 1D diffusion on DNA, folding coupled to binding, RNA folding, and so on.

6. A CASE EXAMPLE: ONE-STATE DOWNHILL FOLDING

In this section we discuss one-state downhill folding in the protein BBL as example of the application of fast SM-FRET to a problem of relevance in protein folding. One-state folding is the extreme manifestation of the downhill scenario in which the folding free energy surface exhibits a single well that continuously shifts from native to unfolded values as denaturing stress raises (Muñoz, 2002). Therefore, at the denaturation midpoint a one-state folder populates an ensemble of partially unstructured conformations rather than the 50–50 populations of folded and unfolded states of barrier-crossing two-state folding (Muñoz, 2002). The one-state scenario is intimately connected to fast folding since its barrier-less process results in microsecond dynamics.

The first identified example of one-state folding was BBL, a small α-helical domain (Garcia-Mira et al., 2002). BBL folds–unfolds in only ~1.5 μs at its denaturation temperature (~325 K) (Li et al., 2009) and shows noncooperative unfolding according to several thermodynamic tests, including probe-dependent unfolding (Garcia-Mira et al., 2002), differential scanning calorimetry (Muñoz & Sanchez-Ruiz, 2004), double-perturbation experiments (Oliva & Muñoz, 2004), and atom-by-atom unfolding via nuclear magnetic resonance (Sadqi, Fushman, & Muñoz, 2006). Those combined results firmly established the one-state character of BBL (Muñoz, 2007; Muñoz et al., 2016). One-state folding remained controversial (Ferguson et al., 2005), however, because it was incompatible with the conventional biochemical paradigm by which single domains ought to fold via a two-state process (Jackson, 1998). In principle, one-state folding is an ideal problem for SM-FRET experiments, which could directly resolve whether under partially denaturing conditions BBL populates two structurally distinct states (with high and low E) or a single one with intermediate E. The difficulty resides in reaching the time resolution required to avoid dynamic averaging, which could blur a bimodal distribution into a broad unimodal FEH, especially given the extremely fast folding of this protein. Dye photophysics had to be minimized too as it could result in additional peaks that could be mistaken with folding species (Cerminara, Campos, Ramanathan, & Muñoz, 2013). To make matters harder, one-state downhill folding proteins tend to exhibit marginal stability (Muñoz et al., 2016) and BBL, with a folding free energy of merely 6–8 kJ/mol (~2–3 RT) (Ferguson et al., 2005; Naganathan, Perez-Jimenez, Sanchez-Ruiz, & Muñoz, 2005), was no exception. Therefore, the fluorescent probes had to be introduced on BBL without perturbing its folding in the slightest way.

The labeling of BBL was done extending the protein termini with short flexible linkers and attaching the dyes to the linker ends (Fig. 11A); chemical denaturation experiments in bulk showed the same stability for the labeled and unlabeled versions (Liu et al., 2012) (Fig. 11B). SM-FRET experiments were done in free diffusion using CW excitation and a designed cocktail of photoprotectors and triplet quenchers (see Section 3.2). In parallel, the folding time of BBL was increased from 1.5 μs measured at 325 K (Li et al., 2009) to approximately 200 μs by performing the experiments at very low temperature (268 K). The photon count rates in these experiments were slightly higher than 1 MHz which permitted to bin the data every 50 μs with thresholds of >50 photons. The resulting FEHs showed a single peak that shifted its position from a maximum at $E \approx 0.8$ in native conditions (no denaturant) to a maximum at $E \approx 0.4$ in highly denaturing conditions (Liu et al., 2012). At the denaturation midpoint the distribution remained unimodal, showing its

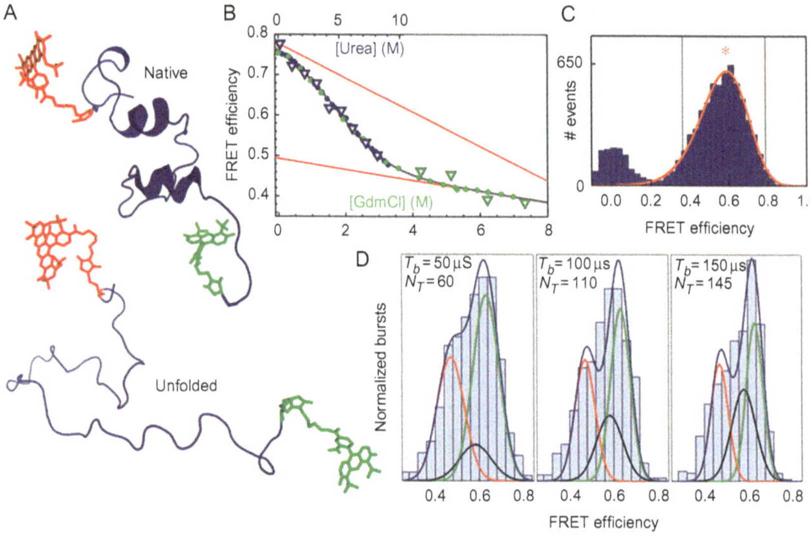

Fig. 11 Investigating the one-state downhill folding scenario in the protein BBL by SM-FRET. (A) Labeling of the BBL protein ends using flexible linkers; (B) bulk chemical denaturation experiments of unlabeled (*solid circles*) and labeled (*empty triangles*) BBL; (C) FEH obtained at the BBL denaturation midpoint (6.1 *M* urea); (D) FEHs at the chemical denaturation midpoint obtained with different time bins and photon thresholds.

maximum at intermediate values ($E \approx 0.6$) as expected for the one-state folding scenario (Fig. 11C). The E distribution was broader than shot noise, however, hinting at the existence of an underlying molecular distribution in slow kinetic exchange relative to the binning time. This idea was confirmed performing an autocorrelation analysis of the free diffusion trajectories that rendered the same 200 μs exponential relaxation observed by laser T-jump, providing a first experimental determination of the folding–unfolding dynamics of a protein by SM-FRET (Campos et al., 2011). Since the folding dynamics of BBL were only fourfold slower than the 50 μs time bins, a dynamically blurred two-state distribution was still a plausible explanation for the unimodal FEH (Fig. 11C). However, the authors could eliminate that possibility by generating a series of FEH with varying binning time and threshold to shift the contributions from shot noise and dynamic averaging. For instance, FEHs obtained with bins of 100 or 150 μs maintained a single intermediate peak, whereas the theoretical prediction for a model of two interconverting states was for distinctly bimodal FEHs due to the much higher photon thresholds allowed in these cases (Liu et al., 2012) (Fig. 11D). The quantitative analysis of FEH as a function of binning time was thus pivotal to rule out the possibility of a hidden two-state model, buttressing the case of BBL as a one-state downhill folding protein.

7. CONCLUDING REMARKS

Protein folding research is in earnest need of modern experimental methods to resolve the folding pathways and mechanisms followed by individual protein molecules in their search for the native structure. SM-FRET is, in principle, an ideal technique to meet this challenge because it recapitulates the conventional chemical denaturation experiments at the single-molecule level. The problem is that the timescales of the relevant folding motions are in the microsecond range, whereas conventional SM-FRET only reaches millisecond times. This is so because the temporal resolution of these experiments ultimately depends on the maximal photon count rates that can be obtained from a single molecule.

Over the last years a combination of technical improvements and practical approaches has been developed that push the boundaries of SM-FRET experiments to finally reach the sought-after microsecond resolution. Such developments are multivariate, but the most important advance came from the identification of photochemical strategies for quickly recovering an actively fluorescent molecule from the long-lived dark states that tend to form under high irradiance and severely limit photon emission. Combining this strategy with simple physical–chemical procedures to slow down the dynamics of protein folding led to the first microsecond resolution SM-FRET experiments on fast folding proteins (Campos et al., 2011). Other highly influencing factors are the type of laser excitation source; the collection efficiency of the instrument; the availability of brighter, more resilient dyes for protein labeling; and the implementation of analytical procedures that reduce the reliance on shot noise statistics.

The wise combination of all those factors gives us the option today of performing microsecond SM-FRET experiments to probe the conformational dynamics of proteins during folding and binding. In parallel, major research efforts are going on in all of those fronts with the overarching goal of further increasing the resolution of protein SM-FRET experiments. With all those incremental developments coming together, we should soon expect to be able to overcome the 1 μs limit that will permit to fully resolve the conformational transitions of individual proteins while they fold and function.

ACKNOWLEDGMENTS

This work was funded in part through grants CSD2009-00088 and BIO2011-28092 from the Spanish Ministry of Economy and Competitiveness, and grant ERC-2012-ADG-323059 from the European Research Council. V.M. and Z.W. also acknowledge support from the W.M. Keck Foundation.

REFERENCES

Abramowitz, M., Spring, K. R., Keller, H. E., & Davidson, M. W. (2002). Basic principles of microscope objectives. *Biotechniques*, *33*, 772–774. 776–778, 780–781.

Aitken, C. E., Marshall, R. A., & Puglisi, J. D. (2008). An oxygen scavenging system for improvement of dye stability in single-molecule fluorescence experiments. *Biophysical Journal*, *94*, 1826–1835.

Altman, R. B., Zheng, Q., Zhou, Z., Terry, D. S., Warren, J. D., & Blanchard, S. C. (2012). Enhanced photostability of cyanine fluorophores across the visible spectrum. *Nature Methods*, *9*, 428–429.

Ansari, A., Jones, C. M., Henry, E. R., Hofrichter, J., & Eaton, W. A. (1992). The role of solvent viscosity in the dynamics of protein conformational changes. *Science*, *256*, 1796–1798.

Antonik, M., Felekyan, S., Gaiduk, A., & Seidel, C. A. (2006). Separating structural heterogeneities from stochastic variations in fluorescence resonance energy transfer distributions via photon distribution analysis. *Journal of Physical Chemistry B*, *110*, 6970–6978.

Axelrod, D. (1990). Total internal reflection fluorescence at biological surfaces. In J. K. Foskett & S. Grinstein (Eds.), *Noninvasive techniques in cell biology* (pp. 93–127). New York: Wiley-Liss.

Axelrod, D. (2003). Total internal reflection fluorescence microscopy in cell biology. *Methods in Enzymology*, *361*, 1–33.

Aznauryan, M., Nettels, D., Holla, A., Hofmann, H., & Schuler, B. (2013). Single-molecule spectroscopy of cold denaturation and the temperature-induced collapse of unfolded proteins. *Journal of the American Chemical Society*, *135*, 14040–14043.

Bahar, I., Lezon, T. R., Yang, L. W., & Eyal, E. (2010). Global dynamics of proteins: Bridging between structure and function. *Annual Review of Biophysics*, *39*, 23–42.

Banerjee, P. R., Mitrea, D. M., Kriwacki, R. W., & Deniz, A. A. (2016). Asymmetric modulation of protein order-disorder transitions by phosphorylation and partner binding. *Angewandte Chemie (International Ed. in English)*, *55*, 1675–1679.

Bartlett, A. I., & Radford, S. E. (2009). An expanding arsenal of experimental methods yields an explosion of insights into protein folding mechanisms. *Nature Structural & Molecular Biology*, *16*, 582–588.

Bernas, T., Zarebski, M., Dobrucki, J. W., & Cook, P. R. (2004). Minimizing photobleaching during confocal microscopy of fluorescent probes bound to chromatin: Role of anoxia and photon flux. *Journal of Microscopy*, *215*, 281–296.

Best, R. B., & Hummer, G. (2010). Coordinate-dependent diffusion in protein folding. *Proceedings of the National Academy of Sciences of the United States of America*, *107*, 1088–1093.

Boukobza, E., Sonnenfeld, A., & Haran, G. (2001). Immobilization in surface-tethered lipid vesicles as a new tool for single biomolecule spectroscopy. *Journal of Physical Chemistry B*, *105*, 12165–12170.

Boutureira, O., & Bernardes, G. J. (2015). Advances in chemical protein modification. *Chemical Reviews*, *115*, 2174–2195.

Bowman, G. R., Voelz, V. A., & Pande, V. S. (2011). Atomistic folding simulations of the five-helix bundle protein lambda(6–85). *Journal of the American Chemical Society*, *133*, 664–667.

Brooks, C. L., 3rd. (1998). Simulations of protein folding and unfolding. *Current Opinion in Structural Biology*, *8*, 222–226.

Brustad, E. M., Lemke, E. A., Schultz, P. G., & Deniz, A. A. (2008). A general and efficient method for the site-specific dual-labeling of proteins for single molecule fluorescence resonance energy transfer. *Journal of the American Chemical Society*, *130*, 17664–17665.

Bryngelson, J. D., Onuchic, J. N., Socci, N. D., & Wolynes, P. G. (1995). Funnels, pathways, and the energy landscape of protein folding: A synthesis. *Proteins*, *21*, 167–195.

Campos, L. A., Liu, J., Wang, X., Ramanathan, R., English, D. S., & Muñoz, V. (2011). A photoprotection strategy for microsecond-resolution single-molecule fluorescence spectroscopy. *Nature Methods*, *8*, 143–146.

Campos, L. A., Sadqi, M., Liu, J., Wang, X., English, D. S., & Muñoz, V. (2013). Gradual disordering of the native state on a slow two-state folding protein monitored by single-molecule fluorescence spectroscopy and NMR. *Journal of Physical Chemistry B*, *117*, 13120–13131.

Cellmer, T., Henry, E. R., Hofrichter, J., & Eaton, W. A. (2008). Measuring internal friction of an ultrafast-folding protein. *Proceedings of the National Academy of Sciences of the United States of America*, *105*, 18320–18325.

Cerminara, M., Campos, L. A., Ramanathan, R., & Muñoz, V. (2013). Slow proton transfer coupled to unfolding explains the puzzling results of single-molecule experiments on BBL, a paradigmatic downhill folding protein. *PLoS One*, *8*, e78044.

Chung, H. S., & Eaton, W. A. (2013). Single-molecule fluorescence probes dynamics of barrier crossing. *Nature*, *502*, 685–688.

Chung, H. S., Gopich, I. V., McHale, K., Cellmer, T., Louis, J. M., & Eaton, W. A. (2011). Extracting rate coefficients from single-molecule photon trajectories and FRET efficiency histograms for a fast-folding protein. *Journal of Physical Chemistry A*, *115*, 3642–3656.

Chung, H. S., Louis, J. M., & Eaton, W. A. (2009). Experimental determination of upper bound for transition path times in protein folding from single-molecule photon-by-photon trajectories. *Proceedings of the National Academy of Sciences of the United States of America*, *106*, 11837–11844.

Chung, H. S., Louis, J. M., & Eaton, W. A. (2010). Distinguishing between protein dynamics and dye photophysics in single-molecule FRET experiments. *Biophysical Journal*, *98*, 696–706.

Chung, H. S., McHale, K., Louis, J. M., & Eaton, W. A. (2012). Single-molecule fluorescence experiments determine protein folding transition path times. *Science*, *335*, 981–984.

Cooper, D. R., Dolino, D. M., Jaurich, H., Shuang, B., Ramaswamy, S., Nurik, C. E., et al. (2015). Conformational transitions in the glycine-bound GluN1 NMDA receptor LBD via single-molecule FRET. *Biophysical Journal*, *109*, 66–75.

Cooper, D., Uhm, H., Tauzin, L. J., Poddar, N., & Landes, C. F. (2013). Photobleaching lifetimes of cyanine fluorophores used for single-molecule Forster resonance energy transfer in the presence of various photoprotection systems. *Chembiochem*, *14*, 1075–1080.

Deniz, A. A., Laurence, T. A., Beligere, G. S., Dahan, M., Martin, A. B., Chemla, D. S., et al. (2000). Single-molecule protein folding: Diffusion fluorescence resonance energy transfer studies of the denaturation of chymotrypsin inhibitor 2. *Proceedings of the National Academy of Sciences of the United States of America*, *97*, 5179–5184.

De Sancho, D., Doshi, U., & Muñoz, V. (2009). Protein folding rates and stability: How much is there beyond size? *Journal of the American Chemical Society*, *131*, 2074–2075.

De Sancho, D., & Muñoz, V. (2011). Integrated prediction of protein folding and unfolding rates from only size and structural class. *Physical Chemistry Chemical Physics*, *13*, 17030–17043.

Dinner, A. R., Sali, A., Smith, L. J., Dobson, C. M., & Karplus, M. (2000). Understanding protein folding via free-energy surfaces from theory and experiment. *Trends in Biochemical Sciences, 25*, 331–339.

Dittrich, P. S., & Schwille, P. (2001). Photobleaching and stabilization of fluorophores used for single-molecule analysis with one- and two-photon excitation. *Applied Physics B, 73*, 829–837.

Eaton, W. A. (1999). Searching for "downhill scenarios" in protein folding. *Proceedings of the National Academy of Sciences of the United States of America, 96*, 5897–5899.

Englander, S. W., & Mayne, L. (2014). The nature of protein folding pathways. *Proceedings of the National Academy of Sciences of the United States of America, 111*, 15873–15880.

Ferguson, N., Sharpe, T. D., Schartau, P. J., Sato, S., Allen, M. D., Johnson, C. M., et al. (2005). Ultra-fast barrier-limited folding in the peripheral subunit-binding domain family. *Journal of Molecular Biology, 353*, 427–446.

Ferreon, A. C., & Deniz, A. A. (2011). Protein folding at single-molecule resolution. *Biochimica et Biophysica Acta, 1814*, 1021–1029.

Fung, A., Li, P., Godoy-Ruiz, R., Sanchez-Ruiz, J. M., & Muñoz, V. (2008). Expanding the realm of ultrafast protein folding: gpW, a midsize natural single-domain with alpha + beta topology that folds downhill. *Journal of the American Chemical Society, 130*, 7489–7495.

Gambin, Y., & Deniz, A. A. (2010). Multicolor single-molecule FRET to explore protein folding and binding. *Molecular Biosystems, 6*, 1540–1547.

Garcia-Mira, M. M., Sadqi, M., Fischer, N., Sanchez-Ruiz, J. M., & Muñoz, V. (2002). Experimental identification of downhill protein folding. *Science, 298*, 2191–2195.

Gell, C., Brockwell, D., & Smith, A. (2006). *Handbook of single molecule fluorescence spectroscopy*. Oxford: Oxford University Press.

Gelman, H., & Gruebele, M. (2014). Fast protein folding kinetics. *Quarterly Reviews of Biophysics, 47*, 95–142.

Glembockyte, V., Lincoln, R., & Cosa, G. (2015). Cy3 photoprotection mediated by Ni2+ for extended single-molecule imaging: Old tricks for new techniques. *Journal of the American Chemical Society, 137*, 1116–1122.

Gopich, I. V., & Szabo, A. (2003). Single-macromolecule fluorescence resonance energy transfer and free-energy profiles. *Journal of Physical Chemistry B, 107*, 5058–5063.

Gopich, I., & Szabo, A. (2005). Theory of photon statistics in single-molecule Förster resonance energy transfer. *Journal of Chemical Physics, 122*, 014707.

Gopich, I. V., & Szabo, A. (2007). Single-molecule FRET with diffusion and conformational dynamics. *Journal of Physical Chemistry B, 111*, 12925–12932.

Gopich, I. V., & Szabo, A. (2009). Decoding the patterns of photon colors in sm-FRET. *Journal of Physical Chemistry B, 113*, 10965–10973.

Gopich, I. V., & Szabo, A. (2011). Theory of single-molecule FRET efficiency histograms. In T. Komatsuzaki, M. Kawakami, S. Takahashi, H. Yang, & R. J. Silbey (Eds.), *Single-molecule biophysics* (pp. 245–297). Hoboken, NJ: John Wiley & Sons, Inc.

Grewer, C., & Brauer, H.-D. (1994). Mechanism of the triplet-state quenching by molecular oxygen in solution. *Journal of Physical Chemistry, 98*, 4230–4235.

Gust, A., Zander, A., Gietl, A., Holzmeister, P., Schulz, S., Lalkens, B., et al. (2014). A starting point for fluorescence-based single-molecule measurements in biomolecular research. *Molecules, 19*, 15824–15865.

Ha, T., Enderle, T., Ogletree, D. F., Chemla, D. S., Selvin, P. R., & Weiss, S. (1996). Probing the interaction between two single molecules: Fluorescence resonance energy transfer between a single donor and a single acceptor. *Proceedings of the National Academy of Sciences of the United States of America, 93*, 6264–6268.

Ha, T., Ting, A. Y., Liang, J., Caldwell, W. B., Deniz, A. A., Chemla, D. S., et al. (1999). Single-molecule fluorescence spectroscopy of enzyme conformational dynamics and cleavage mechanism. *Proceedings of the National Academy of Sciences of the United States of America, 96*, 893–898.

Haran, G. (2012). How, when and why proteins collapse: The relation to folding. *Current Opinion in Structural Biology, 22*, 14–20.

Henry, E. R., Best, R. B., & Eaton, W. A. (2013). Comparing a simple theoretical model for protein folding with all-atom molecular dynamics simulations. *Proceedings of the National Academy of Sciences of the United States of America, 110*, 17880–17885.

Henzler-Wildman, K., & Kern, D. (2007). Dynamic personalities of proteins. *Nature, 450*, 964–972.

Hess, S. T., & Webb, W. W. (2002). Focal volume optics and experimental artifacts in confocal fluorescence correlation spectroscopy. *Biophysical Journal, 83*, 2300–2317.

Heyman, N. S., & Burt, J. M. (2008). Hindered diffusion through an aqueous pore describes invariant dye selectivity of Cx43 junctions. *Biophysical Journal, 94*, 840–854.

Hoffmann, A., Nettels, D., Clark, J., Borgia, A., Radford, S. E., Clarke, J., et al. (2011). Quantifying heterogeneity and conformational dynamics from single molecule FRET of diffusing molecules: Recurrence analysis of single particles (RASP). *Physical Chemistry Chemical Physics, 13*, 1857–1871.

Hofmann, H. (2014). Single-molecule spectroscopy of unfolded proteins and chaperonin action. *Biological Chemistry, 395*, 689–698.

Holzmeister, P., Gietl, A., & Tinnefeld, P. (2014). Geminate recombination as a photoprotection mechanism for fluorescent dyes. *Angewandte Chemie (International Ed. in English), 53*, 5685–5688.

Huang, F., Ying, L., & Fersht, A. R. (2009). Direct observation of barrier-limited folding of BBL by single-molecule fluorescence resonance energy transfer. *Proceedings of the National Academy of Sciences of the United States of America, 106*, 16239–16244.

Inanami, T., Terada, T. P., & Sasai, M. (2014). Folding pathway of a multidomain protein depends on its topology of domain connectivity. *Proceedings of the National Academy of Sciences of the United States of America, 111*, 15969–15974.

Jackson, S. E. (1998). How do small single-domain proteins fold? *Folding & Design, 3*, R81–R91.

Jager, M., Michalet, X., & Weiss, S. (2005). Protein-protein interactions as a tool for site-specific labeling of proteins. *Protein Science, 14*, 2059–2068.

Jager, M., Nir, E., & Weiss, S. (2006). Site-specific labeling of proteins for single-molecule FRET by combining chemical and enzymatic modification. *Protein Science, 15*, 640–646.

Jas, G. S., Eaton, W. A., & Hofrichter, J. (2001). Effect of viscosity on the kinetics of α-helix and β-hairpin formation. *Journal of Physical Chemistry B, 105*, 261–272.

Jones, C. M., Henry, E. R., Hu, Y., Chan, C. K., Luck, S. D., Bhuyan, A., et al. (1993). Fast events in protein folding initiated by nanosecond laser photolysis. *Proceedings of the National Academy of Sciences of the United States of America, 90*, 11860–11864.

Jung, S., & Dickson, R. M. (2009). Hidden Markov analysis of short single molecule intensity trajectories. *Journal of Physical Chemistry B, 113*, 13886–13890.

Kahra, D., Kovermann, M., Low, C., Hirschfeld, V., Haupt, C., Balbach, J., et al. (2011). Conformational plasticity and dynamics in the generic protein folding catalyst SlyD unraveled by single-molecule FRET. *Journal of Molecular Biology, 411*, 781–790.

Kalinin, S., Sisamakis, E., Magennis, S. W., Felekyan, S., & Seidel, C. A. (2010). On the origin of broadening of single-molecule FRET efficiency distributions beyond shot noise limits. *Journal of Physical Chemistry B, 114*, 6197–6206.

Kellner, R., Hofmann, H., Barducci, A., Wunderlich, B., Nettels, D., & Schuler, B. (2014). Single-molecule spectroscopy reveals chaperone-mediated expansion of substrate protein. *Proceedings of the National Academy of Sciences of the United States of America, 111*, 13355–13360.

Kuzmenkina, E. V., Heyes, C. D., & Nienhaus, G. U. (2005). Single-molecule Forster resonance energy transfer study of protein dynamics under denaturing conditions. *Proceedings of the National Academy of Sciences of the United States of America*, *102*, 15471–15476.

Kuzmenkina, E. V., Heyes, C. D., & Nienhaus, G. U. (2006). Single-molecule FRET study of denaturant induced unfolding of RNase H. *Journal of Molecular Biology*, *357*, 313–324.

Lakowicz, J. R. (1983). *Principles of fluorescence spectroscopy*. New York: Plenum.

Lamboy, J. A., Kim, H., Dembinski, H., Ha, T., & Komives, E. A. (2013). Single-molecule FRET reveals the native-state dynamics of the IkappaBalpha ankyrin repeat domain. *Journal of Molecular Biology*, *425*, 2578–2590.

Laurence, T. A., Kong, X., Jager, M., & Weiss, S. (2005). Probing structural heterogeneities and fluctuations of nucleic acids and denatured proteins. *Proceedings of the National Academy of Sciences of the United States of America*, *102*, 17348–17353.

Lee, T. H. (2009). Extracting kinetics information from single-molecule fluorescence resonance energy transfer data using hidden Markov models. *Journal of Physical Chemistry B*, *113*, 11535–11542.

Lee, T. C., Kang, M., Kim, C. H., Schultz, P. G., Chapman, E., Deniz, A. A., et al. (2016). Dual unnatural amino acid incorporation and click-chemistry labeling to enable single-molecule FRET studies of p97 folding. *ChemBioChem*, *17*(11), 981–984.

Li, P., Oliva, F. Y., Naganathan, A. N., & Muñoz, V. (2009). Dynamics of one-state downhill protein folding. *Proceedings of the National Academy of Sciences of the United States of America*, *106*, 103–108.

Lindhoud, S., Pirchi, M., Westphal, A. H., Haran, G., & van Mierlo, C. P. (2015). Gradual folding of an off-pathway molten globule detected at the single-molecule level. *Journal of Molecular Biology*, *427*, 3148–3157.

Liu, J., Campos, L. A., Cerminara, M., Wang, X., Ramanathan, R., English, D. S., et al. (2012). Exploring one-state downhill protein folding in single molecules. *Proceedings of the National Academy of Sciences of the United States of America*, *109*, 179–184.

Liu, B., Mazouchi, A., & Gradinaru, C. C. (2010). Trapping single molecules in liposomes: Surface interactions and freeze-thaw effects. *Journal of Physical Chemistry B*, *114*, 15191–15198.

Liu, Y., Park, J., Dahmen, K. A., Chemla, Y. R., & Ha, T. (2010). A comparative study of multivariate and univariate hidden Markov modelings in time-binned single-molecule FRET data analysis. *Journal of Physical Chemistry B*, *114*, 5386–5403.

Marcinowski, M., Holler, M., Feige, M. J., Baerend, D., Lamb, D. C., & Buchner, J. (2011). Substrate discrimination of the chaperone BiP by autonomous and cochaperone-regulated conformational transitions. *Nature Structural & Molecular Biology*, *18*, 150–158.

Mattheyses, A. L., Simon, S. M., & Rappoport, J. Z. (2010). Imaging with total internal reflection fluorescence microscopy for the cell biologist. *Journal of Cell Science*, *123*, 3621–3628.

Merchant, K. A., Best, R. B., Louis, J. M., Gopich, I. V., & Eaton, W. A. (2007). Characterizing the unfolded states of proteins using single-molecule FRET spectroscopy and molecular simulations. *Proceedings of the National Academy of Sciences of the United States of America*, *104*, 1528–1533.

Michalet, X., Ingargiola, A., Colyer, R. A., Scalia, G., Weiss, S., Maccagnani, P., et al. (2014). Silicon photon-counting avalanche diodes for single-molecule fluorescence spectroscopy. *IEEE Journal of Selected Topics in Quantum Electronics*, *20*, 38044201–380442020.

Michalet, X., Weiss, S., & Jager, M. (2006). Single-molecule fluorescence studies of protein folding and conformational dynamics. *Chemical Reviews*, *106*, 1785–1813.

Moosa, M. M., Ferreon, A. C., & Deniz, A. A. (2015). Forced folding of a disordered protein accesses an alternative folding landscape. *ChemPhysChem*, *16*, 90–94.

Muller, B. K., Zaychikov, E., Brauchle, C., & Lamb, D. C. (2005). Pulsed interleaved excitation. *Biophysical Journal, 89*, 3508–3522.

Muñoz, V. (2002). Thermodynamics and kinetics of downhill protein folding investigated with a simple statistical mechanical model. *International Journal of Quantum Chemistry, 90*, 1522–1528.

Muñoz, V. (2007) Conformational dynamics and ensembles in protein folding. *Annual Review of Biophysics and Biomolecular Structure, 36*, 395–412.

Muñoz, V. (2014). A simple theoretical model goes a long way in explaining complex behavior in protein folding. *Proceedings of the National Academy of Sciences of the United States of America, 111*, 15863–15864.

Muñoz, V., Campos, L. A., & Sadqi, M. (2016). Limited cooperativity in protein folding. *Current Opinion in Structural Biology, 36*, 58–66.

Muñoz, V., & Cerminara, M. (2016). When fast is better: Protein folding fundamentals and mechanisms from ultrafast approaches. *Biochemical Journal, 473*, 2545–2559.

Muñoz, V., & Eaton, W. A. (1999). A simple model for calculating the kinetics of protein folding from three-dimensional structures. *Proceedings of the National Academy of Sciences of the United States of America, 96*, 11311–11316.

Muñoz, V., & Sanchez-Ruiz, J. M. (2004). Exploring protein-folding ensembles: A variable-barrier model for the analysis of equilibrium unfolding experiments. *Proceedings of the National Academy of Sciences of the United States of America, 101*, 17646–17651.

Naganathan, A. N., Doshi, U., & Muñoz, V. (2007). Protein folding kinetics: Barrier effects in chemical and thermal denaturation experiments. *Journal of the American Chemical Society, 129*, 5673–5682.

Naganathan, A. N., & Muñoz, V. (2005). Scaling of folding times with protein size. *Journal of the American Chemical Society, 127*, 480–481.

Naganathan, A. N., Perez-Jimenez, R., Sanchez-Ruiz, J. M., & Muñoz, V. (2005). Robustness of downhill folding: Guidelines for the analysis of equilibrium folding experiments on small proteins. *Biochemistry, 44*, 7435–7449.

Nettels, D., Muller-Spath, S., Kuster, F., Hofmann, H., Haenni, D., Ruegger, S., et al. (2009). Single-molecule spectroscopy of the temperature-induced collapse of unfolded proteins. *Proceedings of the National Academy of Sciences of the United States of America, 106*, 20740–20745.

Nir, E., Michalet, X., Hamadani, K. M., Laurence, T. A., Neuhauser, D., Kovchegov, Y., et al. (2006). Shot-noise limited single-molecule FRET histograms: Comparison between theory and experiments. *Journal of Physical Chemistry B, 110*, 22103–22124.

Oikawa, H., Kamagata, K., Arai, M., & Takahashi, S. (2015). Complexity of the folding transition of the B domain of protein A revealed by the high-speed tracking of single-molecule fluorescence time series. *Journal of Physical Chemistry B, 119*, 6081–6091.

Oliva, F. Y., & Muñoz, V. (2004). A simple thermodynamic test to discriminate between two-state and downhill folding. *Journal of the American Chemical Society, 126*, 8596–8597.

Onuchic, J. N., Nymeyer, H., Garcia, A. E., Chahine, J., & Socci, N. D. (2000). The energy landscape theory of protein folding: Insights into folding mechanisms and scenarios. *Advances in Protein Chemistry, 53*, 87–152.

Pal, P., Lesoine, J. F., Lieb, M. A., Novotny, L., & Knauf, P. A. (2005). A novel immobilization method for single protein spFRET studies. *Biophysical Journal, 89*, L11–L13.

Prigozhin, M. B., & Gruebele, M. (2013). Microsecond folding experiments and simulations: A match is made. *Physical Chemistry Chemical Physics, 15*, 3372–3388.

Ramanathan, R., & Muñoz, V. (2015). A method for extracting the free energy surface and conformational dynamics of fast-folding proteins from single molecule photon trajectories. *Journal of Physical Chemistry B, 119*, 7944–7956.

Rasnik, I., McKinney, S. A., & Ha, T. (2006). Nonblinking and long-lasting single-molecule fluorescence imaging. *Nature Methods, 3*, 891–893.

Rhoades, E., Cohen, M., Schuler, B., & Haran, G. (2004). Two-state folding observed in individual protein molecules. *Journal of the American Chemical Society, 126*, 14686–14687.
Rhoades, E., Gussakovsky, E., & Haran, G. (2003). Watching proteins fold one molecule at a time. *Proceedings of the National Academy of Sciences of the United States of America, 100*, 3197–3202.
Rieger, R., Kobitski, A., Sielaff, H., & Nienhaus, G. U. (2011). Evidence of a folding intermediate in RNase H from single-molecule FRET experiments. *ChemPhysChem, 12*, 627–633.
Rosenkranz, T., Schlesinger, R., Gabba, M., & Fitter, J. (2011). Native and unfolded states of phosphoglycerate kinase studied by single-molecule FRET. *ChemPhysChem, 12*, 704–710.
Roy, R., Hohng, S., & Ha, T. (2008). A practical guide to single-molecule FRET. *Nature Methods, 5*, 507–516.
Sadqi, M., Fushman, D., & Muñoz, V. (2006). Atom-by-atom analysis of global downhill protein folding. *Nature, 442*, 317–321.
Sanchez-Medina, C., Sekhar, A., Vallurupalli, P., Cerminara, M., Muñoz, V., & Kay, L. E. (2014). Probing the free energy landscape of the fast-folding gpW protein by relaxation dispersion NMR. *Journal of the American Chemical Society, 136*, 7444–7451.
Schuler, B., & Eaton, W. A. (2008). Protein folding studied by single-molecule FRET. *Current Opinion in Structural Biology, 18*, 16–26.
Schuler, B., & Hofmann, H. (2013). Single-molecule spectroscopy of protein folding dynamics—Expanding scope and timescales. *Current Opinion in Structural Biology, 23*, 36–47.
Schuler, B., Lipman, E. A., & Eaton, W. A. (2002). Probing the free-energy surface for protein folding with single-molecule fluorescence spectroscopy. *Nature, 419*, 743–747.
Schuler, B., Soranno, A., Hofmann, H., & Nettels, D. (2016). Single-molecule FRET spectroscopy and the polymer physics of unfolded and intrinsically disordered proteins. *Annual Review of Biophysics, 45*, 207–231.
Seidel, R., & Dekker, C. (2007). Single-molecule studies of nucleic acid motors. *Current Opinion in Structural Biology, 17*, 80–86.
Seo, M. H., Han, J., Jin, Z., Lee, D. W., Park, H. S., & Kim, H. S. (2011). Controlled and oriented immobilization of protein by site-specific incorporation of unnatural amino acid. *Analytical Chemistry, 83*, 2841–2845.
Shammas, S. L., Crabtree, M. D., Dahal, L., Wicky, B. I., & Clarke, J. (2016). Insights into coupled folding and binding mechanisms from kinetic studies. *Journal of Biological Chemistry, 291*, 6689–6695.
Shaw, D. E., Maragakis, P., Lindorff-Larsen, K., Piana, S., Dror, R. O., Eastwood, M. P., et al. (2010). Atomic-level characterization of the structural dynamics of proteins. *Science, 330*, 341–346.
Sherman, E., & Haran, G. (2006). Coil-globule transition in the denatured state of a small protein. *Proceedings of the National Academy of Sciences of the United States of America, 103*, 11539–11543.
Sisamakis, E., Valeri, A., Kalinin, S., Rothwell, P. J., & Seidel, C. A. (2010). Accurate single-molecule FRET studies using multiparameter fluorescence detection. *Methods in Enzymology, 475*, 455–514.
Smiley, R. D., & Hammes, G. G. (2006). Single molecule studies of enzyme mechanisms. *Chemical Reviews, 106*, 3080–3094.
Song, L., Varma, C. A., Verhoeven, J. W., & Tanke, H. J. (1996). Influence of the triplet excited state on the photobleaching kinetics of fluorescein in microscopy. *Biophysical Journal, 70*, 2959–2968.
Spicer, C. D., & Davis, B. G. (2014). Selective chemical protein modification. *Nature Communications, 5*, 4740.
Stockmar, F., Kobitski, A. Y., & Nienhaus, G. U. (2016). Fast folding dynamics of an intermediate state in RNase H measured by single-molecule FRET. *Journal of Physical Chemistry B, 120*, 641–649.

Stryer, L., & Haugland, R. P. (1967). Energy transfer: A spectroscopic ruler. *Proceedings of the National Academy of Sciences of the United States of America, 58*, 719–726.
Suzuki, Y., Noel, J. K., & Onuchic, J. N. (2011). A semi-analytical description of protein folding that incorporates detailed geometrical information. *Journal of Chemical Physics, 134*, 245101.
Torella, J. P., Holden, S. J., Santoso, Y., Hohlbein, J., & Kapanidis, A. N. (2011). Identifying molecular dynamics in single-molecule FRET experiments with burst variance analysis. *Biophysical Journal, 100*, 1568–1577.
Turro, N. J., Ramamurthy, V., & Scaiano, J. C. (2010). *Modern molecular photochemistry of organic molecules.* Sausalito, CA: University Science Books.
Vale, R. D. (2008). Microscopes for fluorimeters: The era of single molecule measurements. *Cell, 135*, 779–785.
van der Velde, J. H., Oelerich, J., Huang, J., Smit, J. H., Aminian Jazi, A., Galiani, S., et al. (2016). A simple and versatile design concept for fluorophore derivatives with intramolecular photostabilization. *Nature Communications*, 7, 10144.
Voelz, V. A., Jager, M., Yao, S., Chen, Y., Zhu, L., Waldauer, S. A., et al. (2012). Slow unfolded-state structuring in Acyl-CoA binding protein folding revealed by simulation and experiment. *Journal of the American Chemical Society, 134*, 12565–12577.
Vogelsang, J., Kasper, R., Steinhauer, C., Person, B., Heilemann, M., Sauer, M., et al. (2008). A reducing and oxidizing system minimizes photobleaching and blinking of fluorescent dyes. *Angewandte Chemie (International Ed. in English), 47*, 5465–5469.
Wang, L., Brock, A., Herberich, B., & Schultz, P. G. (2001). Expanding the genetic code of *Escherichia coli. Science, 292*, 498–500.
Wang, Z., & Lu, H. P. (2015). Probing single-molecule protein spontaneous folding-unfolding conformational fluctuation dynamics: The multiple-state and multiple-pathway energy landscape. *Journal of Physical Chemistry B, 119*, 6366–6378.
Weinkam, P., Zimmermann, J., Romesberg, F. E., & Wolynes, P. G. (2010). The folding energy landscape and free energy excitations of cytochrome c. *Accounts of Chemical Research, 43*, 652–660.
Weiss, S. (1999). Fluorescence spectroscopy of single biomolecules. *Science, 283*, 1676–1683.
Widengren, J., Chmyrov, A., Eggeling, C., Lofdahl, P. A., & Seidel, C. A. (2007). Strategies to improve photostabilities in ultrasensitive fluorescence spectroscopy. *Journal of Physical Chemistry A, 111*, 429–440.
Yang, W. Y., & Gruebele, M. (2003). Folding at the speed limit. *Nature, 423*, 193–197.
Young, T. S., Ahmad, I., Yin, J. A., & Schultz, P. G. (2010). An enhanced system for unnatural amino acid mutagenesis in *E. coli. Journal of Molecular Biology, 395*, 361–374.
Zheng, Q., Jockusch, S., Zhou, Z., Altman, R. B., Warren, J. D., Turro, N. J., et al. (2012). On the mechanisms of cyanine fluorophore photostabilization. *Journal of Physical Chemistry Letters, 3*, 2200–2203.
Zheng, Q., Juette, M. F., Jockusch, S., Wasserman, M. R., Zhou, Z., Altman, R. B., et al. (2014). Ultra-stable organic fluorophores for single-molecule research. *Chemical Society Reviews, 43*, 1044–1056.
Zhuang, X. (2005). Single-molecule RNA science. *Annual Review of Biophysics and Biomolecular Structure, 34*, 399–414.
Zimmermann, J. L., Nicolaus, T., Neuert, G., & Blank, K. (2010). Thiol-based, site-specific and covalent immobilization of biomolecules for single-molecule experiments. *Nature Protocols, 5*, 975–985.

CHAPTER FIFTEEN

Single-Molecule Multicolor FRET Assay for Studying Structural Dynamics of Biomolecules

S. Lee*, Y. Jang*, S.-J. Lee†, S. Hohng†,‡,§,1

*Advanced Photonics Research Institute, Gwangju Institute of Science and Technology, Gwangju, Republic of Korea

†National Center for Creative Research Initiatives, Seoul National University, Seoul, Republic of Korea

‡Institute of Applied Physics, Seoul National University, Seoul, Republic of Korea

§Seoul National University, Seoul, Republic of Korea

[1]Corresponding author: e-mail address: shohng@gmail.com

Contents

Abstract

Over the last 2 decades, single-molecule Forster resonance energy transfer (FRET) has been widely used to address important questions in molecular biology. However, a conventional approach based on a single donor–acceptor pair is not powerful enough to study complex biological systems. To address this challenge, single-molecule multicolor FRET techniques have been developed. In this chapter, we present practical considerations required for the successful implementation of single-molecule multicolor FRET in the laboratory.

Methods in Enzymology, Volume 581
ISSN 0076-6879
http://dx.doi.org/10.1016/bs.mie.2016.08.015

© 2016 Elsevier Inc.
All rights reserved.

1. INTRODUCTION

During the last 2 decades, single-molecule techniques have made important contributions to the mechanistic understanding of various biomolecular systems (Bustamante, Cheng, & Mejia, 2011; Kapanidis & Strick, 2009; Tinoco & Gonzalez, 2011; Walter & Bustamante, 2014). Among the techniques, single-molecule Forster resonance energy transfer (FRET) has emerged as a potent tool for studying molecular interactions and conformational dynamics of biomolecules in real time (Ha et al., 1996; Roy, Hohng, & Ha, 2008; Schuler & Eaton, 2008). FRET is the energy transfer phenomenon from a donor dye to an acceptor dye via dipole–dipole interaction (Förster, 1948). It requires sufficient amount of spectral overlap between the emission of donor dye and the absorption of acceptor dye. Since the energy transfer efficiency of FRET strongly depends on the interdye distance in the range of a several nanometers, it can be used for observing interdye distances very accurately. Therefore, when biomolecules are tagged with FRET probes, we can study both the intra- and intermolecular dynamics of biomolecules in real time by tracking changes in the FRET efficiency.

While conventional single-molecule FRET has been popularly used over the last 2 decades, it also suffers from limitations in the amount of available information; in conventional single-molecule FRET measurements, only single interdye distances can be measured. As more complex biological systems are studied, the capability to probe multiple distances simultaneously is frequently needed to visualize unambiguously complex dynamics of biological systems. This technical challenge has been addressed by realizing single-molecule multicolor FRET technique by various groups (Clamme & Deniz, 2005; DeRocco, Anderson, Piehler, Erie, & Weninger, 2010; Hohng, Lee, Lee, & Jo, 2014; Stein, Steinhauer, & Tinnefeld, 2011). In the initial demonstration, single-molecule three-color FRET using one donor (Cy3) and two acceptors (Cy5 and Cy5.5) was realized and used to study the correlated motions of different segments of a DNA Holliday junction in real time (Hohng, Joo, & Ha, 2004). This first demonstration of single-molecule three-color FRET experiment, however, had a limitation that three interdye distances could not be determined simultaneously. To overcome this technical problem, the alternating laser excitation (ALEX) technique was used in single-molecule multicolor FRET experiments (Lee et al., 2007; Lee, Lee, & Hohng, 2010;

Ross et al., 2007). Recently, the single-molecule four-color FRET technique was developed as the most advanced form among the currently available methods (Lee, Lee, Ragunathan, et al., 2010), allowing the determination of six interdye FRET efficiencies in real time. In this chapter, we provide a general overview of single-molecule multicolor FRET and practical considerations required for the successful implementation of the technique in the laboratory.

2. EXPERIMENTAL DESIGN

In this section, we provide a detailed description of the setup.

2.1 Dye Choice

As fluorescent probes for single-molecule measurements, the following properties are desirable (Kapanidis & Weiss, 2002). They should be (1) bright enough for single-molecule detection, (2) photostable to allow long observation time, (3) commercially available in a bioconjugatable form, and (4) excitable in the 450–750 nm wavelength range, where commonly used detectors in single-molecule fluorescence microscopy show the high detection efficiency.

For multicolor FRET measurements, an ideal set of FRET probes should have appreciable spectral overlap between donor emissions and acceptor absorptions for each FRET pair, whereas their absorption and emission spectra should be well separated to incorporated ALEX technique and reduce bleedthrough between detection channels. It is also desirable that they have similar quantum yields and extinction coefficients so that the FRET ranges of each pair are comparable.

Among dyes in the visible range, blue dyes usually have poor photostability (Lee, Lee, & Hohng, 2010). Furthermore, they frequently exhibit spontaneous spectral shift during measurements. Among candidates of blue dyes, we recommend the use of Cy2 as a blue dye (Lee, Lee, Ragunathan, et al., 2010). According to our test measurements (Table 1 and Fig. 1), its emission and absorption spectra are clearly separated from those of Cy3. The spontaneous spectral shift of Cy2 was not as severe other blue dyes. As shown in Fig. 2, when we used a Cy2–Cy3–Cy5–Cy7 dye quartet, their emission and absorption spectra were well separated, and the FRET ranges of each FRET pair were comparable.

Table 1 Comparison of General Specifications and Photostability of Blue Dyes

	Bleaching Time (s)		SNR		Average Intensity		Abs Peak (nm)	Emi Peak (nm)	Extinction Coefficient (M^{-1} cm^{-1})	Quantum Yield
	(gloxy)	(PCD)	(gloxy)	(PCD)	(gloxy)	(PCD)				
Cy2	3.8	6.7	8.4	10.8	1000	1017	489	506	150,000	0.12
Alexa488	14.5	20.8	8.7	9.6	1241	1133	495	519	71,000	0.92
Atto488	35.1	44.9	5.5	6.8	675	589	493	518	70,000	
Dylight488	12.1	13.1	8.7	9.9	1173	1023	510	523	90,000	0.8

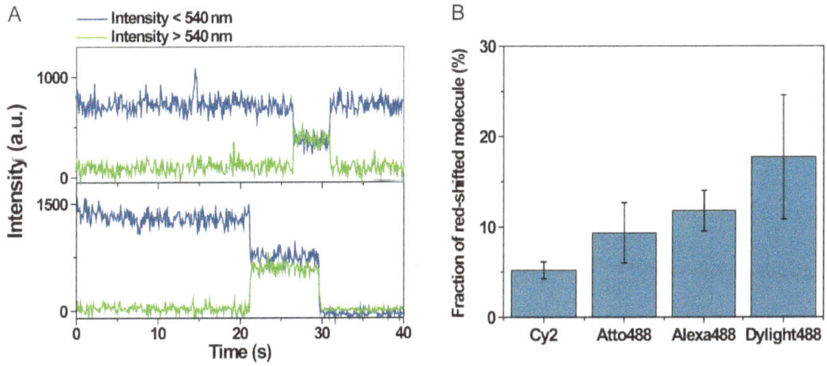

Fig. 1 Emission spectral red-shift behavior of blue dyes. (A) Representative intensity time trace of Alexa488 fluorescence emission observed both in the Cy2 detection channel (*blue line*) and Cy3 detection channel (*green line*). Similar behavior was observed in all blue dyes tested. (B) The fraction of dyes showing emission red-shift behavior in the tested blue dyes. The fractions are 5.2% for Cy2, 9.1% for Atto488, 11.7% for Alexa488, and 17.7% for Dylight488, respectively.

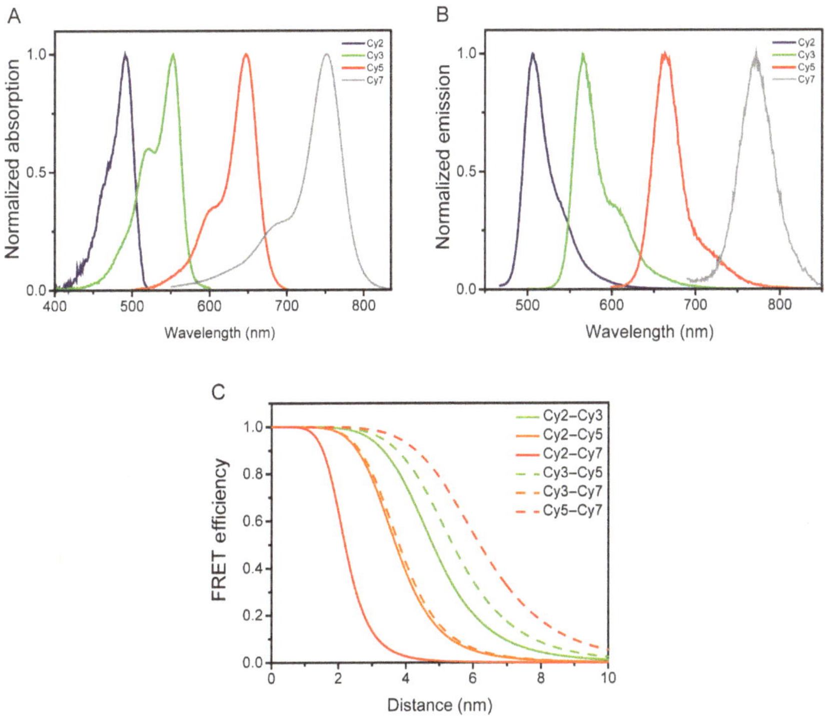

Fig. 2 Emission and absorption spectra of the dye quartet and FRET efficiencies for each dye pair as a function of distance. (A) Absorption spectra of Cy2, Cy3, Cy5, and Cy7. (B) Emission spectra of the dye quartet. (C) Six interdye FRET efficiencies between four quartet dyes as a function of interdye distances using calculated Ro values.

2.2 ALEX Technique for Determining Multiple FRET Efficiencies

The ALEX technique was initially introduced to distinguish donor-only species from donor–acceptor species with low FRET efficiency (Hohlbein, Craggs, & Cordes, 2014; Kapanidis et al., 2004). When the ALEX technique is incorporated into single-molecule multicolor FRET experiments, it can provide the missing information required to determine all FRET efficiencies (Lee, Kapanidis, et al., 2007; Lee, Lee, & Hohng, 2010; Ross et al., 2007). In four-color FRET experiments, three independent lasers (473 nm blue laser for Cy2 excitation, 532 nm green laser for Cy3 excitation, 635 nm red laser for Cy5 excitation) are used to excite each donor dye alternatively. The procedure for determining six FRET efficiencies using three-color ALEX technique is described briefly later (Fig. 3).

As shown in Fig. 3A, there exist six FRET interactions in the dye quartet (E_{12}: FRET efficiency from Cy2 to Cy3, E_{13}: FRET efficiency from Cy2 to Cy5, E_{14}: FRET efficiency from Cy2 to Cy7, E_{23}: FRET efficiency from Cy3 to Cy5, E_{24}: FRET efficiency from Cy3 to Cy7, E_{34}: FRET efficiency from Cy5 to Cy7).

Step 1 Under red laser excitation of Cy5, there exists only one FRET interaction between Cy5 and Cy7. Thus, FRET efficiency between Cy5 and Cy7 (E_{34}) is easily determined as in a conventional FRET scheme.

Step 2 Under green laser excitation of Cy3, however, there exist three FRET interactions. In this case, since E_{34} is already obtained in Step 1, the remaining FRET efficiencies (E_{23} and E_{24}) can be calculated just using three fluorescence signals detected under the excitation.

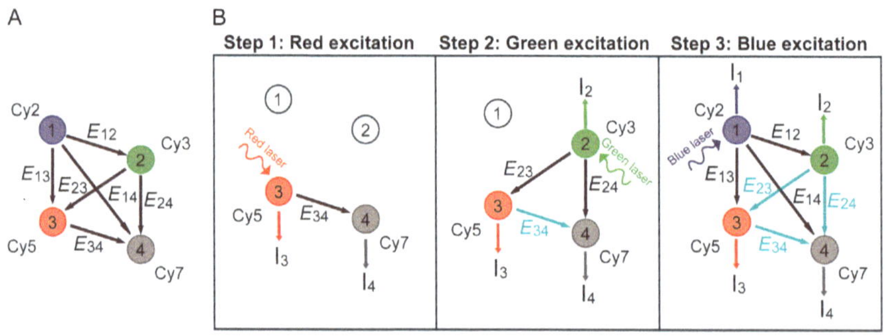

Fig. 3 Brief description of FRET determination using ALEX technique. (A) Interaction diagram of four dyes. (B) A step-by-step guide to determine FRET interactions through three alternative laser excitations.

Step 3 Under blue laser excitation of Cy2, likewise, the remaining FRET efficiencies among the dye quartet (E_{12}, E_{13}, and E_{14}) can be determined on the basis of the three FRET efficiencies that were previously obtained.

A detailed calculation of six FRET efficiencies is described in Section 3.3.

2.3 Optical Setup

To detect single-molecule fluorescence signals, the background signal needs to be effectively reduced. For that purpose, two types of fluorescence microscopies, confocal microscopy and total internal reflection (TIR) microscopy, have been commonly used. Because each microscopy has its own pros and cons (details of the pros and cons are described later), these two types of microscopies are complementary. We describe detailed information on how to build the microscopes.

2.3.1 Confocal Microscopy

In confocal microscopy (Ha, Chemla, Enderle, & Weiss, 1997), a laser beam is tightly focused using a high NA objective lens. A small excitation volume created in such a way is good for the reduction of background noise using a pinhole. As a detector, an avalanche photodiode is usually used. Since it is faster than video-rate camera usually used in TIR microscopy, confocal has an advantage of high temporal resolution compared to TIR microscopy when the photoblinking problems of dyes in the submillisecond range is solved.

Single-molecule four-color FRET in confocal microscopy (Lee, Lee, Ragunathan, et al., 2010) was built by modifying a commercial inverted microscope (TE2000-U, Nikon). Fig. 4 shows a schematic diagram of the setup including the light path and optical components of the instrument. The inverted microscope is equipped with a three-dimensional piezoelectric positioner stage (LP-100, MadCityLabs) for the sample scanning. To utilize three-color ALEX technique in single-molecule four-color FRET measurements, three excitation lasers, 473- and 532-nm diode-pump solid-state lasers (Blues TM50, Cobolt; Excelsior-CDRH, Spectra-Physics) and a 633-nm HeNe laser (HRP050, Thorlabs), are used as excitation sources. Electrooptic modulators (EOMs, 350-50, Conoptics) for each laser are used to switch three excitation lasers alternatively on sub-millisecond timescale. EOMs for each laser achieve high on/off ratio

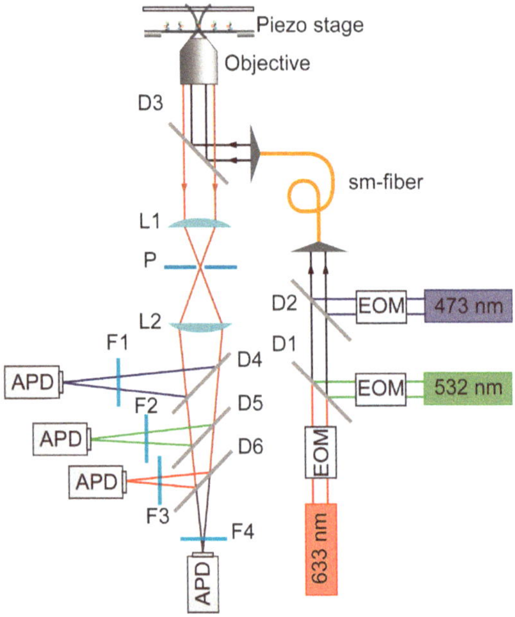

Fig. 4 Schematic diagram of single-molecule four-color FRET setup realized in confocal microscopy. The identities of optical components are described in Section 2.3.1.

(>200:1) using the combination with polarizers. To facilitate the colocalization of three excitation lasers, they are combined using two dichroic mirrors (D1, z532bcm, Chroma; D2, z488rdc, Chroma), and then coupled into a single-mode fiber (460HP, Thorlabs). To excite molecules in the sample chamber, these combined lasers are guided through the back port of the microscope and an objective lens. To create a small volume laser spot for the excitation of a fluorescent molecule and efficiently collect fluorescence emissions from the molecule under the spot, an oil-immersion objective lens with a high numerical aperture (NA = 1.4, UPLASAPO 100X, Olympus) is used. Three fluorescence emissions collected through the objective lens are focused into a 100-μm pinhole (P100S, Thorlabs). The pinhole is used to block the out-of-focus fluorescence, resulting in reduced background noise. To eliminate the effect of chromatic aberration among four fluorescence emissions of different colors, achromatic lenses (L1 and L2, LAO-90.0–25.0/078, CVI) are used. The fluorescence signals are separated by three dichroic mirrors (D4, 540dcxt, Chroma; D5, 640dcxr, Chroma; D6, 740dcxr, Chroma) and filtered through four band-pass filters (F1, FF01–512/25-25, Semrock; F2, HQ580/60m-2p,

Chroma; F3, HQ680/60m-2p, Chroma; F4, HQ790/80m, Chroma). Finally, four fluorescence emissions of the dye quartet (Cy2, Cy3, Cy5, and Cy7) are measured by using four avalanche photo diodes (APD, SPCM-AQRH-14, Perkin Elmer).

To obtain single-molecule fluorescence time traces of immobilized molecules, a fluorescence image is first taken by scanning the confocal spot in the sample plane, and a position of the target molecule is then determined from the image (Fig. 5). After placing the spot on this position, the four fluorescence intensities from one molecule are monitored at one time. Data acquisition of the fluorescence intensities is performed using a counter/timer DAQ board (NI PCI-6602, National Instruments) and synchronized with switching of three lasers by using common trigger lines of the DAQ board. TTL pulses generated by the DAQ board are connected to the inputs of EOM controller for switching three excitation lasers alternatively. The whole setup is run by a custom program written in Visual C++ (program available upon request).

2.3.2 TIR Microscopy

In TIR microscopy (Axelrod, 2001; Roy et al., 2008), an evanescent field excitation generated by TIR of light selectively illuminates fluorescent molecules within a few hundred nanometers (~100–200 nm) from the surface

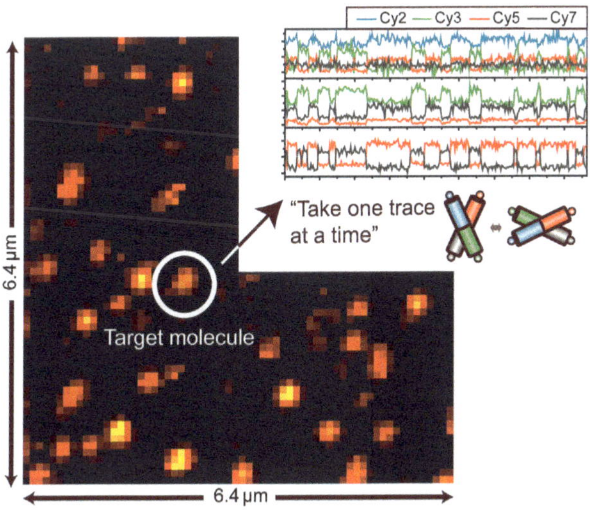

Fig. 5 Extraction of single-molecule fluorescence time trace from a confocal scanning image. Positions of target molecules are determined through manual selection in the scanning image.

and significantly reduces background noise. Compared to confocal microscopy, the setup using an electron-multiplying CCD camera for wide-field imaging significantly increases the data yield by observing hundreds of fluorescent molecules simultaneously with millisecond time resolution. Hence, this is particularly useful for studying irreversible or rare biological reactions.

To achieve TIR of excitation lasers at the quartz–water interface, two types of TIR (objective-type TIR or prism-type TIR) can be adopted in single-molecule fluorescence microscopy. Though objective-type TIR can be used for implementing single-molecule four-color FRET microscopy, we prefer prism-type TIR because real-time buffer-injection exchange experiments can be reliably done only in the prism-type TIRF; in objective-type TIR, when the buffer flow is introduced into the chamber, the imaging surface on the thin coverslip bows instantaneously, and thus the image is defocused during buffer exchange.

A prism-type microscope for single-molecule four-color FRET (Lee, Lee, Ragunathan, et al., 2010) was built based on a commercial inverted microscope (IX71, Olympus). Fig. 6 illustrates a schematic diagram of the setup containing the light path and optical components of the instrument. By modifying a sample stage of the inverted microscope, we hold a prism (Pellin-Broca fused silica prisms, 325–1206, EKSMA) on top of the sample chamber for TIR generating an evanescent field. As with the confocal microscopy, three excitation lasers (473-nm, Blues TM50, Cobolt; 532-nm, Compass215M, Coherent; 640-nm, Cube640-100C, Coherent) are used as excitation sources. The intensities of three lasers are adjusted using a neutral density filter (ND, NDC-50C-4M, Thorlabs) or the combination of a half waveplate ($\lambda/2$, QWPO, CVI) and a polarizing beam splitter (PBS, PBSH-450-700-100, CVI). To switch three excitation lasers alternatively, mechanical shutters (LS-3, Uniblitz) for each laser are used. In TIRF microscopy, the relatively slow switching time of a mechanical shutter ($\sim$1 ms) does not cause any technical problems because the practical time resolution of the setup is limited by the frame rate of an EM-CCD camera. To ensure that three excitation lasers excite the same sample area, three excitation lasers are combined through a dichroic mirror (D1, z488rdc, Chroma) and a polarizing beam splitter. These excitation lasers are focused into the prism using a lens (125 mm focal length, LA1986-A, Thorlabs), and thus the location and area of the excitation can be easily adjusted by moving the position of the lens. The fluorescence signals from the dye quartet are collected through a long working distance (0.28 mm) and high numerical

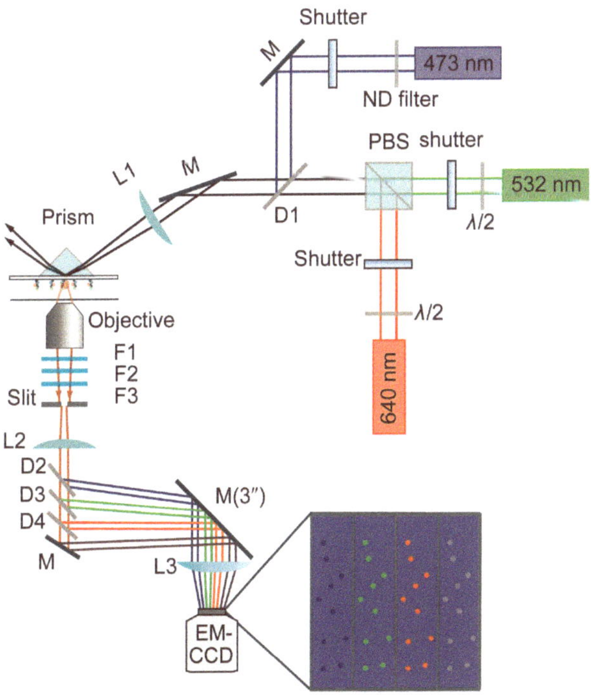

Fig. 6 Schematic diagram of single-molecule four-color FRET setup realized in TIRF microscopy. The identities of optical components are described in Section 2.3.2.

aperture (NA 1.2) water immersion objective (UPlanSApo 60XW, Olympus), and go through a long-pass filter (F1, LP02-473RU-25, Semrock) and two notch filters (F2, NF03-532E-25, Semrock; F3, NF03-633E-25, Semrock) to filter out laser scattering. To obtain fluorescence images for the dye quartet simultaneously using one EM-CCD camera, the fluorescence signals are separated by three dichroic mirrors (D2, T540dcxr-UF1, D3, T635lpxr-UF1, Chroma; D4, T740lpxr-UF1, Chroma) and focus on different areas of an electron-multiplying charged-coupled device (EM-CCD, iXon Ultra897, Andor Technology) in a quad-view scheme. These split images are cut into one-quarter the size of the CCD sensor chip using a vertical slit (M-SV-0.5, Newport) about 1 mm in width. To eliminate the effect of chromatic aberration among four fluorescence emissions of different colors, achromatic lenses (L2, LAO-120.0–40.0/075, CVI; L3, LAO-260.1–50.0/075, CVI) are used. Due to the pair of lenses, this

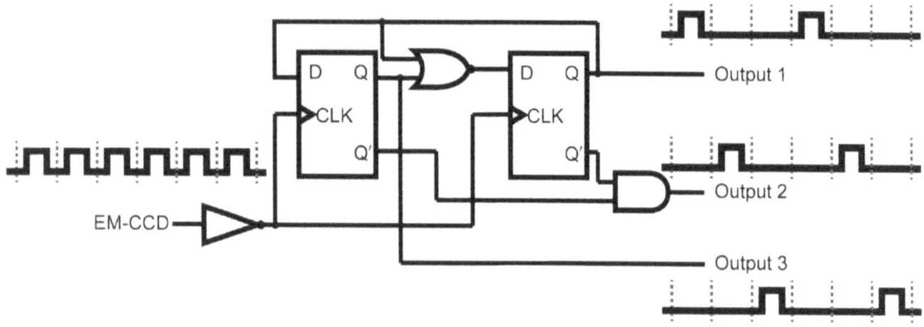

Fig. 7 Diagram of electronic circuit for synchronizing three lasers switching and image acquisition. Inverter, D flip-flop, NOR, and AND are used for the electronic circuit.

microscope practically has a magnification of 130, and thus the observation area is about 15 μm × 60 μm.

Raw fluorescence movies from the EM-CCD camera are recorded in real time and saved to the hard disk directly using home-written Visual C++ software (program available upon request). To synchronize the switching of three excitation lasers with the image acquisition, the "fire" output (pulse signals generated in every frame) from the EM-CCD camera is used as a common trigger line of a counter/timer DAQ board (NI PCI-6602, National Instruments). Three sequential TTL pulses from the DAQ board are coupled to three TTL inputs of a mechanical shutter controller (VMM-D4, Uniblitz) for switching three excitation lasers alternatively. Optionally, a simple electronic circuit based on flip-flops can be also used to synchronize the laser switching and the image acquisition instead of the counter/timer DAQ board. Fig. 7 shows a circuit diagram for the synchronization using the electronic circuit. The "fire" output from the EM-CCD camera is connected to the clock inputs of two D flip-flops (74F374, Motorola), and a combination of flip-flops and other logic gates (NOT, 74HC04, Motorola; AND, 74LS08, Motorola; NOR, 74LS02) generates three sequential pulse trains to switch alternatively three excitation lasers synchronously. The whole setup is operated by a custom program written in Visual C++ (program available upon request).

3. DATA ANALYSIS

As mentioned in Section 2.3, single-molecule multicolor fluorescence signals are acquired in real time using home-written Visual C++ programs. While confocal microscopy directly takes the time trajectories of multicolor fluorescence signals from one molecule at a time, TIR microscopy obtains the raw fluorescence movies of a large number of molecules. To extract individual multicolor fluorescence time traces from the movies, therefore, post-processing of images is required (Fig. 8). To determine accurate FRET efficiencies between multiple FRET pairs, data correction for background noise, leakage between detection channels, and direct excitation of acceptors is essentially required. In this section, we provide a general overview of the data analysis for single-molecule multicolor FRET experiments including image processing, data correction, and FRET calculation.

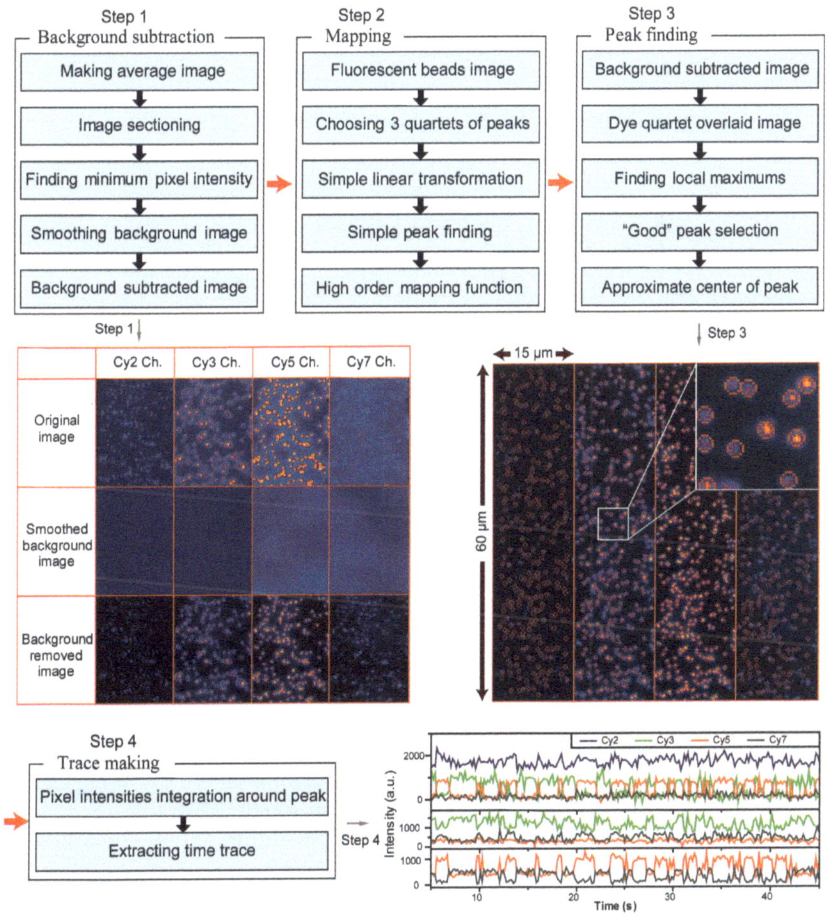

Fig. 8 Detailed procedure of the image processing to extract single-molecule four-color fluorescence time trace.

3.1 Image Processing Algorithms

To process the raw image data and finally extract single-molecule time trajectories from the recorded movies, we use automated scripts written in IDL (the script available upon request). Details of the image processing algorithms used in our laboratories are described later.

The first step in extracting single-molecule time trajectories from the raw movie data is background subtraction. First, we make an average image of first 10 frames from the recorded movie and split the image into four distinct spectral channels for dye quartet. Because in wide-field imaging the background noise level in each pixel is not uniform throughout the imaging area, we divide the imaging area (corresponding to 128×512 pixels detection area) into 8×32 small sections. By taking a minimum pixel intensity for each section, we determine the level of background noise for each section and construct a background image. In this image, there is the discontinuity problem occurring at the boundaries between divided sections. Hence, image smoothing is performed and thus background smoothed image is created. By subtracting this image from the original image, we finally construct background subtracted image. As a second step, the four spit images for quartet dyes are overlaid using a mapping function. Due to spherical or chromatic aberrations and incomplete alignment of optical setup, simple linear transformation is not enough for accurate mapping of dye quartet images. To create the mapping function, we first obtain four split images of surface-immobilized fluorescent beads, which are visible in all detection channels. Then, we manually select three peaks from the Cy2 channel and their corresponding peaks from the other channels. Using those selected peaks, a linear transformation function between the four detection channels is determined. By using this linear transformation, more corresponding peaks are automatically found from each detection channel. These peaks are used to create a high-order polynomial mapping function. In our laboratories, we use a third-order polynomial function for accurate mapping, which requires at least 16 peaks from each image to determine all 32 coefficients of the polynomial function. As a third step, we find single-molecule fluorescence peaks from an overlaid image of dye quartet channels. First, we combine the dye quartet images derived in Step 1 using the mapping function created in Step 2. In this overlaid image, we find many local maximum points as candidates for "true" single-molecule peaks. To select "good" peaks among the candidates, we establish several selection criteria including the appropriate intensity and size of peaks. In addition, quadruply colocalized spots are selected to exclude partially labeled species (except in special experimental designs). Next we find an approximate center of the selected peak by fitting the intensity profile around the peak to a single Gaussian function. As a final step, we extract single-molecule multicolor fluorescence time traces from the recorded movies. To obtain true dye quartet intensities for each single molecule, intensities from 7×7 pixels around the peak derived above are integrated for every frame.

3.2 Data Correction

Before FRET analysis, all data are corrected as follows. This data correction step is performed by a custom program written in Matlab (program available upon request). To generalize the discussion of correction steps, we use indices to indicate FRET probes and detection channels in the increasing order of emission peaks. So, in this paper, the first, the second, the third dyes, and the fourth dyes correspond to Cy2 (or Alexa488), Cy3, Cy5, and Cy7, respectively. Thus the following symbols indicate:

I_i^{raw}, raw intensity of the ith detection channel after background subtraction;

I_i^{eff}, fluorescence intensity of the ith dye detected in the ith detection channel; and

I_i^{tot}, total fluorescence intensity of the ith dye detected in the all detection channels.

The first data correction step is background subtraction. Next we correct the bleedthrough between detection channels. We define l_{ij} as the bleedthrough parameter of the ith dye into the jth detection channel. For example, l_{12} denotes the ratio of Cy2 (or Alexa488) fluorescence signal detected in the second detection channel to that of Cy2 (or Alexa488) signal detected in the first detection channel. Specifically, l_{1i} can be determined from the ratios of I''_i/I''_1 when every dye is photobleached except Cy2 (or Alexa488). The other coefficients can be determined in the same way.

By assuming no bleedthrough from the longer wavelength dyes into the shorter wavelength detection channels, we get the following expressions:

$$I_1^{\text{raw}} = I_1^{\text{eff}} \tag{1}$$

$$I_2^{\text{raw}} = l_{12} I_1^{\text{eff}} + I_2^{\text{eff}} \tag{2}$$

$$I_3^{\text{raw}} = l_{13} I_1^{\text{eff}} + l_{23} I_2^{\text{eff}} + I_3^{\text{eff}} \tag{3}$$

$$I_4^{\text{raw}} = l_{14} I_1^{\text{eff}} + l_{24} I_2^{\text{eff}} + l_{34} I_3^{\text{eff}} + I_4^{\text{eff}} \tag{4}$$

By solving the equations above, we obtain the following equations.

$$I_1^{\text{eff}} = I_1^{\text{raw}} \tag{5}$$

$$I_2^{\text{eff}} = I_2^{\text{raw}} - l_{12} I_1^{\text{eff}} = I_2^{\text{raw}} - l_{12} I_1^{\text{raw}} \tag{6}$$

$$I_3^{\text{eff}} = I_3^{\text{raw}} - l_{13} I_1^{\text{eff}} - l_{23} I_2^{\text{eff}} = I_3^{\text{raw}} - l_{13} I_1^{\text{raw}} - l_{23}\left(I_2^{\text{raw}} - l_{12} I_1^{\text{raw}}\right) \tag{7}$$

$$\begin{aligned} I_4^{\text{eff}} &= I_4^{\text{raw}} - l_{14} I_1^{\text{eff}} - l_{24} I_2^{\text{eff}} - l_{34} I_3^{\text{eff}} \\ &= I_4^{\text{raw}} - l_{14} I_1^{\text{raw}} - l_{24}\left(I_2^{\text{raw}} - l_{12} I_1^{\text{raw}}\right) \\ &\quad - l_{34}\left[I_3^{\text{raw}} - l_{13} I_1^{\text{raw}} - l_{23}\left(I_2^{\text{raw}} - l_{12} I_1^{\text{raw}}\right)\right] \end{aligned} \tag{8}$$

Next, we obtain I_i^{tot} by adding back the bleedthrough signals into the proper detection channels as follows.

$$I_1^{\text{tot}} = I_1^{\text{eff}} + l_{12} I_1^{\text{eff}} + l_{13} I_1^{\text{eff}} + l_{14} I_1^{\text{eff}} = (1 + l_{12} + l_{13} + l_{14}) I_1^{\text{raw}} \tag{9}$$

$$I_2^{\text{tot}} = I_2^{\text{eff}} + l_{23} I_2^{\text{eff}} + l_{24} I_2^{\text{eff}} = (1 + l_{23} + l_{24})\left(I_2^{\text{raw}} - l_{12} I_1^{\text{raw}}\right) \tag{10}$$

$$I_3^{\text{tot}} = I_3^{\text{eff}} + l_{34} I_3^{\text{eff}} = (1 + l_{34})\left[I_3^{\text{raw}} - l_{34} I_2^{\text{raw}} - (l_{13} - l_{12} l_{23}) I_1^{\text{raw}}\right] \tag{11}$$

$$\begin{aligned} I_4^{\text{tot}} = I_4^{\text{eff}} &= I_4^{\text{raw}} - l_{34} I_3^{\text{raw}} - (l_{24} - l_{23} l_{34}) I_2^{\text{raw}} \\ &\quad - (l_{14} - l_{12} l_{24} - l_{13} l_{34} + l_{12} l_{23} l_{34}) I_1^{\text{raw}} \end{aligned} \tag{12}$$

To calculate the FRET efficiency of each FRET pair correctly, we should consider the differences in emission quantum yields of fluorophores, and the detection efficiencies of the detection channels, which can be corrected as follows, a step called the gamma correction.

$$I_1^* = I_1^{\text{tot}} \tag{13}$$

$$I_2^* = \gamma_{12} I_2^{\text{tot}} \tag{14}$$

$$I_3^* = \gamma_{13} I_3^{\text{tot}} \tag{15}$$

$$I_4^* = \gamma_{14} I_4^{\text{tot}} \tag{16}$$

where γ_{ij} is the ratio of $\Delta \widetilde{I_i}$ to $\Delta \widetilde{I_j}$, during a conformational change or photobleaching and I_i* is the intensity of the ith dye after gamma correction. After gamma correction, the intensity of each fluorophore is normalized by the intensity of first fluorophore. Therefore only γ_{12}, γ_{13}, γ_{14} terms are needed for gamma correction.

The final correction step is the subtraction of direct excitation of acceptors. In FRET experiments, it is ideal for each laser to excite only one fluorophore that is expected to behave as a donor in the excitation condition. This is a good approximation for green laser excitation as the absorption of Cy5, and Cy7 at 532-nm are just 3% and 2% of their peak values, respectively. However, the absorption of Cy3 at 473-nm and Cy7 at 633- or 640-nm is not negligible. Therefore, the Cy3 intensity I_2 at 473-nm laser excitation and the Cy7 intensity I_4 at 640-nm laser excitation contain contributions from both FRET and the direct absorption of the excitation. The direct excitation of acceptor dyes must be corrected to determine a reliable FRET efficiency. Based on the absorption spectrum, the ratio of the Cy3 intensity at 473-nm laser excitation to the intensity of 532-nm laser excitation is a constant. Therefore, we define the direct excitation coefficient at 473-nm laser d_b which is the ratio of $I_{2,b}{}^*$ to $I_{2,g}{}^*$ after the photobleaching of Cy2. This coefficient refers to the intensity ratio of Cy3 at 473- and 532-nm lasers. We can then remove the direct excitation part of Cy3 intensity at 473-nm laser excitation by following the equations below. It is important to note that the direct excitation coefficient must be determined in each measurement because the coefficient also changes when the intensity of laser changes.

$$I_{2,b} = I^*_{2,b} - d_b I^*_{2,g} \tag{17}$$

For the direct excitation of Cy7 by a red laser, we use a different method. We assume that the ratio of the direct absorption of Cy7 to the absorption of Cy5 is conserved after gamma correction, and thus Cy7 fluorescence due to direct excitation is proportional to the sum of Cy5 and Cy7 fluorescence at 640-nm laser excitation. Because the Cy7 fluorescence when Cy5 is photobleached comes only from the direct excitation of Cy7, we can obtain the 640-nm laser direct excitation coefficient d_r which is defined as the ratio of $I^*_{4,r}$ after photobleaching to $I^*_{3,r} + I^*_{4,r}$ before Cy5 photobleaching. Then the direct excitation of Cy7 at 640-nm laser excitation can be determined by following equation:

$$I_{4,r} = I^*_{4,r} - d_r \left(I^*_{3,r} + I^*_{4,r} \right) \tag{18}$$

3.3 FRET Calculation

By using the data corrected as explained earlier, six FRET efficiencies are calculated as follows. We use a custom-made program written in Matlab to calculate and display these FRET efficiencies (the program available upon request). Interaction diagrams of four fluorophores are shown in Fig. 3A. The identities of symbols are: G_i, the excitation rate of the ith dye; k_i, intrinsic decay rate of the ith dye; k_{ij}, the rate of FRET from the ith dye to the jth dye; $R_{0,ij}$, the Förster distance of the corresponding FRET pair; R_{ij}, the distance between the ith dye to the jth dye. In this notation, the FRET efficiency from the ith dye to the jth dye is expressed as follows. We assumed that following intensities were already corrected as the correction steps described above.

$$E_{ij} = \frac{k_{ij}}{k_i + k_{ij}} = \frac{1}{1 + \left(R_{ij}/R_{0,ij}\right)^6} \tag{19}$$

From the rate equations, fluorescence intensities at red excitation are written as follows:

$$I_{3,r} = \frac{k_3}{k_3 + k_{34}} G_3 \tag{20}$$

$$I_{4,r} = \frac{k_{34}}{k_3 + k_{34}} G_3 \tag{21}$$

In rigorous forms, the quantum yield and the detection efficiency of the first dye should be multiplied to the right side of the above equations, but these common proportional factors are omitted here for simplicity.

By solving the equations earlier, we obtain E_{34} as follows:

$$E_{34} = \frac{1}{k_3/k_{34} + 1} = \frac{1}{I_{3,r}/I_{4,r} + 1} = \frac{I_{4,r}}{I_{3,r} + I_{4,r}} \tag{22}$$

At green excitation, fluorescence intensities of Cy3, Cy5, and Cy7 are written as follows:

$$I_{2,g} = \frac{k_2}{k_2 + k_{23} + k_{24}} G_2 \tag{23}$$

$$I_{3,g} = \frac{k_3}{k_3 + k_{34}} \frac{k_{23}}{k_2 + k_{23} + k_{24}} G_2 \tag{24}$$

$$I_{4,g} = \left(\frac{k_{24}}{k_2 + k_{23} + k_{24}} + \frac{k_{34}}{k_3 + k_{34}} \frac{k_{23}}{k_2 + k_{23} + k_{24}} \right) G_2 \tag{25}$$

By solving the equations above and using E_{34} obtained at 640-nm laser excitation, we obtain E_{23} and E_{24} as follows:

$$E_{23} = \frac{k_{23}/k_2}{1 + k_{23}/k_2} = \frac{I_{3,g}}{(1 - E_{34})I_{2g} + I_{3,g}} \tag{26}$$

$$E_{24} = \frac{-E_{34}I_{3,g} + (1 - E_{34})I_{4,g}}{(1 - E_{34})I_{2,g} - E_{34}I_{3,g} + (1 - E_{34})I_{4,g}} \tag{27}$$

At blue excitation, fluorescence intensities of all dyes are written as follows:

$$I_{1,b}=\frac{k_1}{k_1+k_{12}+k_{13}+k_{14}}G_1 \quad (28)$$

$$I_{2,b}=\frac{k_2}{k_2+k_{23}+k_{24}}\frac{k_{12}}{k_1+k_{12}+k_{13}+k_{14}}G_1 \quad (29)$$

$$I_{3,b}=\frac{k_3}{k_3+k_{34}}\left[\frac{k_{13}}{k_1+k_{12}+k_{13}+k_{14}}+\frac{k_{23}}{k_2+k_{23}+k_{24}}\frac{k_{12}}{k_1+k_{12}+k_{13}+k_{14}}\right]G_1 \quad (30)$$

$$I_{4,b}=\left[\frac{k_{14}}{k_1+k_{12}+k_{13}+k_{14}}+\frac{k_{24}}{k_2+k_{23}+k_{24}}\frac{k_{12}}{k_1+k_{12}+k_{13}+k_{14}}+\frac{k_{34}}{k_3+k_{34}}\left(\frac{k_{13}}{k_1+k_{12}+k_{13}+k_{14}}+\frac{k_{23}}{k_2+k_{23}+k_{24}}\frac{k_{12}}{k_1+k_{12}+k_{13}+k_{14}}\right)\right]G_1 \quad (31)$$

By solving the equations above and using E_{ij}'s obtained at red and green excitations, we obtain the remaining FRET efficiencies as follows:

$$E_{12}=\frac{(1-E_{23}E_{24})I_{2,b}}{(1-E_{23})(1-E_{24})I_{1,b}+(1-E_{23}E_{24})I_{2,b}} \quad (32)$$

$$E_{13}=\frac{-E_{23}(1-E_{34})I_{2,b}+(1-E_{23})I_{3,b}}{(1-E_{23})(1-E_{34})I_{1,b}-E_{23}(1-E_{34})I_{2,b}+(1-E_{23})I_{3,b}} \quad (33)$$

$$E_{14}=\frac{-E_{24}(1-E_{23})(1-E_{34})I_{2,b}-E_{34}(1-E_{23})(1-E_{24})I_{3,b}+(1-E_{23})(1-E_{24})(1-E_{34})I_{4,b}}{(1-E_{23})(1-E_{24})(1-E_{34})(I_{1,b}+I_{4,b})-E_{24}(1-E_{23})(1-E_{34})I_{2,b}-E_{34}(1-E_{23})(1-E_{24})I_{3,b}} \quad (34)$$

4. SAMPLE PREPARATION

Appropriate sample preparation is crucial for successful single-molecule multicolor FRET experiments. To obtain unambiguous data, specific and high-efficiency dye-labeling, appropriate surface preparation and sample assembly are essentially required. In this section, we describe general considerations and protocols for sample preparation to carry out single-molecule multicolor FRET measurements.

4.1 Dye Labeling

A variety of conjugation strategies are available for nucleic acids and proteins (Selvin & Ha, 2007). The most common strategies for labeling DNA include (1) phosphoramidite chemistry to incorporate directly a fluorescent probe to the backbone and (2) amine reactive ($-NH_2$) chemistry to label an amine-modified base. Fluorescent dyes may be incorporated internally into the nucleic acid chain backbone utilizing phosphoramidite chemistry during synthesis. In some cases where the DNA backbone should not be disrupted, an amine-modified base (thymine with a C_6 amino linker), instead of a fluorescent dye, is inserted into the specific location of interest, which can react later with the *N*-hydroxysuccinimide (NHS) ester form of the fluorescent dyes. In general, oligonucleotides with a C_6 amino linker purchased from a commercial source are labeled with NHS– ester form of dyes in 0.2 *M* sodium tetraborate solution adjusted to pH 8.5 for 3–6 h at room temperature or at 4° overnight. After the labeling reaction, labeled DNA is washed several times with cold EtOH and resuspended with storage buffer. Several reactive groups are involved in labeling proteins with a fluorescent probe. They include (1) N-terminal or lysine amine group ($-NH_2$), (2) cysteine thiol group (–SH), and (3) unnatural amino acid with ketone group (=CO). When structural information is available for the target proteins, the specific location for labeling can be chosen. One of the most popular strategies to label proteins site-specifically is to target cysteine residues, either naturally occurring or introduced by site-directed mutagenesis, via maleimide coupling. The major drawback to maleimide coupling for site-specific labeling of proteins is to mutate all the naturally occurring cysteine residues into serines or alanines, except for the specific cysteine residue of interest. In some cases, the activity or stability of proteins may be affected by those mutations, thereby making the resultant mutant proteins undesirable for further experiments. Thus, it is recommended that many locations for labeling should be tested for its activity or stability of proteins.

4.2 Surface Preparation

In general, a polymer-passivated surface coated with polyethyleneglycol (PEG) is recommended to remove nonspecific surface adsorption of proteins and prevent the potential surface interactions with nucleic acids and beads, although biotinylated bovine serum albumin is also used to adsorb to the glass surface for immobilization of molecules via streptavidin (neutravidin) protein sandwiched in the middle. A precleaned surface-activated slide is aminosilanized and reacted with the N-hydroxysuccinimide (NHS) ester-modified PEG. The detailed protocols are available from previous studies (Roy et al., 2008; Selvin & Ha, 2007). The general protocol for the PEGylated surface preparation involves three steps including (1) precleaning and surface activation, (2) aminosilanization of the surface, and (3) PEGylation, which coats the amino-modified surface with PEG–NHS esters. For the later assembly of a flow chamber, a pair of inlet/outlet holes in the glass (quartz) slide is generated. After a series of rinsing with deionized purified water, detergent (Alconox) and acetone through sonication, the surface of the slides and coverslips are activated in a high concentration of KOH (10 *M*) to improve the aminosilanization result and thereby enhancing the PEGylation efficiency. Aminosilanization is performed in the solution containing methanol (150 mL), acetic acid (7.5 mL), and amino silane (1.5 mL). During PEGylation, a minute fraction (~3%) of biotin–PEG–NHS ester (Bio-PEG-SC, Laysan Bio) is mixed with regular PEG–NHS ester (mPEG-SC, Laysan Bio).

4.3 Flow Cell Assembly and Preparation of the Experiment

Air-tight single- or multichannel sample chambers are prepared by sandwiching double-sided tape between a precleaned quartz slide and a glass coverslip together with pieces of double-sided tape and sealing with epoxy as necessary. Two strips of double-sided tape are attached on the PEG-coated quartz slide bordering a single channel. Then the glass coverslip is placed on top of the slide with the PEG-coated surface facing toward the slide. To prevent the leakage of air or solution into the chamber, the tape adhesion is reinforced by pressing the coverslip with a pipette tip and the two openings of the channel are sealed with epoxy. A syringe or pump is connected to the chamber through tubing and the solution is introduced into a chamber. Molecular oxygen is a source of highly reactive oxygen species that cause photobleaching of fluorescent dyes. The most general enzymatic oxygen scavenging system is a mix of 0.8% glucose, 0.1 mg/mL glucose oxidase, 0.02 mg/mL catalase, 1.5 m*M* Trolox, a triplet-quenching agent (Rasnik, McKinney, & Ha, 2006). Glucose oxidase should be added just prior to initiation of imaging to ensure that the solution pH does not drop dramatically as a result of gluconic acid production (Shi, Lim, & Ha, 2010).

5. APPLICATIONS TO BIOLOGICAL SYSTEMS

Single-molecule multicolor FRET provides a general method to study complex dynamics of biological systems by observing multiple inter-dye distances in real time. In particular, the capability is very useful in investigating correlations between two or more independent reactions (such as molecular interactions or intramolecular motions) of the biological systems. Hence, the single-molecule multicolor FRET technique has been used to solve the several important biological problems (Lee, Koh, Han, & Kim, 2007; Munro et al., 2010; Ratzke, Berkemeier, & Hugel, 2012). For example, Roy, Kozlov, Lohman, and Ha (2009) observed single-stranded binding (SSB) protein movement on single-stranded DNA using single-molecule three-color FRET. They labeled a donor dye (Cy3) onto SSB protein and two different acceptor dyes (Cy5 and Cy5.5) onto each end of single-stranded DNA. Anticorrelated fluctuation of FRET efficiencies for each dye pair clearly showed spontaneous SSB diffusion between two ends of single-stranded DNA. By using two-color ALEX technique and the Cy3/Cy5/Cy7 dye trio which significantly improved single-molecule three-color FRET, Kim et al. (2013) simultaneously monitored both the conformational changes of maltose-binding protein and ligand (maltose) binding. The maltose-binding protein was labeled with Cy3 and Cy5 to observe internal motion of the protein, and the ligand was labeled with Cy7 to observe the ligand binding/dissociation events. In this study, they revealed the preference of ligand binding and dissociation for two different conformations of the protein. Kim et al. (2014) also used the Cy3/Cy5/Cy7 dye trio for single-molecule three-color FRET to observe RNA dynamics modulated by a ribosomal protein S4. They labeled the ribosomal protein S4 with Cy3 and the two helices of ribosomal RNA with Cy5 and Cy7. Three FRET efficiencies between the trio dyes were unambiguously determined and thus clearly revealed complex motions of the RNA helices in S4-RNA complex. A dual FRET pair scheme using single-molecule four-color FRET enables us to observe the correlation between two spatially separated events (beyond the FRET range). Using this capability, Lee, Lee, Ragunathan, et al. (2010) simultaneously monitored docking or propagation events at both ends of RecA synaptic complex during RecA-mediated DNA strand exchange. They labeled two ends of homologous double-stranded DNA with Alexa488 and Cy3 and two ends of RecA filament with Cy5 and Cy7. In this study, they clearly revealed the existence of diverse paths of strand exchange initiation.

6. CONCLUSIONS

In this chapter, we have presented a general overview of the single-molecule multicolor FRET technique and practical considerations for the development of the setup. By considering detailed information of the setup described earlier, this technique can be easily implemented in any laboratory with experience in single-molecule FRET experiments. Thus, in the future, single-molecule multicolor FRET will be widely used to reveal previously unattainable details about complex dynamics of various biological systems by monitoring multiple distances. However, there is still room for technical improvement. First, the information content of single-molecule FRET experiments can be further increased through a combination of the multi-color FRET technique with single-molecule mechanical manipulation techniques (Hohng et al., 2007; Lee, Kim, & Hong, 2010; Long, Parks, Bagshaw, & Stone, 2013; Shroff et al., 2005; Tarsa et al., 2007). We recently combined three-color FRET with optical tweezers (Lee & Hohng, 2013). Second, as mentioned earlier, technical improvements of fluorophores used in the multicolor FRET experiments (such as development of spectrally distinct, bright, and long-lasting fluorophores) are crucially required to increase data yield (Juette et al., 2014; Zheng et al., 2014). Recently, Altman et al. (2012) developed a self-healing strategy to enhance the photostability of organic fluorophores by covalent attachment of triplet-state quenchers such as Trolox, and the strategy was extended to cyanine fluorophores across the visible and near-infrared spectrum (Altman et al., 2012). Finally, protein-labeling strategies with high fidelity and the orthogonal labeling capability will enable a broad range of applications of single-molecule multicolor FRET experiments (Kim et al., 2013; Shi et al., 2012).

ACKNOWLEDGMENTS

This work was supported by a grant from the National R&D Program for Cancer Control of Ministry for Health and Welfare of Korea (1520090), the National Research Foundation of Korea (NRF) Grant funded by the Korean Government (MSIP) (Nos. 2015R1C1A1A01055164 and 2016R1A5A1007318), the GIST Research Institute to S.L., and by Creative Research Initiatives (Physical Genetics Laboratory, 2009-0081562) of National Research Foundation of Korea to S.H.

REFERENCES

Altman, R. B., Terry, D. S., Zhou, Z., Zheng, Q., Geggier, P., Kolster, R. A., et al. (2012). Cyanine fluorophore derivatives with enhanced photostability. *Nature Methods*, *9*(1), 68–71.

Altman, R. B., Zheng, Q., Zhou, Z., Terry, D. S., Warren, J. D., & Blanchard, S. C. (2012). Enhanced photostability of cyanine fluorophores across the visible spectrum. *Nature Methods*, *9*(5), 428–429.

Axelrod, D. (2001). Total internal reflection fluorescence microscopy in cell biology. *Traffic*, *2*(11), 764–774.

Bustamante, C., Cheng, W., & Mejia, Y. X. (2011). Revisiting the central dogma one molecule at a time. *Cell*, *144*(4), 480–497.

Clamme, J. P., & Deniz, A. A. (2005). Three-color single-molecule fluorescence resonance energy transfer. *ChemPhysChem*, *6*(1), 74–77.

DeRocco, V., Anderson, T., Piehler, J., Erie, D. A., & Weninger, K. (2010). Four-color single molecule fluorescence with noncovalent dye labeling to monitor dynamic multimolecular complexes. *Biotechniques*, *49*(5), 807–816.

Förster, T. (1948). Zwischenmolekulare Energiewanderung und Fluoreszenz. *Annalen der Physik*, *437*(2), 55–75.

Ha, T., Chemla, D. S., Enderle, T., & Weiss, S. (1997). Single molecule spectroscopy with automated positioning. *Applied Physics Letters*, *70*(6), 782–784.

Ha, T., Enderle, T., Ogletree, D. F., Chemla, D. S., Selvin, P. R., & Weiss, S. (1996). Probing the interaction between two single molecules: Fluorescence resonance energy transfer between a single donor and a single acceptor. *Proceedings of the National Academy of Sciences of the United States of America*, *93*(13), 6264–6268.

Hohlbein, J., Craggs, T. D., & Cordes, T. (2014). Alternating-laser excitation: Single-molecule FRET and beyond. *Chemical Society Reviews*, *43*(4), 1156–1171.

Hohng, S., Joo, C., & Ha, T. (2004). Single-molecule three-color FRET. *Biophysical Journal*, *87*(2), 1328–1337.

Hohng, S., Lee, S., Lee, J., & Jo, M. H. (2014). Maximizing information content of single-molecule FRET experiments: Multi-color FRET and FRET combined with force or torque. *Chemical Society Reviews*, *43*(4), 1007–1013.

Hohng, S., Zhou, R., Nahas, M. K., Yu, J., Schulten, K., Lilley, D. M., et al. (2007). Fluorescence-force spectroscopy maps two-dimensional reaction landscape of the holliday junction. *Science*, *318*(5848), 279–283.

Juette, M. F., Terry, D. S., Wasserman, M. R., Zhou, Z., Altman, R. B., Zheng, Q., et al. (2014). The bright future of single-molecule fluorescence imaging. *Current Opinion in Chemical Biology*, *20*, 103–111.

Kapanidis, A. N., Lee, N. K., Laurence, T. A., Doose, S., Margeat, E., & Weiss, S. (2004). Fluorescence-aided molecule sorting: Analysis of structure and interactions by alternating-laser excitation of single molecules. *Proceedings of the National Academy of Sciences of the United States of America*, *101*(24), 8936–8941.

Kapanidis, A. N., & Strick, T. (2009). Biology, one molecule at a time. *Trends in Biochemical Sciences*, *34*(5), 234–243.

Kapanidis, A. N., & Weiss, S. (2002). Fluorescent probes and bioconjugation chemistries for single-molecule fluorescence analysis of biomolecules. *The Journal of Chemical Physics*, *117*(24), 10953–10964.

Kim, H., Abeysirigunawarden, S. C., Chen, K., Mayerle, M., Ragunathan, K., Luthey-Schulten, Z., et al. (2014). Protein-guided RNA dynamics during early ribosome assembly. *Nature*, *506*(7488), 334–338.

Kim, E., Lee, S., Jeon, A., Choi, J. M., Lee, H. S., Hohng, S., et al. (2013). A single-molecule dissection of ligand binding to a protein with intrinsic dynamics. *Nature Chemical Biology*, *9*(5), 313–318.

Kim, J., Seo, M. H., Lee, S., Cho, K., Yang, A., Woo, K., et al. (2013). Simple and efficient strategy for site-specific dual labeling of proteins for single-molecule fluorescence resonance energy transfer analysis. *Analytical Chemistry*, *85*(3), 1468–1474.
Lee, S., & Hohng, S. (2013). An optical trap combined with three-color FRET. *Journal of the American Chemical Society*, *135*(49), 18260–18263.
Lee, N. K., Kapanidis, A. N., Koh, H. R., Korlann, Y., Ho, S. O., Kim, Y., et al. (2007). Three-color alternating-laser excitation of single molecules: Monitoring multiple interactions and distances. *Biophysical Journal*, *92*(1), 303–312.
Lee, M., Kim, S. H., & Hong, S. C. (2010). Minute negative superhelicity is sufficient to induce the B-Z transition in the presence of low tension. *Proceedings of the National Academy of Sciences of the United States of America*, *107*(11), 4985–4990.
Lee, N. K., Koh, H. R., Han, K. Y., & Kim, S. K. (2007). Folding of 8-17 deoxyribozyme studied by three-color alternating-laser excitation of single molecules. *Journal of the American Chemical Society*, *129*(50), 15526–15534.
Lee, S., Lee, J., & Hohng, S. (2010a). Single-molecule three-color FRET with both negligible spectral overlap and long observation time. *PLoS One*, *5*(8), e12270.
Lee, J., Lee, S., Ragunathan, K., Joo, C., Ha, T., & Hohng, S. (2010b). Single-molecule four-color FRET. *Angewandte Chemie, International Edition*, *49*(51), 9922–9925.
Long, X., Parks, J. W., Bagshaw, C. R., & Stone, M. D. (2013). Mechanical unfolding of human telomere G-quadruplex DNA probed by integrated fluorescence and magnetic tweezers spectroscopy. *Nucleic Acids Research*, *41*(4), 2746–2755.
Munro, J. B., Altman, R. B., Tung, C. S., Cate, J. H., Sanbonmatsu, K. Y., & Blanchard, S. C. (2010). Spontaneous formation of the unlocked state of the ribosome is a multistep process. *Proceedings of the National Academy of Sciences of the United States of America*, *107*(2), 709–714.
Rasnik, I., McKinney, S. A., & Ha, T. (2006). Nonblinking and longlasting single-molecule fluorescence imaging. *Nature Methods*, *3*(11), 891–893.
Ratzke, C., Berkemeier, F., & Hugel, T. (2012). Heat shock protein 90's mechanochemical cycle is dominated by thermal fluctuations. *Proceedings of the National Academy of Sciences of the United States of America*, *109*(1), 161–166.
Ross, J., Buschkamp, P., Fetting, D., Donnermeyer, A., Roth, C. M., & Tinnefeld, P. (2007). Multicolor single-molecule spectroscopy with alternating laser excitation for the investigation of interactions and dynamics. *The Journal of Physical Chemistry B*, *111*(2), 321–326.
Roy, R., Hohng, S., & Ha, T. (2008). A practical guide to single-molecule FRET. *Nature Methods*, *5*(6), 507–516.
Roy, R., Kozlov, A. G., Lohman, T. M., & Ha, T. (2009). SSB protein diffusion on single-stranded DNA stimulates RecA filament formation. *Nature*, *461*(7267), 1092–1097.
Schuler, B., & Eaton, W. A. (2008). Protein folding studied by single-molecule FRET. *Current Opinion in Structural Biology*, *18*(1), 16–26.
Selvin, P. R., & Ha, T. (2007). *Single molecule techniques: A laboratory manual*. Cold Spring Harbor, NY: Cold Spring Harbor Laboratory Press.
Shi, X., Jung, Y., Lin, L. J., Liu, C., Wu, C., Cann, I. K., et al. (2012). Quantitative fluorescence labeling of aldehyde tagged proteins for single-molecule imaging. *Nature Methods*, *9*(5), 499–503.
Shi, X., Lim, J., & Ha, T. (2010). Acidification of the oxygen scavenging system in single-molecule fluorescence studies: In situ sensing with a ratiometric dual-emission probe. *Analytical Chemistry*, *82*(14), 6132–6138.
Shroff, H., Reinhard, B. M., Siu, M., Agarwal, H., Spakowitz, A., & Liphardt, J. (2005). Biocompatible force sensor with optical readout and dimensions of 6 nm^3. *Nano Letters*, *5*(7), 1509–1514.

Stein, I. H., Steinhauer, C., & Tinnefeld, P. (2011). Single-molecule four-color FRET visualizes energy-transfer paths on DNA Origami. *Journal of the American Chemical Society, 133*(12), 4193–4195.

Tarsa, P. B., Brau, R. R., Barch, M., Ferrer, J. M., Freyzon, Y., Matsudaira, P., et al. (2007). Detecting force-induced molecular transitions with fluorescence resonant energy transfer. *Angewandte Chemie, International Edition, 46*(12), 1999–2001.

Tinoco, I., Jr., & Gonzalez, R. L., Jr. (2011). Biological mechanisms, one molecule at a time. *Genes & Development, 25*(12), 1205–1231.

Walter, N. G., & Bustamante, C. (2014). Introduction to single molecule imaging and mechanics: Seeing and touching molecules one at a time. *Chemical Reviews, 114*(6), 3069–3071.

Zheng, Q., Juette, M. F., Jockusch, S., Wasserman, M. R., Zhou, Z., Altman, R. B., et al. (2014). Ultra-stable organic fluorophores for single-molecule research. *Chemical Society Reviews, 43*(4), 1044–1056.

CHAPTER SIXTEEN

A Multicolor Single-Molecule FRET Approach to Study Protein Dynamics and Interactions Simultaneously

M. Götz[1], P. Wortmann[1], S. Schmid, T. Hugel[2]
Institute of Physical Chemistry, University of Freiburg, Freiburg, Germany
[2]Corresponding author: e-mail address: th@pc.uni-freiburg.de

Contents

[1] Contributed equally to this work.

Methods in Enzymology, Volume 581
ISSN 0076-6879
http://dx.doi.org/10.1016/bs.mie.2016.08.024
© 2016 Elsevier Inc.
All rights reserved.

Abstract

Single-molecule Förster resonance energy transfer (smFRET) is a versatile tool for studying biomolecules in a quantitative manner. Multiple conformations within and interactions between biomolecules can be detected and their kinetics can be determined. Thus, smFRET has become an essential tool in enzymology. Ordinary two-color smFRET experiments can provide only limited insight into the function of biological systems, which commonly consist of more than two components. A complete understanding of complex multicomponent biological systems requires correlated information on conformational rearrangements on the one hand and transient interactions with binding partners on the other. Multicolor smFRET experiments enable the direct observation of such correlated dynamics and interactions. Here we demonstrate the power and limitations of multicolor smFRET experiments including the description of a multicolor smFRET setup and data analysis. A general analytical procedure for multicolor smFRET data is presented and applied to the multicomponent heat shock protein 90 system. This allows us to identify microscopic states in transient complexes. Conformational dynamics and nucleotide binding are simultaneously detected, which is impossible using two-color smFRET. Additionally, their correlation is quantified using 3D ensemble hidden Markov analysis, in and out of equilibrium. This method is perfectly suited for protein systems that are much more sophisticated than previously studied DNA-based systems. By extending the application to biologically relevant systems, multicolor smFRET comes of age and provides a unique mechanistic insight into protein machines.

1. INTRODUCTION

In living organisms many proteins work in conjunction with other molecules as multicomponent protein machines that regulate cellular processes. Prominent examples are the ribosome (Steitz, 2008; Uemura et al., 2010), the transcription machinery (Hahn, 2004), and the heat shock response system (Vabulas, Raychaudhuri, Hayer-Hartl, & Hartl, 2010). The components of these complexes have to transiently adopt different conformational states or structural rearrangements to fulfill their task. When coupled with an external energy source, these complexes can provide directionality in a functional cycle and therefore maintain the steady state in a cell, the prerequisite for life.

When it comes to studying these protein machines, their complexity usually requires the system to be simplified by partial reconstitution in vitro. Single-molecule methods have proven to be a powerful tool for probing the complex behavior of biological molecules and their interactions, unbiased by ensemble averaging (Deniz, Mukhopadhyay, & Lemke, 2008; Juette et al., 2014; Weiss, 1999). With single-molecule Förster resonance energy transfer (smFRET) conformational dynamics and interactions of proteins can be studied quantitatively and in real time. For this purpose, dyes are attached at defined positions in the protein system and the time-resolved FRET efficiency is determined. The experiments can be conducted under near-physiological conditions in thermodynamic equilibrium (i.e., with no net flow of matter or energy) or out of equilibrium. Moreover, measurements using a total internal reflection fluorescence (TIRF) setup with fixed samples allow the succession of states to be extracted, providing unique kinetic and thermodynamic information in the range of milliseconds to minutes (Weiss, 1999). Since most biological systems consist of more than two interaction partners, standard two-color smFRET experiments are insufficient to capture the full complexity of these systems.

Multicolor smFRET measurements add a new dimension to the technique of smFRET and help to tackle the complexity of multicomponent protein systems (Hohng, Joo, & Ha, 2004; Lee et al., 2010; Person, Stein, Steinhauer, Vogelsang, & Tinnefeld, 2009; Ratzke, Berkemeier, & Hugel, 2012; Ratzke, Hellenkamp, & Hugel, 2014). Three or more components labeled with fluorophores build up a network of FRET pairs and allow direct observation of the conformational dynamics and different association states of a multicomponent complex.

In the following, we present a state-of-the-art multicolor smFRET experiment. In addition, we show how a combination with 3D ensemble hidden Markov analysis yields the correlation of processes in a multicomponent system with high reliability. The power of the method is demonstrated on the heat shock protein 90 (Hsp90) system. Finally, we discuss the potential and the limitations of its application to different systems.

1.1 Hsp90 as a Model System for Multicolor smFRET on Proteins

Hsp90 plays a key role in the regulation of many essential cellular functions such as signaling, gene regulation, and protein homeostasis (Gong et al., 2009; Zhao et al., 2005). It is overexpressed upon heat shock and is a part of the cellular stress response. It can prevent protein misfolding and aggregation (Jakob, Lilie, Meyer, & Buchner, 1995). Hsp90 is the key protein of a huge multicomponent machinery, which consists of more than 20 cochaperones that transiently interact with Hsp90 and regulate its function (Taipale, Jarosz, & Lindquist, 2010).

Hsp90 is a highly conserved homodimer with each protomer comprising three domains, commonly referred to as N-terminal (ND), middle (MD), and C-terminal domain (CD). The ND is capable of ATP hydrolysis, while the CD is the dimerization domain (Ali et al., 2006). Surprisingly, the ATPase rate is very slow (about 1 ATP/dimer/min) and is one order of magnitude slower than the slowest structural rearrangements observed with smFRET so far (Mickler, Hessling, Ratzke, Buchner, & Hugel, 2009).

X-ray structures provide evidence of at least one open and one N-terminally closed conformation (Ali et al., 2006; Dollins, Warren, Immormino, & Gewirth, 2007; Lavery et al., 2014; Shiau, Harris, Southworth, & Agard, 2006). Time-resolved smFRET studies reveal that Hsp90 exists in four (kinetically) distinguishable states (Mickler et al., 2009; Schmid, Götz, & Hugel, 2016). In addition, the dynamic nature of Hsp90 suggests altered nucleotide affinity for different conformational states. The binding of nucleotides to Hsp90 has been studied extensively with ensemble experiments (McLaughlin et al., 2006; McLaughlin, Ventouras, Lobbezoo, & Jackson, 2004; Olesen et al., 2015; Panaretou et al., 1998; Prodromou et al., 1997; Richter et al., 2006; Scheibel et al., 1997). The inability of these ensemble experiments to distinguish conformations prevents correlated kinetic analysis.

This makes Hsp90 a perfect case study for demonstrating the applicability of multicolor smFRET to complex protein systems. Here, we monitor Hsp90's N-terminal opening and closing (by attaching one fluorophore to each protomer) and at the same time the binding of a fluorescently labeled AMP-PNP, a nonhydrolyzable analog of ATP.

2. THEORETICAL BACKGROUND

2.1 Two-Color FRET

FRET is an energy transfer phenomenon requiring a dipole–dipole coupling between two fluorophores, where energy is transferred from a donor dye to an acceptor dye when they are in close proximity (generally within 3–10 nm). The efficiency of this process is proportional to the inverse of the sixth power of the distance between the two fluorophores. Thus, it is strongly distance-dependent in a range that is relevant for protein dynamics and protein complex rearrangements (Fig. 1A).

Three properties of the dyes influence the FRET efficiency: interdye distance, spectral overlap, and relative orientation of the transition dipole moments. The relation between the efficiency of the energy transfer and the interdye distance r is given by Eq. (1):

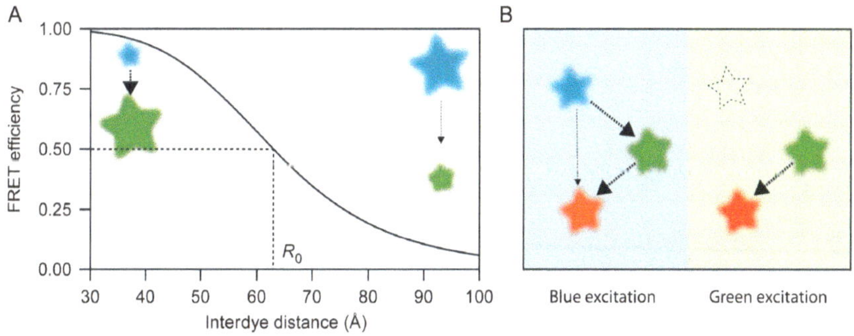

Fig. 1 Nonradiative energy transfer by FRET. (A) The FRET efficiency is strongly distance dependent and therefore often called "molecular ruler." (B) Schematics of a three-color FRET system with possible energy transfer paths upon excitation of the *blue* dye *(left)* and the *green* dye (*right*). The information from both excitations is necessary for a complete description of the relative position of the dyes in this system.

$$E = \frac{\text{energy transfer}}{\text{donor excitation}} = \frac{k_{ET}}{k_f + k_{ET} + \sum k_i} = \frac{1}{1 + \left(\frac{r}{R_0}\right)^6} \tag{1}$$

With the radiative (k_f) and the nonradiative (k_i) deexcitation rates, the FRET rate k_{ET} and the Förster radius R_0 which represents the distance where 50% of the energy is transferred to the acceptor dye.

As shown in Eq. (2), the Förster radius depends on the spectral overlap integral of the donor fluorescence $F_D(\lambda)$ and the acceptor absorption $\varepsilon_A(\lambda)$, over the wavelength λ, the quantum yield of the donor Q_D, the refractive index of the medium n, and the relative dipole orientation factor of the fluorophores κ^2, which can be assumed to be equal to 2/3 for freely rotating dyes (Lakowicz, 2006). All other factors can be determined in independent experiments, and it is usually expected that they do not change within a single experiment.

$$R_0 = \frac{9000\ \ln(10)}{128\pi^5 N_A} \frac{\kappa^2 Q_D}{n^4} \int F_D(\lambda)\varepsilon_A(\lambda)\lambda^4 d\lambda \tag{2}$$

Experimentally, in two-color smFRET TIRF experiments, the FRET efficiency is accessible from the intensities measured in the acceptor and donor channel (I_{acc} and I_{donor}, respectively):

$$E = \frac{k_{ET}}{k_f + k_{ET}} = \frac{I_{acc}}{I_{donor} + I_{acc}} \tag{3}$$

2.2 Multicolor FRET

In multicolor FRET measurements, the energy from one donor fluorophore is transferred to different acceptor fluorophores. The rates for the transfers to all other fluorophores are independent, and the signal in each detection channel results from the sum of all energy transfers to the respective acceptor fluorophore (Fig. 1B). We therefore use the partial fluorescence PF_{em}^{ex} as a proxy for the energy transfer between two fluorophores, where *ex* is the excitation channel and *em* is the emission channel. For simplification, we refer to the corresponding excitation and emission channel with the same identifier (i.e., "blue," "green," "orange," or "red"). We then define the partial fluorescence as:

$$PF_b^a = \frac{I_b^a}{\sum_{i=a}^{c} I_i^a} \tag{4}$$

with I_{em}^{ex} being the intensity in channel *em* after excitation with color *ex* and *c* being the acceptor with the longest wavelength. In case of two-color experiments, the *PF* is the FRET efficiency.

Since in multicolor FRET experiments a donor dye may transfer energy to different acceptor dyes, the relative position of all dyes is not accessible from excitation of one dye alone. Alternating laser excitation (ALEX) overcomes this limitation.

3. EXPERIMENTAL PROCEDURE

3.1 TIRF-Based Multicolor smFRET Setup

The construction scheme of a multicolor (up to four) single-molecule FRET microscope optimized for the dyes Atto488, Atto550, Atto594, and Atto647N (Atto-Tec) is depicted in Fig. 2.

The setup is custom-built to achieve maximal photon sensitivity while maintaining flexibility to adapt to different experiments. All components are mounted on a vibration-damped optical table (RS2000, Newport). The beams of four lasers at the wavelengths 473 nm (Cobolt Blues 50 mW, Cobolt), 532 nm (Compass 215M 75 mW, Coherent Inc.), 594 nm (Cobolt Mambo 100 mW, Cobolt), and 637 nm (LPM635-25C 25 mW, Newport) are expanded and collimated individually by Keplerian telescopes and subsequently overlaid using dichroic mirrors. An iris selects the central part of the beam to obtain homogeneous illumination. An acousto-optical tunable filter (AOTFnC-Vis, AAOptics) is used to control

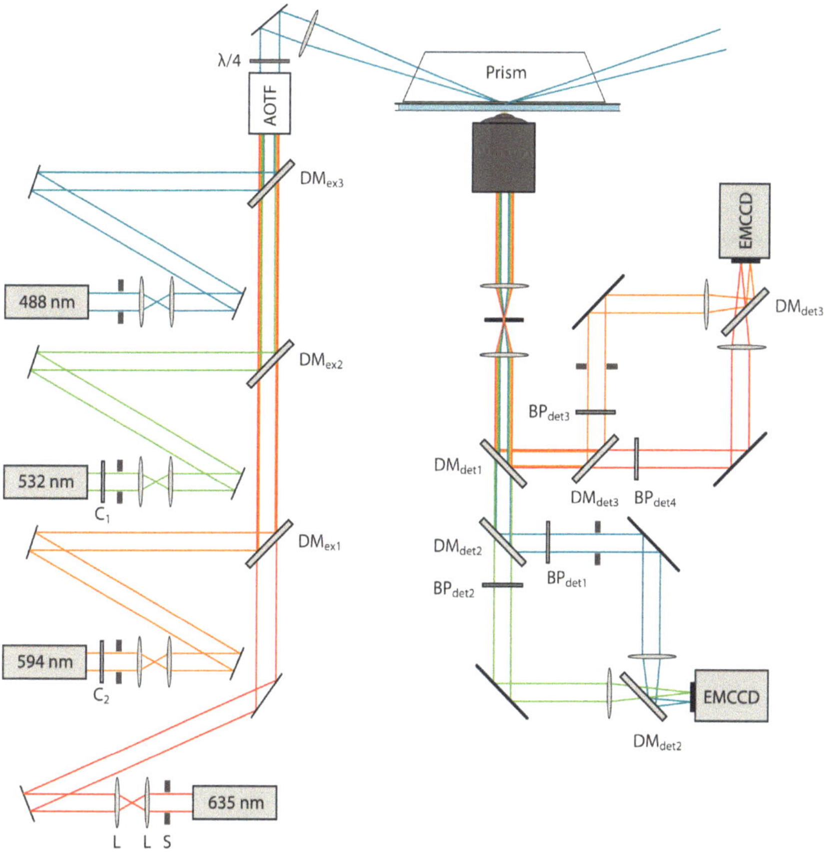

Fig. 2 Scheme of the multicolor smFRET setup. Shown are the excitation (*left*) and detection (*right*) paths. *S*, mechanical shutter; *C*, clean-up filter (C_1: ZET532/10, C_2: ZET 594/10); *DM*, dichroic mirrors (DM_{ex1}: ZQ594RDC, DM_{ex2}: 570DCXR, DM_{ex3}: ZT491RDC, DM_{det1}: T 600 LPXR, DM_{det2}: H 560 LPXR superflat, DM_{det3}: HC BS R635); *BP*, bandpass filter (BP_{det1}: 525/40 BrightLine HC, BP_{det2}: 586/20 BrightLine HC, BP_{det3}: 631/36 BrightLine HC, BP_{det4}: 700/75 ET Bandpass). All filters and dichroic mirrors are purchased from AHF Analysentechnik.

the excitation power and allows fast switching of the different excitation colors. Additional mechanical shutters (SH1, Thorlabs) in each beam ensure that the excitation is blocked completely for measurements on a slower time scale (>20 ms). The combined excitation beam is circularly polarized by a λ/4 plate (AQWP05M-600, Thorlabs) and focused on the sample by an achromatic lens via a quartz prism. This allows for an excitation with total internal reflection (TIR) illumination. The emitted fluorescence is collected by a high-NA oil immersion objective with 100× magnification (CFI Apo

TIRF 100×, Nikon) and a tube lens positioned as close as possible to the objective. An adjustable optical slit (Owis GmbH) is placed in the intermediate image plane to adjust the field of view and to remove off-axis beams. The light is collimated again and spectral regions are separated into the different detection channels with dichroic mirrors (Fig. 3). Each channel is focused with a separate lens onto the chip of the respective EMCCD camera (iXon Ultra 897, Andor Technology Ltd.). The final resolution of the setup is 162, 162, 158, and 159 nm per pixel for the detection channels in the order of increasing wavelength. Bandpass filters decrease the leakage of fluorescence from other dyes. Mechanical shutters in the blue and orange detection channels (Uniblitz VS35, Vincent Associates) can block excitation light from the corresponding excitation laser. One camera is used per two detection channels that are imaged side by side. The orientation of the camera was chosen in a way that channels do not overlap during readout. In the detection paths achromatic aspheric lenses with 50.8 mm diameter and 200 mm focal length (Qioptiq) are used in order to collect as much light as possible and to prevent aberrations, e.g., distortions in off-center regions of the image.

Shutters, AOTF, and cameras are synchronized by TTL signals from a digital I/O card (PCIe-6535, National Instruments) that is controlled by PC software developed with Labview (National Instruments).

The sample is studied in a custom-built flow-chamber, which is a sandwich of a functionalized quartz slide, Tegaderm film (3M) sprayed with

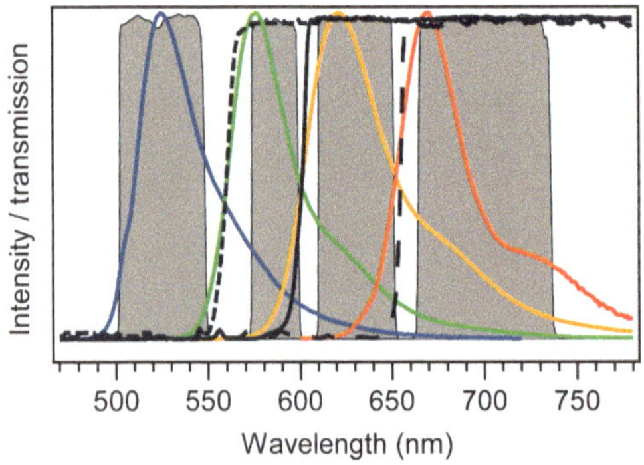

Fig. 3 The emission spectra of the utilized fluorescent dyes (*left* to *right*) Atto488 (*blue*), Atto550 (*green*), Atto594 (*orange*), Atto647N (*red*), and the transmission spectra of the bandpass filter (*gray*) and the transmission spectra of the dichroic mirrors (*bold black lines*). Spectral information provided by Atto-Tec and AHF Analysentechnik.

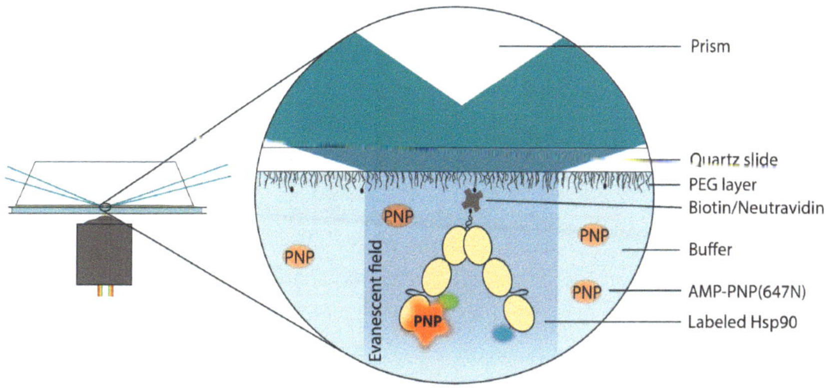

Fig. 4 Detailed view of the measurement chamber. Hsp90 is bound to the surface via biotin-Neutravidin. The quartz slide is passivated with a layer of methoxy-/biotin-PEG. 25 n*M* labeled nucleotide AMP-PNP(647N) is free in solution. Fluorophores are excited in the evanescent field at the quartz/buffer interface and fluorescence is detected with an objective.

adhesive (Photomount, 3M), and a coverslip (Fig. 4). The sample is immobilized on the surface of the quartz slide via biotin-Neutravidin binding. Interaction partners are flushed into the chamber subsequently. A syringe pump (PHD22/2000, Harvard Apparatus) enables buffer exchange.

3.2 Correction for Setup and Dye Properties

Due to experimental shortcomings the measured fluorescence intensities depend not only on the energy transfer but also on fluorophore and setup properties. In order to obtain the true energy transfer efficiency between two fluorophores, the measured intensities are corrected for the following systematic deviations to yield the corrected intensity $({}^{corr}I_{em}^{ex})$:

1. A fraction of the photons emitted by the donor dye are detected in the channel for acceptor fluorescence. This is commonly referred to as leakage and corrected by the leakage factor ℓ.
2. The fluorescent dyes have different quantum yields and the detection of photons from different dyes depends on the detection efficiency of the setup in each detection channel. This is adjusted with the gamma factor γ.
3. Upon donor excitation a small fraction of photons excite the acceptor dye directly. This is commonly referred to as direct excitation with the corresponding correction factor d.

In a two-color FRET experiment all these effects can be accounted for. To gain exact structural information, the correction factors have to be carefully determined to give the unbiased intensities:

$$^{corr}I_b^a = I_b^a - \ell I_a^a - dI_b^b \tag{5}$$

$$^{corr}I_a^a = \gamma I_a^a \tag{6}$$

These correction factors can be obtained from stoichiometry vs FRET efficiency (SE) plots based on experiments with ALEX (Kapanidis et al., 2005), resulting in accurate FRET efficiencies and opening the opportunity to use smFRET as a tool in structural biology.

This correction method is not generally suited for multicolor FRET experiments, because the full ALEX information might not be accessible anymore. Directly exciting a labeled species that is free in solution at concentrations above 10 n*M* is not applicable due to a high background signal from molecules that are not immobilized to the surface. Nevertheless, correcting the recorded intensities is advantageous for the discrimination between different populations and the comparison of multicolor FRET experiment with two-color experiments. Therefore, we describe a correction procedure for multicolor smFRET in the following.

Correction factors for apparent leakage (*lk*) and apparent gamma (*ag*) can be calculated from single-molecule traces that show an acceptor bleaching event. The correction factors are determined independently with a high-FRET dsDNA sample for each dye in combination with the acceptor having the longest excitation wavelength (Atto488–Atto647N, Atto550–Atto647N, Atto594–Atto647N). Properties of the used Atto dyes do not change significantly between DNA and protein conjugates in our hands. This has to be checked for every dye, because differing fluorescence quantum yields or excitation/emission spectra between DNA and protein conjugates will in turn change the determined correction factors.

The calibration sample is fixed to the surface via biotin-Neutravidin coupling and measured in TNM buffer (5 m*M* Tris pH 7.5, 5 m*M* NaCl, 20 m*M* $MgCl_2$) and 2 m*M* trolox. ALEX is used to exclude photophysical artifacts such as acceptor blinking, and the obtained intensity traces are corrected manually for offset. The leakage of the donor dye into every possible acceptor channel is calculated from all data points in the recorded FRET traces where the acceptor dye bleached but the donor is still fluorescent ($^{\dagger}I$, c.f. Fig. 5A):

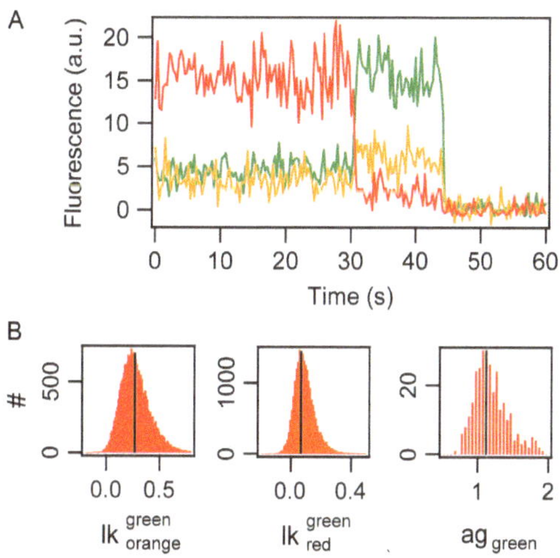

Fig. 5 Obtaining correction factors from a calibration sample with acceptor dye bleaching. (A) Representative trace of a DNA high-FRET sample labeled with Atto550 and Atto647N, with consecutive acceptor bleaching ($t \approx 30$ s) and donor bleaching ($t \approx 45$ s). Detected intensities for the *green*, *orange*, and *red* detection channel are shown. (B) Histograms for the correction factors from about 1000 traces. The median of the distribution for leakage and gamma factor (*black bar*) is used for all following experiments.

$$lk_b^a = {}^{\dagger}I_b^a / {}^{\dagger}I_a^a \tag{7}$$

The median of the leakage histogram is used as the apparent leakage factor (Fig. 5B). After correction for leakage, the apparent gamma factor is determined from the same set of traces. It is calculated by dividing the change of fluorescence in the acceptor channel by the change of fluorescence in the donor channel upon bleaching of the acceptor dye:

$$ag_a = \frac{\Delta I_c^a}{\Delta I_a^a} = \frac{I_c^a - {}^{\dagger}I_c^a}{{}^{\dagger}I_a^a - I_a^a} \tag{8}$$

Here, c is the detection channel for Atto647N. The median of the apparent gamma factor distribution is used as an apparent correction factor. The apparent gamma factor for an arbitrary FRET pair combination is then approximated by the ratio between the two apparent gamma factors of the dyes measured in combination with Atto647N.

The determined apparent correction factors lk and ag include a small contribution from the direct excitation and are therefore dependent on

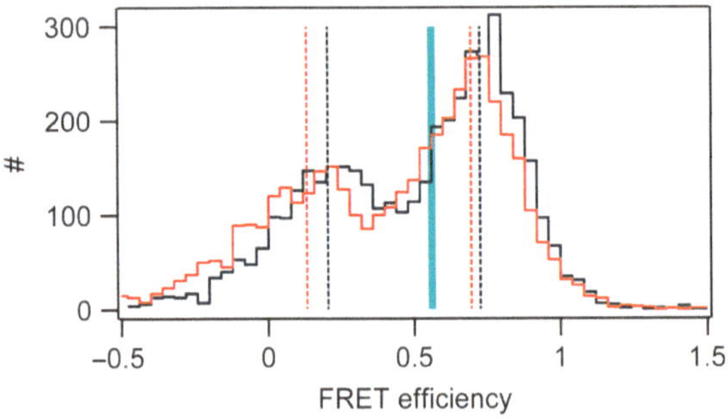

Fig. 6 Comparison of correction procedures for two-color smFRET data. The identical data set of Hsp90 (61C-Atto488 385C-Atto550) in the presence of 250 μ*M* AMP-PNP is corrected either via a SE plot (*red*) or via fixed correction factors (*black*), see main text for details. The peak centers of a Gaussian fit are indicated with *dashed lines* (FRET efficiencies 0.13/0.70 for SE plot and 0.21/0.73 for fixed correction factors). The FRET efficiency of the used DNA calibration sample is shown as *bold blue line*. There are slight deviations in the low FRET range, but the populations are well separated in both cases.

the FRET efficiency. As a result, the corrected FRET efficiency is less accurate (Fig. 6). The deviation increases with the difference between the FRET efficiencies of the studied sample and the calibration standard. Nevertheless, the deviation has negligible influence on the subsequent analysis as multi-color FRET experiments do generally not aim to provide exact distance information from the data.

Finally, the corrected intensity in each channel is obtained by:

$$^{corr}I_b^a = \left(I_b^a - \sum_{i=a}^{b-1} lk_i^a I_i^a \right) \bullet ag_b \tag{9}$$

with the sum over all detection channels with shorter wavelength than the channel that is corrected and $ag_b = 1$ for the channel with the longest wavelength. The PF_{em}^{ex} in a multicolor experiment is then calculated by Eq. (10):

$$PF_b^a = \frac{^{corr}I_b^a}{\sum_{i=a}^{c} {^{corr}I_i^a}} \tag{10}$$

3.3 Surface Functionalization

Specific immobilization of the protein of interest on the quartz slide is achieved by biotin-Neutravidin linkage. Both the slide surface and the protein are biotinylated and cross-linked by Neutravidin (Thermo Fischer Scientific).

Nonspecific binding is prevented by PEGylation of the surface (3 kDa biotin-PEG-NHS, 5 kDa methoxy-PEG-NHS, Rapp Biopolymere). In the following the protocol for surface passivation and functionalization is provided:

Cleaning

1. Sonicate quartz slides 3× in 2% Hellmanex (Hellma Analytics) solution for 10 min and wash 5× with ultrapure water in between and afterward.
2. Incubate slides for 2 h in Piranha solution (3:1 vol% H_2SO_4:H_2O_2) at 60°C.
3. Repeat step 1.
4. Incubate slides for 2 h in RCA (5:1:1 H_2O:H_2O_2:NH_4OH) at 60°C.
5. Rinse slides 5 × with ultrapure water.

Functionalization

1. Dry slides with a stream of nitrogen and incubate them 15 min in acetone.
2. Incubate slides with a 1:200 dilution of Vectabond (Vector Laboratories) in acetone for 5 min.
3. Rinse slides 5 × with ultrapure water.
4. Dissolve 3 mg biotin-PEG and 80 mg methoxy-PEG in 600 μL 100 m*M* sodium hydrogen carbonate.
5. Add a droplet of 50 μL of PEG solution on each slide, place a clean cover slip on top, and incubate overnight at 8°C in humid atmosphere.
6. Rinse slides 5 × with ultrapure water, dry them with a stream of nitrogen, and store them in a dry and light-protected place.

Surface passivation and biotinylation can also be achieved by other methods, e.g., using bovine serum albumin (BSA) and biotin-BSA. Passivation of the glass coverslip that is the counterpart of the sample chamber did not influence our data quality, although it could influence the concentration of diffusing biomolecules.

3.4 Construct, Expression, and Purification

The studied Hsp90 construct is depicted in Fig. 7 and comprises an N-terminal HisTag, a SUMO sequence, the baker's yeast Hsp90 (UniProt ID P02829), a C-terminal zipper (that keeps the protein dimeric at picomolar concentration; Mickler et al., 2009), and optionally an AviTag (a unique

Fig. 7 The protein construct. Yeast Hsp90 is expressed as a fusion protein with an N-terminal His- and Sumo-tag (both are cleaved during purification) and C-terminal zipper followed by an optional biotinylation site (AviTag) for the Hsp90 385C mutant.

15 amino acid peptide tag, Avidity LLC) for specific in vivo biotinylation. As wild-type yeast Hsp90 contains no cysteines, cysteine mutants allow for specific attachment of one fluorescent dye per protomer. They are created by site-directed mutagenesis (QuikChange Lightning, Agilent).

If native cysteines are present in the protein of investigation their reactivity toward maleimides has to be checked. In case they are reactive, either mutagenesis of those into structurally similar residues (serine or alanine) or amino acids occurring at the same position in homologues is an option. Alternatively, a nonnatural amino acid that reacts orthogonally to cysteines can be used for labeling (Plass, Milles, Koehler, Schultz, & Lemke, 2011).

The constructs are contained in pET28 vectors and expressed from BL21 Star (DE3) cells (Thermo Fischer Scientific). For in vivo biotinylation, the biotin ligase (BirA) is coexpressed from pBirAcm (Avidity LLC) following the manufacturer's instructions. In short, TB medium is supplemented with 0.5% glucose and appropriate antibiotics. Expression is induced at an OD_{600} of 0.7 by adding 1 m*M* IPTG and 50 μ*M* D-biotin (from a 5 m*M* stock, in warm 10 m*M* bicine buffer pH 8.3, filter-sterilized). After induction for 3 h at 37°C, cells are harvested by centrifugation, washed with phosphate-buffered saline (PBS), and pelleted again. Cell pellets are stored at −20°C until purification.

The enzymatic biotinylation can also be done in vitro after expression. N-terminal biotinylation via NHS chemistry is also possible, but not recommended due to lower efficiency and possible unspecific side products of the reaction.

For purification, the cells are resuspended in 30 mL PBS and lysed with a Cell Disrupter (Constant Systems) at 1.6 kbar. Cell debris is pelleted by centrifugation at 25,000 × *g* at 4°C for 30 min (Avanti JXN-26, Beckman Coulter). After filtration (Filtropur S 0.45, Sarstedt) the solution is applied to a 5 mL HisTrap HP (GE Healthcare) and eluted by a linear gradient from 0 to 500 m*M* imidazole in 100 m*M* sodium phosphate, 300 m*M* NaCl pH 8.0 at 8°C. Protein-containing fractions are pooled and dialyzed into the imidazole-free buffer overnight in the presence of 1/100 mol SENP protease. This protease cuts off the N-terminal His-SUMO sequence, leaving the native, tag-free protein (Panavas, Sanders, & Butt, 2009). The solution is again applied to the HisTrap column and the flow-through is collected and diluted 1:3 with ddH_2O to decrease the ionic strength. After applying to a HiTrap Q 5 mL (GE Healthcare), the protein is eluted with a linear gradient from 50 m*M* to 1 *M* NaCl in 40 m*M* Hepes pH 7.5. Hsp90 fractions are pooled and concentrated using centrifugal filters with a 50 kDa molecular weight cut-off (Amicon Ultra, Merck Millipore). Finally, the protein is applied to a HiLoad 16/600 Superdex200 (GE Healthcare) and eluted with 40 m*M* Hepes, 200 m*M* KCl pH 7.5. Peak fractions are again pooled and concentrated to 50–100 μ*M*. Aliquots are snap-frozen in liquid nitrogen and stored at −80°C until further usage. Expression and purification is monitored by SDS-PAGE and final sample purity usually exceeds 95%.

3.5 Protein Labeling

Hsp90 61C (the cysteine mutant where the wild-type amino acid at position 61 is replaced) is labeled with Atto488, and Hsp90 385C is labeled with Atto550. Cysteine mutants of the proteins are labeled with the maleimide-dye derivatives according to the following protocol:

1. Add 5 m*M* Tris(2-carboxyethyl)phosphine (TCEP) pH 7 to the protein to reduce the cysteine. Incubate 30 min at room temperature.
2. TCEP is removed by repeatedly washing the protein in a centrifugal filter at 4°C with ice-cold 1 × PBS pH 6.7, and the final concentration of the protein is adjusted to about 50 μ*M*.
3. Add a twofold excess of maleimide-dye (dissolved in DMSO at millimolar concentration) and incubate at room temperature in the dark for 1 h.
4. Free dye is removed using gel filtration spin columns (PD-10, GE Healthcare).
5. Use immediately or freeze aliquots.

The labeling position has to be chosen with care. If available, structural information from crystal or NMR structure should be considered. Preferred positions are in surface-exposed loops, away from interaction interfaces and not buried in a pocket so that the accessible volume of the dye is large.

3.6 Monomer Exchange

In order to create heterodimers of Hsp90 labeled with Atto488 at amino acid position 61 and Atto550 at amino acid position 385, a 10:1 mixture of the variants with a total protein concentration below 10 μ*M* is incubated at 47°C for 30 min. This destabilizes the C-terminal zipper and enables exchange of the monomers (Mickler et al., 2009). Possible aggregates are removed by centrifugation at 14,000 × *g* at 4°C for 30 min. Exchanging with a 10 × excess of 61C-Atto488 will result in mainly heterodimers observed in the measurement chamber, as immobilization on the surface is achieved by biotin-Neutravidin interaction and only 385C-Atto550 is biotinylated.

3.7 Data Acquisition

First the chamber is flushed with buffer, the excitation beam is positioned in the field of view, and the chamber is slowly moved along the complete detection range of the slide with a laser power of about 10 mW in order to bleach any fluorescent contaminants on the surface. Then, a Neutravidin solution (0.25 mg/mL) is flushed into the chamber and incubated for 5 min. Unbound Neutravidin is flushed out with buffer and remaining surface functionalization defects are blocked by flushing a 0.5 mg/mL BSA solution through the chamber. Biotinylated and labeled Hsp90 is flushed into the chamber at rising concentrations until a sufficient surface density is reached, which is usually the case at a concentration of 1–10 p*M*. Unbound protein is flushed out with buffer. Labeled AMP-PNP (AMP-γ-(6-Aminohexyl)-PNP-Atto647N, Jena Bioscience) is flushed into the chamber at a concentration of 25 n*M* in degassed measurement buffer (40 m*M* Hepes, 150 m*M* KCl, 10 m*M* $MgCl_2$ pH 7.5) with 0.5 mg/mL BSA and in a second experiment with additional 250 μ*M* unlabeled ATP. After a few minutes, the same solution is flushed into the chamber again to ensure correct concentration which may have decreased due to unspecific surface binding.

Camera settings are optimized for single-molecule recording (3.3 μs vertical shift speed, normal vertical clock voltage, 17 MHz 16-bit horizontal read out, preamplification gain 3, gain of electron multiplier 1000, sensor cooled to −95°C).

When coupled to protein, the used Atto dyes do not show significant blinking on the time scale of the experiment (70 ms illumination per frame). The most common scavenging system consisting of glucose oxidase, catalase, and trolox (Blanchard, Kim, Gonzalez, Puglisi, & Chu, 2004) does not lead to prolonged fluorescence traces. In order to avoid artifacts due to unspecific protein interactions, we also do not use any other oxygen-scavenging system, which usually contain proteins in the concentration of about 20 mg/mL (Swoboda et al., 2012).

To ensure all fluorescent-labeled interaction partners are present and to help discrimination of states, ALEX is used. The protein is fixed to the surface, the dyes Atto488 and Atto550 are excited alternatingly for 70 ms with the blue or green lasers, and the readout of the CCD chip takes 30 ms. The Atto647N-labeled nucleotide is not excited with the red laser, because at a concentration of 25 n*M* the very high background from dyes not attached to the surface renders this signal useless. For an Atto647N-labeled species in solution with a concentration above 100 n*M*, even direct excitation with the green laser at 532 nm becomes problematic.

4. DATA PROCESSING

The raw data from the EMCCDs (i.e., frames of 512 × 512 pixels of the fluorescence intensities in 16-bit gray scale) are stored as TIFF stacks. Analysis is done with scripts written in Igor Pro (Wavemetrics). The movies are automatically split into the respective detection channels when loaded according to predefined settings.

4.1 Channel Alignment

A precise mapping of the coordinates of one channel to all other channels is necessary to get the fluorescence intensities of all dyes attached to one protein in a multicolor FRET experiment. This is achieved by a calibration with fluorescent beads (Fluospheres amine-modified, 0.2 μm, yellow-green fluorescent, Thermo Fischer Scientific) that show fluorescence emission in the spectral range of all four detection channels of our setup upon excitation with the blue laser. After each measurement, a 1:1000 dilution of the beads in ddH_2O is flushed into the chamber and a movie is recorded while driving stepwise through the chamber.

Bead positions are detected in the calibration movie by searching for the brightest spots, and the central position is determined from a Gaussian fit to the intensity profile. The bead positions in the different channels are linked by image registration of the recorded calibration movies. A distance cut-off is applied to exclude incorrectly assigned beads. The coordinates of beads that are found in all four channels are saved, and the mapping offset in x- and y-direction is each fitted with a 2D polynomial of degree three.

4.2 Extraction of Time Traces

Individual molecules are identified in each movie by searching for the brightest spots in the sum of 5 frames of a certain channel. The central position of the spot is calculated for all channels using the channel mapping. For each channel the intensity traces are obtained by a 3 × 3 binning of the pixels around the central coordinates in each frame. The binning corresponds to an area of approximately 480 × 480 nm for the given magnification of the setup and was determined empirically to give the best signal-to-noise ratio. Intensity traces are corrected for an offset originating from the dark current of the EMCCD and background fluorescence. This is done by automatically subtracting a background time trace that is the mean intensity of the 200 darkest pixels in each channel. For each molecule, a joint raw intensity trace is calculated as the sum of all traces with the same excitation color. Traces are selected according to the following criteria: The joint raw intensity trace should have a roughly flat plateau, a single bleaching step, and the appropriate channel intensities have to show anticorrelated behavior. Please note that a flat plateau can only be expected in the joint intensity traces if the channels have similar apparent gamma factors, since the traces are not corrected at this stage.

As the aim of this work is to characterize nucleotide binding to Hsp90, we select only traces exhibiting intermittent red fluorescence after blue or green excitation, indicating the presence of Atto647N-labeled nucleotide. About 200 spots in each detection channel of a movie are tested for single-molecule behavior.

5. DATA ANALYSIS

In this chapter, the data analysis and interpretation of two multicolor smFRET experiments are demonstrated. In total, 20 movies of each experiment are processed. In these experiments, Hsp90 labeled with Atto488 at amino acid position 61 at one protomer and Atto550 at amino acid position 385 at the other is studied in the presence of 25 n*M* AMP-PNP(647N) and 25 n*M* AMP-PNP(647N) + 250 μ*M* ATP, respectively. 128 and 194 molecules are identified and evaluated subsequently.

In our analysis, we assume that nucleotide-coupled Atto647N does not bleach within the duration of a trace since it is the most photostable of the three dyes. Moreover, AMP-PNP is not hydrolyzable by Hsp90. Thus, a loss of Atto647N fluorescence is treated as unbinding of the labeled nucleotide. Also, we do not observe hydrolysis of nucleotide directly since hydrolysis of unlabeled ATP does not result in a specific conformational change for the used labeling positions.

5.1 Calculation of Partial Fluorescence Traces

All intensity traces of one molecule are displayed, and the local intensity offset in each channel is adjusted by subtracting the individual average intensity of a time range where all dyes are already bleached. The time range that shows FRET is manually selected, and correction factors for leakage and apparent gamma are applied according to Eq. (9). Finally, the *PF* traces are calculated according to Eq. (10). This results in PF_{green}^{blue}, PF_{red}^{blue}, and PF_{red}^{green} representing the energy transfer from Atto488 to Atto550, Atto488 to Atto647N after excitation of Atto488 (possibly via Atto550), and Atto550 to Atto647N after excitation of Atto550.

These *PF*s are shown in a 3D scatter plot (Fig. 8A and Video available in http://dx.doi.org/10.1016/bs.mie.2016.08.024) with projections onto the

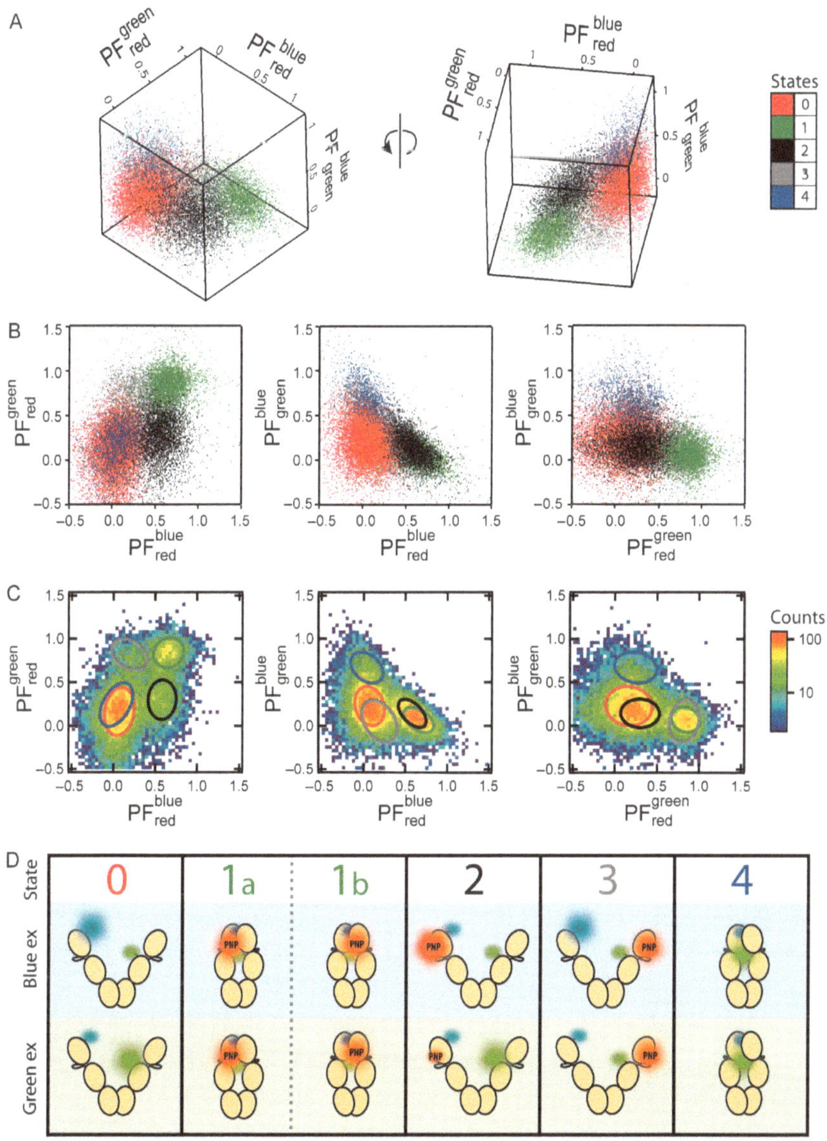

Fig. 8 State allocation for the three-dimensional smFRET data for Hsp90 with AMP-PNP(647N). (A) The 3D scatter plot of the calculated *PF*s shown from two perspectives with the *color code* representing the states allocated by 3D ensemble HMM analysis. (B) 2D projections of the 3D data shown in A. (C) 2D projections of histograms and fits onto the three different *PF* planes. (D) Pictograms of the assigned states after *blue and green* excitation (*ex*). States 1a and 1b are indistinguishable in the presented experiment.

respective *PF* planes shown in Fig. 8B or accumulated in a 3D histogram. 2D projections of this histogram (Fig. 8C) have proven to be useful to distinguish different states.

5.2 State Separation

In two-color experiments Hsp90 labeled with dyes at the opposing amino acid positions 61 and 385 has been found to show two FRET efficiencies, attributed to the "closed" and "open" conformation of Hsp90 (Mickler et al., 2009). Therefore, two populations in the PF_{green}^{blue} signal located at $PF_{red}^{blue} \approx 0$ and $PF_{red}^{green} \approx 0$ in the three-color experiment are expected. Theoretically, each Hsp90 dimer has two binding sites for nucleotides. The dissociation constant K_D of Hsp90 for AMP-PNP binding is at least one order of magnitude higher than the AMP-PNP concentration used in the experiment (25 n*M*). This simplifies analysis since the binding of two AMP-PNP is very unlikely and is therefore neglected. Thus, two populations representing the open conformation of Hsp90 with AMP-PNP(647N) bound either to the 61C-Atto488-labeled protomer or to the 385C-Atto550-labeled protomer are expected. In addition, two indistinguishable populations representing the closed conformation with one nucleotide bound to the dimer should occur.

Indeed, in the 3D scatter plot of the observed *PF*s, five populations can be discriminated, as shown in Fig. 8A. Although states may overlap in a particular 2D projection, the 3D information allows to separate them all. The expected conformations are depicted in Fig. 8D and can be assigned as follows (numbering of states is arbitrary). Three populations representing the nucleotide-bound states are observed in the PF_{red}^{green} vs PF_{red}^{blue} projection (Fig. 8C). These are the state 1 (green), state 2 (black), and state 3 (gray). The fourth peak near the origin results from overlapping populations of nucleotide-free Hsp90 in the open (state 0, purple) and the closed conformation (state 4, blue). These two populations differ in their PF_{green}^{blue} and thus are well separated in the PF_{green}^{blue} vs PF_{red}^{blue} projection. As expected, the equilibrium between Hsp90's open and closed conformation without nucleotide bound (states 0 and 4) is similar to the distribution from two-color experiments in the absence of nucleotides with the open conformation being populated with a probability of about 90%.

Depending on the system under study, different approaches provide optimal state allocation. First, for well-separated states, a free 3D ensemble hidden Markov modeling (HMM) based on the *PF* trace can be applied to identify the underlying states and optimize the corresponding (e.g., Gaussian) probability densities. Second, the original fluorescence traces are a possible alternative because *PF* traces contain unfavorable spikes. However, this benefit comes at the expense of constant signal levels throughout the data set since the fluorescence intensities vary from molecule to molecule. A third possibility is to separate populations by fitting the *PF* histogram with the sum of multidimensional Gaussians.

Here, the 3D histogram is fit with the sum of five 3D Gaussians to determine the exact position of the five states. For this fit to converge properly, the position and width of the Gaussians are free, and only the relative population p_i of each state i has to be held constant. The relative populations are determined beforehand from the 2D projections of the histograms by drawing free-hand polygons that represent the location of the state in the respective 2D projection (Fig. 9). By exploiting all three 2D projections, the relative population of all five states can be estimated.

This procedure was repeated three times for each data set to check the reproducibility of the state allocation. All fits have similar χ^2 values and differ only insignificantly in position and width of the five states. The fits yield the central position and the covariance matrix of the Gaussians that describe each state in the subsequent HMM analysis.

Alternatively, if the number of states is not evident from the data, parsimony criteria like the Bayesian information criterion (BIC) or likelihood ratio tests can provide a way to identify the most appropriate number of states.

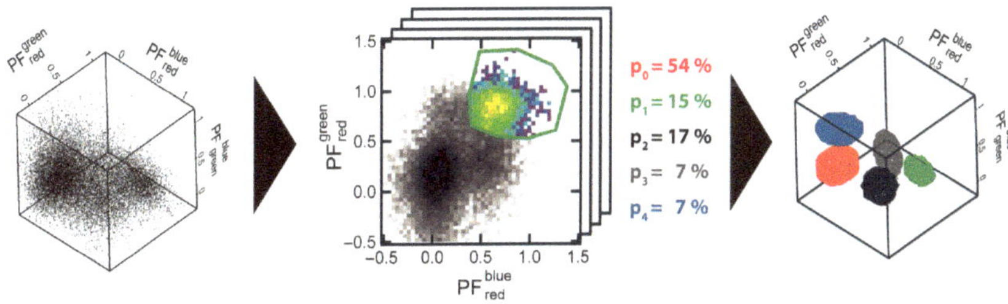

Fig. 9 State separation workflow for the three-dimensional smFRET data on Hsp90 with AMP-PNP(647N). First, the 3D scatter plot of the *PFs* is accumulated in a histogram and state populations are estimated from the 2D projections. Then, the sum of five Gaussians is fit to the 3D histogram with the position and width as free parameters and the predetermined relative populations held fixed. The resulting Gaussians are shown as isosurface at half-maximal probability density (accounts for about 30% of the volume). *Color scale* as in Fig. 8A.

5.3 Kinetic Analysis With 3D Ensemble HMM

Recovery of kinetic information from multicolor smFRET traces is impeded by the following four issues:

1. The true multidimensionality of the signal, i.e., the different conformations of Hsp90 can bind nucleotide.
2. The low signal-to-noise ratio, a challenge common to most single-molecule techniques.
3. The broad FRET efficiency distribution of each state, which is a typical feature of Hsp90 and other protein systems (Kahra et al., 2011; Santoso et al., 2010).
4. Conformational changes in protein systems can occur on a broad range of time scales. Classical dwell time analysis misses primarily long dwells and thus systematically overestimates the extracted rates (Schmid et al., 2016).

To overcome these shortcomings and to provide an unbiased way of state allocation and kinetic rate extraction HMM of FRET trajectories has been established in the field (McKinney, Joo, & Ha, 2006). In general, an HMM describes a system as a succession of states that are not directly observable (i.e., hidden). Each state has a probability distribution over the observable signal range (e.g., the *PF*). The transitions are treated as Markov processes, which means that the progression of states is memoryless. Thus, the probability of a certain transition depends only on the current state. A time-discrete HMM is described by a vector of start probabilities with elements $\boldsymbol{\pi_i}$ per state $\boldsymbol{i}$, the transition matrix $\boldsymbol{A}$ containing the transition probabilities $\boldsymbol{a_{ij}}$ between all hidden states, and state-specific emission probabilities $\boldsymbol{b_i}$ that link the hidden states to the observables. Here, the $\boldsymbol{b_i}$'s are described with continuous Gaussian probability density functions. After the maximum-likelihood estimate (MLE) for the parameters of the HMM given the data set is derived, the most likely sequence of hidden states (called the Viterbi path) can be compiled (c.f. Fig. 10A and B for exemplary intensity, *PF* traces, and the determined Viterbi path).

Implementation of our ensemble HMM approach is based on Fink (2014). Data points with $PF < -1$ or $PF > 2$ are assigned the minimal emission probabilities for all states, which prevents artificial transitions at these time points.

One ensemble HMM was optimized over the full data set with $\boldsymbol{b_i}$'s fixed to the parameters from the Gaussian fit and the resulting transition matrix and Viterbi path were subsequently analyzed. As a quality control, the

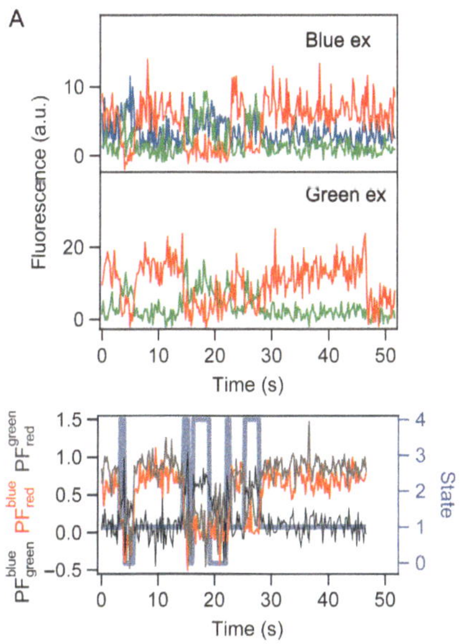

Fig. 10 Two exemplary single-molecule traces (A) and (B) for Hsp90 in the presence of 25 n*M* AMP-PNP(647N) are detailed. In the *top* row the fluorescence traces of Atto488 (*blue*), Atto550 (*green*), and Atto647N (*red*) after *blue* excitation are shown, in the *middle* row the fluorescence of Atto550 (*green*) and Atto647N (*red*) after *green* excitation is depicted. At the *bottom* the resulting *PFs* (PF^{blue}_{green} in *black*, PF^{blue}_{red} in *red*, PF^{green}_{red} in *gray*) and the Viterbi path (*bold blue*) are shown.

Viterbi path is displayed with the *PF* traces and checked for plausibility with regard to state allocation and transitions for all molecules of the data set.

For each of the two data sets the HMM optimization was run with the three different $\boldsymbol{b}_i$'s from the separate fits. The results differ only slightly in their production probability and produced an overall similar transition matrix (Fig. 11A and B).

In order to get a measure for the uncertainty of the extracted transition probabilities the 95% confidence interval (CI) for each transition is determined (Giudici, Ryden, & Vandekerkhove, 2000; Greenfeld, Pavlichin, Mabuchi, & Herschlag, 2012). For this purpose, the likelihood profile of the parameter space around the MLE is calculated for each transition probability. The parameter of interest is varied with all other parameters being fixed to the MLE, yielding the test model λ'. A likelihood ratio test of the likelihood $L(\lambda' \mid O)$ given the data set O is performed according to

$$LR = 2 \cdot (\ln L(\lambda_{MLE} \mid O) - \ln L(\lambda' \mid O)) \tag{11}$$

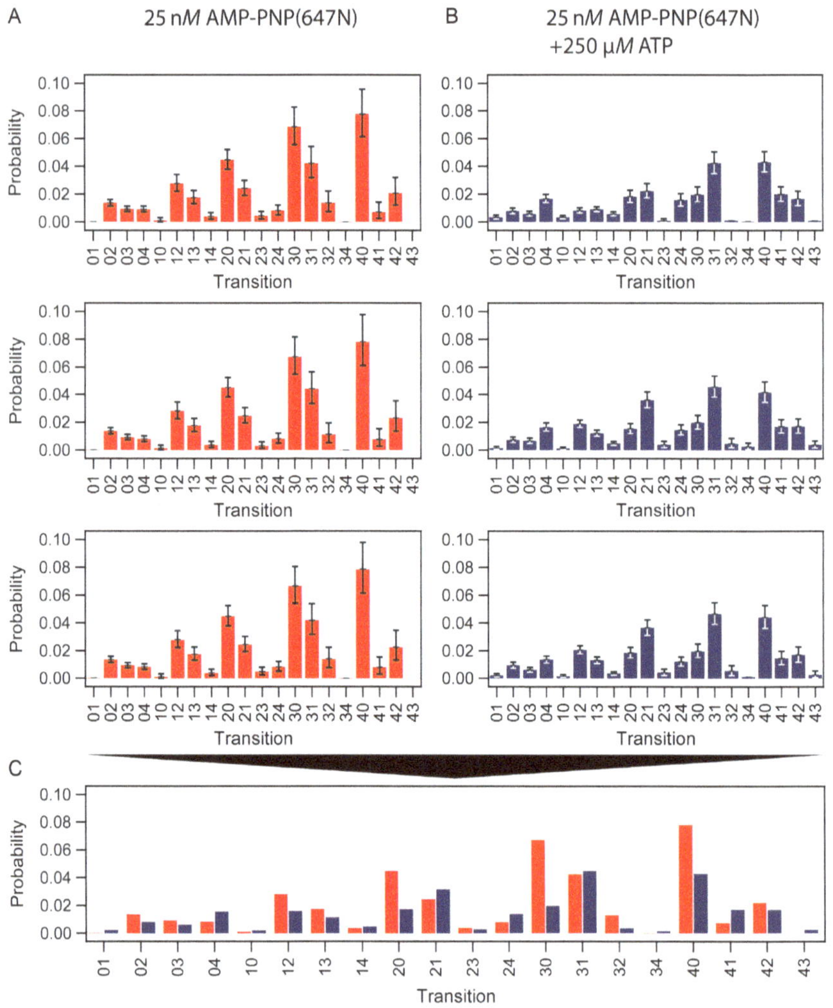

Fig. 11 Transition probabilities for all possible transitions between the five different states in the absence (A) and presence (B) of ATP. The results of three independent HMM optimizations based on different Gaussian population fits are almost identical. The data sets comprise 128 (A) and 194 (B) molecules, respectively. (C) Comparison of the mean transition probabilities from (A) and (B).

The confidence bound for the parameter is reached when LR exceeds 3.841, the 95% quantile of a χ^2-distribution with one degree of freedom.

The CI can be compared to the deviation of the three independent fits for each data set. We find that the deviation of the fits is typically smaller than the CI (c.f. also Fig. 11A and B).

5.4 Data Interpretation and Limitations of Data Analysis of Hsp90

In total, about 1000 transitions are found in each data set. The number of assigned state transitions is shown exemplarily for one HMM run of one data set in Fig. 12. All theoretically possible transitions are allocated, and the number of forward and backward transitions is approximately the same. This implies that the system works at thermodynamic equilibrium, even though detailed balance (i.e., the number of forward and backward transitions between two states is the same) is not enforced. Fig. 11A and B shows that the presence of hydrolyzable ATP does not change the overall picture, indicating that there is no directionality directly coupled to the hydrolysis. The transitions between the states $1 \rightleftharpoons 4$, $(2/3) \rightleftharpoons 4$, and $0 \rightleftharpoons 1$ occur only rarely in both data sets. These transitions could also be an artifact of the finite time resolution and might be caused by two successive transitions within one acquisition cycle. Additional experiments with faster sampling rate can overcome this limitation.

In the presence of additional ATP, the probability of AMP-PNP(647N)-bound Hsp90 being in the closed state is higher (data not shown). At 250 μ*M* ATP the probability for the Hsp90 dimer binding nucleotides in both binding pockets is increased. HMM analysis reveals that the presence of ATP leads to overall slower state transitions of Hsp90, as shown in Fig. 11C. This suggests an increased energy barrier between conformations when two nucleotides are bound to the dimer. Interestingly, the dissociation

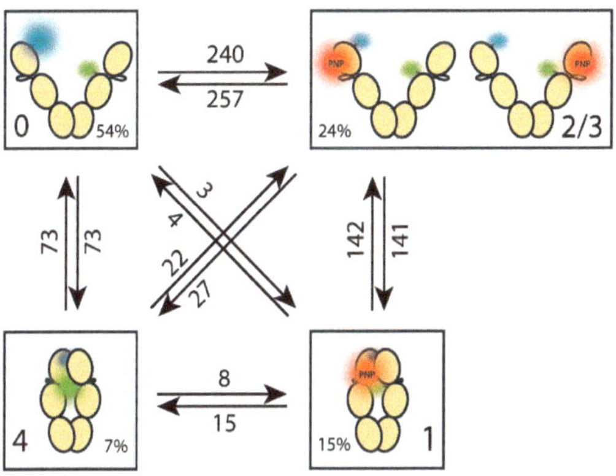

Fig. 12 Summary of the results for the correlated conformational and nucleotide binding dynamics of Hsp90 in the presence of 25 n*M* AMP-PNP(647N). Next to the *arrows* the number of transitions assigned by HMM is given. In the *boxes* the relative population of the different states is indicated.

of AMP-PNP(647N) from the open Hsp90 conformation is slowed down in the presence of ATP, which could hint toward a communication between the two binding sites through the whole protein dimer. Surprisingly, the transition from the open, AMP-PNP(647N)-bound states 2/3 to the closed, AMP-PNP(647N)-bound state 1 seems unaffected by the presence of additional ATP.

For a more detailed analysis, further refinements of the model can be included. For example, transition probabilities that are ruled out (either because the transition does not occur in the data set or because they can be excluded due to a priori knowledge) can be set to zero in the transition matrix. Also, detailed balance can be enforced if no external energy source is present (Greenfeld et al., 2012).

Our data nicely demonstrate the increased resolution of multicolor smFRET that leads to the observation of additional states inaccessible in two-color FRET experiments. The discrimination of more states results in additional transitions, which in turn requires a larger data set for sufficient statistics. A data set of more than 100 molecules (a number that gives reasonable statistics in a two-color experiment) is not large enough to resolve all transitions, especially if there are transitions that occur only rarely. We can correlate the open/close transition of Hsp90 with nucleotide binding. However, here we lost the ability to discriminate states that share the same FRET efficiency but differ kinetically. Namely the closed conformation of Hsp90 with no nucleotide bound is actually a combination of at least two states (Mickler et al., 2009; Schmid et al., 2016). The size of the data set also has a strong influence on the precision of the kinetic analysis by HMM. This is nicely demonstrated by Fig. 11A and B, showing much smaller CIs for the data set that includes about 50% more traces.

Additionally, our data on Hsp90 serve as a good example of the increased complexity of multicomponent experiments. In principle, the simple system of a protein dimer that can independently bind up to two nucleotides and undergo an open/close transition results in eight different states. We reduce this number by appropriate experimental conditions. First, we exclude simultaneous binding of two labeled nucleotides. Second, two states share the same signal. Notably, with only one more interaction partner (e.g., a cochaperone) a four-color experiment could result in up to 16 different states. Moreover, the signal-to-noise ratio decreases with an increasing number of simultaneous signals. Thus, states that are well defined in a two-color experiment can get indistinguishable in a more complex experiment.

In consequence, we believe that most questions concerning correlated dynamics of complex systems can be solved by a set of three-color smFRET experiments with reasonable effort.

6. CONCLUSION

Multicolor single-molecule FRET is an essential method to study multicomponent machineries in or out of equilibrium. Microscopic states that are hidden in two-color experiments can be revealed, and correlations between different conformational changes and interactions can be determined.

The presented general framework is suited for a multicolor experiment with an arbitrary number of colors. We focus on the discussion of a three-color smFRET experiment, which serves as an illustrative example. Three-color smFRET can be applied to the protein machinery Hsp90 to observe nucleotide binding and conformational rearrangements simultaneously. Transition probabilities between the different states can be extracted from the data by applying 3D ensemble HMM analysis, which reveals a correlation between the presence of nucleotide and state transitions of Hsp90. This information is inaccessible in a two-color FRET experiment. Furthermore, CIs for the transition probabilities are used as a measure of the uncertainty.

A major difference between the presented work and previously published multicolor smFRET experiments is the nature of the investigated system. Dynamic DNA or RNA systems, such as Holliday junctions, exhibit transitions between well-defined states. In contrast, protein systems display dynamics on a wide range of time scales between states that are highly flexible themselves. Such dynamics already reduce the signal-to-noise ratio and challenge data analysis. In addition, extracting the complete set of transition probabilities is difficult because of the limited temporal bandwidth in smFRET experiments due to photobleaching. Here, we show that—if applied carefully and with reasonable constraints—these obstacles can be overcome with multicolor smFRET using multidimensional state allocation together with 3D ensemble hidden Markov analysis.

ACKNOWLEDGMENTS

We thank Joanna Urban for helpful comments on the manuscript. This work is funded by the German Research Foundation (INST 39/969-1) and the European Research Council through the ERC Grant Agreement n. 681891.

REFERENCES

Ali, M. M., Roe, S. M., Vaughan, C. K., Meyer, P., Panaretou, B., Piper, P. W., et al. (2006). Crystal structure of an Hsp90-nucleotide-p23/Sba1 closed chaperone complex. *Nature*, *440*(7087), 1013–1017.

Blanchard, S. C., Kim, H. D., Gonzalez, R. L., Puglisi, J. D., & Chu, S. (2004). tRNA dynamics on the ribosome during translation. *Proceedings of the National Academy of Sciences of the United States of America*, *101*(35), 12893–12898.

Deniz, A. A., Mukhopadhyay, S., & Lemke, E. A. (2008). Single-molecule biophysics: At the interface of biology, physics and chemistry. *Journal of the Royal Society Interface*, *5*(18), 15–45.

Dollins, D. E., Warren, J. J., Immormino, R. M., & Gewirth, D. T. (2007). Structures of GRP94-nucleotide complexes reveal mechanistic differences between the hsp90 chaperones. *Molecular Cell*, *28*(1), 41–56.

Fink, G. A. (2014). *Markov models for pattern recognition: From theory to applications* (2nd ed.). London: Springer.

Giudici, P., Ryden, T., & Vandekerkhove, P. (2000). Likelihood-ratio tests for hidden Markov models. *Biometrics*, *56*(3), 742–747.

Gong, Y. C., Kakihara, Y., Krogan, N., Greenblatt, J., Emili, A., Zhang, Z., et al. (2009). An atlas of chaperone-protein interactions in *Saccharomyces cerevisiae*: Implications to protein folding pathways in the cell. *Molecular Systems Biology*, *5*, 275.

Greenfeld, M., Pavlichin, D. S., Mabuchi, H., & Herschlag, D. (2012). Single Molecule Analysis Research Tool (SMART): An integrated approach for analyzing single molecule data. *PLoS One*, 7(2), e30024.

Hahn, S. (2004). Structure and mechanism of the RNA polymerase II transcription machinery. *Nature Structural & Molecular Biology*, *11*(5), 394–403.

Hohng, S., Joo, C., & Ha, T. (2004). Single-molecule three-color FRET. *Biophysical Journal*, *87*(2), 1328–1337.

Jakob, U., Lilie, H., Meyer, I., & Buchner, J. (1995). Transient interaction of Hsp90 with early unfolding intermediates of citrate synthase—Implications for heat-shock in-vivo. *Journal of Biological Chemistry*, *270*(13), 7288–7294.

Juette, M. F., Terry, D. S., Wasserman, M. R., Zhou, Z., Altman, R. B., Zheng, Q., et al. (2014). The bright future of single-molecule fluorescence imaging. *Current Opinion in Chemical Biology*, *20*, 103–111.

Kahra, D., Kovermann, M., Low, C., Hirschfeld, V., Haupt, C., Balbach, J., et al. (2011). Conformational plasticity and dynamics in the generic protein folding catalyst SlyD unraveled by single-molecule FRET. *Journal of Molecular Biology*, *411*(4), 781–790.

Kapanidis, A. N., Laurence, T. A., Lee, N. K., Margeat, E., Kong, X., & Weiss, S. (2005). Alternating-laser excitation of single molecules. *Accounts of Chemical Research*, *38*(7), 523–533.

Lakowicz, J. R. (2006). *Principles of fluorescence spectroscopy* (3rd ed.). New York: Springer.

Lavery, L. A., Partridge, J. R., Ramelot, T. A., Elnatan, D., Kennedy, M. A., & Agard, D. A. (2014). Structural asymmetry in the closed state of mitochondrial Hsp90 (TRAP1) supports a two-step ATP hydrolysis mechanism. *Molecular Cell*, *53*(2), 330–343.

Lee, J., Lee, S., Ragunathan, K., Joo, C., Ha, T., & Hohng, S. (2010). Single-molecule four-color FRET. *Angewandte Chemie International Edition*, *49*(51), 9922–9925.

McKinney, S. A., Joo, C., & Ha, T. (2006). Analysis of single-molecule FRET trajectories using hidden Markov modeling. *Biophysical Journal*, *91*(5), 1941–1951.

McLaughlin, S. H., Sobott, F., Yaol, Z. P., Zhang, W., Nielsen, P. R., Grossmann, J. G., et al. (2006). The co-chaperone p23 arrests the Hsp90 ATPase cycle to trap client proteins. *Journal of Molecular Biology*, *356*(3), 746–758.

McLaughlin, S. H., Ventouras, L. A., Lobbezoo, B., & Jackson, S. E. (2004). Independent ATPase activity of Hsp90 subunits creates a flexible assembly platform. *Journal of Molecular Biology*, *344*(3), 813–826.

Mickler, M., Hessling, M., Ratzke, C., Buchner, J., & Hugel, T. (2009). The large conformational changes of Hsp90 are only weakly coupled to ATP hydrolysis. *Nature Structural & Molecular Biology*, *16*(3), 281–286.

Olesen, S. H., Ingles, D. J., Zhu, J. Y., Martin, M. P., Betzi, S., Georg, G. I., et al. (2015). Stability of the human Hsp90-p50(Cdc37) chaperone complex against nucleotides and Hsp90 inhibitors, and the influence of phosphorylation by casein kinase 2. *Molecules*, *20*(1), 1643–1660.

Panaretou, B., Prodromou, C., Roe, S. M., O'Brien, R., Ladbury, J. E., Piper, P. W., et al. (1998). ATP binding and hydrolysis are essential to the function of the Hsp90 molecular chaperone in vivo. *EMBO Journal*, *17*(16), 4829–4836.

Panavas, T., Sanders, C., & Butt, T. R. (2009). SUMO fusion technology for enhanced protein production in prokaryotic and eukaryotic expression systems. *Methods in Molecular Biology*, *497*, 303–317.

Person, B., Stein, I. H., Steinhauer, C., Vogelsang, J., & Tinnefeld, P. (2009). Correlated movement and bending of nucleic acid structures visualized by multicolor single-molecule spectroscopy. *Chemphyschem*, *10*(9–10), 1455–1460.

Plass, T., Milles, S., Koehler, C., Schultz, C., & Lemke, E. A. (2011). Genetically encoded copper-free click chemistry. *Angewandte Chemie International Edition*, *50*(17), 3878–3881.

Prodromou, C., Roe, S. M., OBrien, R., Ladbury, J. E., Piper, P. W., & Pearl, L. H. (1997). Identification and structural characterization of the ATP/ADP-binding site in the Hsp90 molecular chaperone. *Cell*, *90*(1), 65–75.

Ratzke, C., Berkemeier, F., & Hugel, T. (2012). Heat shock protein 90's mechanochemical cycle is dominated by thermal fluctuations. *Proceedings of the National Academy of Sciences of the United States of America*, *109*(1), 161–166.

Ratzke, C., Hellenkamp, B., & Hugel, T. (2014). Four-colour FRET reveals directionality in the Hsp90 multicomponent machinery. *Nature Communications*, *5*, 1–9.

Richter, K., Moser, S., Hagn, F., Friedrich, R., Hainzl, O., Heller, M., et al. (2006). Intrinsic inhibition of the Hsp90 ATPase activity. *Journal of Biological Chemistry*, *281*(16), 11301–11311.

Santoso, Y., Joyce, C. M., Potapova, O., Le Reste, L., Hohlbein, J., Torella, J. P., et al. (2010). Conformational transitions in DNA polymerase I revealed by single-molecule FRET. *Proceedings of the National Academy of Sciences of the United States of America*, *107*(2), 715–720.

Scheibel, T., Neuhofen, S., Weikl, T., Mayr, C., Reinstein, J., Vogel, P. D., et al. (1997). ATP-binding properties of human Hsp90. *Journal of Biological Chemistry*, *272*(30), 18608–18613.

Schmid S., Götz M., & Hugel, T. (2016). Single-molecule analysis beyond dwell times: Demonstration and assessment in and out of equilibrium, *Biophysical Journal 111*, 1375–1384. http://dx.doi.org/10.1016/j.bpj.2016.08.023.

Shiau, A. K., Harris, S. F., Southworth, D. R., & Agard, D. A. (2006). Structural analysis of *E. coli* hsp90 reveals dramatic nucleotide-dependent conformational rearrangements. *Cell*, *127*(2), 329–340.

Steitz, T. A. (2008). A structural understanding of the dynamic ribosome machine. *Nature Reviews Molecular Cell Biology*, *9*(3), 242–253.

Swoboda, M., Henig, J., Cheng, H. M., Brugger, D., Haltrich, D., Plumere, N., et al. (2012). Enzymatic oxygen scavenging for photostability without pH drop in single-molecule experiments. *ACS Nano*, *6*(7), 6364–6369.

Taipale, M., Jarosz, D. F., & Lindquist, S. (2010). HSP90 at the hub of protein homeostasis: Emerging mechanistic insights. *Nature Reviews Molecular Cell Biology*, *11*(7), 515–528.

Uemura, S., Aitken, C. E., Korlach, J., Flusberg, B. A., Turner, S. W., & Puglisi, J. D. (2010). Real-time tRNA transit on single translating ribosomes at codon resolution. *Nature, 464*(7291), 1012–1017.

Vabulas, R. M., Raychaudhuri, S., Hayer-Hartl, M., & Hartl, F. U. (2010). Protein folding in the cytoplasm and the heat shock response. *Cold Spring Harbor Perspectives in Biology, 2*(12), a004390.

Weiss, S. (1999). Fluorescence spectroscopy of single biomolecules. *Science, 283*(5408), 1676–1683.

Zhao, R. M., Davey, M., Hsu, Y. C., Kaplanek, P., Tong, A., Parsons, A. B., et al. (2005). Navigating the chaperone network: An integrative map of physical and genetic interactions mediated by the Hsp90 chaperone. *Cell, 120*(5), 715–727.

CHAPTER SEVENTEEN

Interferometric Scattering Microscopy for the Study of Molecular Motors

J. Andrecka*, Y. Takagi†, K.J. Mickolajczyk‡,§, L.G. Lippert¶, J.R. Sellers†, W.O. Hancock‡,§, Y.E. Goldman¶, P. Kukura*,[1]

*Physical and Theoretical Chemistry Laboratory, University of Oxford, Oxford, United Kingdom
†Laboratory of Molecular Physiology, National Heart, Lung and Blood Institute, National Institutes of Health, Bethesda, MD, United States
‡Pennsylvania State University, University Park, PA, United States
§Intercollege Graduate Degree Program in Bioengineering, Pennsylvania State University, University Park, PA, United States
¶Pennsylvania Muscle Institute, Perelman School of Medicine, University of Pennsylvania, Philadelphia, PA, United States
[1]Corresponding author: e-mail address: philipp.kukura@chem.ox.ac.uk

Contents

Abstract

Our understanding of molecular motor function has been greatly improved by the development of imaging modalities, which enable real-time observation of their motion at the single-molecule level. Here, we describe the use of a new method, interferometric scattering microscopy, for the investigation of motor protein dynamics by attaching and tracking the motion of metallic nanoparticle labels as small as 20 nm diameter. Using myosin-5, kinesin-1, and dynein as examples, we describe the basic assays, labeling strategies, and principles of data analysis. Our approach is relevant

Methods in Enzymology, Volume 581
ISSN 0076-6879
http://dx.doi.org/10.1016/bs.mie.2016.08.016

© 2016 Elsevier Inc.
All rights reserved.

not only for motor protein dynamics but also provides a general tool for single-particle tracking with high spatiotemporal precision, which overcomes the limitations of single-molecule fluorescence methods.

1. INTRODUCTION

Single-particle tracking (SPT) is a powerful tool for studying micro- and nanoscopic motions of cells, organelles, and molecules. Since the advent of video microscopy, there has been a continuous improvement in sensitivity, speed, precision, and spatial resolution, leading to our current ability to track single molecules on millisecond timescales and with nanometer precision. Early video microscopes used optical density, phase, or interference to provide contrast for visualizing large cargoes and natural tracks for motor proteins, such as mitochondria, and microtubules (MTs) (Allen, Metuzals, Tasaki, Brady, & Gilbert, 1982; Horio & Hotani, 1986; Howard, Hudspeth, & Vale, 1989; Vale et al., 1985). To detect motions of individual proteins, polymer microspheres (beads) were attached to individual motor molecules and imaged (Schnapp, Gelles, & Sheetz, 1988) or stabilized within an infrared optical trap (Rief et al., 2000; Svoboda, Schmidt, Schnapp, & Block, 1993). The average step size of kinesin-1, for instance, was determined to be 8 nm, matching the spacing of tubulin dimers along the MT axis (Svoboda et al., 1993). Optical trap experiments found that myosin-5a, which walks along actin filaments, exhibits ~36 nm steps (Mehta et al., 1999), the largest step size in the myosin superfamily.

In contrast to attaching beads, labeling single-protein molecules with fluorescent probes renders them clearly visible by fluorescence microscopy (Funatsu, Harada, Tokunaga, Saito, & Yanagida, 1995; Sakamoto, Amitani, Yokota, & Ando, 2000). The traditional spatial resolution limits in optical microscopy (including fluorescence) are related to diffraction of light, which results in the smallest objects producing an image whose size is equal to approximately half the wavelength of the light. This distance is much greater than the step size of molecular motors, such as the 8 nm steps of kinesin and the 5–40 nm steps, or power strokes, of myosin motors. Several investigators realized that when only one fluorescent object is present in the emitting volume, the center of the ~250 nm wide image of the single molecule, termed the point spread function, could be localized with very high precision (Thompson, Larson, & Webb, 2002; Yildiz et al., 2003). High-resolution tracking of myosin-5a and kinesin-1 (Forkey, Quinlan, Shaw, Corrie, & Goldman, 2003; Snyder, Sakamoto, Hammer, Sellers, & Selvin, 2004; Warshaw et al., 2005; Yildiz, Tomishige, Vale, & Selvin, 2004) found that these motors move in a "hand-over-hand" fashion, while recent studies on dynein showed comparatively erratic stepping behavior, demonstrating both "hand-over-hand" and "inchworm" mechanisms (DeWitt, Chang, Combs, & Yildiz, 2012; Qiu et al., 2012; Reck-Peterson et al., 2006).

As is apparent from the development of differential interference contrast and then fluorescence microscopy for SPT, continuous improvements to temporal and spatial resolution have led to deeper insights into the mechanisms of molecular machines responsible for cellular motion; many problems, however, remain unresolved. Outstanding questions in our understanding of molecular motor function include the details of myosin dynamics, while each step is being produced, how the two heads of kinesin coordinate their chemomechanical cycles to achieve processive stepping, and the determinants of dynein's irregular stepping pattern. Here, we present detailed protocols for tracking different molecular motors by interferometric scattering microscopy (iSCAT), which enables orders of magnitude improved spatiotemporal resolution compared to single-molecule fluorescence and, thereby, provides a route to addressing questions such as those outlined earlier.

2. INTERFEROMETRIC SCATTERING MICROSCOPY

An iSCAT image results from the interference between a reference light, commonly provided by a weak reflection at the sample surface, and light scattered by an object in the medium. A typical iSCAT setup is based on an inverted microscope (Fig. 1A), where the sample is illuminated by a coherent light source passing through a beamsplitter, which enables any photons scattered or reflected by the sample (Fig. 1B) to be separated from the incident light beam and subsequently imaged onto a digital camera. As a result, the experimental setup is almost identical to those used for the various implementations of interference reflection microscopy (Curtis, 1964; Filler & Peuker, 2000), with the important distinction that a coherent light source is used, which enhances the interferometric contrast (Dulin, Barland, Hachair, & Pedaci, 2014).

The most critical aspect to iSCAT performance is the precise measurement and removal of any static imaging background, as it determines imaging sensitivity and precision. When controlled appropriately, iSCAT

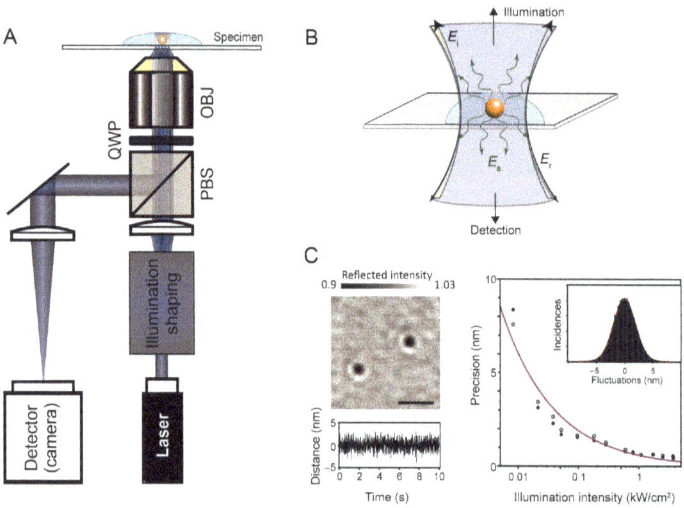

Fig. 1 Principles and performance of iSCAT for single-particle tracking. (A) Experimental setup. *OBJ*, microscope objective; *PBS*, polarizing beam splitter; *QWP*, quarter-wave plate. Illumination shaping refers to different schemes for illuminating the sample without generating excessive speckle such as slim field (Piliarik & Sandoghdar, 2014) or rapid beam scanning (Kukura et al., 2009). (B) Operating principle of iSCAT detection. Light reflected at the sample/glass interface is collected along with any scattered light from the object of interest. E_i, incident electric field; E_r, reflected electric field; E_s, scattered electric field. (C) Localization of 20 nm gold particles in the shot-noise-limited regime. *Left*: flat-field-corrected iSCAT image of 20 nm gold nanoparticles immobilized on a glass coverslip in water. Scale bar: 1 μm. The *bottom trace* shows the fluctuation in the distance between two nanoparticles as a function of time sampled at 1000 frames per second, caused solely by shot noise in the measurement. *Right*: localization precision as a function of illumination intensity for 20 nm gold nanoparticles. An increase in illumination intensity improves the localization precision for a given exposure time. *Open* and *filled circles*: localization precision in *x* and *y*, respectively. The *line* depicts the expected behavior for shot-noise-limited imaging.

has been shown to detect and image single, unlabeled proteins (Ortega Arroyo et al., 2014; Piliarik & Sandoghdar, 2014) and to track 20 nm gold labels at up to 500 kHz frame rates with nm precision (Andrecka et al., 2015; Lin, Chang, & Hsieh, 2014; Mickolajczyk et al., 2015). Interferometric detection leads to a lesser drop in scattering contrast with decreasing particle size compared to dark-field microscopy, simplifying the detection of very weak scatterers. In comparison to fluorescence, iSCAT suffers from much poorer background rejection ability, but in exchange can completely remove the need for any fluorescent labeling, including labeling of actin filaments and microtubules, which themselves generate clear iSCAT signals

(Andrecka, Arroyo, Lewis, Cross, & Kukura, 2016; Ortega Arroyo et al., 2014). As a result, the relevant single-molecule assays are in many instances easier to implement as they require fewer labeling steps. Most importantly, however, iSCAT provides access to experimental approaches and timescales that are inaccessible to single molecule fluorescence-based studies, either due to the underlying photophysics or due to the difficulty associated with removing fluorescence background (Arroyo & Kukura, 2016; Ortega-Arroyo & Kukura, 2012).

3. METHODS AND PROTOCOLS

3.1 Flow Cell Preparation

In all experiments described later we used the same type of flow cell, which is common in studies of molecular motors with a variety of methods. Briefly, we clean a 50 × 25-mm, #1.5, coverslip (VWR) by washing with ultrapure water (MilliQ) followed by ethanol and again MilliQ water. After drying, passing the coverslip through a blue flame removes dust and moisture. Proper washing is important to remove any dirt adhered to the coverslip surface, which generates iSCAT signal. Importantly, however, iSCAT detection does not require special substrates, such as quartz glass, and the cleaning procedure is straightforward compared to single-molecule fluorescence measurements, where special care must be taken to minimize any residual fluorescence background. The final flow channel consists of the cleaned microscope slide and a 25 × 25-mm coverslip connected by double-sided sticky tape. The volume of the chamber is on the order of ~15–20 μL.

3.2 Preparation and Immobilization of Cytoskeletal Filaments

One of the strengths of iSCAT in the context of molecular motor studies is its ability to visualize cytoskeletal filaments such as actin or microtubules without the need for any labels. When illuminated with blue light (<500 nm), an actin filament generates ~1% iSCAT contrast, while a single microtubule produces ~8% contrast, comparable with a 20-nm gold nanoparticle (Andrecka et al., 2016; Ortega Arroyo et al., 2014). Complete immobilization of actin and microtubules on the glass surface is paramount to success when running single-molecule motor assays with high spatiotemporal precision. Poorly immobilized filaments will introduce additional fluctuations, which can reduce the ability to detect steps. In this section we describe how actin filaments and microtubules are obtained and attached to the glass surface.

3.2.1 Actin

Required materials

- Globular (G) actin can be prepared as described (Pardee & Spudich, 1982) or purchased from Cytoskeleton
- Polymerization buffer: 10 m*M* imidazole pH 7.3, 50 m*M* KCl, 1 m*M* $MgCl_2$, 2 m*M* DTT, 3 m*M* ATP, 1 m*M* EGTA
- Motility buffer (MB): 20 m*M* MOPS pH 7.3, 5 m*M* $MgCl_2$, 0.1 m*M* EGTA
- Phosphate-buffered saline (PBS buffer) pH 7.4
- Bovine serum albumin (BSA)
- Poly(ethylene glycol)–poly-L-lysine branch copolymer (PEG–PLL) from Surface Solutions SuSoS, Switzerland

To polymerize and immobilize actin

1. Prepare 20 μ*M* filamentous (F) actin in polymerization buffer and incubate on ice for 2 h.
2. Rinse the chamber (prepared as described in Section 3.1) with 1 mg/mL solution of PEG–PLL dissolved in PBS and incubate for 30 min.
3. Wash the chamber with MB before adding actin solution.
4. Dilute actin 20–50 times in MB, load into the chamber, and incubate for 2–5 min. Inspect the chamber under the iSCAT microscope and add more actin if necessary.
5. Wash the cell with MB and then block with BSA (2 mg/mL in MB) for 5 min.

   NOTE: This procedure includes a minor modification of an established method (Dunn & Spudich, 2011). The polymer (PEG–PLL) is usually used to passivate the surface. In this protocol, however, PEG–PLL provides weak interactions with glass that immobilize actin. Additional blocking with BSA is required to avoid nonspecific binding of myosin and gold to the surface. Alternative methods, such as actin binding to nitrocellulose, are not recommended due to high iSCAT background of the nitrocellulose. Using biotinylated actin- and neutravidin-coated surfaces fails when gold nanoparticles are attached to myosin using the same moiety.

   TIP: If actin binding is insufficiently strong, plasma cleaning the cover glass before assembling the flow cells improves the attachment.

3.2.2 Microtubules

Required materials

- Free tubulin can be extracted from bovine or porcine brain as previously described (Ozeki et al., 2009) purchased directly from commercial

sources (Cytoskeleton), or expressed in yeast cells (Johnson, Ayaz, Huddleston, & Rice, 2011)
- Full-length *Drosophila* KHC R210A rigor mutant (see later)
- BRB80 buffer: 80 m*M* PIPES pH to 6.8 with KOH, 1 m*M* EGTA, 1 m*M* $MgCl_2$
- Casein (Sigma Aldrich, Cat. No. C7078)
- Taxol (Sigma Aldrich, Cat. No. T1912) dissolved in DMSO

To polymerize microtubules

1. Prepare a solution of 1.6 mg/mL free tubulin, 4 m*M* $MgCl_2$, 1 m*M* GTP, and 1:25 DMSO in BRB80 buffer, and incubate at 37°C for 30 min.
2. Stabilize by adding taxol to 10 μ*M*.
3. Dilute to working concentration (~1:80) in BRB80 plus 10 μ*M* taxol and 0.5 mg/mL casein.

Recent work has shown that a rigor form of kinesin-1 (full-length *Drosophila* KHC R210A) is highly effective in microtubule immobilization (Mickolajczyk et al., 2015). The tail domain of full-length kinesin binds strongly to casein-treated surfaces, while the head domains bind tightly to microtubules. Rigor kinesin binds microtubules more tightly than tubulin antibodies, improving microtubule immobilization. Cross-linking microtubules with glutaraldehyde is another effective method for immobilizing microtubules, but it greatly reduces the number of kinesin landing events likely due to damage to the microtubules (Turner, Chang, Fang, Cuomo, & Murphy, 1996).

To immobilize microtubules

1. Create a flow cell as described in Section 3.1.
2. Prepare casein solution (final 0.5 mg/mL in BRB80) and flow into the flow cell. Let sit for 5–10 min.
3. Dilute rigor mutant to 200 n*M* in casein solution plus 1 μ*M* ATP, and add to the flow cell, washing out the previous solution. Incubate for 5 min to adsorb motors to the surface.
4. Wash any unbound rigor mutant out of the flow cell using casein solution.
5. Load diluted microtubules (see earlier) into the flow cell. Incubate for 5 min.
6. Wash excess microtubules out of the flow well using casein solution and observe microtubules under the iSCAT microscope.

NOTE: The rigor mutant requires at least 1 μ*M* ATP with equimolar $MgCl_2$ at all times for proper function.

TIP: If there are multiple overlapping microtubules or clumps, then a further microtubule dilution is necessary. If microtubules appear to wobble, a higher concentration or a new preparation of rigor kinesin may be necessary.

3.3 Labeling Motor Proteins With Gold Nanoparticles

Attachment of nanoparticle labels to kinesin, myosin, and dynein motors has to date been achieved via biotin–streptavidin chemistry (Andrecka et al., 2015; Dunn & Spudich, 2007; Isojima, Iino, Niitani, Noji, & Tomishige, 2016; Mickolajczyk et al., 2015; Schneider, Glaser, Berndt, & Diez, 2013; Schneider, Korten, Walter, & Diez, 2015). One method of introducing biotin into kinesin is to create a cysteine-lite mutant, in which solvent-exposed cysteines are mutated out of the motor domain and a single cysteine is kept to take advantage of sulfhydryl chemistry for attaching a maleimide-functionalized label. This method has the advantage of direct, covalent attachment of the biotin and a minimal linker size, but has the disadvantage that cysteine-lite kinesin mutants exhibit markedly different mechanochemistry from wild type (Andreasson et al., 2015). For myosin-5a, a native or engineered cysteine residue in purified calmodulin can be labeled with a dye or gold nanoparticle and then exchanged for endogenous calmodulin in the myosin-5a (Dunn & Spudich, 2007; Sakamoto et al., 2000). However, in this method, it is difficult to control which of the six possible calmodulins is labeled. Since many existing constructs of kinesin and myosin possess an N- or C-terminal GFP moiety, another option is to use biotinylated antibodies against GFP (Andrecka et al., 2015; Schneider et al., 2013).

An alternative way of attaching a gold nanoparticle, which was used for the kinesin and myosin work here, is to fuse an Avi-Tag (GLNDIFEAQKIEWHE) to the motor and to biotinylate the protein with the biotin ligase BirA (Avidity) (Schatz, 1993). The constructs used to efficiently attach gold nanoparticles are described later.

3.3.1 Myosin

Mouse myosin-5a HMM was expressed in the presence of calmodulin and purified as described (Wang et al., 2000). The N-terminus was modified by the addition of a nucleotide sequence encoding an Avi-Tag peptide (Andrecka et al., 2015). Myosin was biotinylated at 4°C, overnight and dialyzed against 10 m*M* MOPS pH 7.3 buffer containing 0.1 m*M* EGTA, 1 m*M* DTT, and 100 m*M* KCl to remove excess of biotin. Before labeling, the myosin sample was diluted in MB containing 40 m*M* KCl, 5 m*M* DTT,

0.1 mg/mL BSA, and 5 μ*M* calmodulin. Gold nanoparticles conjugated with streptavidin were purchased from BBI (UK) and directly mixed with biotinylated myosin-5a in a 4:1, gold particle to myosin molecule ratio, consistent with one or zero myosin molecules per gold particle and only one motor domain labeled per myosin dimer. The mixture was incubated on ice for at least 15 min (sample volume 50 μL, final concentration of myosin 300 p*M*). The myosin–gold conjugate sample was then diluted in the MB: 20 m*M* MOPS pH 7.3, 5 m*M* $MgCl_2$, 0.1 m*M* EGTA supplemented with 20 m*M* KCl, 5 m*M* DTT, 0.1 mg/mL BSA, 1–1000 μ*M* ATP and loaded into the flow cell prepared as described earlier.

3.3.2 Kinesin

Kinesin-1 (*Drosophila* KHC truncated at 560) was prepared with an N-terminal Avi-Tag and coexpressed with BirA in BL21(DE3) bacteria, which resulted in in vivo biotinylation of the motor (Mickolajczyk et al., 2015). Motors were purified by affinity chromatography using a C-terminal His-tag. The biotinylation percentage was kept low by waiting until 6 h after induction by isopropyl β-D-1-thiogalactopyranoside before adding biotin to avoid doubly biotinylating the homodimer motor. The extent of biotinylation was quantified using the colorimetric HABA assay (ThermoFisher, Cat No. 28005). Motors were diluted in imaging solution (0.5 mg/mL casein, 10 μ*M* taxol, 20 m*M* glucose, 20 μg/mL glucose oxidase, 8 μg/mL catalase, 0.2 mg/mL BSA, 1:200 β-mercaptoethanol, 2 m*M* ATP-$MgCl_2$ in BRB80) and mixed with streptavidin-coated 30 nm gold nanoparticles (Mickolajczyk et al., 2015).

NOTE: In order to ensure the presence of only one biotinylated motor per bead, a titration experiment was performed where the number of landing events per linear micron of microtubule is measured as a function of molar ratio of motors to nanoparticles (NPs). The assumption is that one motor is sufficient for a particle to bind to a microtubule and more motors on a bead do not increase the probability of binding. By Poisson statistics, this plot will be linear if there is truly one motor per NP (Block, Goldstein, & Schnapp, 1990), and will plateau when there are multiple motors per bead. Thus, any motor:gold ratio from the linear regime is acceptable for single-molecule assays. In our experiment we used the single-molecule landing rate which showed a linear relationship up to 6:1 motor-to-gold ratio, justifying the use of these molar ratios with the expectation of only one motor per nanoparticle. Kinesin concentrations are often obtained through absorbance at 280 nm; however, only a subset of total protein is active kinesin, and only a further subset is active and biotinylated, meaning that there is actually a superstoichiometric amount of gold in the linear regime. Each new batch of kinesin will have a different linear regime for the number of landing events per micron of microtubule.

3.3.3 Dynein

Tail-truncated (331 kDa) dynein mini-dimer (GST dimerized) from *Saccharomyces cerevisiae* (Reck-Peterson et al., 2006) was biotinylated in vivo at AAA5 and AAA6 by inserting biotinylation sites at A3666 and P4032 in the dynein ring (Kulman, Satake, & Harris, 2007). Dynein was diluted in dynein lysis buffer (30 m*M* HEPES pH 7.4, 2 m*M* Mg acetate, 1 m*M* EGTA, 10% glycerol, 1.25 mg/mL casein, 1 m*M* DTT, 20 μ*M* taxol) and mixed with streptavidin-coated 30 nm gold nanoparticles in a 1:5, gold particle to dynein molecule ratio. The mixture was incubated on ice for at least 15 min. The dynein–gold conjugate sample was then diluted in the lysis buffer supplemented with oxygen scavenger system (300 μg/mL glucose oxidase (Sigma), 120 μg/mL catalase (Sigma), 0.4%, w/v glucose) and loaded into the flow cell.

3.4 Data Acquisition and Analysis

3.4.1 Image Operations: Flat-Fielding and Background Subtraction

The raw images from an iSCAT microscope contain static features resulting from nonuniform illumination, nonuniform pixel response, fixed-pattern noise, spurious back reflections, unwanted interference between back reflections, and imperfections of optical elements. These features are time invariant, do not change with lateral sample displacements, and can therefore be removed. We call the process of isolating and subsequently removing these signatures "flat-fielding" (Arroyo, Cole, & Kukura, 2016). First, the sample is translated laterally while recording a stack of 100 images. Second, a median image is generated by computing the median for each pixel across the stack of acquired images. Any subsequently recorded images are divided by this median image to account for illumination inhomogeneities.

TIP: The focus position of the sample must remain constant during acquisition of the flat-field stack of images, as some background signatures are focus dependent. In addition, the sample must be translated sufficiently quickly so that constant sample features are not imprinted onto the median image.

NOTE: Flat-fielding can only correct for truly static background features as the median image is acquired over several seconds.

Despite the "flat-fielding" process, intrinsic sample features still contribute to the scattering background that limits the achievable signal-to-noise ratio (SNR). In the case of motor proteins, such background originates from actin filaments or microtubules, along with any inactive motors and other surface-bound objects. This scattering background can overwhelm the signal of interest, especially for microtubules, which produce ~8% iSCAT signal, comparable to 20 nm gold nanoparticles (Fig. 3). If the scattering background remains constant, however, it can be removed by background subtraction, as follows: from a recorded stack of images, a single median image is generated by averaging the entire stack (Figs. 2 and 3, insets). This median image contains only static features intrinsic to the sample (background). Subtracting this median image from each image in the acquired stack yields an image stack dominated by shot noise and the signal of the gold-labeled translocating motor.

NOTE: If a motor protein tends to pause or stops altogether, the median filter approach will fail unless static images are removed during the median image processing. The background image should only contain immobile filaments, therefore, it is necessary to select a substack of images before a particle binds to a filament or after it detaches. In this case a median (or average) image can be calculated and subtracted from the remaining stack to reveal a moving particle.

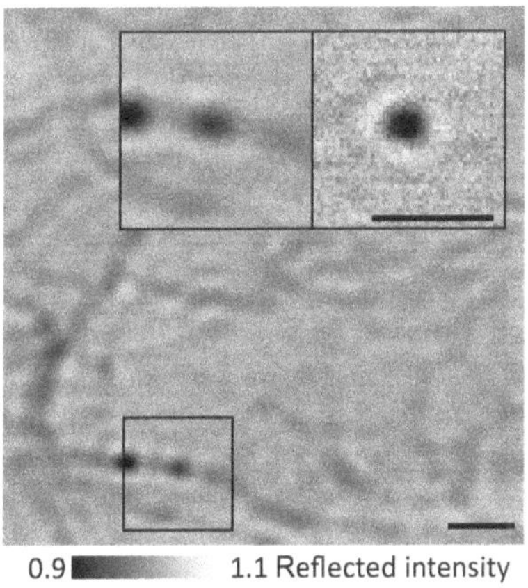

Fig. 2 iSCAT assay for gold-labeled myosin-5a tracking. Flat-field-corrected image showing actin filaments and some 20 nm gold-labeled myosin-5a molecules bound to actin. Inset: zoom of the indicated region showing 20 nm gold-labeled myosin-5a bound to actin (*left*) and the same region after background subtraction, which removes all static features including the actin filament and any immobile particles (*right*). Scale bars: 1 μm.

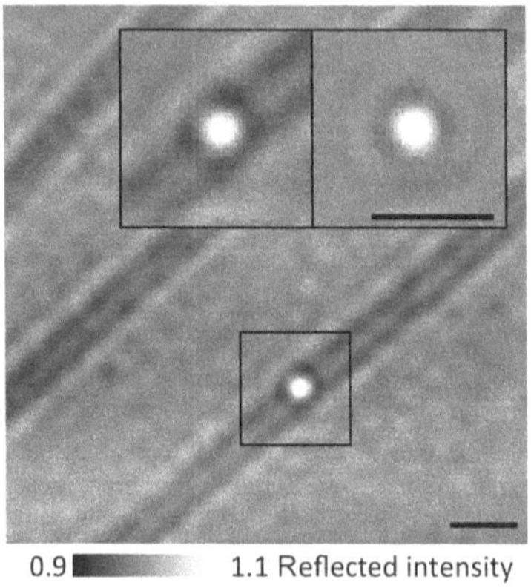

Fig. 3 iSCAT imaging of gold-labeled dynein. 30 nm gold-labeled dynein bound to a microtubule. In this assay, a gold label produces a positive iSCAT contrast when placed into focus (*bright*) due to the increased distance between the label and the coverglass surface, in contrast to myosin-5a head-labeled gold, which invariably appeared *dark* (Fig. 2). Scale bars: 1 μm.

3.4.2 Fitting Diffraction-Limited Spots

From the processed images, diffraction-limited spots can be identified by a sequence of pattern recognition operations that output the location of a candidate feature to the nearest integer pixel value. First, each image is filtered using a Gaussian kernel with a size defined by the full width at half maximum of the diffraction-limited spots (~200 nm). This Gaussian blurring step artificially enhances the SNR of such features by smoothing out high-frequency noise caused by shot statistics. Next, a nonmaximum suppression algorithm identifies the local extrema within the optical resolution of the imaging system (Neubeck & Van Gool, 2006). In practice, selecting only pixel values exceeding twice the standard deviation of the entire image in the absence of outliers removes local maxima attributed to noise. Finally, candidate particles are segmented into regions of interest corresponding to approximately 1 μm^2.

Subpixel localization information can be obtained from the candidate features, present in each segmented region of interest, by a nonlinear least square fit to an elliptical 2D Gaussian with a constant offset expressed by the following functional form:

$$G(x, y) = C \exp\left(-\left(\frac{(x-x_o)^2}{2{s_x}^2} + \frac{(y-y_o)^2}{2{s_y}^2}\right)\right) + B$$

where C represents the iSCAT contrast, x_o and y_o the center of mass location along each axis, s_x and s_y are proportional to the width of the Gaussian along each axis, and B a constant baseline. The nonlinear least square fit is performed by the iterative Levenberg–Marquardt algorithm, which requires an adequate initial guess for each of the adjustable parameters. This can be determined by a faster localization algorithm developed by Parthasarathy, termed radial symmetry centers (Neubeck & Van Gool, 2006). The advantage of this approach over iterative ones is that it is model free and requires a single iteration, which leads to an improvement in the execution time by two orders of magnitude, all while achieving similar performance to model-based techniques.

NOTE: The above fitting routine represents only one possible tracking algorithm. We also used open-source Fluorescent Image Evaluation Software for Tracking (FIESTA) to track gold-labeled kinesin (Ruhnow, Zwicker, & Diez, 2011).

3.4.3 Step-Finding Algorithms

Traditional step-finding algorithms analyze time vs distance traces in order to find time points where instantaneous and significant distance changes (steps) occur. In order to turn *XY* trajectories into time–distance traces that are amenable to step-finding analysis, the trajectory can simply be rotated such that the standard deviation in a given direction is minimized (Mickolajczyk et al., 2015). This rotation changes the coordinate system from *XY* axes to filament longitudinal and lateral axes and allows for more direct interpretation of how a motor is moving along the known structure of the filament.

Step-finding algorithms can broadly be split into two classes, model free and model dependent. Model-free algorithms make no assumptions about the data and simply perform significance testing to determine points where statistically significant changes in the data occur (Chen, Deffenbaugh, Anderson, & Hancock, 2014; Deffenbaugh, Chen, Anderson, & Hancock, 2015; Kalafut & Visscher, 2008). This class of algorithms is preferable when the goal of the study is to directly measure the size or number of steps present within a trace. A drawback of model-free algorithms is that they can suggest the existence of nonexisting step data if the significance level is set too low, and can miss steps if the significance level is set too high. The model-free detector algorithm, developed for counting the number of steps in GFP photobleaching data (Chen et al., 2014), has been applied with success to molecular motor data (Mickolajczyk et al., 2015). Model-dependent algorithms incorporate known information (a model) into the decision-making process for step finding (Kerssemakers et al., 2006; Milescu, Yildiz, Selvin, & Sachs, 2006a, 2006b; Mullner, Syed, Selvin, & Sigworth, 2010; Syed, Mullner, Selvin, & Sigworth, 2010). Example models might be the known lattice spacing of the filament or the known number of steps in a trace. Model-dependent algorithms are less-prone to over-fitting and are often the better choice when the goal of the study is to measure steps duration rather than step sizes. However, it is critical to justify the use of a model-dependent algorithm with prior experiments.

4. APPLICATIONS

4.1 Revealing the Motion of the Detached Head of Myosin-5a

For the movement of myosin-5a along actin, a body of biophysical work has led to a hand-over-hand stepping model in which the two heads alternate in leading, an average displacement of the striding head is 74 nm, and the detached head reaches the next binding site after Brownian rotation around a free joint at the head–head junction (Beausang, Shroder, Nelson, & Goldman, 2013; Dunn & Spudich, 2007; Forkey et al., 2003; Sakamoto, Webb, Forgacs, White, & Sellers, 2008; Veigel, Wang, Bartoo, Sellers, & Molloy, 2002; Warshaw et al., 2005; Yildiz et al., 2003). The last decade of research has been focused on understanding how unidirectional movement is achieved in the presence of random motion of the detached head (Karagiannis, Ishii, & Yanagida, 2014; Sellers & Veigel, 2006; Vale, 2003). Despite the variety of techniques with which myosin's motion has been characterized, it has remained beyond reach to directly visualize individual head stepping in real time.

In order to resolve the detached head movement a 20-nm gold label was attached to the N-terminus of the myosin-5a head and we tracked its motion in an iSCAT microscope. According to the Brownian search model, we expected do detect periods of increased positional fluctuations of the labeled head between the 74 nm steps, which were previously interpreted as a signature of Brownian search (Dunn & Spudich, 2007). Surprisingly, instead of random motion of the bead between the two binding sites, we revealed a transient state with a center of mass just over half way between the two binding sites and offset by 40 nm perpendicular to the actin filament (Fig. 4A and B). The myosin head repeatedly moved back and forth between the transient state and the next actin-binding site (Andrecka et al., 2015). An important implication of our data is that the appearance of the transient state seems inconsistent with a purely rotational diffusion type of search. The position of the transient state, together with signatures of constrained diffusion, suggests that myosin side-steps along actin in a combined twisting and leaning motion, much more like a drawing compass rather than a freely jointed swivel (Fig. 4C).

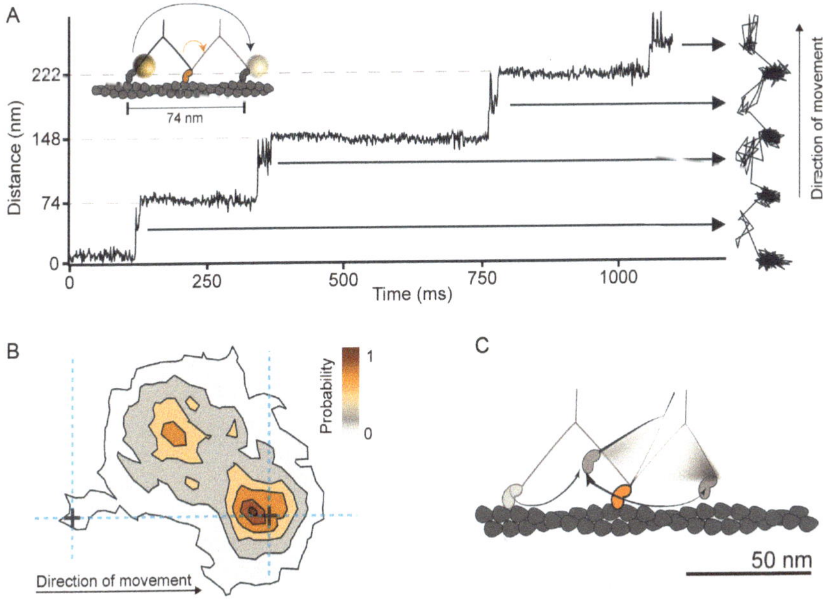

Fig. 4 Myosin-5a head motion during processive stepping. (A) iSCAT tracking results: distance traveled as a function of time for a single-motor domain labeled with a 20-nm gold particle at its N-terminus and the corresponding 2D-trajectory, which is shown to the right of the time trace. The *arrows* indicate ~40 nm off-axis side position of the unbound motor domain (transient state). *Inset*: schematic of myosin-5a labeling and movement. Imaging speed: 1000 frames per second. (B) Contour map of a two-dimensional histogram with a 10×10-nm^2 bin width obtained from the transient state of the unbound head *x–y* trajectories (number of steps = 486). All contributing steps were aligned and those to the right of the filament when viewed in the direction of motion were mirrored. (C) The data lead to a hand-over-hand spinning model (shown schematically), in which each step proceeds via a single, spatially constrained transient state of the detached head (Andrecka et al., 2015).

4.2 Microtubule Motors

4.2.1 The Kinesin-1 Mechanochemical Cycle

The kinesin superfamily includes 45 genes in humans and is responsible for a huge diversity of tasks in the cell, including vesicle transport, microtubule dynamics control, and mitotic spindle organization (Hirokawa & Takemura, 2004). We recently applied iSCAT to solve long-standing questions regarding kinesin mechanochemistry by directly visualizing the dynamics of each step of the canonical kinesin-1 motor (Mickolajczyk et al., 2015). Attaching 30 nm gold nanoparticle to a single-motor domain of processive kinesin-1 dimers (Fig. 5A) enabled us to follow the motion of the motor over time. By identifying and quantifying the amount of time that

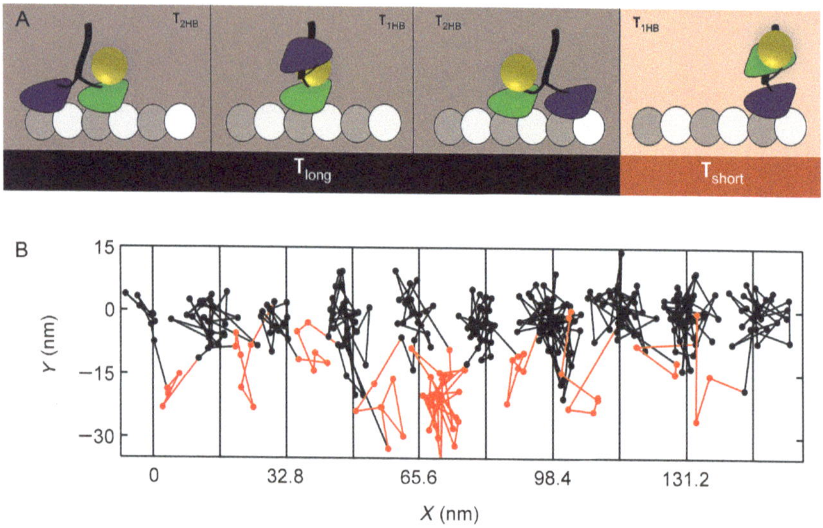

Fig. 5 Example results of kinesin tracking with iSCAT. (A) Schematic showing detectable transitions in the two-step cycle of kinesin with only one motor domain tagged. The entire step (including both one- and two-head-bound durations) of the unlabeled head as well as the two-heads-bound duration of the labeled head precedes the first translocation of the bead (*black*), and the one-head-bound duration of the labeled head precedes the second translocation of the bead (*red*). (B) Example *XY* trajectory for kinesin-1 at saturating ATP. The amount of time that the motor spends on the microtubule and off the microtubule is easily determined. *Black* and *red* coloring corresponds to (A). *Vertical lines* show available microtubule-binding sites. Imaging speed: 1000 frames per second (Mickolajczyk et al., 2015).

kinesin-1 spent with one head bound vs two heads bound at various ATP concentrations, we determined the order and rates of mechanochemical transitions (Fig. 5B). We found the ATP-waiting state to be most similar to a two-heads-bound state, and the one-head-bound state duration was found to be ATP independent with a lifetime of about 8 ms. By using the slowly hydrolyzable ATP analog ATPγS, we further determined that hydrolysis precedes exit from the one-head-bound state, consistent with recent optical trapping studies (Milic, Andreasson, Hancock, & Block, 2014).

4.2.2 Dynein

We also applied iSCAT tracking to the microtubule motor, cytoplasmic dynein from *S. cerevisiae*, brewer's yeast. In contrast to kinesin-1, dynein walks toward the minus end of microtubules (Paschal & Vallee, 1987) and has an irregular stepping pattern, taking both forward and backward steps of variable size as well as moving laterally, switching between different protofilaments that make up the microtubule wall (Reck-Peterson et al., 2006). Dynein has been shown to use both a hand over-hand stepping mechanism, where the leading head alternates with each step, and an inchworm-like mechanism where either the front or rear head steps without passing the other head (DeWitt et al., 2012; Qiu et al., 2012). iSCAT provides the opportunity to characterize in detail this unusual and highly variable stepping mechanism and the events that take place during each step.

Previous single-molecule fluorescence tracking studies of dynein with TIRF microscopy used low ATP concentrations (between 4 and 12 μM) to slow the stepping rate sufficiently to enable detection of individual steps (DeWitt et al., 2012; Qiu et al., 2012; Reck-Peterson et al., 2006). The high spatial and temporal resolution of iSCAT enabled us to track dynein at near saturating ATP concentrations (100 μM), which results in a mean velocity of 36.7 ± 20.4 nm/s.

As described in the protocol, we labeled yeast cytoplasmic dynein with 20–40 nm gold nanoparticles. As in earlier studies (DeWitt et al., 2012; Qiu et al., 2012), we observed a wide distribution of step sizes when we tracked the position of the particle on the dynein motor domain (Fig. 6). With the high spatial and temporal resolution of iSCAT, 16 and 24 nm steps were easily resolved as clusters of points in the lateral position of the tracked gold nanoparticle (Fig. 6), as well as smaller 8 nm steps. Yeast cytoplasmic dynein is highly processive (Gennerich, Carter, Reck-Peterson, & Vale, 2007), and we frequently observed runs >2 μm during the entire 40 s period of recording.

A prominent feature of many of the recordings was large off-axis motion, seen as changes perpendicular to the microtubule axis greater than 50 nm, followed by a return to the original axis of motion (Fig. 6). In some cases dynein appeared to transition between sequences of straight steps and off-axis steps. Off-axis steps were often correlated with substantial changes in iSCAT contrast, presumably due to z-axis motions with protofilament switching. Conversely, traces without off-axis stepping tended to have more constant contrast levels (not shown).

Our results indicate that during processive stepping along microtubules, dynein moves not only parallel to the microscope slide (x and y motions) but also closer and farther away from the slide (in z), consistent with changes to microtubule protofilament and azimuthal position around the microtubule.

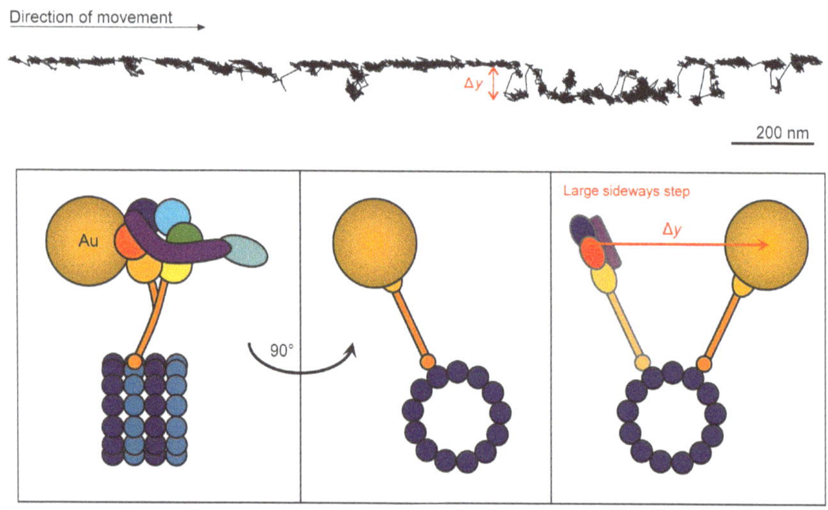

Fig. 6 iSCAT tracking of dynein. Example *x–y* plot of cytoplasmic dynein position along a microtubule exhibiting side deviations from the track (Δ*y*). The schematic presents one possible explanation of the observed trace where the dynein molecule switches microtubule protofilaments. Dynein was tracked at 200 frames per second using a 20-nm gold nanoparticle bound to AAA domains 5 and 6 (indicated by *orange* and *red circles*, respectively).

Further iSCAT studies should be valuable toward understanding the detailed motions of dynein's puzzling transport mechanism.

5. OUTLOOK

We have shown how iSCAT coupled with nanoparticle labels can be used to produce nm-precise and highly time-resolved tracks of several processive motor proteins moving along their tracks in vitro. In addition to the capabilities presented here, there are numerous opportunities for both improving the capabilities of iSCAT for SPT and broadening the scope of iSCAT more generally.

In terms of improving the capabilities for SPT, there are three relatively straightforward avenues: (i) increasing localization precision for smaller labels; (ii) increasing imaging speed; and (iii) 3D tracking. Label size is important for two reasons. First, as imaging speed increases beyond the ms threshold, the observed motion will increasingly be affected by the drag or fluctuations of the bead. Thus, even though nominal localization precision may be on the order of a few nm, the positional fluctuations of the particle will be much larger, as the motions of the label itself are resolved. This effect is minimized for smaller labels, both because the positional fluctuations themselves become smaller and because the label moves more rapidly, causing any such fluctuations to average out until much higher imaging speeds are achieved. This distinction is also important when comparing dark field with interferometric scattering (Isojima et al., 2016), which have similar theoretical SNRs for the same incident power and scattering properties (Ortega-Arroyo & Kukura, 2012). Even though the nominal spatiotemporal precision can be very high (Isojima et al., 2016), large particle sizes require time averaging to remove localization noise caused by label motion, lowering the effective achievable localization precision and imaging speed. Thus, achieving true sub-ms and nm-level precision simultaneously will invariably require the use of labels with diameters of 20 nm or smaller.

Ideally, labels on the order of single fluorescent proteins would be desirable, with the smallest possible linker length. For instance, the high spatiotemporal precision for low incident power for 20 nm labels shown in Fig. 1C would translate into similar imaging performance with 5 nm silver labels and a moderate increase of illumination intensity. The major challenge associated with using labels with diameters on the order of 10 nm or less is purely one of convenience—they become difficult to detect outright due to the rough imaging background generated by microscope cover glass, however, in dynamic measurements where labeled motors bind to their tracks, move, and unbind, this background can be efficiently subtracted, paving the way for future studies with smaller labels and higher imaging speeds without any loss in localization precision. The ultimate limits are likely to be determined by light absorption and the resulting sample heating, although this could be alleviated by pulsed excitation, for example.

Beyond label size, extending localization from two to three dimensions would be valuable for many applications. There is precedent for using iSCAT for 3D tracking, although the precise calibration of the 3D response has generally been challenging and a clear methodology has yet to be developed (Andrecka et al., 2015; Goldfain, Garmann, Jin, Lahini, & Manoharan, 2016; Krishnan, Mojarad, Kukura, & Sandoghdar, 2010). The interferometric nature of the technique, however, lends itself well to very high localization precision along the optical axis on the length scale of a few tens of nm.

In many cases, it is highly advantageous to perform iSCAT and fluorescence experiments simultaneously (Andrecka et al., 2015; Kukura et al., 2009). The compatibility of iSCAT with single-molecule fluorescence imaging makes it possible to use well-established, fluorescent assays and combine them with the advantages of iSCAT imaging. Furthermore, multilabel experiments, where simultaneous tracking is performed, can be achieved with an iSCAT and fluorescence imaging modality (Andrecka et al., 2015). Such approaches could be of particular value for initial experiments on cell surfaces, which may be possible on cellular features that are particularly thin and produce little background scatter, such as filopodia, as opposed to cell bodies.

Finally, it is worth noting that there are ample opportunities for using iSCAT in a completely label-free manner. We initially demonstrated label-free tracking of myosin-5a (Ortega Arroyo et al., 2014), but such approaches could be of particular interest also for dynein tracking, which has a greater molecular weight and thereby produces a stronger iSCAT signal. The advantages here are that the center of mass of the protein can be tracked with very long observation times.

ACKNOWLEDGMENTS

P.K. is supported by an ERC starting grant (NanoScope). J.A. was supported by a Marie Curie Fellowship (330215). K.J.M. and W.O.H. were supported by the NIH R01 GM076476. L.G.L. and Y.E.G. were supported by NIH Grants P01 GM087253 and R01 GM086352. L.G.L was supported by the NIH training grant T32- GM008275. J.R.S. and Y.T. were supported by funds from the NHLBI Intramural Research Program.

REFERENCES

Allen, R. D., Metuzals, J., Tasaki, I., Brady, S. T., & Gilbert, S. P. (1982). Fast axonal-transport in squid giant-axon. *Science, 218*(4577), 1127–1129.

Andreasson, J. O. L., Milic, B., Chen, G. Y., Guydosh, N. R., Hancock, W. O., & Block, S. M. (2015). Examining kinesin processivity within a general gating framework. *eLife, 4*, e07403.

Andrecka, J., Arroyo, J. O., Lewis, K., Cross, R. A., & Kukura, P. (2016). Label-free imaging of microtubules with sub-nm precision using interferometric scattering microscopy. *Biophysical Journal, 110*(1), 214–217.

Andrecka, J., Ortega Arroyo, J., Takagi, Y., de Wit, G., Fineberg, A., MacKinnon, L., … Kukura, P. (2015). Structural dynamics of myosin 5 during processive motion revealed by interferometric scattering microscopy. *eLife, 4*, e05413.

Arroyo, J. O., Cole, D., & Kukura, P. (2016). Interferometric scattering microscopy and its combination with single-molecule fluorescence imaging. *Nature Protocols, 11*(4), 617–633.

Arroyo, J. O., & Kukura, P. (2016). Non-fluorescent schemes for single-molecule detection, imaging and spectroscopy. *Nature Photonics, 10*(1), 11–17.

Beausang, J. F., Shroder, D. Y., Nelson, P. C., & Goldman, Y. E. (2013). Tilting and wobble of myosin V by high-speed single-molecule polarized fluorescence microscopy. *Biophysical Journal, 104*, 1263–1273.

Block, S. M., Goldstein, L. S. B., & Schnapp, B. J. (1990). Bead movement by single kinesin molecules studied with optical tweezers. *Nature, 348*(6299), 348–352.

Chen, Y., Deffenbaugh, N. C., Anderson, C. T., & Hancock, W. O. (2014). Molecular counting by photobleaching in protein complexes with many subunits: Best practices and

application to the cellulose synthesis complex. *Molecular Biology of the Cell*, *25*(22), 3630–3642.
Curtis, A. S. G. (1964). Mechanism of adhesion of cells to glass—Study by interference reflection microscopy. *Journal of Cell Biology*, *20*(2), 199–215.
Deffenbaugh, N. C., Chen, Y. L., Anderson, C. T., & Hancock, W. O. (2015). Novel step detection algorithms for photobleaching analysis of protein complexes with many subunits. *Biophysical Journal*, *108*(2), 438a.
DeWitt, M. A., Chang, A. Y., Combs, P. A., & Yildiz, A. (2012). Cytoplasmic dynein moves through uncoordinated stepping of the AAA+ ring domains. *Science*, *335*(6065), 221–225.
Dulin, D., Barland, S., Hachair, X., & Pedaci, F. (2014). Efficient illumination for microsecond tracking microscopy. *PLoS One*, *9*(9), e107335.
Dunn, A. R., & Spudich, J. A. (2007). Dynamics of the unbound head during myosin V processive translocation. *Nature Structural & Molecular Biology*, *14*, 246–248.
Dunn, A. R., & Spudich, J. A. (2011). Single-molecule gold-nanoparticle tracking. *Cold Spring Harbor Protocols*, *2011*(12), 1498–1506.
Filler, T. J., & Peuker, E. T. (2000). Reflection contrast microscopy (RCM): A forgotten technique? *Journal of Pathology*, *190*(5), 635–638.
Forkey, J. N., Quinlan, M. E., Shaw, M. A., Corrie, J. E., & Goldman, Y. E. (2003). Three-dimensional structural dynamics of myosin V by single-molecule fluorescence polarization. *Nature*, *422*, 399–404.
Funatsu, T., Harada, Y., Tokunaga, M., Saito, K., & Yanagida, T. (1995). Imaging of single fluorescent molecules and individual ATP turnovers by single myosin molecules in aqueous-solution. *Nature*, *374*(6522), 555–559.
Gennerich, A., Carter, A. P., Reck-Peterson, S. L., & Vale, R. D. (2007). Force-induced bidirectional stepping of cytoplasmic dynein. *Cell*, *131*(5), 952–965.
Goldfain, A. M., Garmann, R. F., Jin, Y., Lahini, Y., & Manoharan, V. N. (2016). Dynamic measurements of the position, orientation, and DNA content of individual unlabeled bacteriophages. *Journal of Physical Chemistry B*, *120*(26), 6130–6138.
Hirokawa, N., & Takemura, R. (2004). Kinesin superfamily proteins and their various functions and dynamics. *Experimental Cell Research*, *301*(1), 50–59. http://dx.doi.org/10.1016/j.yexcr.2004.08.010.
Horio, T., & Hotani, H. (1986). Visualization of the dynamic instability of individual microtubules by dark-field microscopy. *Nature*, *321*(6070), 605–607.
Howard, J., Hudspeth, A. J., & Vale, R. D. (1989). Movement of microtubules by single kinesin molecules. *Nature*, *342*(6246), 154–158.
Isojima, H., Iino, R., Niitani, Y., Noji, H., & Tomishige, M. (2016). Direct observation of intermediate states during the stepping motion of kinesin-1. *Nature Chemical Biology*, *12*(4), 290–297.
Johnson, V., Ayaz, P., Huddleston, P., & Rice, L. M. (2011). Design, overexpression, and purification of polymerization-blocked yeast alpha beta-tubulin mutants. *Biochemistry*, *50*(40), 8636–8644.
Kalafut, B., & Visscher, K. (2008). An objective, model-independent method for detection of non-uniform steps in noisy signals. *Computer Physics Communications*, *179*(10), 716–723.
Karagiannis, P., Ishii, Y., & Yanagida, T. (2014). Molecular machines like myosin use randomness to behave predictably. *Chemical Reviews*, *114*, 3318–3334.
Kerssemakers, J. W. J., Munteanu, E. L., Laan, L., Noetzel, T. L., Janson, M. E., & Dogterom, M. (2006). Assembly dynamics of microtubules at molecular resolution. *Nature*, *442*(7103), 709–712.
Krishnan, M., Mojarad, N., Kukura, P., & Sandoghdar, V. (2010). Geometry-induced electrostatic trapping of nanometric objects in a fluid. *Nature*, *467*(7316), 692–695.

Kukura, P., Ewers, H., Muller, C., Renn, A., Helenius, A., & Sandoghdar, V. (2009). High-speed nanoscopic tracking of the position and orientation of a single virus. *Nature Methods*, *6*(12), 923–927.
Kulman, J. D., Satake, M., & Harris, J. E. (2007). A versatile system for site-specific enzymatic biotinylation and regulated expression of proteins in cultured mammalian cells. *Protein Expression and Purification*, *52*(2), 320–328.
Lin, Y. H., Chang, W. L., & Hsieh, C. L. (2014). Shot-noise limited localization of single 20 nm gold particles with nanometer spatial precision within microseconds. *Optics Express*, *22*, 9159–9170.
Mehta, A. D., Rock, R. S., Rief, M., Spudich, J. A., Mooseker, M. S., & Cheney, R. E. (1999). Myosin-V is a processive actin-based motor. *Nature*, *400*, 590–593.
Mickolajczyk, K. J., Deffenbaugh, N. C., Arroyo, J. O., Andrecka, J., Kukura, P., & Hancock, W. O. (2015). Kinetics of nucleotide-dependent structural transitions in the kinesin-1 hydrolysis cycle. *Proceedings of the National Academy of Sciences of the United States of America*, *112*(52), E7186–E7193.
Milescu, L. S., Yildiz, A., Selvin, P. R., & Sachs, F. (2006a). Extracting dwell time sequences from processive molecular motor data. *Biophysical Journal*, *91*(9), 3135–3150.
Milescu, L. S., Yildiz, A., Selvin, P. R., & Sachs, F. (2006b). Maximum likelihood estimation of molecular motor kinetics from staircase dwell-time sequences. *Biophysical Journal*, *91*(4), 1156–1168.
Milic, B., Andreasson, J. O. L., Hancock, W. O., & Block, S. M. (2014). Kinesin processivity is gated by phosphate release. *Proceedings of the National Academy of Sciences of the United States of America*, *111*(39), 14136–14140.
Mullner, F. E., Syed, S., Selvin, P. R., & Sigworth, F. J. (2010). Improved hidden Markov models for molecular motors, part 1 basic theory. *Biophysical Journal*, *99*(11), 3684–3695.
Neubeck, A., & Van Gool, L. (2006). Efficient non-maximum suppression. In *Proceedings of the 18th international conference on pattern recognition, Vol. 3*. (pp. 850–855).
Ortega Arroyo, J., Andrecka, J., Spillane, K. M., Billington, N., Takagi, Y., Sellers, J. R., & Kukura, P. (2014). Label-free, all-optical detection, imaging, and tracking of a single protein. *Nano Letters*, *14*, 2065–2070.
Ortega-Arroyo, J., & Kukura, P. (2012). Interferometric scattering microscopy (iSCAT): New frontiers in ultrafast and ultrasensitive optical microscopy. *Physical Chemistry Chemical Physics*, *14*(45), 15625–15636.
Ozeki, T., Verma, V., Uppalapati, M., Suzuki, Y., Nakamura, M., Catchmark, J. M., & Hancock, W. O. (2009). Surface-bound casein modulates the adsorption and activity of kinesin on SiO2 surfaces. *Biophysical Journal*, *96*(8), 3305–3318.
Pardee, J. D., & Spudich, J. A. (1982). Purification of muscle actin. *Methods in Cell Biology*, *24*, 271–289.
Paschal, B. M., & Vallee, R. B. (1987). Retrograde transport by the microtubule-associated protein MAP 1C. *Nature*, *330*(6144), 181–183.
Piliarik, M., & Sandoghdar, V. (2014). Direct optical sensing of single unlabelled proteins and super-resolution imaging of their binding sites. *Nature Communications*, *5*, 4495.
Qiu, W., Derr, N. D., Goodman, B. S., Villa, E., Wu, D., Shih, W., & Reck-Peterson, S. L. (2012). Dynein achieves processive motion using both stochastic and coordinated stepping. *Nature Structural & Molecular Biology*, *19*(2), 193–200.
Reck-Peterson, S. L., Yildiz, A., Carter, A. P., Gennerich, A., Zhang, N., & Vale, R. D. (2006). Single-molecule analysis of dynein processivity and stepping behavior. *Cell*, *126*(2), 335–348.
Rief, M., Rock, R. S., Mehta, A. D., Mooseker, M. S., Cheney, R. E., & Spudich, J. A. (2000). Myosin-V stepping kinetics: A molecular model for processivity. *Proceedings of the National Academy of Sciences of the United States of America*, *97*, 9482–9486.

Ruhnow, F., Zwicker, D., & Diez, S. (2011). Tracking single particles and elongated filaments with nanometer precision. *Biophysical Journal*, *100*(11), 2820–2828.

Sakamoto, T., Amitani, I., Yokota, E., & Ando, T. (2000). Direct observation of processive movement by individual myosin V molecules. *Biochemical and Biophysical Research Communications*, *272*(2), 586–590.

Sakamoto, T., Webb, M. R., Forgacs, E., White, H. D., & Sellers, J. R. (2008). Direct observation of the mechanochemical coupling in myosin Va during processive movement. *Nature*, *455*(7209), 128–132.

Schatz, P. J. (1993). Use of peptide libraries to map the substrate-specificity of a peptide-modifying enzyme—A 13 residue consensus peptide specifies biotinylation in Escherichia coli. *Bio/Technology*, *11*(10), 1138–1143.

Schnapp, B. J., Gelles, J., & Sheetz, M. P. (1988). Nanometer-scale measurements using video light-microscopy. *Cell Motility and the Cytoskeleton*, *10*(1–2), 47–53.

Schneider, R., Glaser, T., Berndt, M., & Diez, S. (2013). Using a quartz paraboloid for versatile wide-field TIR microscopy with sub-nanometer localization accuracy. *Optics Express*, *21*(3), 3523–3539.

Schneider, R., Korten, T., Walter, W. J., & Diez, S. (2015). Kinesin-1 motors can circumvent permanent roadblocks by side-shifting to neighboring protofilaments. *Biophysical Journal*, *108*(9), 2249–2257.

Sellers, J. R., & Veigel, C. (2006). Walking with myosin V. *Current Opinion in Cell Biology*, *18*, 68–73.

Snyder, G. E., Sakamoto, T., Hammer, J. A., 3rd., Sellers, J. R., & Selvin, P. R. (2004). Nanometer localization of single green fluorescent proteins: Evidence that myosin V walks hand-over-hand via telemark configuration. *Biophysical Journal*, *87*(3), 1776–1783.

Svoboda, K., Schmidt, C. F., Schnapp, B. J., & Block, S. M. (1993). Direct observation of kinesin stepping by optical trapping interferometry. *Nature*, *365*(6448), 721–727.

Syed, S., Mullner, F. E., Selvin, P. R., & Sigworth, F. J. (2010). Improved hidden Markov models for molecular motors, part 2 extensions and application to experimental data. *Biophysical Journal*, *99*(11), 3696–3703.

Thompson, R. E., Larson, D. R., & Webb, W. W. (2002). Precise nanometer localization analysis for individual fluorescent probes. *Biophysical Journal*, *82*(5), 2775–2783.

Turner, D., Chang, C. Y., Fang, K., Cuomo, P., & Murphy, D. (1996). Kinesin movement on glutaraldehyde-fixed microtubules. *Analytical Biochemistry*, *242*(1), 20–25.

Vale, R. D. (2003). Myosin V motor proteins: Marching stepwise towards a mechanism. *Journal of Cell Biology*, *163*(3), 445–450.

Vale, R. D., Schnapp, B. J., Mitchison, T., Steuer, E., Reese, T. S., & Sheetz, M. P. (1985). Different axoplasmic proteins generate movement in opposite directions along microtubules in vitro. *Cell*, *43*(3), 623–632.

Veigel, C., Wang, F., Bartoo, M. L., Sellers, J. R., & Molloy, J. E. (2002). The gated gait of the processive molecular motor, myosin V. *Nature Cell Biology*, *4*, 59–65.

Wang, F., Chen, L., Arcucci, O., Harvey, E. V., Bowers, B., Xu, Y., … Sellers, J. R. (2000). Effect of ADP and ionic strength on the kinetic and motile properties of recombinant mouse myosin V. *Journal of Biological Chemistry*, *275*, 4329–4335.

Warshaw, D. M., Kennedy, G. G., Work, S. S., Krementsova, E. B., Beck, S., & Trybus, K. M. (2005). Differential labeling of myosin V heads with quantum dots allows direct visualization of hand-over-hand processivity. *Biophysical Journal*, *88*, L30–L32.

Yildiz, A., Forkey, J. N., McKinney, S. A., Ha, T., Goldman, Y. E., & Selvin, P. R. (2003). Myosin V walks hand-over-hand: Single fluorophore imaging with 1.5-nm localization. *Science*, *300*, 2061–2065.

Yildiz, A., Tomishige, M., Vale, R. D., & Selvin, P. R. (2004). Kinesin walks hand-over-hand. *Science*, *303*(5658), 676–678.

CHAPTER EIGHTEEN

Enzyme Kinetics in Femtoliter Arrays

P. Mogalisetti, D.R. Walt[1]
Tufts University, Medford, MA, United States
[1]Corresponding author: e-mail address: David.Walt@tufts.edu

Contents

Abstract

Over the last decade, femtoliter arrays have been used as a simple and robust way to encapsulate and monitor the kinetics of single enzyme molecules. Encapsulating individual enzyme molecules within a femtoliter-sized reaction chamber does not require immobilization of the enzyme molecules or fluorescent tagging of the enzyme molecules, which offers the unique advantage of observing unmodified single enzyme molecules free in solution. Several fascinating details about enzyme kinetics have been revealed using these femtoliter arrays, which were unattainable from traditional ensemble experiments. Here, we discuss various considerations to take into account when developing single-molecule enzyme assays in femtoliter arrays and the advantages and disadvantages of various protocols.

1. INTRODUCTION AND BACKGROUND

The advent and development of single-molecule measurements over the last 2 decades have enabled a new understanding of the dynamics and kinetics of enzymatic catalysis (Liebherr & Gorris, 2014; Lu, 2014; Lu, Xun, & Xie, 1998). Two fluorescence-based methods for observing these

Methods in Enzymology, Volume 581
ISSN 0076-6879
http://dx.doi.org/10.1016/bs.mie.2016.08.017
© 2016 Elsevier Inc.
All rights reserved.

reactions are typically used to detect and study single enzyme molecule catalysis. One method is based on the ability to observe directly single fluorophore molecules. A fluorescent-tagged enzyme or a fluorescent enzymatic substrate/product is used in conjunction with an immobilized enzyme on a surface to observe the single enzyme molecules and their interaction with substrate using either total internal reflection microscopy or confocal microscopy. This method allows researchers to study millisecond time-scale dynamics of enzyme activity and conformational dynamics of the enzyme molecules (Chen, Hu, Vorpagel, & Lu, 2003; De Cremer et al., 2007; Funatsu, Harada, Tokunaga, Saito, & Yanagida, 1995; Hatzcbeakis et al., 2006; Kuznetsova et al., 2008; Lerch, Rigler, & Mikhailov, 2005; Min et al., 2005; Santoso et al., 2010; Wang & Lu, 2010). In this method, single fluorophore (and hence single enzyme molecule) detection is achieved by reducing the observation volume (the region of sample that is excited), which eliminates the background signal from the bulk solution while maximizing the collection of photons from the fluorophore.

An alternative method is to reduce the observation volume physically by trapping single enzyme molecules in femtoliter reaction containers. In this method, instead of direct detection of fluorescent-tagged single enzyme molecules, indirect detection is performed by measuring the amount and rate of fluorescent product generated due to enzymatic catalysis. The background signal is eliminated because the enzyme is physically separated from the bulk solution and the fluorescent product does not get diluted by the bulk solution. Typically, a large array of femtoliter-sized reaction containers is used to trap single molecules by simply using a highly dilute enzyme solution. When the number of reaction containers is much greater than the number of enzyme molecules, the probability of trapping more than one enzyme in a container is very low. The single enzyme molecule within the container catalyzes many turnovers of a fluorogenic substrate to a fluorescent product that builds up to a detectable concentration in the small reaction volume. This method of single-molecule enzyme isolation has the advantage of observing enzyme molecules that are free in solution as well as the ability to observe the enzyme molecules for long periods of time, which is difficult with most optical methods because of photobleaching. In addition, the high-throughput array format enables many single enzyme molecules to be measured simultaneously.

Arrays of small reaction containers (zeptoliter–femtoliter) have been generated using optical fiber bundles (Rissin & Walt, 2006), polydimethylsiloxane

(PDMS) (Rondelez et al., 2005), fused silica slides (Ehrl, Liebherr, & Gorris, 2013; Liebherr, Renner, & Gorris, 2014), water in oil droplets (Arayanarakool, Shui, Kengen, van den Berg, & Eijkel, 2013), and electrophoresis capillaries (Craig & Nichols, 2008) among other approaches (Piwonski, Goomanovsky, Bensimon, Horovitz, & Haran, 2012). In these systems, single enzyme molecules are encapsulated within the small reaction containers using a dilute enzyme solution such that the number of reaction containers is much larger than the number of enzyme molecules present. Poisson statistics describe the probability of occurrence of a rare event in a large number of trials and are used to estimate the probability of finding an enzyme within the reaction chambers (Fig. 1). The equation $P_\mu(x) = e^{-\mu}\frac{\mu^x}{x!}$, gives the probability $P_\mu(x)$ of finding x enzyme molecules in a well given the ratio of enzyme molecules to wells is μ. In order to reduce the probability of trapping more than one enzyme molecule in a reaction container, the ratio of enzyme molecules to wells μ should be low (<0.1). A low value of μ also means that the majority of the reaction chambers are empty and only a small fraction of the containers encapsulate single enzyme molecules. As a result, a large number of reaction containers are required to ensure that a statistically significant number of single enzyme molecules are encapsulated and monitored. Therefore, the ability to generate a large number of reaction chambers in a high-throughput manner is essential for single-molecule studies, as demonstrated in the methods mentioned earlier. In this chapter, we will focus on the methods for developing single enzyme assays using the optical fiber bundle platform.

Microwells etched onto the tip of optical fiber bundles have been used for a wide variety of applications (Gorris, Blicharz, & Walt, 2007). The optical fibers have a core and a cladding with different refractive indices as well as different etching properties. As a result, when the fiber bundles are submerged in an acid solution, the core etches faster than the cladding, resulting in a femtoliter-sized cylindrical well at the tip of each fiber (Pantano & Walt, 1996). Using the high-density femtoliter array etched onto the cross section of optical fiber bundles, several interesting features about enzyme catalysis have been uncovered over the last decade. For example, individual enzyme molecules from the same enzyme population were shown to have distinct activities from each other in earlier literature (Craig, Arriaga, Wong, Lu, & Dovichi, 1996; Craig & Dovichi, 1998; Nichols, Shadabi, & Craig, 2009; Xue & Yeung, 1995). However, these experiments only examined a few enzyme molecules at a time, making it difficult to measure a

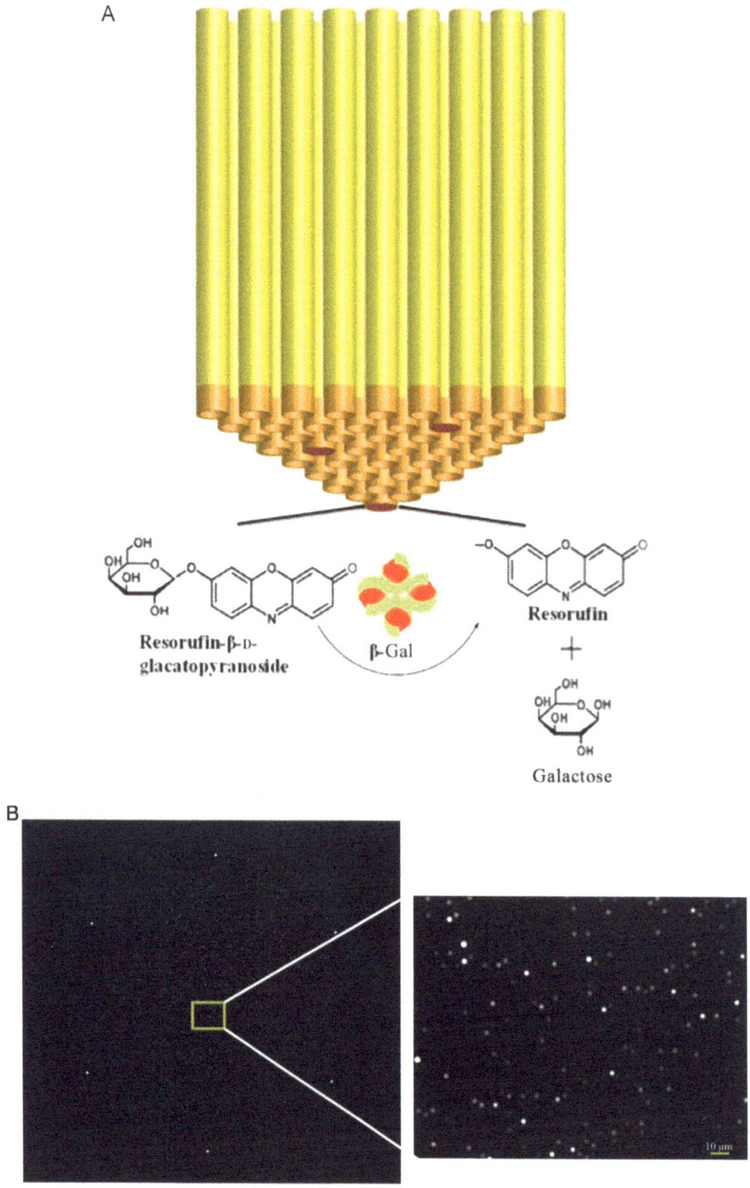

Fig. 1 (A) Schematic showing array of femtoliter wells at the tip of an optical fiber bundle. To isolate single enzyme molecules in the femtoliter array, the array is filled with an extremely dilute enzyme solution such that the probability of encapsulating more than one enzyme per well is low. The ratio of the number of enzyme molecules to wells is maintained at 0.1 or lower. The distribution of enzyme molecules in the wells can be described by using Poisson statistics. The probability of a well containing more than one enzyme molecule can be estimated using the Poisson distribution $P_\mu(x) = e^{-\mu}\frac{\mu^x}{x!}$, where $P_\mu(x)$ is the probability of a well containing x enzyme molecules when the ratio of enzyme molecules to wells is μ. For example at $\mu = 0.01$ (i.e. one enzyme for every 100 wells), the probabilities of a well containing 0, 1 and more than one enzyme molecule are $P_{0.01}(0) = 0.99$, $P_{0.01}(1) = 0.0099$, and $P_{0.01}(>1) = 1 - P_{0.01}(1) - P_{0.01}(0) = 4.9 \times 10^{-5}$, respectively. Therefore, 99% of the wells do not have any enzyme in them and ~1% of the wells have 1 enzyme (B).

statistically significant number of enzyme molecules. Using the high-density femtoliter arrays, the activities of hundreds of individual β-galactosidase molecules were measured simultaneously for extended times in a single experiment. The activities of the enzyme molecules were found to be different from each other and individual enzyme molecules retained their distinct activity over long periods of time (Rissin, Gorris, & Walt, 2008). The activity of purified β-galactosidase molecules varied as much as 10-fold. Two possible reasons for this observed variation have been proposed: conformational changes between individual enzyme molecules or chemical variations in the enzyme, such as variations in primary sequence due to errors in transcription and translation or oxidative damage. Recently, the conformation of single enzyme molecules was probed using thermal pulses (Rojek & Walt, 2014). Enzyme molecules were shown to randomly change their activity after a thermal pulse. Thermal pulses provide enough energy for an enzyme molecule to overcome an activation barrier and convert from one local energy minimum to another with a different random conformational state, resulting in a different activity. The femtoliter arrays were also used to study the inhibition of β-galactosidase (Gorris, Rissin, & Walt, 2007; Mogalisetti, Gorris, Rojek, & Walt, 2014) at the single-molecule level as well as to elucidate the mechanism of horseradish peroxidase (Gorris & Walt, 2009). The high density of the femtoliter arrays also enables accurate counting of the number of enzyme molecules in solution. Using this feature, the lab developed ultrasensitive digital ELISAs for detecting immunocomplexes for a variety of protein biomarkers (Gaylord, Abdul-Aziz, & Walt, 2015; Gaylord et al., 2015; Rissin et al., 2010; Schubert et al., 2015; Schubert, Walter, Manesse, & Walt, 2016; Warren et al., 2014; Wu, Milutinovic, & Walt, 2015). These single-molecule arrays (coined Simoa) are now commercialized by Quanterix Corporation (Kan et al., 2012; Wilson et al., 2016). Recently, we used the single-molecule counting technique to determine the stoichiometry of enzyme fragment complementation in β-galactosidase (Mogalisetti & Walt, 2015). In this chapter, we provide a detailed description of the protocols for generating the femtoliter arrays using optical fiber bundles, we discuss various factors to take into consideration while developing an assay to detect single enzyme molecules in the femtoliter array, and we provide a brief discussion on the data analysis.

2. PREPARATION OF FEMTOLITER ARRAY

Optical fiber-based femtoliter arrays are fabricated by acid etching optical fiber bundles purchased from Schott North America, Inc.

(Elmsford, NY). The silica-based optical fiber bundles consist of many individual optical fibers with micron-sized diameters bundled together. In our laboratory, fiber bundles consisting of 50,000 fibers of 4.8 μm diameter each are most commonly used for trapping single enzyme molecules. Due to the differential etching properties of the core and the cladding of the individual fibers, when the fiber bundle is submerged in an acid solution, the core etches faster than the cladding resulting in a well at the tip of each individual optical fiber, with the same diameter as the fiber core and the depth depending on the etching time. Etching the 4.8 μm diameter fiber bundles in 0.025N HCl for 1′55″ results in an array of 50,000 cylindrical wells with 4.8 μm diameter and 2.9 μm depth (46 fL volume).

Materials for femtoliter well array preparation

Fiber bundle (Schott North America, Inc.)
Allied High Tech Multiprep system polisher (item number 15-2000) and fiber bundle fixture (item number 40-31200)
Diamond lapping films (Allied High Tech)—30, 15, 6, 6, 3, and 1 μm
Two 20 mL scintillation vials
Magnetic stir bars
1N HCl (Sigma-Aldrich)
DI water

Protocol for femtoliter well array preparation

1. The fiber bundles are cut to approximately 5 cm in length.
2. The fiber bundles are polished on both sides on the polisher using diamond lapping films of consecutively decreasing roughness (30, 15, 6, 6, 3, and 1 μm) to generate a smooth surface on the distal end of the fiber bundle.
3. In a 20 mL scintillation vial, 19.5 mL DI water, and 0.5 mL HCl are mixed using a magnetic stir bar for 5 min.
4. The polished fiber bundles are dipped into the HCl solution for 1′55″.
5. The fiber bundles are immediately transferred into a scintillation vial filled with DI water and stirred for 1 min.
6. The fibers are air dried and stored until use.

3. SEALING METHODS

In order to observe single enzyme molecules within the femtoliter wells, it is essential to seal the femtoliter wells so that the individual enzyme molecules, as well as the contents of the enzymatic reaction (the substrate and product), are confined within the individual femtoliter wells. If the femtoliter wells are not sealed, an enzyme molecule can escape from the femtoliter well to the bulk sample solution or to a neighboring well, making it difficult to track the activity of that enzyme molecule. More often, even if the enzyme is still confined to the femtoliter well, the product of the reaction (usually a fluorophore indicating the extent of the reaction) can leak into the adjacent wells. Since the product concentration in the wells is measured to determine the activity of the enzyme, leaking of product into neighboring wells can result in an inaccurate estimate of enzyme activity.

To seal the contents of the enzyme reaction within the femtoliter wells, a mechanical sealing approach is commonly used (Rissin et al., 2008). A layer of silicone gasket is mechanically pressed against the femtoliter wells with substantial pressure, which is maintained for the duration of the experiment (Fig. 2). While mechanical sealing is straightforward, it limits the total duration of the experiment. The solution within the femtoliter wells can evaporate because the silicone allows water vapor to permeate. Consequently, silicone sealing limits the reaction time to 30 min or less.

For longer observation periods, an alternative method of sealing is to use a layer of oil on the femtoliter wells (Kathryn, Jason, Douglas, & David, 2015; Zhang, Nie, Etson, Wang, & Walt, 2012) (Fig. 3). First, the cladding of the femtoliter wells is modified with a hydrophobic layer. To coat only the cladding of the femtoliter wells and not the interior, a contact printing method was used to chemically attach n-octadecyltrichlorosilane (OTS) to the cladding. OTS-modified fibers are then incubated in the enzyme–substrate mixture to fill the wells. The fiber bundle is then immersed in a solution of oil and spun. The spinning generates a shear force between the oil and the fiber bundle, which removes the excess aqueous solution from the fiber bundle and results in a layer of oil on the wells (Fig. 3).

Materials for mechanical sealing

Etched fiber bundles

0.01 Nonreinforced gloss silicone sheeting (Specialty Manufacturing Inc.)

Glass slide

Z-direction platform

Microscope stage adaptor for the fiber bundle (custom made or Thorlabs part number SM1PT)

β-Galactosidase from *Escherichia coli* grade VII (Sigma) lyophilized powder dissolved to 2 μ*M* concentration in 1 × PBS, 1 m*M* $MgCl_2$ buffer, and 50% glycerol

Resorufin β-D-galactosidase (Sigma) dissolved in DMSO to 100 m*M* stock

1 × PBS, 1 m*M* $MgCl_2$, and pH 7.4

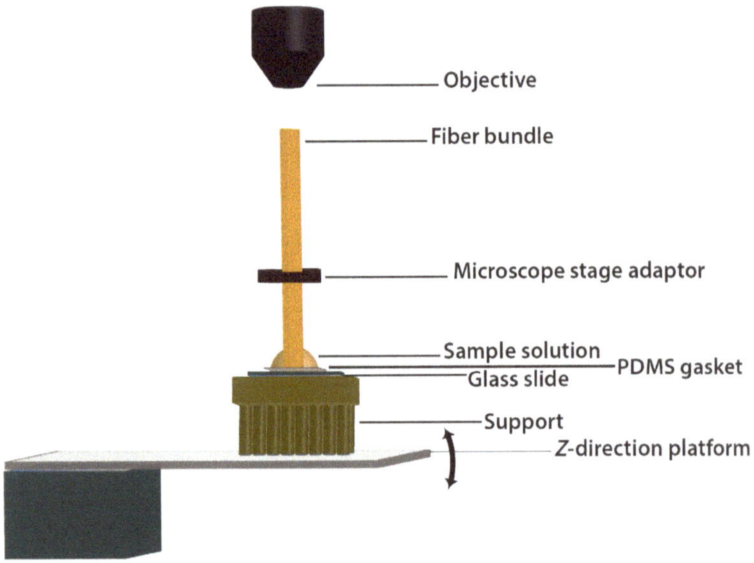

Fig. 2 The figure illustrates the mechanical sealing method to seal the enzymatic reaction contents in the optical fiber wells. A polished, etched, and surface modified (see Section 3 on surface modification of femtoliter wells) optical fiber bundle is firmly fixed to the microscope stage with the etched end facing away from the objective. A simple custom microscope stage adaptor that can firmly hold the fiber bundle is used for this purpose. The fiber bundle is brought into focus and centered in the field of view. A manually operated *z*-direction platform (vertical translation stage), which can travel at least 20 mm is placed beneath the microscope stage. A 25 μL drop of sample solution containing the enzyme and substrate is placed on a small piece of silicone sheet (PDMS gasket) that is bonded to a small glass slide, which is then placed on a small sample support block. This assembly is then placed on the *z*-direction platform. Mechanical sealing is achieved by simply translating the *z*-platform toward the fiber bundle. Once the fiber bundle is in contact with the PDMS, the *z*-platform is further gradually translated in the vertical direction to apply pressure on the PDMS gasket. The fiber bundle can be imaged immediately after sealing, after readjusting the focus.

Protocol for mechanical sealing of sample into femtoliter well array

1. The enzyme stock is diluted to 3.6 n*M* in the reaction buffer.
2. The substrate solution is prepared by diluting 1 μL of the substrate stock in 9 μL of DMSO to give 10 m*M* substrate solution. The 10 μL substrate solution is then diluted in 990 μL of reaction buffer.
3. The glass slide is rinsed with DI water and ethanol, dried with a blowgun and placed on the *z*-direction platform placed under the microscope stage.

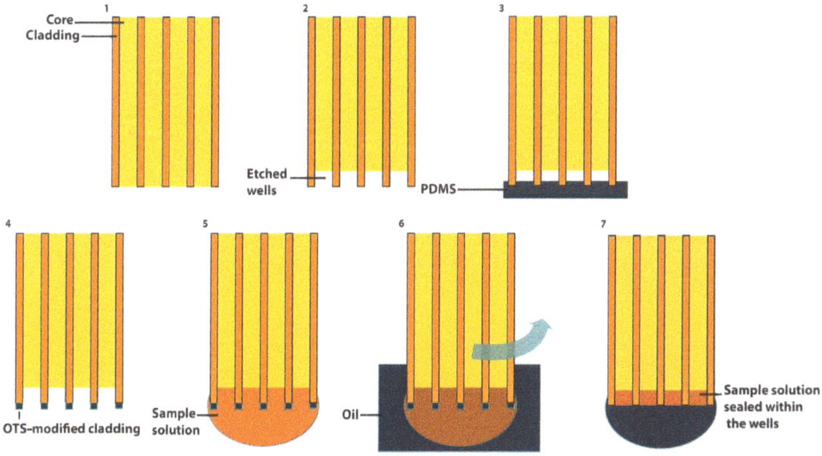

Fig. 3 This figure illustrates the oil sealing method to confine the enzymatic reaction contents within the optical fiber bundle wells (Zhang et al., 2012). (1) The core and the cladding of the optical fibers within the bundle. (2) After etching the fiber bundle in 0.025N HCl, the core etches faster than the cladding resulting in femtoliter-sized wells at the tips of the optical fibers. (3) OTS is spin coated onto a layer of PDMS, as explained in the protocol for oil sealing. The etched fiber bundle is then brought into contact with the PDMS layer for 1 min. (4) After stamping with OTS, only the cladding is hydrophobic while the interiors of the wells are still hydrophilic. (5) When the wells are filled with the enzyme–substrate solution, a drop of excess solution remains on the tip of the fiber bundle. (6) The fiber is then dipped into a reservoir of mineral oil and spun at high speed. Only the aqueous solution in the wells remains after spinning. (7) The oil forms a layer on the wells, making contacts with the cladding, thus trapping the aqueous solution within the wells. The fiber bundle can be removed from the oil reservoir and imaged on the microscope for several hours.

4. Approximate 1 cm^2 piece of silicone is rinsed thoroughly in soap and DI water and cleaned with lint free water to remove dust and residual water droplets. The silicone is then placed on the glass slide.
5. The fiber bundle is fixed to the microscope stage using an adaptor that can hold the fiber with the etched end facing down.
6. 1 μL of 3.6 n*M* enzyme is mixed with 1 mL of substrate solution. 25 μL of the enzyme–substrate solution is placed on the silicone sheet on the glass slide.
7. The *z*-direction stage is moved upward until the fiber is in contact with the enzyme solution followed by the PDMS gasket. Once the fiber is in contact with the PDMS, the *z*-direction stage is further moved upward to press against the fiber with substantial pressure (approximately 3.8×10^4 kg/m^2).
8. The proximal end of the fiber is brought into focus and imaged.

Materials for oil sealing

Etched fiber bundles
Glass slide
Slygard 184 silicone elastomer base and curing agent
n-Octadecyltrichlorosilane (OTS) (Sigma)
Anhydrous Toluene (Sigma)
Methanol (Sigma)
Enzyme–substrate mixture (prepared just before sealing)
Stir bar
Light mineral oil (Sigma)
Drill press
Adaptor to fix the fiber to the drill press

Protocol for oil sealing

1. 50 mL of PDMS prepolymer is prepared by mixing 10:1 ratio of elastomer base and curing agent, centrifuged to remove the bubbles, poured into a 150 mm diameter Petri dish and cured on a stable bench surface overnight.
2. The PDMS is placed in the oven at 80°C for 30 min and a small piece (approximately 0.5 cm × 4 cm) of PDMS is cut and bonded on to a glass slide.
3. A fresh solution of 2 m*M* OTS is prepared in anhydrous toluene by mixing 1 μL of OTS into 1.7 mL of toluene.
4. The OTS is immediately spin coated onto the PDMS followed by gentle air drying of the PDMS with a blow gun for 1 min.
5. The etched fiber bundle is immediately brought into contact with the spin-coated PDMS for 1 min. The contact angle is maintained at 90 degrees using a *z*-axis microscope stage to mount the fiber bundle.
6. The fiber bundle is mildly sonicated for 1 min in methanol and for 1 min in toluene.
7. The bundle is then dried using an air gun and then dried in the oven at 80°C for 20 min. The OTS functionalized fibers can be stored for at least 1 week for later use.
8. The OTS functionalized fiber bundle is first gently sonicated in the reaction buffer for 1 min, and then transferred to the enzyme–substrate solution for 5 min with stirring.
9. The fiber bundle is attached to a drill press using an adaptor and spun in light mineral oil for 30 s.
10. The oil-sealed fiber is then placed on the microscope and imaged.

To determine the quality of sealing—i.e. is there any leakage of product from 1 fL well to another—a fluorescence recovery after photobleaching type experiment can be performed (Fig. 4). The femtoliter wells are first filled with a fluorophore solution and sealed using either mechanical or

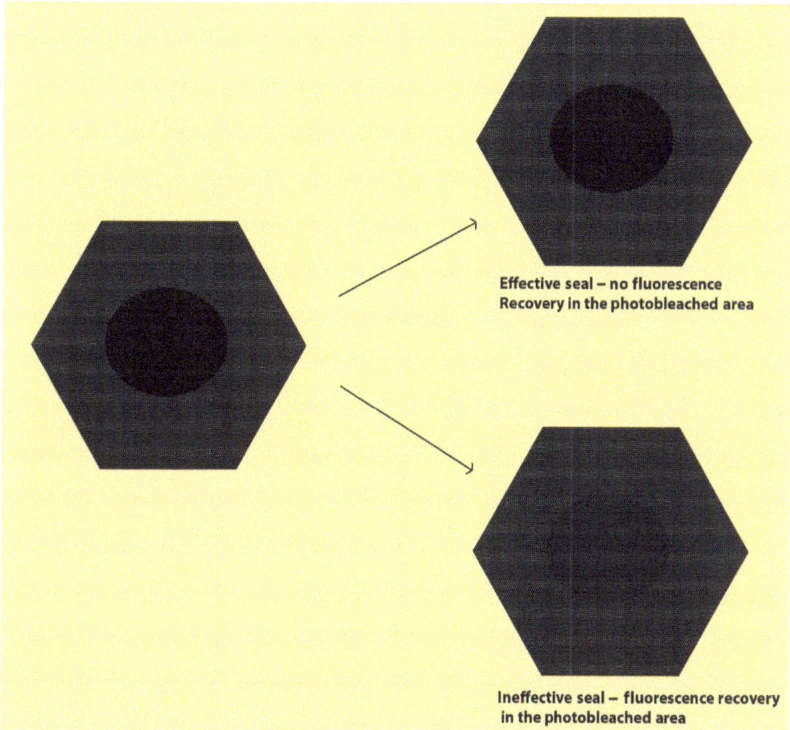

Fig. 4 The schematic illustrates a fluorescence recovery after photobleaching method to test the integrity of sealing either by mechanical or by oil sealing. A solution of fluorophore (for example, 10 μ*M* of resorufin) is first sealed within the fiber bundle wells. Using a field stop, a small region of the fiber bundle is exposed to excitation light for about 2 min until a well-defined photobleached area can be distinguished (bundle shown on *left*). After exposure, when the fiber bundle is imaged in the fluorescence channel after 30 min, the photobleached area still remains well defined and no increase in fluorescence is observed if the fiber bundle is well sealed. On the other hand, if the fiber bundle is not well sealed (either due to insufficient pressure in mechanical sealing or ineffective functionalization of fiber bundles with OTS in oil sealing), then the fluorescence intensity within the photobleached area gradually increases, indicating that the fluorophore is able to diffuse from the nonphotobleached area to the photobleached area.

oil sealing. The fiber bundle is mounted on the microscope and brought into focus. Then, using a field aperture iris, a small portion of the fiber bundle is photobleached by illuminating with the excitation wavelength for 30 min. As a result, there was less fluorescence signal from the photobleached part of the fiber bundle. The fluorescence of this region is then monitored over time using time-lapse imaging. If the fluorophore solution is properly sealed, then the bleached region remains dark; if the seal leaks, then the fluorescence of the photobleached area will increase over time due to the fluorophore leaking in from the nonbleached area.

4. SURFACE MODIFICATION OF FEMTOLITER WELLS

Enzyme molecules encapsulated within the optical fiber wells are exposed to the negatively charged glass surface of the femtoliter reaction well and to the PDMS or oil "seal" of the wells. Glass surfaces are notorious for inactivating or modifying the activities of enzyme molecules; therefore, passivating the femtoliter well surface is crucial to observe enzyme activities similar to bulk enzyme activities. Several passivation strategies can be adopted depending on the enzyme of interest and finding the optimal passivation method is often a trial-and-error process. The first passivation strategy is to include nonspecific binding agents such as BSA and Tween20 in the reaction buffer as long as they do not affect the activity of the enzyme (Gorris, Rissin, et al., 2007; Mogalisetti et al., 2014; Rissin et al., 2008). The second strategy is to suspend the femtoliter well array in a stirred solution of a blocking buffer such as protein-free blocking buffers or Superblock for an optimal amount of time followed by suspending in reaction buffer for 5 min to remove excess blocking agent (Kathryn et al., 2015). The final strategy is to use a covalent modification using a silane with an appropriate end group (Rojek & Walt, 2014). This strategy is particularly useful for thermal studies, where protein-based blocking can result in aggregation at high temperatures.

Materials for surface modification of femtoliter wells with silane

Etched fibers

Silane—(n-cyclohexylaminopropyl)trimethoxy silane, (3-aminopropyl) trimethoxy silane, or PEG silanes)

Anhydrous ethanol (Sigma)

Scintillation vials

Stir bars

Protocol for covalent modification of femtoliter wells with silane

1. The etched fibers are first treated with plasma for 5 min.
2. At the same time a 5% (v/v) solution of water in ethanol is prepared under inert conditions, either in a glove box or a scintillation vial filled with N_2.
3. 10–30 mg/mL of silane is added to the ethanol–water mixture.
4. The plasma-cleaned fibers are immediately immersed into the silane solution and stirred for 30 min.
5. The fiber bundles are then immersed in a vial containing ethanol and stirred for 20 min.
6. The fiber bundles are then vacuum desiccated for 1 h.
7. The fiber bundles are placed in a falcon tube with a rubber stopper and filled with N_2. The falcon tube is then placed in the oven at 80°C for 30 min. The fibers can be stored in the falcon tube for 1 week until further use.

The effectiveness of a passivation method can be tested by comparing the average activity of all the enzyme molecules observed in the femtoliter well array to the bulk enzyme activity. If the passivation is ineffective, the single-molecule activity will be significantly lower than the bulk activity. In addition, a seal and release experiment can be performed if mechanical sealing is employed, to see if the enzyme is bound to the femtoliter well surface (Fig. 5). In this experiment, the enzyme solution is first sealed within the array and the activity is imaged for 2.5 min (Fig. 5A). The seal is then released by pulling the fiber bundle away from the PDMS gasket containing the solution. While the fiber bundle is still immersed in the sample solution, it is no longer in contact with the PDMS so the enzyme molecules are free to diffuse from one well to another. If the enzyme molecules are bound to the surface, they remain in the same well as before. After 2 min the fiber bundle is sealed again with substrate and imaged for another 2 min. If the enzyme molecules are bound to the surface, the same wells will light up as before. If the enzyme molecules are free in solution, different wells will light up as shown in Fig. 5B.

5. SUBSTRATE SELECTION AND SAMPLE PREPARATION

In order to achieve the sensitivity required to detect a single enzyme molecule, fluorogenic substrates are generally used. To measure enzyme kinetics, the concentration of the substrate should be well above the Michaelis constant of the enzyme–substrate pair. In addition, since a very small reaction volume is entrapped in the femtoliter wells, depletion of the substrate must be considered when selecting the substrate concentration.

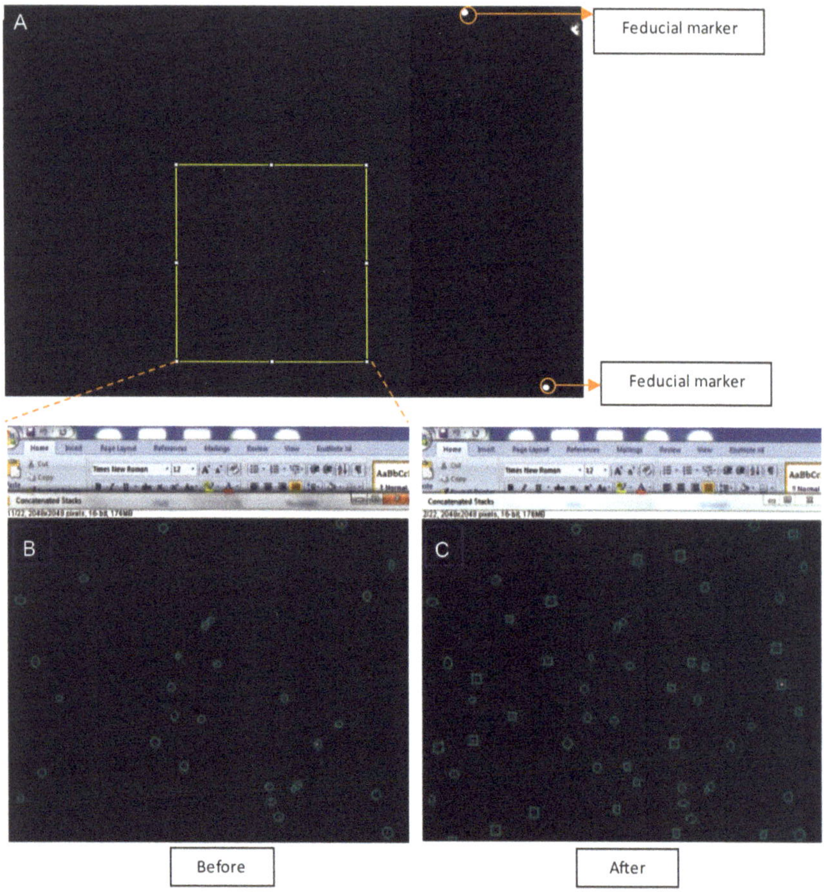

Fig. 5 Seal and release experiments were performed to demonstrate that the enzyme molecules are not bound to the surface of the fiber wells. The fibers were etched and stirred in the reaction buffer (1 × PBS, 1 m*M* $MgCl_2$, 0.005% Tween20, and 0.005% BSA) for 1 h. 0.036 p*M* of the enzyme is sealed within the fiber wells and imaged for 2.5 min (A). A small portion of the fiber bundle is shown in (B). The *blue circles* indicate wells that show activity. The seal is then released to allow the enzyme molecules to diffuse out of the fiber wells. The enzyme is once again sealed within the fiber wells and imaged for 2.5 min (C). The enzyme molecules now appear in different wells (indicated by the *blue rectangles*). Note that the *blue circles* do not show any activity. If the enzyme molecules are bound to the fiber surface they are not free to diffuse out the wells. As a result, the same wells that were active in (B) (the *blue circles*), should show activity in (C), which is not the case. Thus, the seal and release experiment demonstrates that enzymes do not stick to the fiber wells. The fiducial markers shown in (A) are used to align the images to ensure that the position of wells is not shifted.

For example, at 100 μM substrate concentration, there are 3.01×10^6 substrate molecules in a 50 fL well. If the enzyme has a substrate turnover rate of 250 s^{-1}, within 20 min only 10% of the substrate is depleted, ensuring that there is still a large excess of substrate available.

To capture single enzyme molecules and to minimize the possibility of capturing more than one molecule in a well, the ratio of the wells to enzyme molecules is maintained at more than 10:1. For 50,000 total wells of 46 fL volume each, the number of enzyme molecules is 5000 or lower, which means the concentration is $\frac{5000}{N \times 46\text{fL} \times 50,000 \text{ wells}} = 3.6\text{p}M$, where N is Avogadro's number. At 3.6 pM, the number of wells containing more than 1 enzyme is less than 1%.

6. MICROSCOPE SETUP AND IMAGING

The fiber bundle is fixed to the stage of an upright Olympus (Tokyo, Japan) microscope (BX-61) using a custom fiber holder. With mechanical sealing, it is important to ensure that the fiber bundle is tightly fixed to the microscope stage. To reduce evaporation effects, the etched end of the fiber bundle is fixed facing down and away from the objective. Twenty-five microliters of the sample is placed on a slide containing a PDMS gasket. A second stage that can be moved in the z-direction is used to hold the sample slide. To load the sample in the femtoliter wells, the sample is brought into contact with the femtoliter wells for 30 s by moving up the sample stage. After 30 s, the PDMS gasket is then brought into contact with the etched end of the fiber bundle by moving the sample stage several microns further in the z-direction until the gasket is pressed against the fiber bundle. Time-lapse images of the fiber bundle are taken immediately after sealing using a CCD or a CMOS camera.

The exposure time and the lamp intensity must be chosen to have a sufficient signal-to-noise ratio but at the same time must minimize photobleaching of the accumulated product. Photobleaching of the product results in a lower estimate of the activity of enzyme molecules. A long light exposure time in combination with high intensity excitation light will increase the signal as well as the photobleaching rate of the product. In practice, a low lamp intensity is used to reduce photobleaching effects and a long detector exposure time is used for a good signal-to-noise ratio. A very short time interval between images can also contribute to photobleaching of the product due to repeated exposure. To observe single molecules of β-galactosidase, time-lapse imaging is done at 10% of lamp intensity (USH-103D short arc mercury lamp from Ushio (Tokyo, Japan)) using a neutral-density filter and 2 s of exposure.

7. IMAGE AND DATA ANALYSIS

The time-lapse microarray images are processed using a MATLAB algorithm. First a black and white well mask image is generated, where the white regions correspond to the geometric centers of the femtoliter wells. Using the well mask, the intensities of femtoliter wells in each frame are extracted. These intensities are used to compute the percent intensity increase (%*I*) over the course of the experiment, where $\%I = \frac{Final\,Intensity - Initial\,Intensity}{Initial\,Intensity} \times 100$. Since the majority of the wells do not contain an enzyme molecule and only a small fraction of the wells contain a molecule (~1% at 100:1 well to enzyme ratio), a histogram of %*I* is typically a normal distribution with a small fraction of outliers. Because most of the wells do not contain an enzyme, the normal distribution peak corresponds to wells that do not contain any enzyme and the outliers correspond to the wells containing an enzyme molecule. The mean value of this normal distribution peak is the percent change in intensity due to the uncatalyzed background hydrolysis of the substrate solution. The outliers correspond to the wells that contain an enzyme molecule and have a high percent increase in intensity compared to the background. To identify the femtoliter wells that contain single enzyme molecules within the entire array, a threshold is applied to %*I*. To determine the threshold, a control experiment is performed, where the substrate solution is sealed within the fiber wells and imaged for the same duration as the single-molecule experiment. A Gaussian curve is fit to the histogram of %*I* in this control experiment and the Gaussian fit is used to find the mean and standard deviation for the control. The sum of mean and three times the standard deviation of %*I* is used as the threshold value to determine wells that contain an enzyme. Therefore any well that has a %*I* that is higher than the threshold value is counted as an "active well" (a well containing an enzyme molecule).

Once the active wells are identified, the activity of the individual enzyme molecules can be easily calculated using a calibration factor (c.f.). To determine the c.f., different concentrations of the fluorescent product are sealed within the femtoliter wells and imaged using the identical exposure and lamp conditions as performed during the experiment. The slope of the number of product molecules in the wells vs intensity of the wells is used as the c.f. The c.f. corresponds to the number of fluorescent product molecules required for one unit increase in intensity. The activity of enzyme molecules is calculated using the equation $Activity = \frac{\Delta I \times cf}{\Delta t}$, where ΔI is the change in intensity of the active well and Δt is the duration of the experiment. When the experiment is performed for an extended period of time, since the substrate is depleted over time, the initial reaction rate is used to calculate the activity of the enzyme. ΔI is determined from the intensity change within the first 2 min when less than 2% of the substrate is depleted. To determine the K_m and k_{cat}, first the average activity can be determined at different substrate concentrations. Then the traditional Michaelis–Menten equation can be applied to determine the K_m and k_{cat}.

For accurate determination of enzyme activity, a photobleaching correction is applied. First, the photobleaching decay constant is determined by performing a time-lapse image of 10 μM of fluorophore sealed within the femtoliter wells. The intensity change over time is fit to an exponential decay model $I = I_0 e^{-K_{ph} t}$, where I_0 is the starting intensity, to extract the first-order photobleaching decay constant K_{ph}. The photobleaching correction is applied using the formula $S_t = I_t {}^* K_{ph} + I'_t$, where S_t is the substrate turnover rate at time t, I_t is the intensity at time t and I'_t is the derivative of I.

8. SUMMARY AND CONCLUSION

Femtoliter arrays provide a simple way to encapsulate and study single enzyme molecules. Over the past decade, several methods for generating femtoliter arrays have been published and used to study the kinetics of single enzyme molecules and to elucidate mechanistic details. The main limitation for developing single-molecule enzyme assays in the femtoliter arrays is the lack of available fluorogenic substrates. The development of new fluorogenic substrates for the enzymes of interest is essential for progress in femtoliter array-based single-molecule enzymology. Femtoliter arrays are also limited in the time resolution at which enzyme processes can be studied, thus limiting the method to processes that occur in the minutes to hours time scale. Even with these limitations, femtoliter arrays offer a unique platform for studying catalytic processes at the single-molecule scale.

REFERENCES

Arayanarakool, R., Shui, L., Kengen, S. W. M., van den Berg, A., & Eijkel, J. C. T. (2013). Single-enzyme analysis in a droplet-based micro- and nanofluidic system. *Lab on a Chip, 13*(10), 1955–1962. http://dx.doi.org/10.1039/C3LC41100A.

Chen, Y., Hu, D., Vorpagel, E. R., & Lu, H. P. (2003). Probing single-molecule T4 lysozyme conformational dynamics by intramolecular fluorescence energy transfer. *The Journal of Physical Chemistry B, 107*(31), 7947–7956. http://dx.doi.org/10.1021/jp022406z.

Craig, D. B., Arriaga, E. A., Wong, J. C. Y., Lu, H., & Dovichi, N. J. (1996). Studies on single alkaline phosphatase molecules: Reaction rate and activation energy of a reaction catalyzed by a single molecule and the effect of thermal denaturation the death of an enzyme. *Journal of the American Chemical Society, 118*(22), 5245–5253. http://dx.doi.org/10.1021/ja9540839.

Craig, D. B., & Dovichi, N. J. (1998). Escherichia coli β-galactosidase is heterogeneous with respect to the activity of individual molecules. *Canadian Journal of Chemistry, 76*(5), 623–626. http://dx.doi.org/10.1139/v98-055.

Craig, D. B., & Nichols, E. R. (2008). Continuous flow assay for the simultaneous measurement of the electrophoretic mobility, catalytic activity and its variation over time of individual molecules of Escherichia coli β-galactosidase. *Electrophoresis, 29*(21), 4298–4303. http://dx.doi.org/10.1002/elps.200800482.

De Cremer, G., Roeffaers, M. B. J., Baruah, M., Sliwa, M., Sels, B. F., Hofkens, J., & De Vos, D. E. (2007). Dynamic disorder and stepwise deactivation in a chymotrypsin catalyzed hydrolysis reaction. *Journal of the American Chemical Society, 129*(50), 15458–15459. http://dx.doi.org/10.1021/ja077621d.

Ehrl, B. N., Liebherr, R. B., & Gorris, H. H. (2013). Single molecule kinetics of horseradish peroxidase exposed in large arrays of femtoliter-sized fused silica chambers. *Analyst, 138*(15), 4260–4265. http://dx.doi.org/10.1039/C3AN00809F.

Funatsu, T., Harada, Y., Tokunaga, M., Saito, K., & Yanagida, T. (1995). Imaging of single fluorescent molecules and individual atp turnovers by single myosin molecules in aqueous solution. *Nature, 374*(6522), 555–559. http://dx.doi.org/10.1038/374555a0.

Gaylord, S. T., Abdul-Aziz, S., & Walt, D. R. (2015). Single-molecule arrays for ultrasensitive detection of host immune response to dengue virus infection. *Journal of Clinical Microbiology, 53*(5), 1722–1724. http://dx.doi.org/10.1128/jcm.03487-14.

Gaylord, S. T., Dinh, T. L., Goldman, E. R., Anderson, G. P., Ngan, K. C., & Walt, D. R. (2015). Ultrasensitive detection of ricin toxin in multiple sample matrixes using single-domain antibodies. *Analytical Chemistry, 87*(13), 6570–6577. http://dx.doi.org/10.1021/acs.analchem.5b00322.

Gorris, H. H., Blicharz, T. M., & Walt, D. R. (2007). Optical-fiber bundles. *FEBS Journal, 274*(21), 5462–5470. http://dx.doi.org/10.1111/j.1742-4658.2007.06078.x.

Gorris, H. H., Rissin, D. M., & Walt, D. R. (2007). Stochastic inhibitor release and binding from single-enzyme molecules. *Proceedings of the National Academy of Sciences of the United States of America, 104*(45), 17680–17685. http://dx.doi.org/10.1073/pnas.0705411104.

Gorris, H. H., & Walt, D. R. (2009). Mechanistic aspects of horseradish peroxidase elucidated through single-molecule studies. *Journal of the American Chemical Society, 131*(17), 6277–6282. http://dx.doi.org/10.1021/ja9008858.

Hatzcbeakis, N. S., Engelkamp, H., Velonia, K., Hofkens, J., Christianen, P. C. M., Svendsen, A., … Nolte, R. J. M. (2006). Synthesis and single enzyme activity of a clicked lipase-BSA hetero-dimer. *Chemical Communications*, (19), 2012–2014. http://dx.doi.org/10.1039/B516551B.

Kan, C. W., Rivnak, A. J., Campbell, T. G., Piech, T., Rissin, D. M., Mosl, M., … Duffy, D. C. (2012). Isolation and detection of single molecules on paramagnetic beads using sequential fluid flows in microfabricated polymer array assemblies. *Lab on a Chip*, *12*(5), 977–985. http://dx.doi.org/10.1039/c2lc20744c.

Kathryn, M. M., Jason, S., Douglas, D., & David, R. W. (2015). Catalytic kinetics of single gold nanoparticles observed via optical microwell arrays. *Nanotechnology*, *26*(5), 055704.

Kuznetsova, S., Zauner, G., Aartsma, T. J., Engelkamp, H., Hatzakis, N., Rowan, A. E., … Canters, G. W. (2008). The enzyme mechanism of nitrite reductase studied at single-molecule level. *Proceedings of the National Academy of Sciences of the United States of America*, *105*(9), 3250–3255. http://dx.doi.org/10.1073/pnas.0707736105.

Lerch, H.-P., Rigler, R., & Mikhailov, A. S. (2005). Functional conformational motions in the turnover cycle of cholesterol oxidase. *Proceedings of the National Academy of Sciences of the United States of America*, *102*(31), 10807–10812. http://dx.doi.org/10.1073/pnas.0504995102.

Liebherr, R. B., & Gorris, H. H. (2014). Enzyme molecules in solitary confinement. *Molecules*, *19*(9), 14417–14445. http://dx.doi.org/10.3390/molecules190914417.

Liebherr, R. B., Renner, M., & Gorris, H. H. (2014). A single molecule perspective on the functional diversity of in vitro evolved β-glucuronidase. *Journal of the American Chemical Society*, *136*(16), 5949–5955. http://dx.doi.org/10.1021/ja412379p.

Lu, H. P. (2014). Sizing up single-molecule enzymatic conformational dynamics. *Chemical Society Reviews*, *43*(4), 1118–1143. http://dx.doi.org/10.1039/C3CS60191A.

Lu, H. P., Xun, L., & Xie, X. S. (1998). Single-molecule enzymatic dynamics. *Science*, *282*(5395), 1877–1882. http://dx.doi.org/10.1126/science.282.5395.1877.

Min, W., English, B. P., Luo, G., Cherayil, B. J., Kou, S. C., & Xie, X. S. (2005). Fluctuating enzymes: Lessons from single-molecule studies. *Accounts of Chemical Research*, *38*(12), 923–931. http://dx.doi.org/10.1021/ar040133f.

Mogalisetti, P., Gorris, H. H., Rojek, M. J., & Walt, D. R. (2014). Elucidating the relationship between substrate and inhibitor binding to the active sites of tetrameric [small beta]-galactosidase. *Chemical Science*, *5*, 4467–4473. http://dx.doi.org/10.1039/C4SC01437E.

Mogalisetti, P., & Walt, D. R. (2015). Stoichiometry of the α-complementation reaction of Escherichia coli β-galactosidase as revealed through single-molecule studies. *Biochemistry*, *54*(8), 1583–1588. http://dx.doi.org/10.1021/bi5015024.

Nichols, E. R., Shadabi, E., & Craig, D. B. (2009). Effect of alteration of translation error rate on enzyme microheterogeneity as assessed by variation in single molecule electrophoretic mobility and catalytic activity. *Biochemistry and Cell Biology*, *87*(3), 517–529. http://dx.doi.org/10.1139/O09-010.

Pantano, P., & Walt, D. R. (1996). Ordered nanowell arrays. *Chemistry of Materials*, *8*(12), 2832–2835. http://dx.doi.org/10.1021/cm9603314.

Piwonski, H. M., Goomanovsky, M., Bensimon, D., Horovitz, A., & Haran, G. (2012). Allosteric inhibition of individual enzyme molecules trapped in lipid vesicles. *Proceedings of the National Academy of Sciences of the United States of America*, *109*(22), E1437–E1443. http://dx.doi.org/10.1073/pnas.1116670109.

Rissin, D. M., Gorris, H. H., & Walt, D. R. (2008). Distinct and long-lived activity states of single enzyme molecules. *Journal of the American Chemical Society*, *130*(15), 5349–5353. http://dx.doi.org/10.1021/ja711414f.

Rissin, D. M., Kan, C. W., Campbell, T. G., Howes, S. C., Fournier, D. R., Song, L., … Duffy, D. C. (2010). Single-molecule enzyme-linked immunosorbent assay detects serum proteins at subfemtomolar concentrations. *Nature Biotechnology*, *28*(6), 595–599. http://www.nature.com/nbt/journal/v28/n6/abs/nbt.1641.html—supplementary-information.

Rissin, D. M., & Walt, D. R. (2006). Digital concentration readout of single enzyme molecules using femtoliter arrays and Poisson statistics. *Nano Letters*, *6*(3), 520–523. http://dx.doi.org/10.1021/nl060227d.

Rojek, M. J., & Walt, D. R. (2014). Observing single enzyme molecules interconvert between activity states upon heating. *PLoS One*, *9*(1), e86224. http://dx.doi.org/10.1371/journal.pone.0086224.
Rondelez, Y., Tresset, G., Tabata, K. V., Arata, H., Fujita, H., Takeuchi, S., & Noji, H. (2005). Microfabricated arrays of femtoliter chambers allow single molecule enzymology. *Nature Biotechnology*, *23*(3), 361–365. http://www.nature.com/nbt/journal/v23/n3/suppinfo/nbt1072_S1.html.
Santoso, Y., Joyce, C. M., Potapova, O., Le Reste, L., Hohlbein, J., Torella, J. P., … Kapanidis, A. N. (2010). Conformational transitions in DNA polymerase I revealed by single-molecule fret. *Proceedings of the National Academy of Sciences of the United States of America*, *107*(2), 715–720. http://dx.doi.org/10.1073/pnas.0910909107.
Schubert, S. M., Arendt, L. M., Zhou, W., Baig, S., Walter, S. R., Buchsbaum, R. J., … Walt, D. R. (2015). Ultra-sensitive protein detection via single molecule arrays towards early stage cancer monitoring. *Scientific Reports*, *5*, 11034. http://dx.doi.org/10.1038/srep11034. http://www.nature.com/articles/srep11034—supplementary-information.
Schubert, S. M., Walter, S. R., Manesse, M., & Walt, D. R. (2016). Protein counting in single cancer cells. *Analytical Chemistry*, *88*(5), 2952–2957. http://dx.doi.org/10.1021/acs.analchem.6b00146.
Wang, Y., & Lu, H. P. (2010). Bunching effect in single-molecule T4 lysozyme nonequilibrium conformational dynamics under enzymatic reactions. *The Journal of Physical Chemistry. B*, *114*(19), 6669–6674. http://dx.doi.org/10.1021/jp1004506.
Warren, A. D., Gaylord, S. T., Ngan, K. C., Dumont Milutinovic, M., Kwong, G. A., Bhatia, S. N., & Walt, D. R. (2014). Disease detection by ultrasensitive quantification of microdosed synthetic urinary biomarkers. *Journal of the American Chemical Society*, *136*(39), 13709–13714. http://dx.doi.org/10.1021/ja505676h.
Wilson, D. H., Rissin, D. M., Kan, C. W., Fournier, D. R., Piech, T., Campbell, T. G., … Duffy, D. C. (2016). The Simoa HD-1 analyzer: A novel fully automated digital immunoassay analyzer with single-molecule sensitivity and multiplexing. *Journal of Laboratory Automation*, *21*, 533–547. http://dx.doi.org/10.1177/2211068215589580.
Wu, D., Milutinovic, M. D., & Walt, D. R. (2015). Single molecule array (Simoa) assay with optimal antibody pairs for cytokine detection in human serum samples. *Analyst*, *140*(18), 6277–6282. http://dx.doi.org/10.1039/C5AN01238D.
Xue, Q., & Yeung, E. S. (1995). Differences in the chemical reactivity of individual molecules of an enzyme. *Nature*, *373*(6516), 681–683.
Zhang, H., Nie, S., Etson, C. M., Wang, R. M., & Walt, D. R. (2012). Oil-sealed femtoliter fiber-optic arrays for single molecule analysis. *Lab on a Chip*, *12*(12), 2229–2239. http://dx.doi.org/10.1039/C2LC21113K.

AUTHOR INDEX

Note: Page numbers followed by "*f*" indicate figures, and "*t*" indicate tables.

A

B

C

E

I

L

M

N

O

P

Q

R

S

SUBJECT INDEX

Note: Page numbers followed by "*f*" indicate figures, and "*t*" indicate tables.

E

F

G

N

Q

R

S

T

V

W

X

Y

FSC
www.fsc.org
MIX
Papier | Fördert
gute Waldnutzung
FSC® C083411

Kolibri 360 GmbH
Ferdinand-Jühlke-Straße 7
99095 Erfurt, Deutschland
produktsicherheit@kolibri360.de